Mobile Cellular
Telecommunications

Other McGraw-Hill Communications Books of Interest

Best • PHASE-LOCKED LOOPS, SECOND EDITION
Bush • PRIVATE BRANCH EXCHANGE SYSTEM AND APPLICATIONS
Chorafas • THE COMPLETE LAN REFERENCE
Cooper • COMPUTER AND COMMUNICATIONS SECURITY
Dayton • INTEGRATING DIGITAL SERVICES
Folts • THE MCGRAW-HILL COMPILATION OF OPEN SYSTEMS STANDARDS
Fortier • HANDBOOK OF LAN TECHNOLOGY
Ha • DIGITAL SATELLITE COMMUNICATIONS
Hebrawi • OSI: UPPER LAYER STANDARDS AND PRACTICES
Held, Sarch • DATA COMMUNICATIONS
Heldman • INFORMATION TELECOMMUNICATIONS
Heldman • GLOBAL TELECOMMUNICATIONS
Hughes • DATA COMMUNICATIONS
Inglis • ELECTRONIC COMMUNICATIONS HANDBOOK
Kessler • ISDN, SECOND EDITION
Kessler, Train • METROPOLITAN AREA NETWORKS
Lindberg • DIGITAL BROADBAND NETWORKS AND SERVICES
Logsdon • MOBILE COMMUNICATION SATELLITES
Macario • CELLULAR RADIO
Radicati • ELECTRONIC MAIL
Ranade • INTRODUCTION TO SNA NETWORKING
Ranade • VSAM: PERFORMANCE, DESIGN, AND FINE TUNING
Ranade, Ranade • VSAM: CONCEPTS, PROGRAMMING, AND DESIGN
Rhee • ERROR CORRECTION CODING THEORY
Rohde, Bucher • COMMUNICATION RECEIVERS
Schlar • INSIDE X.25
Simon et al. • SPREAD SPECTRUM COMMUNICATIONS HANDBOOK
Winch • TELECOMMUNICATIONS TRANSMISSION SYSTEMS

To order or receive additional information on these or any other McGraw-Hill titles, in the United States please call 1-800-822-8158. In other countries, contact your local McGraw-Hill representative.

BC14BCZ

Mobile Cellular Telecommunications

Analog and Digital Systems

William C. Y. Lee Vice President

Applied Research and Science
AirTouch Communications Inc. (Formerly Pactel Co.)
Walnut Creek, California

Second Edition

McGraw-Hill, Inc.

New York   San Francisco   Washington, D.C.   Auckland   Bogotá
Caracas   Lisbon   London   Madrid   Mexico City   Milan
Montreal   New Delhi   San Juan   Singapore
Sydney   Tokyo   Toronto

MOBILE CELLULAR TELECOMMUNICATIONS
Analog and Digital Systems, Second Edition
International Editions 1995

Exclusive rights by McGraw-Hill Book Co. – Singapore for manufacture and export. This book cannot be re-exported from the country to which it is consigned by McGraw-Hill.

The first edition of this book was published in 1989 under the title *Mobile Cellular Telecommunications Systems*.

5 6 7 8 9 0 CWP UPE 9 8 7

The sponsoring editor for this book was Stephen S. Chapman, the editing supervisor was Paul R. Sobel, and the production supervisor was Pamela A. Pelton. It was set in Century Schoolbook by PRO-Image Corporation, Techna-Type Div., York, PA.

Library of Congress Cataloging-in-Publication Data

Lee, Williams C. Y.
     Mobile cellular telecommunications : analog and digital systems /
William C.Y. Lee. – 2nd ed.
          p.   cm.
     Rev. ed. of: Mobile cellular telecommunications systems. c1989.
     Includes index.
     ISBN 0-07-038089-9 (alk. paper)
     1. Cellular radio.   I. Lee, William C. Y.   Mobile cellular
telecommunications systems.   II. Title.
TK6570.M6L35   1995
621.3845'6–dc20   94-41512
                                                               CIP

When ordering this title, use ISBN 0-07-113479-4

Printed in Singapore

Contents

Chapter 5. Cell-Site Antennas and Mobile Antenna   157

Chapter 6. Cochannel Interference Reduction   189

Chapter 7. Types of Noncochannel Interference   221

Chapter 8. Frequency Management and Channel Assignment   257

Chapter 9. Handoffs and Dropped Calls   283

Chapter 10. Operational Techniques and Technologies   307

Chapter 11. Switching and Traffic   343

Chapter 12. Data Links and Microwaves   363

Preface to the Second Edition

Since the book *Mobile Cellular Telecommunications Systems* was last published in 1989, a lot of readers have written to me and given me a great deal of encouragement and compliments. In the last five years, the wireless communications field has been rapidly growing. Many new concepts and new systems have been developed. When Mr. Chapman from McGraw Hill asked me to revise this book, I immediately accepted his offer. Because of the rapid growth in the industry, many engineers are joining in this field from the military industry. Many new engineers who are just graduating from school are also joining this field. They need to learn about this field as quick as possible. Also, the engineers who are already in the field may only work on a small specialized area and want to broaden their knowledge. Therefore, when I revised this book, I wanted to aim it towards their needs. Besides adding many sections in the current chapters, such as calculations of near-field propagation, wireless information superhighway, call blocking and call dropping, etc., I also added three chapters. Chapter 15, Digital Cellular Systems describes newly developed digital systems. The three major ones, GSM, North American TDMA, and CDMA are stated in detail. Others such as DECT, PDC, PHS, MIRS, CD2, CDPD, DCS-1800, PCS, etc., are also included. In Chap. 16, Intelligent Cell Concept and Applications, I have introduced the intelligent means to increase capacity and improve system performance. In Chap. 17, Advanced Intelligent Network (AIN), was introduced. Also, the wideband switch for the future wireless information superhighway was also stated.

In the future, intelligence will be applied to cell sites, networks, and systems. These same thoughts used in cellular will also be applied to PCS.

I hope this book will broaden the readers thoughts. The wireless information superhighway will be our future goal but we still have along way to go.

William C.Y. Lee

Preface to the First Edition

As the number of cellular subscribers increases, the interference that will be experienced by the systems will also increase. This means that many large cellular systems will, sooner or later, have to handle interference problems. This is a lucrative field that is ripe for research and that will soon be begging for more advanced applications.

This all-inclusive and self-contained work, consisting of fifteen chapters, is a basic textbook that supports further exploration in a new communications field, cellular communications. Since it is the first in its field, this book may be considered a handbook or building block for future research.

For years it has been my desire to write a book on the technical aspects of cellular systems. Since it is a new field, the theory has to be developed and then verified by experiment. I am seeking to adhere to the progression of learning that I described in *Who's Who in America:*

1. Use mathematics to solve problems.

2. Use physics to interpret results.

3. Use experiments and counterexamples to check outcomes.

4. Use pictures to emphasize important points.

Since I have accumulated many pictures in my mind, I would like to share them with my readers. In this field many new applications and theories have been discovered. Thus, my findings will help the reader to assimilate this new knowledge and accelerate learning time. The many mistakes that have been made in the past in designing cellular systems can now be avoided. Engineers who work in other communication systems will appreciate the many diverse concepts used in cellular systems. The reader should be aware that it is possible to apply the various theories improperly and thereby create many serious problems. I would like to hear from readers about their cellular systems experiences, both successful and unsuccessful.

Overall, I have written this book for technical engineers who would like to explore options in the cellular industry. However, Chapters 1 and 2 are for *executives* and for anyone who would like to familiarize himself or herself with key concepts of the field. Chapter 3 describes the specification of cellular systems. The North American specification works in Canada, the United States, and Mexico, so a cellular phone will work anywhere in this territory because of the standardized specification.

Chapter 4 introduces the point-to-point model I developed over the last 15 years. It can be used as a core to develop many design tools. Chapters 5 and 6 deal with cochannel interference problems, and Chapter 7 deals with non-cochannel interference problems. Chapters 8 through 13 offer detailed material for engineers to solve problems concerning improved system performance. Chapter 14 describes the digital systems which may become the next-generation cellular systems, and presents many key issues in order to alert readers to possible future developments. Chapter 15 highlights some miscellaneous topics related to cellular systems.

Much of my unpublished work is included in these 15 chapters. I welcome feedback from readers about how I can better meet their needs in the second edition of the book.

I have always felt that cellular technology should be openly shared by cellular operators. Competition only occurs in a saturated market, and the cellular market is almost unlimited. Therefore, competition is not to be feared in this early stage of the cellular industry. We need to promote this industry as much as possible by involving more interested engineers and investors.

In the last six years, I have taught 3-day seminars sponsored by George Washington University. I am trying to convince the cellular industry that if we have narrow-minded attitudes and do not share our experiments or knowledge, the whole industry will not advance fast enough and could be replaced by other new industries, such as wireless communications or in-building communications.

Let us join together to allow the cellular industry its optimum potential and set our goal that one day a pocket cellular phone will carry our calls to any place in the world.

William C.Y. Lee

Acknowledgments

I would like to sincerely thank all the engineers who took my seminars sponsored by George Washington University in the last six years and who stimulated many valuable thoughts for this book.

I still and always will remember the valuable advice and encouragement that C. C. Cutler and Frank Blecher gave me at different times in my career. Bell Laboratories and ITT have provided me the opportunity to study mobile cellular systems during a system-development stage. PacTel Cellular Inc., and Cellular Telecommunications Industrial Association have provided me with much information for the system implementation stage. I want to thank my colleagues who have been associated with my work during these two stages. Above all, I want to thank Phil Quigley and Gloria Everett, who provided me the opportunity of joining PacTel Cellular and dealing with cellular technologies.

I am deeply grateful to Ms. Ella Saunders who has been so helpful in typing this manuscript and to George McClure who patiently reviewed it and gave many valuable suggestions.

This is the third book which I have written in my leisure time. Of course, I have tried to convince my family and especially my lovely wife how important this book is for the cellular industry. Although my family does not believe a word of it, they have generously supported me in making this book possible. I am gratefully obliged to my wife, Margaret, and our two daughters, Betty and Lily.

The first edition of this book was published in 1989. In the last five years, the wireless communications field has changed very rapidly. A lot of new material had to be included in this book. During the revising stage of this book AirTouch's Librarian, Ms. Maribeth Eisenmann, enthusiastically gathered new information for my book. Ms. Susan Shaffer was so patient in typing and proofreading the manuscript for the new additional sections in the old chapters and three new chapters.

When attending the PIMRC Conference in Haag, Netherlands, in October 1994, my briefcase was stolen in the train station which contained some of my manuscript that had to be rewritten. Fortunately, most of the material had been duplicated so the publication date did not slip.

Finally, I would like to thank my family and especially my wife, Margaret, for their understanding while writing this second edition and to the readers who are so supportive of my book.

Mobile Cellular
Telecommunications

Introduction to Cellular Mobile Systems

1.1 Why Cellular Mobile Telephone Systems?

1.1.1 Limitations of conventional mobile telephone systems

One of many reasons for developing a *cellular* mobile telephone system and deploying it in many cities is the operational limitations of conventional mobile telephone systems: limited service capability, poor service performance, and inefficient frequency spectrum utilization.

Limited service capability. A conventional mobile telephone system is usually designed by selecting one or more channels from a specific frequency allocation for use in autonomous geographic zones, as shown in Fig. 1.1. The communications coverage area of each zone is normally planned to be as large as possible, which means that the transmitted power should be as high as the federal specification allows. The user who starts a call in one zone has to reinitiate the call when moving into a new zone (see Fig. 1.1) because the call will be dropped. This is an undesirable radio telephone system since there is no guarantee that a call can be completed without a handoff capability.

The handoff is a process of automatically changing frequencies as the mobile unit moves into a different frequency zone so that the conversation can be continued in a new frequency zone without redialing. Another disadvantage of the conventional system is that the number of active users is limited to the number of channels assigned to a particular frequency zone.

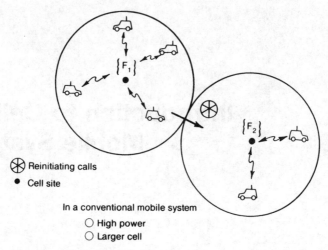

Reinitiating calls
Cell site

In a conventional mobile system
○ High power
○ Larger cell

Figure 1.1 Conventional mobile system.

Poor service performance.   In the past, a total of 33 channels were allocated to three mobile telephone systems: Mobile Telephone Service (MTS), Improved Mobile Telephone Service (IMTS) MJ systems, and Improved Mobile Telephone Service (IMTS) MK systems. MTS operates around 40 MHz and MJ operates at 150 MHz; both provide 11 channels; IMTS MK operates at 450 MHz and provides 12 channels. These 33 channels must cover an area 50 mi in diameter. In 1976, New York City had 6 channels of MJ serving 320 customers, with another 2400 customers on a waiting list. New York City also had 6 channels of MK serving 225 customers, with another 1300 customers on a waiting list. The large number of subscribers created a high blocking probability during busy hours. The actual number of blockings will be shown later. Although service performance was undesirable, the demand was still great. A high-capacity system for mobile telephones was needed.

Inefficient frequency spectrum utilization.   In a conventional mobile telephone system, the frequency utilization measurement M_0 is defined as the maximum number of customers that could be served by one channel at the busy hour. Equation (1.1-1) gives the 1976 New York City data cited earlier.

$$M_0 = \frac{\text{no. of customers}}{\text{channel}} \qquad \text{(conventional systems)} \qquad (1.1\text{-}1)$$

or
$$M_0 = \begin{cases} 53 \text{ customers/channel} & \text{(MJ system)} \\ 37 \text{ customers/channel} & \text{(MK system)} \end{cases}$$

Assume an average calling time of 1.76 min and apply the Erlang B model (lost-calls-cleared conditions). Calculate the blocking probability as follows: Use 6 channels, with each channel serving the two different numbers of customers shown in Eq. (1.1-1). The offered load can then be obtained by Eq. (1.1-2).

$$A = \frac{\text{av calling time (minutes)} \times \text{total customers}}{60 \text{ min}} \quad \text{erlangs} \quad (1.1\text{-}2)$$

$$A_1 = \frac{1.76 \times 53 \times 6}{60} = 9.33 \text{ erlangs} \quad \text{(MJ system)}$$

$$A_2 = \frac{1.76 \times 37 \times 6}{60} = 6.51 \text{ erlangs} \quad \text{(MK system)}$$

Given that the number of channels is 6 and the offered loads are $A_1 = 9.33$ and $A_2 = 6.51$, read from the table in Appendix 1.1 to obtain the blocking probabilities $B_1 = 50$ percent (MJ system) and $B_2 = 30$ percent (MK system), respectively. It is likely that half the initiating calls will be blocked in the MJ system, a very high blocking probability.

If the actual average calling time is greater than 1.76 min, the blocking probability can be even higher. To reduce the blocking probability, we must decrease the value of the frequency spectrum utilization measurement M_0 as shown in Eq. (1.1-1).

As far as frequency spectrum utilization is concerned, the conventional system does not utilize the spectrum efficiently since each channel can only serve one customer at a time in a whole area. A new cellular system that measures the frequency spectrum utilization differently from Eq. (1.1-1) and proves to be efficient is discussed in Sec. 1.1.2.

1.1.2 Spectrum efficiency considerations

A major problem facing the radio communication industry is the limitation of the available radio frequency spectrum. In setting allocation policy, the Federal Communications Commission (FCC) seeks systems which need minimal bandwidth but provide high usage and consumer satisfaction.

The ideal mobile telephone system would operate within a limited assigned frequency band and would serve an almost unlimited number of users in unlimited areas. Three major approaches to achieve the ideal are

1. Single-sideband (SSB), which divides the allocated frequency band into maximum numbers of channels

2. Cellular, which reuses the allocated frequency band in different geographic locations

3. Spread spectrum or frequency-hopped, which generates many codes over a wide frequency band

In 1971, the cellular approach was shown to be a spectrally efficient system.[1] The comparison of an analog cellular approach with other digital cellular system approaches is given in Chap. 15.

1.1.3 Technology, feasibility, and service affordability

In 1971, the computer industry entered a new era. Microprocessors and minicomputers are now used for controlling many complicated features and functions with less power and size than was previously possible. Large-scale integrated (LSI) circuit technology reduced the size of mobile transceivers so that they easily fit into the standard automobile. These achievements were a few of the requirements for developing advanced mobile phone systems and encouraging engineers to pursue this direction.

Another factor was the price reduction of the mobile telephone unit. LSI technology and mass production contribute to reduced cost so that in the near future an average-income family should be able to afford a mobile telephone unit.

On Jan. 4, 1979, the FCC authorized Illinois Bell Telephone Co. (IBT) to conduct a developmental cellular system in the Chicago area and make a limited commercial offering of its cellular service to the public. In addition, American Radio Telephone Service, Inc., (ARTS) was authorized to operate a cellular system in the Washington, D.C.--Baltimore, Md., area. These first systems showed the technological feasibility and affordability of cellular service.

1.1.4 Why 800 MHz?

The FCC's decision to choose 800 MHz was made because of severe spectrum limitations at lower frequency bands. FM broadcasting services operate in the vicinity of 100 MHz. The television broadcasting service starts at 41 MHz and extends up to 960 MHz.

Air-to-ground systems use 118 to 136 MHz; military aircraft use 225 to 400 MHz. The maritime mobile service is located in the vicinity of 160 MHz. Also fixed-station services are allocated portions of the 30- to 100-MHz band. Therefore, it was hard for the FCC to allocate a spectrum in the lower portions of the 30- to 400-MHz band since the services of this band had become so crowded. On the other hand, mobile radio transmission cannot be applied at 10 GHz or above because severe propagation path loss, multipath fading, and rain activity make the medium improper for mobile communications.

Fortunately, 800 MHz was originally assigned to educational TV channels. Cable TV service became a big factor in the mid-70s and shared the load of providing TV channels. This situation opened up the 800-MHz band to some extent, and the FCC allocated a 40-MHz system at 800 MHz to mobile radio cellular systems.

Although 800 MHz is not the ideal transmission medium for mobile radio, it has been demonstrated that a cellular mobile radio system[2,3] that does not go beyond this frequency band can be deployed. Needless to say, the medium of transmitting an 800-MHz signal, although it is workable, is already very difficult. Section 1.6.1 briefly describes the transmission medium.

1.2 History of 800-MHz Spectrum Allocation

In 1958, the Bell System (FCC Docket 11997) proposed a 75-MHz system at 800 MHz, quite a broadband proposal. In 1970, the FCC (Docket 18262)[13] tentatively decided to allocate 75 MHz for a wire-line common carrier. In December 1971 the Bell System assured technical feasibility by showing how a cellular mobile system could be designed.[1] In 1974, the FCC allocated 40 MHz of the spectrum, with one cellular system to be licensed per market. There was considerable uncertainty in predicting the cellular market. However, the FCC strategically placed spectrum reserves totaling 20 MHz in proximity to the cellular allocation.

In 1980, the FCC reconsidered its one-system-per-market strategy and studied the possibility of introducing competition into the previous one-carrier markets. Although cost savings make one cellular system per market attractive, balancing the benefits of economies of scale against the benefits of competition, two licensed carriers per service area was more in line with emerging FCC policies.

Trunking efficiency degradation using two carriers per service area will be discussed in Sec. 1.3. It was the FCC's view that such an approach, while not gaining the full competitive market structure, would provide some competitive advantages. The frequencies will be assigned in 20-MHz groups identified as block A and block B, or called band A and band B.

Two bands serve two different groups in the standard situation: one for wire-line (telephone) companies* and one for non-wire-line (non-

---

* Let telephone companies operate mobile radio telephone systems.

TABLE 1.1   Mobile and Base Transmission Frequencies*

Band†	Mobile	Base	Two systems/market
A	824–835, 845–846.5	869–880, 890–891.5	Non-wire-line‡
B	835–845, 846.5–849	880–890, 891.5–894	Wire-line

*On July 24, 1986, an additional 5 MHz was allocated to each band. Therefore, an additional 83 channels were added to each band. The system accommodating an additional 83 channels may be ready in mid-1988.
†416 channels per band, 30 kHz per channel.
‡Majority are non-wire-line companies.

telephone) companies. Each company designs its own system and divides the area into geographic areas, or cells. Each cell operates within its own bands (see Table 1.1).

Since 30 kHz is the specified bandwidth, each band operating nowadays consists of 333 channels.† How to utilize these limited resources to provide adequate voice quality and service performance to an unrestricted population size presents a challenge.

1.3   Trunking Efficiency

To explore the trunking efficiency degradation inherent in licensing two or more carriers rather than one, compare the trunking efficiency between one cellular system per market operating 666 channels and two cellular systems per market each operating 333 channels. Assume that all frequency channels are evenly divided into seven subareas called *cells*. In each cell, the blocking probability of 0.02 is assumed. Also the average calling time is assumed to be 1.76 min.

Look up the table of Appendix 1.1 with $N_1 = {}^{666}\!/_7 = 95$ and $B = 0.02$ to obtain the offered load $A_1 = 83.1$ and with $N_2 = {}^{333}\!/_7 = 47.5$ and $B = 0.02$ to obtain $A_2 = 38$. Since two carriers each operating 333 channels are considered, the total offered load is $2A_2$. We then realize that

$$A_1 \geq 2A_2 \tag{1.3-1}$$

By converting Eq. (1.3-1) to the number of users who can be served in a busy hour, the average calling time of 1.76 min is introduced. The number of calls per hour served in a cell can be expressed as

$$Q_i = \frac{A \times 60}{1.76} \text{ calls/h} \tag{1.3-2}$$

Then

† Most analyses in this book are based on the present channel numbers of 333 channels per system.

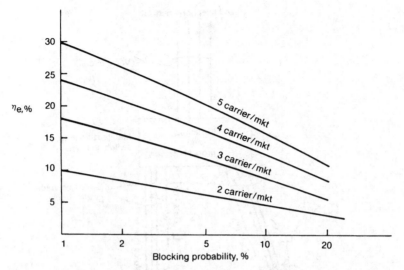

Figure 1.2 Degradation of trunking efficiency—comparing one carrier/market and other-than-one-carrier/market.

$$Q_i = \begin{cases} 2832.95 \text{ calls/h} & (1 \text{ carrier/market}) \\ 1295.45 \times 2 = 2590.9 \text{ calls/h} & (2 \text{ carriers/market}) \end{cases}$$

The trunking efficiency degradation factor can be calculated as

$$\eta_e = \frac{2832.95 - 2590.9}{2832.95} = 8.5\% \qquad (1.3\text{-}3)$$

for a blocking probability of 2 percent. Figure 1.2 shows η_e by comparing one carrier per market with more than one carrier per market situations with different blocking probability conditions. The degradation of trunking efficiency decreases as the blocking probability increases. As the number of carriers per market increases the degradation increases. However, when a high percentage of blocking probability, say more than 20 percent, occurs, the performance of one carrier per market is already so poor that further degradation becomes insignificant, as Fig. 1.2 shows.

For a 2 percent blocking probability, the trunking efficiency of one carrier per market does show a greater advantage when compared to other scenarios.

1.4 A Basic Cellular System

A basic cellular system consists of three parts: a mobile unit, a cell site, and a mobile telephone switching office (MTSO), as Fig. 1.3 shows, with connections to link the three subsystems.

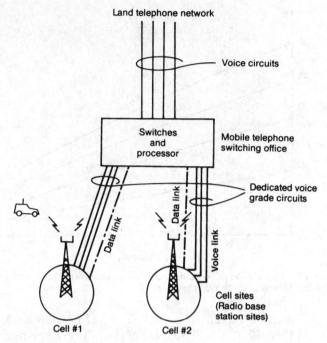

Figure 1.3 Cellular system.

1. *Mobile units.* A mobile telephone unit contains a control unit, a transceiver, and an antenna system.

2. *Cell site.* The cell site provides interface between the MTSO and the mobile units. It has a control unit, radio cabinets, antennas, a power plant, and data terminals.

3. *MTSO.* The switching office, the central coordinating element for all cell sites, contains the cellular processor and cellular switch. It interfaces with telephone company zone offices, controls call processing, and handles billing activities.

4. *Connections.* The radio and high-speed data links connect the three subsystems. Each mobile unit can only use one channel at a time for its communication link. But the channel is not fixed; it can be any one in the entire band assigned by the serving area, with each site having multichannel capabilities that can connect simultaneously to many mobile units.

The MTSO is the heart of the cellular mobile system. Its processor provides central coordination and cellular administration.

The cellular switch, which can be either analog or digital, switches calls to connect mobile subscribers to other mobile subscribers and to the nationwide telephone network. It uses voice trunks similar to telephone company interoffice voice trunks. It also contains data links providing supervision links between the processor and the switch and between the cell sites and the processor. The radio link carries the voice and signaling between the mobile unit and the cell site. The high-speed data links cannot be transmitted over the standard telephone trunks and therefore must use either microwave links or T-carriers (wire lines). Microwave radio links or T-carriers carry both voice and data between the cell site and the MTSO.

1.5 Performance Criteria

There are three categories for specifying performance criteria.

1.5.1 Voice quality

Voice quality is very hard to judge without subjective tests from users' opinions. In this technical area engineers cannot decide how to build a system without knowing the voice quality that will satisfy the users. In military communications, the situation differs: armed forces personnel *must* use the assigned equipment.

For any given commercial communications system, the voice quality will be based upon the following criterion: a set value x at which y percent of customers rate the system voice quality (from transmitter to receiver) as good or excellent, the top two circuit merits (CM) of the five listed below.

CM	Score	Quality scale
CM5	5	Excellent (speech perfectly understandable)
CM4	4	Good (speech easily understandable, some noise)
CM3	3	Fair (speech understandable with a slight effort, occasional repetitions needed)
CM2	2	Poor (speech understandable only with considerable effort, frequent repetitions needed)
CM1	1	Unsatisfactory (speech not understandable)

As the percentage of customers choosing CM4 and CM5 increases, the cost of building the system rises.

The average of the CM scores obtained from all the listeners is called *mean opinion score* (MOS). Usually the toll-quality voice is around MOS ≥ 4.

1.5.2 Service quality

Three items are required for service quality.

1. *Coverage.* The system should serve an area as large as possible. With radio coverage, however, because of irregular terrain configurations, it is usually not practical to cover 100 percent of the area for two reasons:

   a. The transmitted power would have to be very high to illuminate weak spots with sufficient reception, a significant added cost factor.

   b. The higher the transmitted power, the harder it becomes to control interference.

   Therefore, systems usually try to cover 90 percent of an area in flat terrain and 75 percent of an area in hilly terrain. The combined voice quality and coverage criteria in AMPS cellular systems[3] state that 75 percent of users rate the voice quality between good and excellent in 90 percent of the served area, which is generally flat terrain. The voice quality and coverage criteria would be adjusted as per decided various terrain conditions. In hilly terrain, 90 percent of users must rate voice quality good or excellent in 75 percent of the served area. A system operator can lower the percentage values stated above for a low-performance and low-cost system.

2. *Required grade of service.* For a normal start-up system the grade of service is specified for a blocking probability of .02 for initiating calls at the busy hour. This is an average value. However, the blocking probability at each cell site will be different. At the busy hour, near freeways, automobile traffic is usually heavy, so the blocking probability at certain cell sites may be higher than 2 percent, especially when car accidents occur. To decrease the blocking probability requires a good system plan and a sufficient number of radio channels.

3. *Number of dropped calls.* During Q calls in an hour, if a call is dropped and Q − 1 calls are completed, then the call drop rate is 1/Q. This drop rate must be kept low. A high drop rate could be caused by either coverage problems or handoff problems related to inadequate channel availability. How to estimate the number of dropped calls will be described in Chap. 9.

1.5.3 Special features

A system would like to provide as many special features as possible, such as call forwarding, call waiting, voice stored (VSR) box, automatic roaming, or navigation services. However, sometimes the customers may not be willing to pay extra charges for these special services.

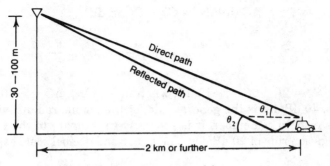

Figure 1.4 Mobile radio transmission model.

1.6 Uniqueness of Mobile Radio Environment

1.6.1 Description of mobile radio transmission medium

The propagation attenuation. In general, the propagation path loss increases not only with frequency but also with distance. If the antenna height at the cell site is 30 to 100 m and at the mobile unit about 3 m, and the distance between the cell site and the mobile unit is usually 2 km or more, then the incident angles of both the direct wave and the reflected wave are very small, as Fig. 1.4 shows. The incident angle of the direct wave is θ_1, and the incident angle of the reflected wave is θ_2. θ_1 is also called the *elevation angle.* The propagation path loss would be 40 dB/dec,[4] where "dec" is an abbreviation of *decade,* i.e., a period of 10. This means that a 40-dB loss at a signal receiver will be observed by the mobile unit as it moves from 1 to 10 km. Therefore C is inversely proportional to R^4.

$$C \propto R^{-4} = \alpha R^{-4} \qquad (1.6\text{-}1)$$

where C = received carrier power
   R = distance measured from the transmitter to the receiver
   α = constant

The difference in power reception at two different distances R_1 and R_2 will result in

$$\frac{C_2}{C_1} = \left(\frac{R_2}{R_1}\right)^{-4} \qquad (1.6\text{-}2a)$$

and the decibel expression of Eq. (1.6-2) is

$$\Delta C \text{ (in dB)} = C_2 - C_1 \text{ (in dB)}$$

$$= 10 \log \frac{C_2}{C_1} = 40 \log \frac{R_1}{R_2} \qquad (1.6\text{-}2b)$$

When $R_2 = 2R_1$, $\Delta C = -12$ dB; when $R_2 = 10R_1$, $\Delta C = -40$ dB.

This 40 dB/dec is the general rule for the mobile radio environment and is easy to remember. It is also easy to compare to the free-space propagation rule of 20 dB/dec. The linear and decibel scale expressions are

$$C \propto R^{-2} \qquad \text{(free space)} \qquad (1.6\text{-}3a)$$

and

$$\Delta C = C_2 \text{ (in dB)} - C_1 \text{ (in dB)}$$

$$= 20 \log \frac{R_1}{R_2} \qquad \text{(free space)} \qquad (1.6\text{-}3b)$$

In a real mobile radio environment, the propagation path-loss slope varies as

$$C \propto R^{-\gamma} = \alpha R^{-\gamma} \qquad (1.6\text{-}4)$$

γ usually lies between 2 and 5 depending on the actual conditions.[5] Of course γ cannot be lower than 2, which is the free-space condition. The decibel scale expression of Eq. (1.6-4) is

$$C = 10 \log \alpha - 10\gamma \log R \qquad \text{dB} \qquad (1.6\text{-}5)$$

Severe fading. Since the antenna height of the mobile unit is lower than its typical surroundings, and the carrier frequency wavelength is much less than the sizes of the surrounding structures, multipath waves are generated. At the mobile unit, the sum of the multipath waves causes a signal-fading phenomenon. The signal fluctuates in a range of about 40 dB (10 dB above and 30 dB below the average signal). We can visualize the nulls of the fluctuation at the baseband at about every half wavelength in space, but all nulls do not occur at the same level, as Fig. 1.5 shows. If the mobile unit moves fast, the rate of fluctuation is fast. For instance, at 850 MHz, the wavelength is roughly 0.35 m (1 ft). If the speed of the mobile unit is 24 km/h (15 mi/h), or 6.7 m/s, the rate of fluctuation of the signal reception at a 10-dB level below the average power of a fading signal is 15 nulls per second (see Sec. 1.6.3).

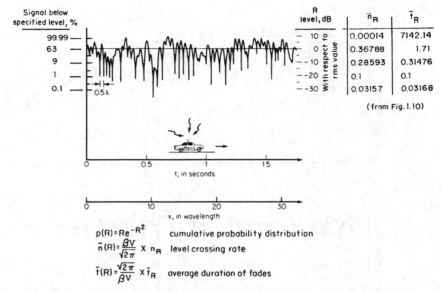

The following equations appear below the figure:

$$p(R) = Re^{-R^2} \quad \text{cumulative probability distribution}$$

$$\bar{n}(R) = \frac{\beta V}{\sqrt{2\pi}} \times n_R \quad \text{level crossing rate}$$

$$\bar{t}(R) = \frac{\sqrt{2\pi}}{\beta V} \times \bar{t}_R \quad \text{average duration of fades}$$

Figure 1.5 A typical fading signal received while the mobile unit is moving. *(Reprint after Lee, Ref. 4, p. 46.)*

1.6.2 Model of transmission medium

A mobile radio signal $r(t)$, illustrated in Fig. 1.6, can be artificially characterized by two components $m(t)$ and $r_0(t)$ based on natural physical phenomena.

$$r(t) = m(t)r_0(t) \tag{1.6-6}$$

The component $m(t)$ is called *local mean, long-term fading,* or *lognormal fading* and its variation is due to the terrain contour between the base station and the mobile unit. The factor r_0 is called *multipath fading, short-term fading,* or *Rayleigh fading* and its variation is due to the waves reflected from the surrounding buildings and other structures. The long-term fading $m(t)$ can be obtained from Eq. (1.6-7a).

$$m(t_1) = \frac{1}{2T} \int_{t_1-T}^{t_1+T} r(t)\, dt \tag{1.6-7a}$$

where $2T$ is the time interval for averaging $r(t)$. T can be determined based on the fading rate of $r(t)$, usually 40 to 80 fades.[5] Therefore, $m(t)$ is the envelope of $r(t)$, as shown in Fig. 1.6a. Equation (1.6-7a) also can be expressed in spatial scale as

$$m(x_1) = \frac{1}{2L} \int_{x_1-L}^{x_1+L} r(x)\, dx \tag{1.6-7b}$$

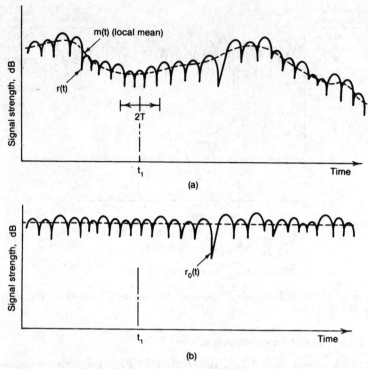

Figure 1.6 A mobile radio signal fading representation. (*a*) A mobile signal fading. (*b*) A short-term signal fading.

The length of $2L$ has been determined to be 20 to 40 wavelengths.[5] Using 36 or up to 50 samples in an interval of 40 wavelengths is an adequate averaging process for obtaining the local means.[4]

The factor $m(t)$ or $m(x)$ is also found to be a log-normal distribution based on its characteristics caused by the terrain contour. The short-term fading r_0 is obtained by

$$r_0 \text{ (in dB)} = r(t) - m(t) \qquad \text{dB} \qquad (1.6\text{-}8)$$

as shown in Fig. 1.6b. The factor $r_0(t)$ follows a Rayleigh distribution, assuming that only reflected waves from local surroundings are the ones received (a normal situation for the mobile radio environment). Therefore, the term Rayleigh fading is often used.

1.6.3 Mobile fading characteristics

Rayleigh fading is also called multipath fading in the mobile radio environment. When these multipath waves bounce back and forth due to the buildings and houses, they form many standing-wave pairs in

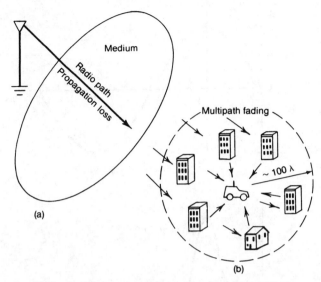

Figure 1.7 A mobile radio environment—two parts. (1) Propagation loss; (2) multipath fading.

space, as shown in Fig. 1.7. Those standing-wave pairs are summed together and become an irregular wave-fading structure. When a mobile unit is standing still, its receiver only receives a signal strength at that spot, so a constant signal is observed. When the mobile unit is moving, the fading structure of the wave in the space is received. It is a multipath fading. The recorded fading becomes fast as the vehicle moves faster.

The radius of the active scatterer region. The mobile radio multipath fading shown in Fig. 1.7 explains the fading mechanism. The radius of the active scatterer region at 850 MHz can be obtained indirectly as shown in Ref. 12. The radius is roughly 100 wavelengths. The active scatterer region always moves with the mobile unit as its center. It means that some houses were inactive scatterers and became active as the mobile unit approached them; some houses were active scatterers and became inactive as the mobile unit drove away from them.

Standing waves expressed in a linear scale and a log scale. We first introduce a sine wave in a log scale.

$$y = 10 \cos \beta x \qquad \text{dB} \qquad (1.6\text{-}9)$$

A log plot of the sine wave of Eq. (1.6-9) is shown in Fig. 1.8a. The linear expression of Eq. (1.6-9) then is shown in Fig. 1.8b. The sym-

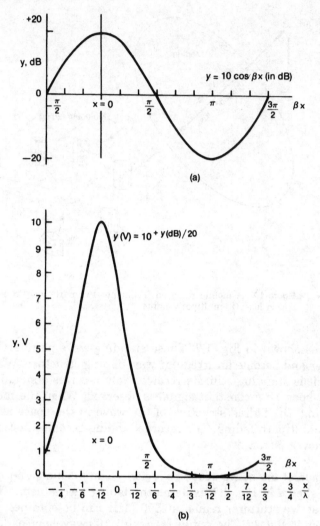

Figure 1.8 The linear plot and the log plot of a sine wave. (*a*) In linear scale; (*b*) in log scale.

metrical waveform in a log plot becomes an unsymmetrical waveform when plotted on a linear scale. It shows that the sine wave waveform in a log scale becomes a completely different waveform when expressed on a linear scale and vice versa. Two sine waves, the incident wave traveling along the x-axis (traveling to the left) and the reflected wave traveling in the opposite direction, can be expressed as

$$e_0 = E_0 e^{j(\omega t + \beta x)} \qquad (1.6\text{-}10)$$

and
$$e_1 = E_1 e^{j(\omega t - \beta x + \delta)} \qquad (1.6\text{-}11)$$

where ω = angular frequency
    β = wave number ($= 2\pi/\lambda$)
    δ = time-phase lead of e_1 with respect to e_0 at $x = 0$

The two waves form a standing-wave pattern.

$$e = e_0 + e_1 = R \cos (\omega t - \delta) \qquad (1.6\text{-}12)$$

where the amplitude R becomes

$$R = \sqrt{(E_0 + E_1)^2 \cos^2 \beta x + (E_0 - E_1)^2 \sin^2 \beta x} \qquad (1.6\text{-}13)$$

We are plotting two cases.

Case 1. $E_0 = 1$, $E_1 = 1$; that is, the reflection coefficient = 1,

$$\text{Standing wave ratio (SWR)} = \frac{E_0 + E_1}{E_0 - E_1} = \infty$$

and
$$R = 2 \cos \beta x \qquad (1.6\text{-}14)$$

Case 2. $E_0 = 1$, $E_1 = 0.5$; that is, the reflection coefficient = 0.5, SWR = 3, and

$$R = \sqrt{(1.5)^2 \cos^2 \beta x + (0.5)^2 \sin^2 \beta x} \qquad (1.6\text{-}15)$$

The linear expression of Eqs. (1.6-14) and (1.6-15) are shown in Fig. 1.9a. The log-scale expression of Eqs. (1.6-14) and (1.6-15) are shown in Fig. 1.9b. The waveform of Fig. 1.9b is the first sign of the fading signal which resembles the real fading signal shown in Fig. 1.5.

First-order and second-order statistics of fading. Fading occurs on the signal reception when the mobile unit is moving. The first-order statistics, such as average power probability cumulative distribution function (CDF) and bit error rate, are independent of time. The second-order statistics, such as level crossing rate, average duration of fades, and word error rate, are time functions or velocity-related functions. The data signaling format is based on these characteristics. The description of the fading characteristic can be found in detail in two books, Refs. 4 and 5.

Some data can be found from Fig. 1.10a, the cumulative distribution function (CDF), and Fig. 1.10b, the level crossing rate. In Fig. 1.10a, the equation of CDF for a Rayleigh fading is used as follows:

$$P(x \le A) = 1 - e^{-A^2/\bar{A}^2} \qquad (1.6\text{-}16)$$

and
$$P(y \le L) = 1 - e^{-L/\bar{L}} \qquad (1.6\text{-}17)$$

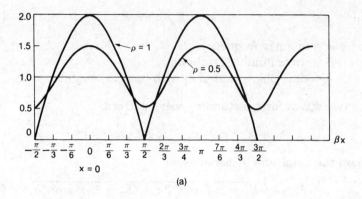

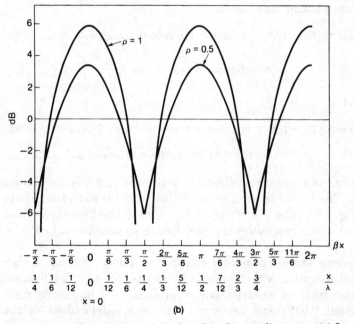

Figure 1.9 The linear plot and the log plot of a standing wave. (*a*) In linear scale; (*b*) in log scale.

where $\overline{A^2}$ and $\overline{L}$ are the mean square value and the average power, respectively. In Fig. 1.10*a*, about 9 percent of the total signal is below a level of -10 dB with respect to average power. In Fig. 1.10*b*, the level crossing rate (lcr) at a level A is

$$\overline{n}(A) = \frac{\beta v}{\sqrt{2\pi}}\, n_R$$

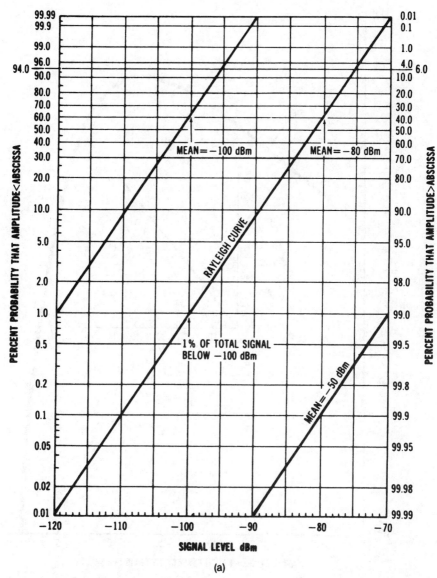

Figure 1.10 Fading characteristics. (*a*) CDF. (*After Lee, Ref. 5, p. 33.*) (*b*) Level crossing rate. (*c*) Average duration of fades.

where n_R is the normalized lcr which is independent of wavelength and the car speed. At a level of -10 dB, $n_0 = 0.3$ can be found from Fig. 1.10*b*. Assume that a signal of 850 MHz is received at a mobile unit with a velocity of 24 km/h (15 mi/h). Then

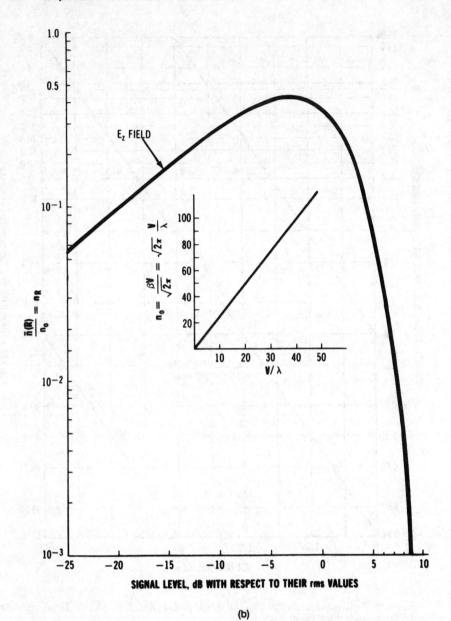

Figure 1.10   *(Continued)*

$$n_0 = \frac{\beta V}{\sqrt{2\pi}} \approx 50$$

and $$\overline{n} = 50 \times 0.3 = 15$$

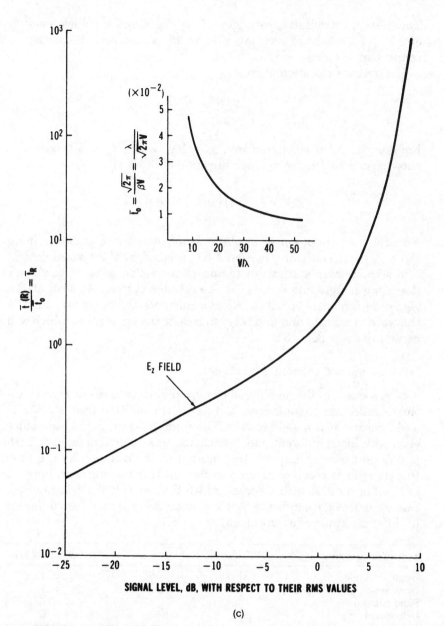

E_z FIELD

SIGNAL LEVEL, dB, WITH RESPECT TO THEIR RMS VALUES

(c)

Figure 1.10 *(Continued)*

Therefore, at a cellular frequency of 800 MHz and a vehicle velocity of 15 mi/h, the level crossing rate is 15 per second. It is easy to remember.

The average duration of fade is

$$\bar{t} = \frac{\text{CDF}}{\bar{t_0}\bar{t_R}} = \frac{\text{CDF}}{\bar{n}} \tag{1.6-19}$$

Equation (1.6-19) is plotted in Fig. 1.10c, where t_0 and t_R are also shown. At -10 dB, the average duration of fades is

$$\bar{t} = \frac{\text{CDF}}{\bar{n}} = 0.0066 \text{ s} = 6.6 \text{ ms}$$

Now the average power level plays an important role in determining the statistics. Therefore, it should be specified by the system design. The second-order statistic of fading phenomenon is most useful for designing a signaling format for the cellular system. As soon as the signaling format is specified, we can calculate the bit error rate and the word error rate and find ways to reduce the error rates, which will be described in Sec. 13.2.

Delay spread and coherence bandwidth

Delay spread. In the mobile radio environment, as a result of the multipath reflection phenomenon, the signal transmitted from a cell site and arriving at a mobile unit will be from different paths, and since each path has a different path length, the time of arrival for each path is different. For an impulse transmitted at the cell site, by the time this impulse is received at the mobile unit it is no longer an impulse but rather a pulse with a spread width that we call the *delay spread*. The measured data indicate that the mean delay spreads are different in different kinds of environment.

Type of environment	Delay spread Δ, μs
Inside the building	<0.1
Open area	<0.2
Suburban area	0.5
Urban area	3

Coherence bandwidth. The coherence bandwidth is the defined bandwidth in which either the amplitudes or the phases of two received signals have a high degree of similarity. The delay spread is a natural phenomenon, and the coherence bandwidth is a defined creation related to the delay spread.

A coherence bandwidth for two fading amplitudes of two received signals is

$$B_c = \frac{1}{2\pi\Delta}$$

A coherence bandwidth for two random phases of two received signals is

$$`B_c' = \frac{1}{4\pi\Delta}$$

1.6.4 Direct wave path, line-of-sight path, and obstructive path

A *direct wave path* is a path clear from the terrain contour. The *line-of-sight path* is a path clear from buildings. In the mobile radio environment, we do not always have a line-of-sight condition.

When a line-of-sight condition occurs, the average received signal at the mobile unit at a 1-mi intercept is higher, although the 40 dB/dec path-loss slope remains the same. It will be described in Sec. 4.2. In this case the short-term fading is observed to be a rician fading.[14] It results from a strong line-of-sight path and a ground-reflected wave combined, plus many weak building-reflected waves.

When an out-of-sight condition is reached, the 40-dB/dec path-loss slope still remains. However, all reflected waves, including ground reflected waves and building-reflected waves, become dominant. The short-term received signal at the mobile unit observes a Rayleigh fading. The Rayleigh fading is the most severe fading.

When the terrain contour blocks the direct wave path, we call it the *obstructive path*. In this situation, the shadow loss from the signal reception can be found by using the knife-edge diffraction curves shown in Sec. 4.7.2.

1.6.5 Noise level in cellular frequency band

The thermal noise kTB at a temperature T of 290 K (17°C) and a bandwidth B of 30 kHz is -129 dBm.* Assume that the received front-end noise is 9 dB, then the noise level is -120 dBm. Now there are two kinds of man-made noise, the ignition noise generated by the vehicles and the noise generated by 800-MHz emissions.

---

* k is Boltzmann's constant, and $kT = -174$ dBm/Hz at $T = 290$ K.

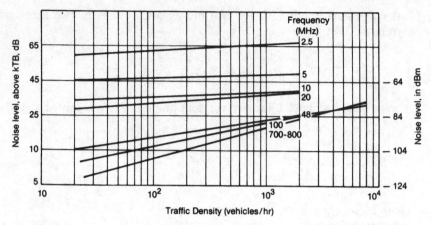

Figure 1.11 Average automotive-traffic-noise power for various traffic densities and frequencies. Detector noise bandwidth 30 kHz at room temperature (17°C). *(After Lee, Ref. 16.)*

The ignition noise. In the past, 800 MHz was not widely used. Therefore, the man-made noise at 800 MHz is merely generated by the vehicle ignition noise.[10] The automotive noise introduced at 800 MHz with a bandwidth of 30 kHz can be deduced from Ref. 15, as shown in Fig. 1.11.

The 800-MHz-emission noise. As a result of the cellular mobile systems operating in all the major cities in the United States and the spurious energy generated outside each channel bandwidth, the early noise data measurements[9,10] are no longer valid. The 800-MHz-emission noise can be measured at an idle channel (a forward voice channel) in the 869- to 894-MHz region while the mobile receiver is operating on a car battery in a no-traffic spot in a city. In this case, no automotive ignition noise is involved, and no cochannel operation is in the proximity of the idle-channel receiver. We found that in some areas the noise level is 2 to 3 dB higher than -120 dBm at the cell sites and 3 to 4 dB higher than -120 dBm at the mobile stations.

1.6.6   Amplifier noise

A mobile radio signal received by a receiving antenna, either at the cell site or at the mobile unit, will be amplified by an amplifier. We would like to understand how the signal is affected by the amplifier noise. Assume that the amplifier has an available power gain g and the available noise power at the output is N_0. The input signal-to-noise

(S/N) ratio is P_s/N_i, the output signal-to-noise ratio is P_o/N_o, and the internal amplifier noise is N_a. Then the output P_o/N_o becomes

$$\frac{P_o}{N_o} = \frac{gP_s}{g(N_i) + N_a} = \frac{P_s}{N_i + (N_a/g)} \qquad (1.6\text{-}20)$$

The noise figure F is defined as

$$F = \frac{\text{maximum possible S/N ratio}}{\text{actual S/N ratio at output}} \qquad (1.6\text{-}21)$$

where the maximum possible S/N ratio is measured when the load is an open circuit. Equation (1.6-21) can be used for obtaining the noise figure of the amplifier.

$$F = \frac{P_s/kTB}{P_o/N_o} = \frac{N_o}{(P_o/P_s)kTB} = \frac{N_o}{g(kTB)} \qquad (1.6\text{-}22)$$

Also substituting Eq. (1.6-20) into Eq. (1.6-22) yields

$$F = \frac{P_s/kTB}{P_s/[N_i + (N_a/g)]} = \frac{N_i + (N_a/g)}{kTB} \qquad (1.6\text{-}23)$$

The term kTB is the thermal noise as described in Sec. 1.6.5. The noise figure is a reference measurement between a minimum noise level due to thermal noise and the noise level generated by both the external and internal noise of an amplifier.

1.7 Operation of Cellular Systems

This section briefly describes the operation of the cellular mobile system from a customer's perception without touching on the design parameters.[17,18] The operation can be divided into four parts and a handoff procedure.

Mobile unit initialization. When a user sitting in a car activates the receiver of the mobile unit, the receiver scans 21 set-up channels which are designated among the 416 channels. It then selects the strongest and locks on for a certain time. Since each site is assigned a different set-up channel, locking onto the strongest set-up channel usually means selecting the nearest cell site. This self-location scheme is used in the idle stage and is user-independent. It has a great advantage because it eliminates the load on the transmission at the cell site for locating the mobile unit. The disadvantage of the self-location scheme is that no location information of idle mobile units appears at each cell site. Therefore, when the call initiates from the land line to a mobile unit, the paging process is longer. Since a large percentage of calls originates at the mobile unit, the use of self-location schemes is justified. After 60 s, the self-location procedure is repeated. In the future, when land-line originated calls increase, a feature called "registration" can be used.

Mobile originated call. The user places the called number into an originating register in the mobile unit, checks to see that the number is correct, and pushes the "send" button. A request for service is sent on a selected set-up channel obtained from a self-location scheme. The cell site receives it, and in directional cell sites, selects the best directive antenna for the voice channel to use. At the same time the cell site sends a request to the mobile telephone switching office (MTSO) via a high-speed data link. The MTSO selects an appropriate voice channel for the call, and the cell site acts on it through the best directive antenna to link the mobile unit. The MTSO also connects the wire-line party through the telephone company zone office.

Network originated call. A land-line party dials a mobile unit number. The telephone company zone office recognizes that the number is mobile and forwards the call to the MTSO. The MTSO sends a paging message to certain cell sites based on the mobile unit number and the search algorithm. Each cell site transmits the page on its own set-up channel. The mobile unit recognizes its own identification on a strong set-up channel, locks onto it, and responds to the cell site. The mobile unit also follows the instruction to tune to an assigned voice channel and initiate user alert.

Call termination. When the mobile user turns off the transmitter, a particular signal (signaling tone) transmits to the cell site, and both sides free the voice channel. The mobile unit resumes monitoring pages through the strongest set-up channel.

Handoff procedure. During the call, two parties are on a voice channel. When the mobile unit moves out of the coverage area of a particular cell site, the reception becomes weak. The present cell site requests a handoff. The system switches the call to a new frequency channel in a new cell site without either interrupting the call or alerting the user. The call continues as long as the user is talking. The user does not notice the handoff occurrences. *Handoff* was first used by the AMPS system, then renamed *handover* by the European systems because the different meanings in English English and American English. Description of handoff will appear in Chap. 9.

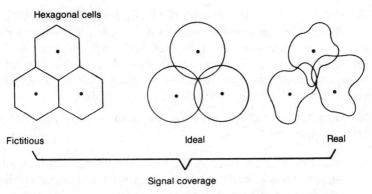

Figure 1.12 Hexagonal cells and the real shapes of their coverages.

1.8 Marketing Image of Hexagonal-shaped Cells

We have to realize that hexagonal-shaped communication cells are artificial and that such a shape cannot be generated in the real world. Engineers draw hexagonal-shaped cells on a layout to simplify the planning and design of a cellular system because it approaches a circular shape that is the ideal power coverage area. The circular shapes have overlapped areas which make the drawing unclear. The hexagonal-shaped cells fit the planned area nicely, as shown in Fig. 1.12, with no gap and no overlap between the hexagonal cells. The ideal cell shapes as well as the real cell shapes are also shown in Fig. 1.12.

A simple mechanism which makes the cellular system implementable based on hexagonal cells will be illustrated in later chapters. Otherwise, a statistical approach will be used in dealing with a real-world situation. Fortunately, the outcomes resulting from these two approaches are very close, yet the latter does not provide a clear physical picture, as shown later. Besides, today these hexagonal-shaped cells have already become a widely promoted symbol for cellular mobile systems. An analysis using hexagonal cells, if it is desired, can easily be adapted by the reader.

1.9 Planning a Cellular System

1.9.1 How to start planning

Assume that the construction permit for a cellular system in a particular market area is granted. The planning stage becomes critical. A great deal of money can be spent and yet poor service may be provided if we do not know how to create a good plan. First, we have to determine two elements: regulations and the market situation.

Regulations. The federal regulations administered by the FCC are the same throughout the United States. The state regulations may be different from state to state, and each city and town may have its own building codes and zoning laws. Become familiar with the rules and regulations. Sometimes waivers need to be applied for ahead of time. Be sure that the plan is workable.

Market situation. There are three tasks to be handled by the marketing department.

1. *Prediction of gross income.* We have to determine the population, average income, business types, and business zones so that the gross income can be predicted.

2. *Understanding competitors.* We also need to know the competitor's situation, coverage, system performance, and number of customers. Any system should provide a unique and outstanding service to overcome the competition.

3. *Decision* of *geographic coverage.* What general area should ultimately be covered? What near-term service can be provided in a limited area? These questions should be answered and the decisions passed on to the engineering department.

1.9.2 The engineer's role

The engineers follow the market decisions by

1. Initiating a cellular mobile service in a given area by creating a plan that uses a minimum number of cell sites to cover the whole area. It is easy for marketing to request but hard for the engineers to fulfill. We will address this topic later.

2. Checking the areas that marketing indicated were important revenue areas. The number of radios (number of voice channels) required to handle the traffic load at the busy hours should be determined.

3. Studying the interference problems, such as cochannel and adjacent channel interference, and the intermodulation products generated at the cell sites, and finding ways to reduce them.

4. Studying the blocking probability of each call at each cell site, and trying to minimize it.

5. Planning to absorb more new customers. The rate at which new customers subscribe to a system can vary depending on the service charges, system performance, and seasons of the year. Engineering has to try to develop new technologies to utilize fully the limited

spectrum assigned to the cellular system. The analysis of spectrum efficiency due to the natural limitations may lead to a request for a larger spectrum.

1.9.3 Finding solutions

Many practical design tools, methods of reducing interference, and ways of solving the blocking probability of call initiation will be introduced in this book.

1.10 Analog Cellular Systems

1.10.1 Cellular systems in the United States

There are 150 major market areas* in the United States where licenses for cellular systems can be granted by the FCC. They have been classified by their populations into five groups. Each group has 30 cities.

1. Top 30 markets—very large cities

2. Top 31 to 60 markets—large-sized cities

3. Top 61 to 90 markets—medium-sized cities

4. Top 91 to 120 markets—below medium-sized cities

5. Top 121 to 150 markets—small-sized cities

Each market area is planned to have two systems. The status of each system in each area of groups 1 to 3 as of December 1985 appears in Appendix 1.2. The specifications of the system are described in detail in Chap. 3. There are 305 MSAs (metropolitan statistical areas) and 482 RSAs (rural statistical areas). Both MSAs and RSAs are listed in Appendix 1.3.

1.10.2 Cellular systems outside the United States

Japan.[6] Nippon Telegraph and Telephone Corporation (NTT) developed an 800-MHz land mobile telephone system and put it into service in the Tokyo area in 1979. The general system operation is similar to the AMPS system. It accesses approximately 40,000 subscribers in 500 cities. It covers 75 percent of all Japanese cities, 25 percent of inhabitable areas, and 60 percent of the population. In Japan, 9 automobile switching centers (ASCs), 51 mobile control stations (MCSs), 465 mo-

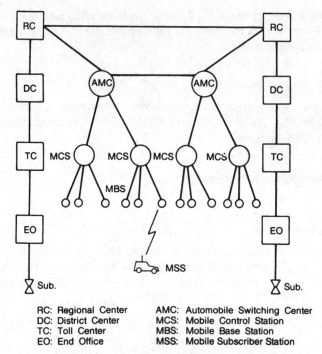

RC: Regional Center   AMC: Automobile Switching Center
DC: District Center   MCS: Mobile Control Station
TC: Toll Center   MBS: Mobile Base Station
EO: End Office   MSS: Mobile Subscriber Station

Figure 1.13 Japanese mobile telephone service network configuration.

bile base stations (MBSs), and 39,000 mobile subscriber stations (MSSs) were in operation as of February 1985.

The Japanese mobile telephone service network configuration is shown in Fig. 1.13. In the metropolitan Tokyo area, about 30,000 subscribers are being served.

The 1985 system operated over a spectrum of 30 MHz. The total number of channels was 600, and the channel bandwidth was 25 kHz. This system comprised an automobile switching center (ASC), a mobile control station (MCS), a mobile base station (MBS), and a mobile subscriber station (MSS). At present there is no competitive situation set up by the government. However, the Japanese Ministry of Post and Telecommunication (MFT) is considering providing a dual competitive situation similar to that in the United States.

United Kingdom.[7]   In June 1982 the government of the United Kingdom announced two competing national cellular radio networks. The UK system is called TACS (Total Access Communications System). The total number of channels was 1000, with a channel bandwidth of

25 kHz per channel. Among them, 600 channels are assigned and 400 are reserved. Two competing cellular network operators, Cellnet and Vodafone, are operating in the United Kingdom. Each network system has only 300 spectral channels. The Cellnet system started operating in January 1985. Cellnet has over 200 cell sites, covering 82 percent of the United Kingdom. Vodaphone, though, which started operations late, has served the same areas as Cellnet.

Canadian system.[8] In 1978, a system called AURORA was designed for the Alberta government telephone (AGT). The system provides provincewide mobile telephone service at 400 MHz. Ongoing developmental work on the AURORA is underway at 800 MHz.

AURORA 400 system. It is aimed at 40,000 subscribers living in an area approximately 1920 km × 960 km. The AURORA 400 system initially has 40 channels and is expected to add an additional 20 channels with frequency reuse and a seven-cell cluster plan. A fully implemented system has 120 cells. The 400-MHz system does not have a handoff capability.

AURORA 800 system. The AURORA 800 system is truly frequency-transparent. By repackaging the radio frequency (RF) sections on the cell site, the mobile unit can be operated on any mobile RF band up to 800 MHz. The handoff capability will be implemented in this system.

Nordic system. This system was built mostly by Scandinavian countries (Denmark, Norway, Sweden, and Finland) in cooperation with Saudi Arabia and Spain and is called the *NMT network*. It is currently a 450-MHz system, but an 800-MHz system will be implemented soon since the frequency-transparent concept as the AURORA 800 system is used to convert the 450-MHz system to the 800-MHz system.

The total bandwidth is 10 MHz, which has 200 channels with a bandwidth of 25 kHz per channel. This system does have handoff and roaming capabilities. It also uses repeaters to increase the coverage in a low traffic area. The total number of subscribers is around 100,000.

European cellular systems.[11] All the present generation of European cellular networks is totally lacking in cross-border compatibility. Besides the United Kingdom and NMT networks, the others include the following.

Benelux-country network. The Netherlands served on their ATF2 network (the same as the NMT 450 network) at the beginning of 1985. It has a nationwide coverage using 50 cell sites with two different cell sizes, 20- and 5-km radii. The capacity of the present system is 15,000 to 20,000 subscribers. Dutch PT&T is using a single Ericsson AXE10 switch. Luxembourg came on air in August 1985. In 1986, Belgium joined the network. It operates at 450 MHz. The network is compatible among the three countries.

France. A direct-dial car telephone operating at 160 MHz can access the system in 10 regional areas. The network serves 10,000 subscribers. By the end of 1984, 450 MHz was in operation. In the meantime Radicom 2000 (digital signaling) was introduced, operating at 200 MHz but with no handoff feature.

Spain. It uses an NMT 450-MHz cellular network introduced in 1982. It was the first cellular system in Europe. The number of cells in service is 13. There are three separate networks operating 104 channels. Each channel bandwidth is 25 kHz.

Austria. A new NMT cellular network called *Autotelefonnetz C* has two mobile switching exchanges and has enough capacity for 30,000 subscribers.

The Austrian PT&T has allocated 222 duplex channels in ranges 451.3 to 455.7 MHz and 461.3 to 465.7 MHz, with a channel bandwidth of 20 kHz.

Although both Austria and Spain are using NMT 450 systems, their systems are not compatible because of different frequency allocations, channel spacings (bandwidth), and protocols by different PT&Ts.

Germany. A full national coverage, including West Berlin, using a C-450 cellular system was installed in September 1985 with 100 cell sites. Another 75 cell sites were completed in mid-1986. Also, Germany and France are working on cross-border compatibility in cellular radio systems and have proposed a CD-900 digital system. The details of the CD-900 system will be described in Chap. 14.

Switzerland. Swiss PT&T decided to install an NMT 900-MHz cellular network that had a capacity of 12,000 subscribers. A pilot scheme with 20 transmitters (cell sites) was installed in the Zurich area in late 1986.

Cellular systems in the rest of the world. Australia is installing a system using Ericsson's AXE-10 switching network and will operate at 800 MHz with 12 sites concentrated in three big cities.

Kuwait's cellular system uses NEC's switches and provides 12 sites. It operates at 800 MHz.

Hong Kong has three systems. The United Kingdom's TACS system is installed with Motorola switches. The United States' AMPS system and Japanese NEC systems were also installed in Hong Kong. It is a very competitive market. All systems are penetrating the markets of both portable sets and car sets.

1.11 Digital Cellular Systems

In 1992 the first digital cellular system, GSM (Special Mobile Group), was deployed in Germany. GSM is a European standard system. In the United States, an NA-TDMA system (IS-54) and a CDMA system (IS-95) have been developed. NA-TDMA was deployed in 1993 and CDMA is planned for deployment in 1995. A Japanese system, PDC (Personal Digital Cellular), was deployed in Osaka in June 1994. All the digital cellular systems and some noncellular digital systems will be described in Chap. 15.

References

1. Bell Laboratories, "High-Capacity Mobile Telephone System Technical Report," December 1971, submitted to FCC.
2. F. H. Blecher, "Advanced Mobile Phone Service," *IEEE Transactions on Vehicular Technology,* Vol. VT-29, May 1980, pp. 238–244.
3. V. H. MacDonald, "The Cellular Concept," *Bell System Technical Journal,* Vol. 58, January 1979, pp. 15–42.
4. W. C. Y. Lee, *Mobile Communications Engineering,* McGraw-Hill Book Co., 1982, p. 104.
5. W. C. Y. Lee, *Mobile Communications Design Fundamentals,* Howard W. Sams & Co., 1986, Chap. 2.
6. T. Kiuchi, K. Tsujimura, and K. Kato, "Progress in the 800-MHz Land Mobile Telephone System in Japan," *35th IEEE Vehicular Technology Conference Record,* Boulder, Colo., May 21–23, 1985, pp. 304–309.
7. David M. Barnes, "The Introduction of Cellular Radio in the United Kingdom," *35th IEEE Vehicular Technology Conference Record,* Boulder, Colo., May 21–23, 1985, pp. 147–152.
8. M. Ali, "The AURORA Cellular Mobile Telephone System," *33rd IEEE Vehicular Technology Conference Record,* Toronto, May 1983, pp. 302–307.
9. *ITT Reference Data for Radio Engineers,* Howard W. Sams & Co., 1970, p. 2–27.
10. A. D. Spaulding and R. T. Disney, "Man-Made Radio Noise, Part I," U.S. Department of Commerce, Office of Technical Services Report 74-38, June 1974.
11. N. Cawthorne, "Cellular Radio—A European Round-up," *Wireless World,* April 1986, pp. 33–36.
12. W. C. Y. Lee, *Mobile Communications Engineering,* McGraw-Hill Book Co., 1982, p. 202.
13. Henry G. Fischer and John W. Willis (eds.), Pike and Fischer Radio Regulations, 2d series, vol. 19, pp. 1681–1728, for a summary of submissions received by the FCC on Dockets 18261 and 18262.

14. W. C. Y. Lee, *Mobile Communications Design Fundamentals,* Howard W. Sams & Co., 1986, pp. 28 and 33.
15. E. N. Skomal, *Man-Made Radio Noise,* Van Nostrand Reinhold Co., 1978, Chap. 2.
16. W. C. Y. Lee, *Mobile Communications Design Fu:.Jamentals,* Howard W. Sams & Co., 1986, p. 224.
17. J. Oetting, "Cellular Mobile Radio—An Emerging Technology," *IEEE Communications Magazine,* Vol. 21, No. 8, November 1983, pp. 10–15.
18. "Advanced Mobile Phone Service," special issue, *Bell System Technical Journal,* Vol. 58, January 1979.

Appendix 1.1
Blocked-Calls-Cleared
(Erlang *B*)

A, erlangs

N	B												
	1.0%	1.2%	1.5%	2%	3%	5%	7%	10%	15%	20%	30%	40%	50%
1	.0101	.0121	.0152	.0204	.0309	.0526	.0753	.111	.176	.250	.429	.667	1.00
2	.153	.168	.190	.223	.282	.381	.470	.595	.796	1.00	1.45	2.00	2.73
3	.455	.489	.535	.602	.715	.899	1.06	1.27	1.60	1.93	2.63	3.48	4.59
4	.869	.922	.992	1.09	1.26	1.52	1.75	2.05	2.50	2.95	3.39	5.02	6.50
5	1.36	1.43	1.52	1.66	1.88	2.22	2.50	2.88	3.45	4.01	5.19	6.60	8.44
6	1.91	2.00	2.11	2.28	2.54	2.96	3.30	3.76	4.44	5.11	6.51	8.19	10.4
7	2.50	2.60	2.74	2.94	3.25	3.74	4.14	4.67	5.46	6.23	7.86	9.80	12.4
8	3.13	3.25	3.40	3.63	3.99	4.54	5.00	5.60	6.50	7.37	9.21	11.4	14.3
9	3.78	3.92	4.09	4.34	4.75	5.37	5.88	6.55	7.55	8.52	10.6	13.0	16.3
10	4.46	4.61	4.81	5.08	5.53	6.22	6.78	7.51	8.62	9.68	12.0	14.7	18.3
11	5.16	5.32	5.54	5.84	6.33	7.08	7.69	8.49	9.69	10.9	13.3	16.3	20.3
12	5.88	6.05	6.29	6.61	7.14	7.95	8.61	9.47	10.8	12.0	14.7	18.0	22.2
13	6.61	6.80	7.05	7.40	7.97	8.83	9.54	10.5	11.9	13.2	16.1	19.6	24.2
14	7.35	7.56	7.82	8.20	8.80	9.73	10.5	11.5	13.0	14.4	17.5	21.2	26.2
15	8.11	8.33	8.61	9.01	9.65	10.6	11.4	12.5	14.1	15.6	18.9	22.9	28.2
16	8.88	9.11	9.41	9.83	10.5	11.5	12.4	13.5	15.2	16.8	20.3	24.5	30.2
17	9.65	9.89	10.2	10.7	11.4	12.5	13.4	14.5	16.3	18.0	21.7	26.2	32.2
18	10.4	10.7	11.0	11.5	12.2	13.4	14.3	15.5	17.4	19.2	23.1	27.8	34.2
19	11.2	11.5	11.8	12.3	13.1	14.3	15.3	16.6	18.5	20.4	24.5	29.5	36.2
20	12.0	12.3	12.7	13.2	14.0	15.2	16.3	17.6	19.6	21.6	25.9	31.2	38.2

Appendix 1.1
Blocked-Calls-Cleared
(Erlang B) (Continued)

						A, erlangs							
						B							
N	1.0%	1.2%	1.5%	2%	3%	5%	7%	10%	15%	20%	30%	40%	50%
21	12.8	13.1	13.5	14.0	14.9	16.2	17.3	18.7	20.8	22.8	27.3	32.8	40.2
22	13.7	14.0	14.3	14.9	15.8	17.1	18.2	19.7	21.9	24.1	28.7	34.5	42.1
23	14.5	14.8	15.2	15.8	16.7	18.1	19.2	20.7	23.0	25.3	30.1	36.1	44.1
24	15.3	15.6	16.0	16.6	17.6	19.0	20.2	21.8	24.2	26.5	31.6	37.8	46.1
25	16.1	16.5	16.9	17.5	18.5	20.0	21.2	22.8	25.3	27.7	33.0	39.4	48.1
26	17.0	17.3	17.8	18.4	19.4	20.9	22.2	23.9	26.4	28.9	34.4	41.1	50.1
27	17.8	18.2	18.6	19.3	20.3	21.9	23.2	24.9	27.6	30.2	35.8	42.8	52.1
28	18.6	19.0	19.5	20.2	21.2	22.9	24.2	26.0	28.7	31.4	37.2	44.4	54.1
29	19.5	19.9	20.4	21.0	22.1	23.8	25.2	27.1	29.9	32.6	38.6	46.1	56.1
30	20.3	20.7	21.2	21.9	23.1	24.8	26.2	28.1	31.0	33.8	40.0	47.7	58.1
31	21.2	21.6	22.1	22.8	24.0	25.8	27.2	29.2	32.1	35.1	41.5	49.4	60.1
32	22.0	22.5	23.0	23.7	24.9	26.7	28.2	30.2	33.3	36.3	42.9	51.1	62.1
33	22.9	23.3	23.9	24.6	25.8	27.7	29.3	31.3	34.4	37.5	44.3	52.7	64.1
34	23.8	24.2	24.8	25.5	26.8	28.7	30.3	32.4	35.6	38.8	45.7	54.4	66.1
35	24.6	25.1	25.6	26.4	27.7	29.7	31.3	33.4	36.7	40.0	47.1	56.0	68.1
36	25.5	26.0	26.5	27.3	28.6	30.7	32.3	34.5	37.9	41.2	48.6	57.7	70.1
37	26.4	26.8	27.4	28.3	29.6	31.6	33.3	35.6	39.0	42.4	50.0	59.4	72.1
38	27.3	27.7	28.3	29.2	30.5	32.6	34.4	36.6	40.2	43.7	51.4	61.0	74.1
39	28.1	28.6	29.2	30.1	31.5	33.6	35.4	37.7	41.3	44.9	52.8	62.7	76.1
40	29.0	29.5	30.1	31.0	32.4	34.6	36.4	38.8	42.5	46.1	54.2	64.4	78.1

41	29.9	30.4	31.0	31.9	33.4	35.6	37.4	39.9	43.6	47.4	55.7	66.0	80.1
42	30.8	31.3	31.9	32.8	34.3	36.6	38.4	40.9	44.8	48.6	57.1	67.7	82.1
43	31.7	32.2	32.8	33.8	35.3	37.6	39.5	42.0	45.9	49.9	58.5	69.3	84.1
44	32.5	33.1	33.7	34.7	36.2	38.6	40.5	43.1	47.1	51.1	59.9	71.0	86.1
45	33.4	34.0	34.6	35.6	37.2	39.6	41.5	44.2	48.2	52.3	61.3	72.7	88.1
46	34.3	34.9	35.6	36.5	38.1	40.5	42.6	45.2	49.4	53.6	62.8	74.3	90.1
47	35.2	35.8	36.5	37.5	39.1	41.5	43.6	46.3	50.6	54.8	64.2	76.0	92.1
48	36.1	36.7	37.4	38.4	40.0	42.5	44.6	47.4	51.7	56.0	65.6	77.7	94.1
49	37.0	37.6	38.3	39.3	41.0	43.5	45.7	48.5	52.9	57.3	67.0	79.3	96.1
50	37.9	38.5	39.2	40.3	41.9	44.5	46.7	49.6	54.0	58.5	68.5	81.0	98.1
51	38.8	39.4	40.1	41.2	42.9	45.5	47.7	50.6	55.2	59.7	69.9	82.7	100.1
52	39.7	40.3	41.0	42.1	43.9	46.5	48.8	51.7	56.3	61.0	71.3	84.3	102.1
53	40.6	41.2	42.0	43.1	44.8	47.5	49.8	52.8	57.5	62.2	72.7	86.0	104.1
54	41.5	42.1	42.9	44.0	45.8	48.5	50.8	53.9	58.7	63.5	74.2	87.6	106.1
55	42.4	43.0	43.8	44.9	46.7	49.5	51.9	55.0	59.8	64.7	75.6	89.3	108.1
56	43.3	43.9	44.7	45.9	47.7	50.5	52.9	56.1	61.0	65.9	77.0	91.0	110.1
57	44.2	44.8	45.7	46.8	48.7	51.5	53.9	57.1	62.1	67.2	78.4	92.6	112.1
58	45.1	45.8	46.6	47.8	49.6	52.6	55.0	58.2	63.3	68.4	79.8	94.3	114.1
59	46.0	46.7	47.5	48.7	50.6	53.6	56.0	59.3	64.5	69.7	81.3	96.0	116.1
60	46.9	47.6	48.4	49.6	51.6	54.6	57.1	60.4	65.6	70.9	82.7	97.6	118.1
61	47.9	48.5	49.4	50.6	52.5	55.6	58.1	61.5	66.8	72.1	84.1	99.3	120.1
62	48.8	49.4	50.3	51.5	53.5	56.6	59.1	62.6	68.0	73.4	85.5	101.0	122.1
63	49.7	50.4	51.2	52.5	54.5	57.6	60.2	63.7	69.1	74.6	87.0	102.6	124.1
64	50.6	51.3	52.2	53.4	55.4	58.6	61.2	64.8	70.3	75.9	88.4	104.3	126.1
65	51.5	52.2	53.1	54.4	56.4	59.6	62.3	65.8	71.4	77.1	89.8	106.0	128.1
66	52.4	53.1	54.0	55.3	57.4	60.6	63.3	66.9	72.6	78.3	91.2	107.6	130.1
67	53.4	54.1	55.0	56.3	58.4	61.6	64.4	68.0	73.8	79.6	92.7	109.3	132.1
68	54.3	55.0	55.9	57.2	59.3	62.6	65.4	69.1	74.9	80.8	94.1	111.0	134.1
69	55.2	55.9	56.9	58.2	60.3	63.7	66.4	70.2	76.1	82.1	95.5	112.6	136.1
70	56.1	56.8	57.8	59.1	61.3	64.7	67.5	71.3	77.3	83.3	96.9	114.3	138.1

Appendix 1.1
Blocked-Calls-Cleared
(Erlang *B*) (*Continued*)

A, erlangs

N	B												
	1.0%	1.2%	1.5%	2%	3%	5%	7%	10%	15%	20%	30%	40%	50%
71	57.0	57.8	58.7	60.1	62.3	65.7	68.5	72.4	78.4	84.6	98.4	115.9	140.1
72	58.0	58.7	59.7	61.0	63.2	66.7	69.6	73.5	79.6	85.8	99.8	117.6	142.1
73	58.9	59.6	60.6	62.0	64.2	67.7	70.6	74.6	80.8	87.0	101.2	119.3	144.1
74	59.8	60.6	61.6	62.9	65.2	68.7	71.7	75.6	81.9	88.3	102.7	120.9	146.1
75	60.7	61.5	62.5	63.9	66.2	69.7	72.7	76.7	83.1	89.5	104.1	122.6	148.0
76	61.7	62.4	63.4	64.9	67.2	70.8	73.8	77.8	84.2	90.8	105.5	124.3	150.0
77	62.6	63.4	64.4	65.8	68.1	71.8	74.8	78.9	85.4	92.0	106.9	125.9	152.0
78	63.5	64.3	65.3	66.8	69.1	72.8	75.9	80.0	86.6	93.3	108.4	127.6	154.0
79	64.4	65.2	66.3	67.7	70.1	73.8	76.9	81.1	87.7	94.5	109.8	129.3	156.0
80	65.4	66.2	67.2	68.7	71.1	74.8	78.0	82.2	88.9	95.7	111.2	130.9	158.0
81	66.3	67.1	68.2	69.6	72.1	75.8	79.0	83.3	90.1	97.0	112.6	132.6	160.0
82	67.2	68.0	69.1	70.6	73.0	76.9	80.1	84.4	91.2	98.2	114.1	134.3	162.0
83	68.2	69.0	70.1	71.6	74.0	77.9	81.1	85.5	92.4	99.5	115.5	135.9	164.0
84	69.1	69.9	71.0	72.5	75.0	78.9	82.2	86.6	93.6	100.7	116.9	137.6	166.0
85	70.0	70.9	71.9	73.5	76.0	79.9	83.2	87.7	94.7	102.0	118.3	139.3	168.0
86	70.9	71.8	72.9	74.5	77.0	80.9	84.3	88.8	95.9	103.2	119.8	140.9	170.0
87	71.9	72.7	73.8	75.4	78.0	82.0	85.3	89.9	97.1	104.5	121.2	142.6	172.0
88	72.8	73.7	74.8	76.4	78.9	83.0	86.4	91.0	98.2	105.7	122.6	144.3	174.0
89	73.7	74.6	75.7	77.3	79.9	84.0	87.4	92.1	99.4	106.9	124.0	145.9	176.0
90	74.7	75.6	76.7	78.3	80.9	85.0	88.5	93.1	100.6	108.2	125.5	147.6	178.0

91	75.6	76.5	77.6	79.3	81.9	86.0	89.5	94.2	101.7	109.4	126.9	149.3	180.0
92	76.6	77.4	78.6	80.2	82.9	87.1	90.6	95.3	102.9	110.7	128.3	150.9	182.0
93	77.5	78.4	79.6	81.2	83.9	88.1	91.6	96.4	104.1	111.9	129.7	152.6	184.0
94	78.4	79.3	80.5	82.2	84.9	89.1	92.7	97.5	105.3	113.2	131.2	154.3	186.0
95	79.4	80.3	81.5	83.1	85.8	90.1	93.7	98.6	106.4	114.4	132.6	155.9	188.0
96	80.3	81.2	82.4	84.1	86.8	91.1	94.8	99.7	107.6	115.7	134.0	157.6	190.0
97	81.2	82.2	83.4	85.1	87.8	92.2	95.8	100.8	108.8	116.9	135.5	159.3	192.0
98	82.2	83.1	84.3	86.0	88.8	93.2	96.9	101.9	109.9	118.2	136.9	160.9	194.0
99	83.1	84.1	85.3	87.0	89.8	94.2	97.9	103.0	111.1	119.4	138.3	162.6	196.0
100	84.1	85.0	86.2	88.0	90.8	95.2	99.0	104.1	112.3	120.6	139.7	164.3	198.0
102	85.9	86.9	88.1	89.9	92.8	97.3	101.1	106.3	114.6	123.1	142.6	167.6	202.0
104	87.8	88.8	90.1	91.9	94.8	99.3	103.2	108.5	116.9	125.6	145.4	170.9	206.0
106	89.7	90.7	92.0	93.8	96.7	101.4	105.3	110.7	119.3	128.1	148.3	174.2	210.0
108	91.6	92.6	93.9	95.7	98.7	103.4	107.4	112.9	121.6	130.6	151.1	177.6	214.0
110	93.5	94.5	95.8	97.7	100.7	105.5	109.5	115.1	124.0	133.1	154.0	180.9	218.0
112	95.4	96.4	97.7	99.6	102.7	107.5	111.7	117.3	126.3	135.6	156.9	184.2	222.0
114	97.3	98.3	99.7	101.6	104.7	109.6	113.8	119.5	128.6	138.1	159.7	187.6	226.0
116	99.2	100.2	101.6	103.5	106.7	111.7	115.9	121.7	131.0	140.6	162.6	190.9	230.0
118	101.1	102.1	103.5	105.5	108.7	113.7	118.0	123.9	133.3	143.1	165.4	194.2	234.0
120	103.0	104.0	105.4	107.4	110.7	115.8	120.1	126.1	135.7	145.6	168.3	197.6	238.0
122	104.9	105.9	107.4	109.4	112.6	117.8	122.2	128.3	138.0	148.1	171.1	200.9	242.0
124	106.8	107.9	109.3	111.3	114.6	119.9	124.4	130.5	140.3	150.6	174.0	204.2	246.0
126	108.7	109.8	111.2	113.3	116.6	121.9	126.5	132.7	142.7	153.0	176.8	207.6	250.0
128	110.6	111.7	113.2	115.2	118.6	124.0	128.6	134.9	145.0	155.5	179.7	210.9	254.0
130	112.5	113.6	115.1	117.2	120.6	126.1	130.7	137.1	147.4	158.0	182.5	214.2	258.0
132	114.4	115.5	117.0	119.1	122.6	128.1	132.8	139.3	149.7	160.5	185.4	217.6	262.0
134	116.3	117.4	119.0	121.1	124.6	130.2	134.9	141.5	152.0	163.0	188.3	220.9	266.0
136	118.2	119.4	120.9	123.1	126.6	132.3	137.1	143.7	154.4	165.5	191.1	224.2	270.0
138	120.1	121.3	122.8	125.0	128.6	134.3	139.2	145.9	156.7	168.0	194.0	227.6	274.0
140	122.0	123.2	124.8	127.0	130.6	136.4	141.3	148.1	159.1	170.5	196.8	230.9	278.0

Appendix 1.1
Blocked-Calls-Cleared
(Erlang B) (Continued)

A, erlangs

N	B												
	1.0%	1.2%	1.5%	2%	3%	5%	7%	10%	15%	20%	30%	40%	50%
142	123.9	125.1	126.7	128.9	132.6	138.4	143.4	150.3	161.4	173.0	199.7	234.2	282.0
144	125.8	127.0	128.6	130.9	134.6	140.5	145.6	152.5	163.8	175.5	202.5	237.6	286.0
146	127.7	129.0	130.6	132.9	136.6	142.6	147.7	154.7	166.1	178.0	205.4	240.9	290.0
148	129.7	130.9	132.5	134.8	138.6	144.6	149.8	156.9	168.5	180.5	208.2	244.2	294.0
150	131.6	132.8	134.5	136.8	140.6	146.7	151.9	159.1	170.8	183.0	211.1	247.6	298.0
152	133.5	134.8	136.4	138.8	142.6	148.8	154.0	161.3	173.1	185.5	214.0	250.9	302.0
154	135.4	136.7	138.4	140.7	144.6	150.8	156.2	163.5	175.5	188.0	216.8	254.2	306.0
156	137.3	138.6	140.3	142.7	146.6	152.9	158.3	165.7	177.8	190.5	219.7	257.6	310.0
158	139.2	140.5	142.3	144.7	148.6	155.0	160.4	167.9	180.2	193.0	222.5	260.9	314.0
160	141.2	142.5	144.2	146.6	150.6	157.0	162.5	170.0	182.5	195.5	225.4	264.2	318.0
162	143.1	144.4	146.1	148.6	152.7	159.1	164.7	172.4	184.9	198.0	228.2	267.6	322.0
164	145.0	146.3	148.1	150.6	154.7	161.2	166.8	174.6	187.2	200.4	231.1	270.9	326.0
166	146.9	148.3	150.0	152.6	156.7	163.3	168.9	176.8	189.6	202.9	233.9	274.2	330.0
168	148.9	150.2	152.0	154.5	158.7	165.3	171.0	179.0	191.9	205.4	236.8	277.6	334.0
170	150.8	152.1	153.9	156.5	160.7	167.4	173.2	181.2	194.2	207.9	239.7	280.9	338.0
172	152.7	154.1	155.9	158.5	162.7	169.5	175.3	183.4	196.6	210.4	242.5	284.2	342.0
174	154.6	156.0	157.8	160.4	164.7	171.5	177.4	185.6	198.9	212.9	245.4	287.6	346.0
176	156.6	158.0	159.8	162.4	166.7	173.6	179.6	187.8	201.3	215.4	248.2	290.9	350.0
178	158.5	159.9	161.8	164.4	168.7	175.7	181.7	190.0	203.6	217.9	251.1	294.2	354.0
180	160.4	161.8	163.7	166.4	170.7	177.8	183.8	192.2	206.0	220.4	253.9	297.5	358.0
182	162.3	163.8	165.7	168.3	172.8	179.8	185.9	194.4	208.3	222.9	256.8	300.9	362.0
184	164.3	165.7	167.6	170.3	174.8	181.9	188.1	196.6	210.7	225.4	259.6	304.2	366.0
186	166.2	167.7	169.6	172.3	176.8	184.0	190.2	198.9	213.0	227.9	262.5	307.5	370.0
188	168.1	169.6	171.5	174.3	178.8	186.1	192.3	201.1	215.4	230.4	265.4	310.9	374.0
190	170.1	171.5	173.5	176.3	180.8	188.1	194.5	203.3	217.7	232.9	268.2	314.2	378.0

192	172.0	173.5	175.4	178.2	182.8	190.2	196.6	205.5	220.1	235.4	271.1	317.5	382.0
194	173.9	175.4	177.4	180.2	184.8	192.3	198.7	207.7	222.4	237.9	273.9	320.9	386.0
196	175.9	177.4	179.4	182.2	186.9	194.4	200.8	209.9	224.8	240.4	276.8	324.2	390.0
198	177.8	179.3	181.3	184.2	188.9	196.4	203.0	212.1	227.1	242.9	279.6	327.5	394.0
200	179.7	181.3	183.3	186.2	190.9	198.5	205.1	214.3	229.4	245.4	282.5	330.9	398.0
202	181.7	183.2	185.2	188.1	192.9	200.6	207.2	216.5	231.8	247.9	285.4	334.2	402.0
204	183.6	185.2	187.2	190.1	194.9	202.7	209.4	218.7	234.1	250.4	288.2	337.5	406.0
206	185.5	187.1	189.2	192.1	196.9	204.7	211.5	221.0	236.5	252.9	291.1	340.9	410.0
208	187.5	189.1	191.1	194.1	199.0	206.8	213.6	223.2	238.8	255.4	293.9	344.2	414.0
210	189.4	191.0	193.1	196.1	201.0	208.9	215.8	225.4	241.2	257.9	296.8	347.5	418.0
212	191.4	193.0	195.1	198.1	203.0	211.0	217.9	227.6	243.5	260.4	299.6	350.9	422.0
214	193.3	194.9	197.0	200.0	205.0	213.0	220.0	229.8	245.9	262.9	302.5	354.2	426.0
216	195.2	196.9	199.0	202.0	207.0	215.1	222.2	232.0	248.2	265.4	305.3	357.5	430.0
218	197.2	198.8	201.0	204.0	209.1	217.2	224.3	234.2	250.6	267.9	308.2	360.9	434.0
220	199.1	200.8	202.9	206.0	211.1	219.3	226.4	236.4	252.9	270.4	311.1	364.2	438.0
222	201.1	202.7	204.9	208.0	213.1	221.4	228.6	238.6	255.3	272.9	313.9	367.5	442.0
224	203.0	204.7	206.8	210.0	215.1	223.4	230.7	240.9	257.6	275.4	316.8	370.9	446.0
226	204.9	206.6	208.8	212.0	217.1	225.5	232.8	243.1	260.0	277.8	319.6	374.2	450.0
228	206.9	208.6	210.8	213.9	219.2	227.6	235.0	245.3	262.3	280.3	322.5	377.5	454.0
230	208.8	210.5	212.8	215.9	221.2	229.7	237.1	247.5	264.7	282.8	325.3	380.9	458.0
232	210.8	212.5	214.7	217.9	223.2	231.8	239.2	249.7	267.0	285.3	328.2	384.2	462.0
234	212.7	214.4	216.7	219.9	225.2	233.8	241.4	251.9	269.4	287.8	331.1	387.5	466.0
236	214.7	216.4	218.7	221.9	227.2	235.9	243.5	254.1	271.7	290.3	333.9	390.9	470.0
238	216.6	218.3	220.6	223.9	229.3	238.0	245.6	256.3	274.1	292.8	336.8	394.2	474.0
240	218.6	220.3	222.6	225.9	231.3	240.1	247.8	258.6	276.4	295.3	339.6	397.5	478.0
242	220.5	222.3	224.6	227.9	233.3	242.2	249.9	260.8	278.8	297.8	342.5	400.9	482.0
244	222.5	224.2	226.5	229.9	235.3	244.3	252.0	263.0	281.1	300.3	345.3	404.2	486.0
246	224.4	226.2	228.5	231.8	237.4	246.3	254.2	265.2	283.4	302.8	348.2	407.5	490.0
248	226.3	228.1	230.5	233.8	239.4	248.4	256.3	267.4	285.8	305.3	351.0	410.9	494.0
250	228.3	230.1	232.5	235.8	241.4	250.5	258.4	269.6	288.1	307.8	353.9	414.2	498.0
	.976	.982	.988	.998	1.014	1.042	1.070	1.108	1.176	1.250	1.428	1.666	2.000

Appendix 1.1
Blocked-Calls-Cleared
(Erlang *B*) (*Continued*)

A, erlangs

| | | | | | | | | *B* | | | | | | |
N	1.0%	1.2%	1.5%	2%	3%	5%	7%	10%	15%	20%	30%	40%	50%
300	277.1 / *.982*	279.2 / *.984*	281.9 / *.990*	285.7 / *1.000*	292.1 / *1.016*	302.6 / *1.044*	311.9 / *1.070*	325.0 / *1.108*	346.9 / *1.174*	370.3 / *1.248*	425.3 / *1.428*	497.5 / *1.668*	598.0 / *2.000*
350	326.2 / *.982*	328.4 / *.988*	331.4 / *.994*	335.7 / *1.004*	342.9 / *1.020*	354.8 / *1.046*	365.4 / *1.070*	380.4 / *1.108*	405.6 / *1.176*	432.7 / *1.250*	496.7 / *1.430*	580.9 / *1.666*	698.0 / *2.000*
400	375.3 / *.986*	377.8 / *.990*	381.1 / *.996*	385.9 / *1.004*	393.9 / *1.018*	407.1 / *1.046*	418.9 / *1.072*	435.8 / *1.110*	464.4 / *1.176*	495.2 / *1.250*	568.2 / *1.428*	664.2 / *1.666*	798.0 / *2.000*
450	424.6 / *.988*	427.3 / *.994*	430.9 / *.998*	436.1 / *1.006*	444.8 / *1.022*	459.4 / *1.048*	472.5 / *1.070*	491.3 / *1.108*	523.2 / *1.176*	557.7 / *1.250*	639.6 / *1.428*	747.5 / *1.668*	898.0 / *2.000*
500	474.0 / *.991*	477.0 / *.994*	480.8 / *1.000*	486.4 / *1.008*	495.9 / *1.022*	511.8 / *1.047*	526.0 / *1.073*	546.7 / *1.110*	582.0 / *1.176*	620.2 / *1.249*	711.0 / *1.429*	830.9 / *1.666*	998.0 / *2.000*
600	573.1 / *.993*	576.4 / *.997*	580.8 / *1.002*	587.2 / *1.010*	598.1 / *1.024*	616.5 / *1.049*	633.3 / *1.073*	657.7 / *1.110*	699.6 / *1.176*	745.1 / *1.250*	853.9 / *1.428*	997.5 / *1.665*	1198. / *2.00*
700	672.4 / *.994*	676.1 / *.998*	681.0 / *1.004*	688.2 / *1.011*	700.5 / *1.025*	721.4 / *1.050*	740.6 / *1.073*	768.7 / *1.110*	817.2 / *1.176*	870.1 / *1.250*	996.7 / *1.433*	1164. / *1.67*	1398. / *2.00*
800	771.8 / *.997*	775.9 / *1.000*	781.4 / *1.004*	789.3 / *1.013*	803.0 / *1.025*	826.4 / *1.050*	847.9 / *1.074*	879.7 / *1.111*	934.8 / *1.172*	995.1 / *1.249*	1140. / *1.42*	1331. / *1.67*	1598. / *2.00*
900	871.5 / *.997*	875.9 / *1.001*	881.8 / *1.006*	890.6 / *1.013*	905.5 / *1.025*	931.4 / *1.046*	955.3 / *1.077*	990.8 / *1.112*	1052. / *1.18*	1120. / *1.25*	1282. / *1.43*	1498. / *1.66*	1798. / *2.00*
1000	971.2 / *.998*	976.0 / *1.000*	982.4 / *1.006*	991.9 / *1.011*	1008. / *1.03*	1036. / *1.05*	1063. / *1.07*	1102. / *1.11*	1170. / *1.18*	1245. / *1.25*	1425. / *1.43*	1664. / *1.67*	1998. / *2.00*
1100	1071.	1076.	1083.	1093.	1111.	1141.	1170.	1213.	1288.	1370.	1568.	1831.	2198.

Appendix 1.2   Status of First 90 Cities (1993) Key: A—Band A Carrier; B—Band B Carrier

MSA no./name		System operators	On-line status	Rank (based on number of subscribers)	Switching equipment
1. New York, NY	B	Nynex Mobile	6/15/84		AT&T
	A	Cellular One	4/5/86	2	Motorola
2. Los Angeles, CA	B	PacTel Cellular	6/13/84		Motorola
	A	LA Cellular Telephone	3/27/87	1	Ericsson
3. Chicago, IL	B	Ameritech Mobile	10/13/83		AT&T
	A	Cellular One	1/3/85	3	Ericsson
4. Philadelphia, PA	B	Bell Atlantic Mobile	7/12/84		AT&T
	A	Metrophone	2/12/86	8	Motorola
5. Detroit, MI	B	Ameritech Mobile	9/21/84		AT&T
	A	Cellular One	7/30/85	5	Ericsson
6. Boston, MA	B	Nynex Mobile	1/1/85		AT&T
	A	Cellular One	1/1/85	11	Motorola
7. San Francisco, CA	B	GTE Mobilnet	4/2/85		Motorola
(ranked with San Jose)	A	Cellular One	9/26/86	9	Ericsson
8. Washington, DC	B	Bell Atlantic Mobile	4/2/84		AT&T
(ranked with Baltimore)	A	Cellular One	12/16/83	4	Motorola
9. Dallas, TX	B	Southwestern Bell Mobile	7/31/84		AT&T
	A	MetroCel	3/1/86	6	Motorola
10. Houston, TX	B	GTE Mobilnet	9/28/84		Motorola
	A	Houston Cellular Telephone	5/16/86	10	Ericsson
11. St. Louis, MO	B	Southwestern Bell Mobile	7/.3/84		AT&T
	A	CyberTel	7/16/84	18	Motorola
12. Miami, FL	B	BellSouth Mobility	5/25/84		AT&T
	A	Cellular One	3/6/87	7	NTI/GE
13. Pittsburgh, PA	B	Bell Atlantic Mobile	12/10/84		AT&T
	A	Cellular One/McCaw		26	Ericsson
14. Baltimore, MD	B	Bell Atlantic Mobile	4/2/84		AT&T
	A	Cellular One	12/16/83	4	Motorola
15. Minneapolis, MN	B	U.S. West	6/6/84		GE/Motorola
	A	Cellular One	7/23/84	19	NTI/GE
16. Cleveland, OH	B	GTE Mobilnet	12/18/84		Motorola
	A	Cellular One	5/31/85	14	NTI/GE
17. Atlanta, GA	B	BellSouth Mobility	9/5/84		AT&T
	A	PacTel Cellular	2/1/88	12	Motorola
18. San Diego, CA	B	PacTel Cellular	8/15/85		AT&T
	A	U.S. West	4/1/86	16	Motorola
19. Denver, CO	B	U.S. West	7/10/84		NTI/GE
	A	Cellular One	11/21/86	21	AT&T/Panasonic

Appendix 1.2   Status of First 90 Cities (1993) Key: A—Band A Carrier; B—Band B Carrier *(Continued)*

MSA no./name		System operators	On-line status	Rank (based on number of subscribers)	Switching equipment
20. Seattle, WA	B	U.S. West	7/12/84		NTI/GE
	A	Cellular One	12/12/85	13	AT&T
21. Milwaukee, WI	B	Ameritech Mobile	8/1/84		AT&T
	A	Cellular One	6/1/84	30	Ericsson
22. Tampa, FL	B	GTE Mobilnet	11/30/84		Motorola
	A	Cellular One	9/25/87	15	Ericsson
23. Cincinnati, OH	B	Ameritech Mobile	11/5/84		AT&T
	A	Cellular One	8/26/86	22	Ericsson
24. Kansas City, MO	B	Southwestern Bell Mobile	8/14/84		Motorola
	A	Cellular One	2/14/86	27	AT&T
25. Buffalo, NY	B	Nynex Mobile	4/16/84		Motorola
	A	Cellular One	6/1/84	35	Ericsson
26. Phoenix, AZ	B	U.S. West	8/15/84		AT&T
	A	Bell Atlantic Mobile	3/1/86	17	Motorola
27. San Jose, CA	B	GTE Mobilnet	4/2/85		Motorola
	A	Cellular One	9/26/86	9	Ericsson
28. Indianapolis, IN	B	GTE Mobilnet	5/3/84		AT&T
	A	Cellular One	2/3/84	20	Ericsson
29. New Orleans, LA	B	BellSouth Mobility	9/1/84		Motorola
	A	Radiofone	9/6/85	36	Motorola
30. Portland, OR	B	GTE Mobilnet	3/5/85		AT&T
	A	Cellular One	7/12/85	25	Ericsson
31. Columbus, OH	B	Ameritech Mobile	5/30/85		NTI/GE
	A	Cellular One	7/1/86	34	NTI
32. Hartford, CT	B	SNET Cellular	1/31/85		AT&T
	A	Bell Atlantic Mobile	10/16/87	37	Motorola
33. San Antonio, TX	B	Southwestern Bell Mobile	1/28/85		AT&T
	A	Cellular One	10/21/86	29	
34. Rochester, NY	B	Rochester Telephone Mobile	6/4/85		AT&T
	A	Genesee Telephone Co.	9/22/86	—	Ericsson
35. Sacramento, CA	B	PacTel Cellular	8/29/85		Motorola
	A	Cellular One	10/30/87	24	Ericsson
36. Memphis, TN	B	BellSouth Mobility	5/1/85		Motorola
	A	Cellular One	12/24/86	38	AT&T
37. Louisville, KY	B	BellSouth Mobility	1/3/85		Motorola
	A	Cellular One	2/15/85	—	AT&T
38. Providence, RI	B	Nynex Mobile	8/22/85		AT&T
	A	Bell Atlantic Mobile	7/8/87	31	Motorola
39. Salt Lake City, UT	B	U.S. West	1/29/85		AT&T
	A	Cellular One	12/17/86	39	Motorola/NEC/NovAtel
40. Dayton, OH	B	Ameritech Mobile	5/31/85		NTI/GE
	A	Cellular One	8/26/86	—	Northern Telecom

Appendix 1.2   Status of First 90 Cities (1993) Key: A—Band A Carrier; B—Band B Carrier (Continued)

MSA no./name		System operators	On-line status	Rank (based on number of subscribers)	Switching equipment
41. Birmingham, AL	B	BellSouth Mobility	9/26/85		Motorola
	A	Cellular One	12/31/86	—	AT&T
42. Bridgeport, CT	B	Southern New England Tele.	5/20/85		AT&T
	A	Bell Atlantic Mobile	11/20/87	—	Motorola
43. Norfolk, VA	B	Cellular One	5/3/85		AT&T
	A	Centel Cellular	11/1/85	33	Motorola
44. Albany, NY	B	Nynex Mobile	6/25/85		NTI/GE
	A	Cellular One	12/3/86		Northern Telecom/
				—	Ericsson
45. Oklahoma City, OK	B	Southwestern Bell Mobile	1/14/85		AT&T
	A	Cellular One	1/17/86	—	AT&T
46. Nashville, TN	B	BellSouth Mobility	6/10/85		Motorola
	A	Cellular One	6/30/87	—	AT&T
47. Greensboro, NC	B	Centel	5/15/85		Motorola
	A	Cellular One	12/27/85		Motorola
48. Toledo, OH	B	Centel Cellular	7/25/85		Motorola
	A	Cellular One/PacTel Cellular	4/15/86	—	Ericsson
49. New Haven, CT	B	Southern New England Tele.	3/4/85		AT&T
	A	Bell Atlantic Mobile	11/20/87	—	Motorola
50. Honolulu, HI	B	GTE Mobilnet	3/26/84		Motorola
	A	Honolulu Cellular Telephone	6/1/86	—	Ericsson
51. Jacksonville, FL	B	BellSouth Mobility	6/12/85		Motorola
	A	Cellular One	8/14/87	40	Ericsson
52. Akron, OH	B	GTE Mobilnet	10/31/85		Motorola
	A	Cellular One	12/15/86		Motorola
53. Syracuse, NY	B	Nynex Mobile	1/24/86		AT&T
	A	Cellular One	12/31/85	—	Motorola
54. Gary, IN	B	Ameritech Mobile	3/11/85		AT&T
	A	Cellular One	4/21/86	—	Ericsson
55. Worcester, MA	B	Nynex Mobile	11/18/85		AT&T
	A	Cellular One	11/18/85	—	Motorola
56. Northeast, PA	B	Cellular Plus	7/2/85		AT&T
	A	Cellular One	12/31/85	—	NTI/GE
57. Tulsa, OK	B	United States Cellular	8/30/85		NEC
	A	Cellular One	5/22/86	—	AT&T
58. Allentown, PA	B	Bell Atlantic Mobile	3/18/85		AT&T
	A	Cellular One	10/18/85	—	NTI/GE
59. Richmond, VA	B	Contel Cellular, Inc.	5/10/85		AT&T
	A	Cellular One	12/2/86	—	Motorola
60. Orlando, FL	B	BellSouth Mobility	2/27/85		Astronet
	A	Cellular One	12/29/86	32	Ericsson
61. Charlotte, NC	B	Alltel	4/15/85		Motorola
	A	Bell Atlantic Mobile	3/1/86	23	Motorola

Appendix 1.2   Status of First 90 Cities (1993) Key: A—Band A Carrier; B—Band B Carrier (*Continued*)

MSA no./name		System operators	On-line status	Rank (based on number of subscribers)	Switching equipment
62. New Brunswick, NJ	B	Nynex Mobile	1/21/87		AT&T
	A	Cellular One	11/1/86	—	Motorola
63. Springfield, MA	B	Nynex Mobile	4/10/87		AT&T
	A	Springfield Cellular Telephone	10/16/87	—	Motorola
64. Grand Rapids, MI	B	Century Cellunet	9/17/86		AT&T
	A	Cellular One	6/18/86	—	Ericsson
65. Omaha, NE	B	Centel Cellular	4/15/85		Motorola
	A	U.S. West Cellular	12/23/85	—	NTI
66. Youngstown, OH	B	Centel Cellular	9/19/85		Motorola
	A	Youngstown Cellular/Wilcom	12/23/85	—	Northern Telecom
67. Greenville, SC	B	Centel Cellular	7/30/86		Motorola
	A	Bell Atlantic Mobile	8/18/86	—	Motorola
68. Flint, MI	B	Ameritech Mobile	7/12/85		AT&T
	A	Cellular One	7/30/85	—	Ericsson
69. Wilmington, DE	B	Bell Atlantic Mobile	3/27/85		AT&T
	A	Cellular One	5/86	—	Motorola
70. Long Branch, NJ	B	Nynex Mobile	2/20/87		AT&T
	A	Cellular One	11/1/86	—	Motorola
71. Raleigh-Durham, NC	B	Centel Cellular	11/11/85		Motorola
	A	Cellular One	9/16/85	—	NTI/GE
72. W. Palm Beach, FL	B	BellSouth Mobility	5/23/85		AT&T
	A	Cellular One	3/6/87	—	
73. Oxnard, CA	B	PacTel Mobile Access	10/30/85		AT&T
	A	Ventura Cellular	9/87	—	Ericsson
74. Fresno, CA	B	Contel Cellular	5/1/86		AT&T
	A	Cellular One	10/23/87	—	Ericsson
75. Austin, TX	B	GTE Mobilnet	9/27/85		Motorola
	A	Cellular One	12/27/85	—	AT&T
76. New Bedford, MA	B	Nynex Mobile	12/9/85		AT&T
	A	Bell Atlantic Mobile	7/8/87	—	Motorola
77. Tucson, AZ	B	U.S. West	8/6/85		NTI/GE
	A	Bell Atlantic Mobile	4/1/86	—	Motorola
78. Lansing, MI	B	Century Cellunet	9/16/87		Motorola
	A	Cellular One	9/17/86	—	Ericsson
79. Knoxville, TN	B	United States Cellular	7/23/85		NEC
	A	Cellular One	12/22/87	—	AT&T
80. Baton Rouge, LA	B	BellSouth Mobility	7/2/85		Motorola
	A	Cellular One	8/1/86	—	Motorola
81. El Paso, TX	B	Contel Cellular	2/25/85		AT&T
	A	Bell Atlantic Mobile	5/2/86	—	Motorola
82. Tacoma, WA	B	U.S. West	4/18/85		NTI/GE
	A	Cellular One	12/12/85	—	AT&T

Appendix 1.2   Status of First 90 Cities (1993) Key: A—Band A Carrier; B—Band B Carrier (*Continued*)

MSA no./name		System operators	On-line status	Rank (based on number of subscribers)	Switching equipment
83. Mobile, AL	B	GTE of Mobile	9/3/85		AT&T
	A	BellSouth Mobility	6/8/87	—	Astronet
84. Harrisburg, PA	B	United Telespectrum	10/18/85		Motorola
	A	Cellular One	9/18/85	—	NTI/GE
85. Johnson City, TN	B	Centel	10/3/85		Motorola
	A	U.S. West	1/22/88	—	AT&T
86. Albuquerque, NM	B	U.S. West	8/13/85		NTI/GE
	A	Bell Atlantic Mobile	11/1/85	—	Motorola
87. Canton, OH	B	GTE Mobilnet	2/25/87		Motorola
	A	Cellular One	2/2/87	—	NTI/GE
88. Chattanooga, TN	B	BellSouth Mobility	8/1/85		Motorola
	A	Cellular One	11/20/87	—	AT&T
89. Wichita, KS	B	Southwestern Bell Mobile	2/11/85		Motorola
	A	Cellular One	1/24/86		Mitsubishi/NEC/
				—	Motorola/Novatel
90. Charleston, SC	B	Centel Cellular	9/11/85		Motorola
	A	Cellular One	1/22/88	—	NTI/GE

Appendix 1.3   List of MSAs and RSAs

1. New York, NY
2. Los Angeles, CA
3. Chicago, IL
4. Philadelphia, PA
5. Detroit, MI
6. Boston, MA-NH
7. San Francisco, CA
8. Washington, DC
9. Dallas, TX
10. Houston, TX
11. St. Louis, MO
12. Miami, FL
13. Pittsburgh, PA
14. Baltimore, MD
15. Minneapolis, MN-WI
16. Cleveland, OH
17. Atlanta, GA
18. San Diego, CA
19. Denver, CO
20. Seattle, WA
21. Milwaukee, WI
22. Tampa, FL
23. Cincinnati, OH
24. Kansas City, MO
25. Buffalo, NY
26. Phoenix, AZ
27. San Jose, CA
28. Indianapolis, IN
29. New Orleans, LA
30. Portland, OR-WA
31. Columbus, OH
32. Hartford, CT
33. San Antonio, TX
34. Rochester, NY
35. Sacramento, CA
36. Memphis, TN
37. Louisville, KY
38. Providence, RI
39. Salt Lake City, UT
40. Dayton, OH
41. Birmingham, AL
42. Bridgeport, CT
43. Norfolk, VA
44. Albany, NY
45. Oklahoma City, OK
46. Nashville, TN
47. Greensboro, NC
48. Toledo, OH
49. New Haven, CT
50. Honolulu, HI
51. Jacksonville, FL
52. Akron, OH
53. Syracuse, NY
54. Gary, IN
55. Worcester, MA
56. Northeast, PA
57. Tulsa, OK
58. Allentown, PA-NJ
59. Richmond, VA
60. Orlando, FL
61. Charlotte, NC
62. New Brunswick, NJ
63. Springfield, MA
64. Grand Rapids, MI
65. Omaha, NE
66. Youngstown, OH
67. Greenville, SC
68. Flint, MI
69. Wilmington, DE-NJ-MD
70. Long Branch, NJ
71. Raleigh-Durham, NC
72. W. Palm Beach, FL
73. Oxnard, CA
74. Fresno, CA
75. Austin, TX
76. New Bedford, MA
77. Tucson, AZ
78. Lansing, MI
79. Knoxville, TN
80. Baton Rouge, LA
81. El Paso, TX
82. Tacoma, WA
83. Mobile, AL
84. Harrisburg, PA
85. Johnson City, TN-VA
86. Albuquerque, NM
87. Canton, OH
88. Chattanooga, TN
89. Wichita, KS
90. Charleston, SC
91. San Juan, PR
92. Little Rock, AR
93. Las Vegas, NV
94. Saginaw Bay–Midland, MI
95. Columbia, SC
96. Fort Wayne, IN
97. Bakersfield, CA
98. Davenport, IA-Mol., IL
99. York, PA
100. Shreveport, LA
101. Beaumont, TX
102. Des Moines, IA
103. Peoria, IL
104. Newport News, VA
105. Lancaster, PA
106. Jackson, MS
107. Stockton, CA
108. Augusta, GA-SC
109. Spokane, WA
110. Huntington-Ashland, WV
111. Vallejo, CA
112. Corpus Christi, TX
113. Madison, WI
114. Lakeland, FL
115. Utica-Rome, NY
116. Lexington-Fayette, KY
117. Colorado Springs, CO
118. Reading, PA
119. Evansville, IN
120. Huntsville, AL
121. Trenton, NJ
122. Binghamton, NY
123. Santa Rosa-Petaluma, CA
124. Santa Barbara, CA
125. Appleton, WI
126. Salinas, CA
127. Pensacola, FL
128. McAllen, TX
129. South Bend, IN
130. Erie, PA
131. Rockford, IL
132. Kalamazoo, MI
133. Manchester-Nashua, NH
134. Atlantic City, NJ
135. Eugene-Springfield, OR
136. Lorain-Elyria, OH
137. Melbourne, FL
138. Macon, GA
139. Montgomery, AL
140. Charleston, WV
141. Duluth, MN
142. Modesto, CA
143. Johnstown, PA
144. Orange County, NY
145. Hamilton, OH
146. Daytona Beach, FL
147. Ponce, PR
148. Salem, OR
149. Fayetteville, NC
150. Visalia-Tulare, CA
151. Poughkeepsie, NY
152. Portland, ME
153. Columbus, GA
154. New London-Norwich, CT
155. Savannah, GA
156. Portsmouth, NH
157. Roanoke, VA
158. Lima, OH
159. Provo-Orem, UT
160. Killeen-Temple, TX
161. Lubbock, TX
162. Brownsville, TX
163. Springfield, MO
164. Fort Myers, FL
165. Fort Smith, AR-OK
166. Hickory, NC
167. Sarasota, FL
168. Tallahassee, FL
169. Mayaguez, PR
170. Galveston, TX
171. Reno, NV
172. Lincoln, NE
173. Biloxi-Gulfport, MS
174. Lafayette, LA
175. Santa Cruz, CA
176. Springfield, IL
177. Battle Creek, MI
178. Wheeling, WV-OH
179. Topeka, KS
180. Springfield, OH
181. Muskegon, MI
182. Fayetteville-Sprngdl., AR
183. Asheville, NC
184. Houma-Thibodaux, LA
185. Terre Haute, IN
186. Green Bay, WI
187. Anchorage, AK
188. Amarillo, TX
189. Racine, WI
190. Boise City, ID
191. Yakima, WA
192. Gainesville, FL
193. Benton Harbor, MI
194. Waco, TX
195. Cedar Rapids, IA
196. Champaign-Urbana, IL
197. Lake Charles, LA
198. St. Cloud, MN
199. Steubenville-Werton, OH
200. Parkersburg-Marietta, WV
201. Waterloo-Cedar Falls, IA
202. Arecibo, PR
203. Lynchburg, VA
204. Aguadilla, PR
205. Alexandria, LA
206. Longview-Marshall, TX
207. Jackson, MI
208. Fort Pierce, FL

Appendix 1.3   List of MSAs and RSAs
(Continued)

209. Clarksville, TN-KY
210. Fort Collins-Loveland, CO
211. Bradenton, FL
212. Bremerton, WA
213. Pittsfield, MA
214. Richland-Kennewick, WA
215. Chico, CA
216. Janesville-Beloit, WI
217. Anderson, IN
218. Wilmington, NC
219. Monroe, LA
220. Abilene, TX
221. Fargo-Moorhead, ND-MN
222. Tuscaloosa, AL
223. Elkhart-Goshen, IN
224. Bangor, ME
225. Altoona, PA
226. Florence, AL
227. Anderson, SC
228. Vineland-Millville, NJ
229. Medford, OR
230. Decatur, IL
231. Mansfield, OH
232. Eau Claire, WI
233. Wichita Falls, TX
234. Athens, GA
235. Petersburg, VA
236. Muncie, IN
237. Tyler, TX
238. Sharon, PA
239. Joplin, MO
240. Texarkana, TX-AR
241. Pueblo, CO
242. Olympia, WA
243. Greeley, CO
244. Kenosha, WI
245. Ocala, FL
246. Dothan, AL
247. Lafayette, IN
248. Burlington, VT
249. Anniston, AL
250. Bloomington-Normal, IL
251. Williamsport, PA
252. Pascagoula, MS
253. Sioux City, IA-NE
254. Redding, CA
255. Odessa, TX
256. Charlottesville, VA
257. Hagerstown, MD
258. Jacksonville, NC
259. State College, PA
260. Lawton, OK

261. Albany, GA
262. Danville, VA
263. Wausau, WI
264. Florence, SC
265. Fort Walton Beach, FL
266. Glens Falls, NY
267. Sioux Falls, SD
268. Billings, MT
269. Cumberland, MD-WV
270. Bellingham, WA
271. Kokomo, IN
272. Gadsden, AL
273. Kankakee, IL
274. Yuba City, CA
275. St. Joseph, MO
276. Grand Forks, ND
277. Sheboygan, WI
278. Columbia, MO
279. Lewiston-Auburn, ME
280. Burlington, NC
281. Laredo, TX
282. Bloomington, IN
283. Panama City, FL
284. Elmira, NY
285. Las Cruces, NM
286. Dubuque, IA
287. Bryan-College Station, TX
288. Rochester, MN
289. Rapid City, SD
290. La Crosse, WI
291. Pine Bluff, AR
292. Sherman-Denison, TX
293. Owensboro, KY
294. San Angelo, TX
295. Midland, TX
296. Iowa City, IA
297. Great Falls, MT
298. Bismarck, ND
299. Casper, WY
300. Victoria, TX
301. Lawrence, KS
302. Enid, OK
303. Aurora-Elgin, IL
304. Joliet, IL
305. Alton-Granite City, IL
307. Franklin, AL-1
308. Jackson, AL-2
309. Lamar, AL-3
310. Bib, AL-4
311. Cleburne, AL-5
312. Washington, AL-6
313. Butler, AL-7

314. Lee, AL-8
315. Wade Hampton, AK-1
316. Bethel, AK-2
317. Haines, AK-3
318. Mohave, AZ-1
319. Coconino, AZ-2
320. Navajo, AZ-3
321. Yuma, AZ-4
322. Gila, AZ-5
323. Graham, AZ-6
324. Madison, AR-1
325. Marion, AR-2
326. Sharp, AR-3
327. Clay, AR-4
328. Cross, AR-5
329. Cleburne, AR-6
330. Pope, AR-7
331. Franklin, AR-8
332. Polk, AR-9
333. Garland, AR-10
334. Hempstead, AR-11
335. Quachita, AR-12
336. Del Norte, CA-1
337. Modoc, CA-2
338. Alpine, CA-3
339. Madera, CA-4
340. San Luis Obispo, CA-5
341. Mono, CA-6
342. Imperial, CA-7
343. Tehama, CA-8
344. Mendocino, CA-9
345. Sierra, CA-10
346. El Dorado, CA-11
347. Kings, CA-12
348. Moffat, CO-1
349. Logan, CO-2
350. Garfield, CO-3
351. Park, CO-4
352. Elbert, CO-5
353. San Miguel, CO-6
354. Saguache, CO-7
355. Kiowa, CO-8
356. Costilla, CO-9
357. Litchfield, CT-1
358. Windham, CT-2
359. Kent, DE-1
360. Collier, FL-1
361. Glades, FL-2
362. Hardee, FL-3
363. Citrus, FL-4
364. Putnam, FL-5
365. Dixie, FL-6

366. Hamilton, FL-7
367. Jefferson, FL-8
368. Calhoun, FL-9
369. Walton, FL-10
370. Monroe, FL-11
371. Whitfield, GA-1
372. Dawson, GA-2
373. Chattooga, GA-3
374. Jasper, GA-4
375. Haralson, GA-5
376. Spalding, GA-6
377. Hancock, GA-7
378. Warren, GA-8
379. Marion, GA-9
380. Bleckley, GA-10
381. Toombs, GA-11
382. Liberty, GA-12
383. Early, GA-13
384. Worth, GA-14
385. Kauai, HI-1
386. Maui, HI-2
387. Hawaii, HI-3
388. Boundary, ID-1
389. Idaho, ID-2
390. Lemhi, ID-3
391. Elmore, ID-4
392. Butte, ID-5
393. Clark, ID-6
394. Jo Daviess, IL-1
395. Bureau, IL-2
396. Mercer, IL-3
397. Adams, IL-4
398. Mason, IL-5
399. Montgomery, IL-6
400. Vermilion, IL-7
401. Washington, IL-8
402. Clay, IL-9
403. Newton, IN-1
404. Kosciusko, IN-2
405. Huntington, IN-3
406. Miami, IN-4
407. Warren, IN-5
408. Randolph, IN-6
409. Owen, IN-7
410. Brown, IN-8
411. Decatur, IN-9
412. Mills, IA-1
413. Union, IA-2
414. Monroe, IA-3
415. Muscatine, IA-4
416. Jackson, IA-5
417. Iowa, IA-6

Appendix 1.3   List of MSAs and RSAs
(Continued)

418. Audubon, IA-7
419. Monona, IA-8
420. Ida, IA-9
421. Humboldt, IA-10
422. Hardin, IA-11
423. Winneshiek, IA-12
424. Mitchell, IA-13
425. Kossuth, IA-14
426. Dickinson, IA-15
427. Lyon, IA-16
428. Cheyenne, KS-1
429. Norton, KS-2
430. Jewell, KS-3
431. Marshall, KS-4
432. Brown, KS-5
433. Wallace, KS-6
434. Trego, KS-7
435. Ellsworth, KS-8
436. Morris, KS-9
437. Franklin, KS-10
438. Hamilton, KS-11
439. Hodgeman, KS-12
440. Edwards, KS-13
441. Reno, KS-14
442. Elk, KS-15
443. Fulton, KY-1
444. Union, KY-2
445. Meade, KY-3
446. Spencer, KY-4
447. Barren, KY-5
448. Madison, KY-6
449. Trimble, KY-7
450. Mason, KY-8
451. Elliott, KY-9
452. Powell, KY-10
453. Clay, KY-11
454. Claiborne, LA-1
455. Morehouse, LA-2
456. De Soto, LA-3
457. Caldwell, LA-4
458. Beauregard, LA-5
459. Iberville, LA-6
460. West Feliciana, LA-7
461. St. James, LA-8
462. Plaquemines, LA-9
463. Oxford, ME-1
464. Somerset, ME-2
465. Kennebec, ME-3
466. Washington, ME-4
467. Garrett, MD-1
468. Kent, MD-2
469. Frederick, MD-3

470. Franklin, MA-1
471. Barnstable, MA-2
472. Gogebic, MI-1
473. Alger, MI-2
474. Emmet, MI-3
475. Cheboygan, MI-4
476. Manistee, MI-5
477. Roscommon, MI-6
478. Newaygo, MI-7
479. Allegan, MI-8
480. Cass, MI-9
481. Tuscola, MI-10
482. Kittson, MN-1
483. Lake of the Woods, MN-2
484. Koochiching, MN-3
485. Lake, MN-4
486. Wilkin, MN-5
487. Hubbard, MN-6
488. Chippewa, MN-7
489. Lac qui Parle, MN-8
490. Pipestone, MN-9
491. Le Sueur, MN-10
492. Goodhue, MN-11
493. Tunica, MS-1
494. Benton, MS-2
495. Bolivar, MS-3
496. Yalobusha, MS-4
497. Washington, MS-5
498. Montgomery, MS-6
499. Leake, MS-7
500. Claiborne, MS-8
501. Copiah, MS-9
502. Smith, MS-10
503. Lamar, MS-11
504. Atchison, MO-1
505. Harrison, MO-2
506. Schuyler, MO-3
507. De Kalb, MO-4
508. Linn, MO-5
509. Marion, MO-6
510. Saline, MO-7
511. Callaway, MO-8
512. Bates, MO-9
513. Benton, MO-10
514. Moniteau, MO-11
515. Maries, MO-12
516. Washington, MO-13
517. Barton, MO-14
518. Stone, MO-15
519. Laclede, MO-16
520. Shannon, MO-17
521. Perry, MO-18

522. Stoddard, MO-19
523. Lincoln, MT-1
524. Toole, MT-2
525. Phillips, MT-3
526. Daniels, MT-4
527. Mineral, MT-5
528. Deer Lodge, MT-6
529. Fergus, MT-7
530. Beaverhead, MT-8
531. Carbon, MT-9
532. Prairie, MT-10
533. Sioux, NE-1
534. Cherry, NE-2
535. Knox, NE-3
536. Grant, NE-4
537. Boone, NE-5
538. Keith, NE-6
539. Hall, NE-7
540. Chase, NE-8
541. Adams, NE-9
542. Cass, NE-10
543. Humboldt, NV-1
544. Lander, NV-2
545. Storey, NV-3
546. Mineral, NV-4
547. White Pine, NV-5
548. Coos, NH-1
549. Carroll, NH-2
550. Hunterdon, NJ-1
551. Ocean, NJ-2
552. Sussex, NJ-3
553. San Juan, NM-1
554. Colfax, NM-2
555. Catron, NM-3
556. Santa Fe, NM-4
557. Grant, NM-5
558. Lincoln, NM-6
559. Jefferson, NY-1
560. Franklin, NY-2
561. Chautauqua, NY-3
562. Yates, NY-4
563. Otsego, NY-5
564. Columbia, NY-6
565. Cherokee, NC-1
566. Yancey, NC-2
567. Ashe, NC-3
568. Henderson, NC-4
569. Anson, NC-5
570. Chatham, NC-6
571. Rockingham, NC-7
572. Northampton, NC-8
573. Camden, NC-9

574. Harnett, NC-10
575. Hoke, NC-11
576. Sampson, NC-12
577. Greene, NC-13
578. Pitt, NC-14
579. Cabarrus, NC-15
580. Divide, ND-1
581. Bottineau, ND-2
582. Barnes, ND-3
583. McKenzie, ND-4
584. Kidder, ND-5
585. Williams, OH-1
586. Sandusky, OH-2
587. Ashtabula, OH-3
588. Mercer, OH-4
589. Hancock, OH-5
590. Morrow, OH-6
591. Tuscarawas, OH-7
592. Clinton, OH-8
593. Ross, OH-9
594. Perry, OH-10
595. Columbiana, OH-11
596. Cimarron, OK-1
597. Harper, OK-2
598. Grant, OK-3
599. Nowata, OK-4
600. Roger Mills, OK-5
601. Seminole, OK-6
602. Beckham, OK-7
603. Jackson, OK-8
604. Garvin, OK-9
605. Haskell, OK-10
606. Clatsop, OR-1
607. Hood River, OR-2
608. Umatilla, OR-3
609. Lincoln, OR-4
610. Coos, OR-5
611. Crook, OR-6
612. Crawford, PA-1
613. McKean, PA-2
614. Potter, PA-3
615. Bradford, PA-4
616. Wayne, PA-5
617. Lawrence, PA-6
618. Jefferson, PA-7
619. Union, PA-8
620. Greene, PA-9
621. Bedford, PA-10
622. Huntingdon, PA-11
623. Lebanon, PA-12
624. Newport, RI-1
625. Oconee, SC-1

Appendix 1.3   List of MSAs and RSAs
(Continued)

626. Laurens, SC-2	678. Piute, UT-6	730. St. Thomas, VI-1
627. Cherokee, SC-3	679. Franklin, VT-1	731. St. Croix, VI-2
628. Chesterfield, SC-4	680. Addison, VT-2	732. Guam
629. Georgetown, SC-5	681. Lee, VA-1	733. American Samoa
630. Clarendon, SC-6	682. Tazewell, VA-2	734. N. Marianas
631. Calhoun, SC-7	683. Giles, VA-3	
632. Hampton, SC-8	684. Bedford, VA-4	
633. Lancaster, SC-9	685. Bath, VA-5	
634. Harding, SD-1	686. Highland, VA-6	
635. Corson, SD-2	687. Buckingham, VA-7	
636. McPherson, SD-3	688. Amelia, VA-8	
637. Marshall, SD-4	689. Greensville, VA-9	
638. Custer, SD-5	690. Frederick, VA-10	
639. Haakon, SD-6	691. Madison, VA-11	
640. Sully, SD-7	692. Caroline, VA-12	
641. Kingsbury, SD-8	693. Clallam, WA-1	
642. Hanson, SD-9	694. Okanogan, WA-2	
643. Lake, TN-1	695. Ferry, WA-3	
644. Cannon, TN-2	696. Grays Harbor, WA-4	
645. Macon, TN-3	697. Kittitas, WA-5	
646. Hamblen, TN-4	698. Pacific, WA-6	
647. Fayette, TN-5	699. Skamania, WA-7	
648. Giles, TN-6	700. Whitman, WA-8	
649. Bledsoe, TN-7	701. Mason, WV-1	
650. Johnson, TN-8	702. Wetzel, WV-2	
651. Maury, TN-9	703. Monongalia, WV-3	
652. Dallam, TX-1	704. Grant, WV-4	
653. Hansford, TX-2	705. Tucker, WV-5	
654. Parmer, TX-3	706. Lincoln, WV-6	
655. Briscoe, TX-4	707. Raleigh, WV-7	
656. Hardeman, TX-5	708. Burnett, WI-1	
657. Jack, TX-6	709. Bayfield, WI-2	
658. Fannin, TX-7	710. Vilas, WI-3	
659. Gaines, TX-8	711. Marinette, WI-4	
660. Runnels, TX-9	712. Pierce, WI-5	
661. Navarro, TX-10	713. Trempealeau, WI-6	
662. Cherokee, TX-11	714. Wood, WI-7	
663. Hudspeth, TX-12	715. Vernon, WI-8	
664. Reeves, TX-13	716. Columbia, WI-9	
665. Loving, TX-14	717. Door, WI-10	
666. Concho, TX-15	718. Park, WY-1	
667. Burleson, TX-16	719. Sheridan, WY-2	
668. Newton, TX-17	720. Lincoln, WY-3	
669. Edwards, TX-18	721. Niobrara, WY-4	
670. Atascosa, TX-19	722. Converse, WY-5	
671. Wilson, TX-20	723. Rincon, PR-1	
672. Chambers, TX-21	724. Adjuntas, PR-2	
673. Box Elder, UT-1	725. Ciales, PR-3	
674. Morgan, UT-2	726. Aibonito, PR-4	
675. Juab, UT-3	727. Ceiba, PR-5	
676. Beaver, UT-4	728. Culebra-7	
677. Carbon, UT-5	729. Viegues, PR-6	

Elements of Cellular Mobile Radio System Design

2.1 General Description of the Problem

Based on the concept of efficient spectrum utilization, the cellular mobile radio system design can be broken down into many elements, and each element can be analyzed and related to the others. The major elements are (1) the concept of frequency reuse channels, (2) the co-channel interference reduction factor, (3) the desired carrier-to-interference ratio, (4) the handoff mechanism, and (5) cell splitting. The purpose of this chapter is to introduce a simple methodology which will enable us to better understand how each element affects a cellular mobile radio system.

Since the limitation in the system is the frequency resource, the challenge is to serve the greatest number of customers with a specified system quality. We may ask ourselves three questions.

1. How many customers can we serve in a busy hour?

2. How many subscribers can we take into our system?

3. How many frequency channels do we need?

2.1.1 Maximum number of calls per hour per cell

To calculate the predicted number of calls per hour per cell Q in each cell, we have to know the size of the cell and the traffic conditions in the cell. The calls per hour per cell is based on how small the theoretical cell size can be. The control of the coverage of small cells is based on technological development.

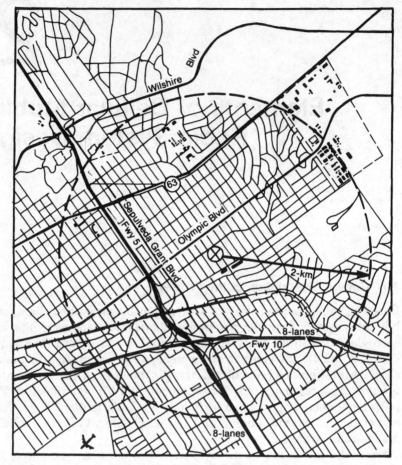

Figure 2.1 To establish the traffic capacity from a geographic map (west Los Angeles).

We assume that the cell can be reduced to a 2-km cell, which means a cell of 2-km radius. A 2-km cell in some areas may cover many highways, and in other areas a 2-km cell may only cover a few highways.

Let a busy traffic area of 12 km radius fit seven 2-km cells. The heaviest traffic cell may cover 4 freeways and 10 heavy traffic streets, as shown in Fig. 2.1. A total length of 64 km of 2 eight-lane freeways, 48 km of 2 six-lane freeways, and 588 km of 43 four-lane roads, including the 10 major roads, are obtained from Fig. 2.1. Assume that the average spacing between cars is 10 m during busy periods. We can determine that the total number of cars is about 70,000. If one-half the cars have car phones, and among them eight-tenths will make a call (η_c = 0.8) during the busy hour, there are 28,000 calls per hour, based on an average of one call per car if that car phone is used.

The maximum predicted number of calls per hour per a 2-km cell Q is derived from the above scenario. It may be an unrealistic case. However, it demonstrates how we can calculate Q for different scenarios and apply this method to finding the different Q in different geographic areas.

2.1.2 Maximum number of frequency channels per cell

The maximum number of frequency channels per cell N is closely related to an average calling time in the system. The standard user's calling habits may change as a result of the charging rate of the system and the general income profile of the users. If an average calling time T is 1.76 min and the maximum calls per hour per cell Q_i is obtained from Sec. 2.1.1, then the offered load can be derived as

$$A = \frac{Q_i T}{60} \quad \text{erlangs} \tag{2.1-1}$$

Assume that the blocking probability is given, then we can easily find the required number of radios in each cell.[1]

Example 2.1 Let the maximum calls per hour Q_i in one cell be 3000 and an average callng time T be 1.76 min. The blocking probability B is 2 percent. Then we may use Q from Eq. (2.1-1) to find the offered load A.

$$A = \frac{3000 \times 1.76}{60} = 88$$

With the blocking probability B = 2 percent, the maximum number of channels can be found from Appendix 1.1 as N = 100.

Example 2.2 If we let Q_i = 28,000 calls per cell per hour, based on one scenaro shown in Sec. 2.1.1, B = 2 percent, and T = 1.76 min, how many radio channels are needed? The offered load A is obtained as

$$A = \frac{28,000 \times 1.76}{60} = 821$$

Inserting the above known figures into the table of Appendix 1.1, we find that N = 820 channels per cell.

Example 2.3 If there are 50 channels in a cell to handle all the calls and the average is 100 s per call, how many calls can be handled in this cell with a blocking probability of 2 percent? Since N = 50 and B = 2 percent, the offered load can be found from Appendix 1.1 as

$$A = 40.3$$

The number of calls per hour in a cell is

$$Q_i = \frac{40.3 \times 3600}{100} = 1451 \text{ calls per hour}$$

Example 2.4 If the maximum number of calls per hour per cell is 1451 and there is a seven-cell reuse pattern* in the system ($K = 7$), and assuming that $B = 2$ percent and $T = 100$ s as in Example 2.3, then $N = 50$ as indicated. The total number of required channels for a $K = 7$ reuse system is

$$N_t = 50 \times 7 = 350 \text{ radios}$$

If a large area is covered by 28 cells, $K_t = 28$; the total number of customers $M_t = \Sigma_{i=1}^{K_t} M_i$ in the system increases. Therefore, we may assume that the number of subscribers per cell M_i is somehow related to the percentage of car phones used in the busy hours (η_c) and the number of calls per hour per cell Q_i as

$$M_i = f(Q_i, \eta_c) \tag{2.1-2}$$

where the value Q_i is a function of the blocking probability B, the average calling time T, and the number of channels N.

$$Q_i = f(B, T, N) \tag{2.1-3}$$

If the $K = 7$ frequency reuse pattern is used, the total number of required channels in the system is $N_t = 7 \times N$. We must realize that it is the maximum number of calls per cell Q_i that determines the total required channels N_t, not the total number of subscribers M_t. In this case ($K_t = 30$ and $K = 7$), the total number of channels N_t has been used four times in the system.

2.2   Concept of Frequency Reuse Channels

A radio channel consists of a pair of frequencies, one for each direction of transmission that is used for full-duplex operation. A particular radio channel, say F_1, used in one geographic zone to call a cell, say C_1, with a coverage radius R can be used in another cell with the same coverage radius at a distance D away.

Frequency reuse is the core concept of the cellular mobile radio system. In this frequency reuse system, users in different geographic locations (different cells) may simultaneously use the same frequency

---

* Its pattern is shown in Fig. 2.3 and described in Sec. 2.2.1.

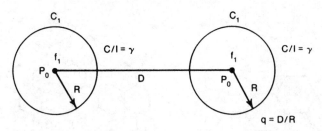

Figure 2.2 The ratio of D/R.

channel (see Fig. 2.2). The frequency reuse system can drastically increase the spectrum efficiency, but if the system is not properly designed, serious interference may occur. Interference due to the common use of the same channel is called *cochannel interference* and is our major concern in the concept of frequency reuse.

2.2.1 Frequency reuse schemes

The frequency reuse concept can be used in the time domain and the space domain. Frequency reuse in the time domain results in the occupation of the same frequency in different time slots. It is called *time-division multiplexing* (TDM). Frequency reuse in the space domain can be divided into two categories.

1. Same frequency assigned in two different geographic areas, such as AM or FM radio stations using the same frequency in different cities.

2. Same frequency repeatedly used in a same general area in one system[2]—the scheme is used in cellular systems. There are many cochannel cells in the system. The total frequency spectrum allocation is divided into K frequency reuse patterns, as illustrated in Fig. 2.3 for $K = 4, 7, 12,$ and 19.

2.2.2 Frequency reuse distance[2,3,7,8]

The minimum distance which allows the same frequency to be reused will depend on many factors, such as the number of cochannel cells in the vicinity of the center cell, the type of geographic terrain contour, the antenna height, and the transmitted power at each cell site.

The frequency reuse distance D can be determined from

$$D = \sqrt{3K}\,R \qquad\qquad (2.2\text{-}1)$$

Where K is the frequency reuse pattern shown in Fig. 2.2, then

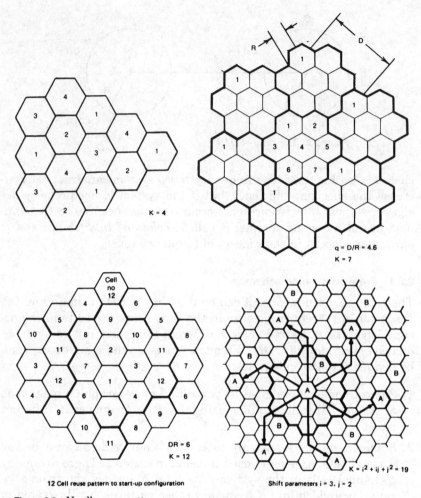

Figure 2.3 N-cell reuse pattern.

$$D = \begin{cases} 3.46R & K = 4 \\ 4.6R & K = 7 \\ 6R & K = 12 \\ 7.55R & K = 19 \end{cases}$$

If all the cell sites transmit the same power, then K increases and the frequency reuse distance D increases. This increased D reduces the chance that cochannel interference may occur.

Theoretically, a large K is desired. However, the total number of allocated channels is fixed. When K is too large, the number of channels assigned to each of K cells becomes small. It is always true that if the total number of channels in K cells is divided as K increases, trunking inefficiency results. The same principle applies to spectrum inefficiency: if the total number of channels are divided into two network systems serving in the same area, spectrum inefficiency increases, as mentioned in Sec. 1.3.

Now the challenge is to obtain the smallest number K which can still meet our system performance requirements. This involves estimating cochannel interference and selecting the minimum frequency reuse distance D to reduce cochannel interference. The smallest value of K is $K = 3$, obtained by setting $i = 1$, $j = 1$ in the equation $K = i^2 + ij + j^2$ (see Fig. 2.3).

2.2.3 Number of customers in the system

When we design a system, the traffic conditions in the area during a busy hour are some of the parameters that will help determine both the sizes of different cells and the number of channels in them.

The maximum number of calls per hour per cell is driven by the traffic conditions at each particular cell. After the maximum number of frequency channels per cell has been implemented in each cell, then the maximum number of calls per hour can be taken care of in each cell. Now, take the maximum number of calls per hour in each cell Q_i and sum them over all cells. Assume that 60 percent of the car phones will be used during the busy hour, on average, one call per phone ($\eta_c = 0.6$) if that phone is used. The total allowed subscriber traffic M_t can then be obtained.

Example 2.5 During a busy hour, the number of calls per hour Q_i for each of 10 cells is 2000, 1500, 3000, 500, 1000, 1200, 1800, 2500, 2800, 900. Assume that 60 percent of the car phones will be used during this period ($\eta_c = 0.6$) and that one call is made per car phone. Summing over all Q_i gives the total Q_t

$$Q_t = \sum_{i=1}^{10} Q_i = 17{,}200 \text{ calls per hour}$$

Since $\eta_c = 0.6$, the number of customers in the system is

$$M_t = \frac{17{,}200}{0.6} = 28{,}667$$

2.3 Cochannel Interference Reduction Factor

Reusing an identical frequency channel in different cells is limited by cochannel interference between cells, and the cochannel interference can become a major problem. Here we would like to find the minimum frequency reuse distance in order to reduce this cochannel interference.

Assume that the size of all cells is roughly the same. The cell size is determined by the coverage area of the signal strength in each cell. As long as the cell size is fixed, cochannel interference is independent of the transmitted power of each cell. It means that the received threshold level at the mobile unit is adjusted to the size of the cell. Actually, cochannel interference is a function of a parameter q defined as

$$q = \frac{D}{R} \tag{2.3-1}$$

The parameter q is the cochannel interference reduction factor. When the ratio q increases, cochannel interference decreases. Furthermore, the separation D in Eq. (2.3-1) is a function of K_I and C/I,

$$D = f(K_I, C/I) \tag{2.3-2}$$

where K_I is the number of cochannel interfering cells in the first tier and C/I is the received carrier-to-interference ratio at the desired mobile receiver.[3]

$$\frac{C}{I} = \frac{C}{\sum\limits_{k=1}^{K_I} I_k} \tag{2.3-3}$$

In a fully equipped hexagonal-shaped cellular system, there are always six cochannel interfering cells in the first tier, as shown in Fig. 2.4; that is, $K_I = 6$. The maximum number of K_I in the first tier can be shown as six (i.e., $2\pi D/D \approx 6$). Cochannel interference can be experienced both at the cell site and at mobile units in the center cell. If the interference is much greater, then the carrier-to-interference ratio C/I at the mobile units caused by the six interfering sites is (on the average) the same as the C/I received at the center cell site caused by interfering mobile units in the six cells. According to both the reciprocity theorem and the statistical summation of radio propagation, the two C/I values can be very close. Assume that the local noise is much less than the interference level and can be neglected. C/I then can be expressed, from Eq. (1.6-4), as

$$\frac{C}{I} = \frac{R^{-\gamma}}{\sum\limits_{k=1}^{K_I} D_k^{-\gamma}} \tag{2.3-4}$$

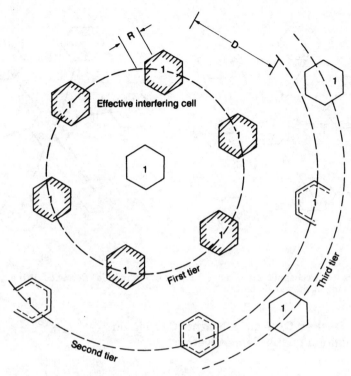

Figure 2.4 Six effective interfering cells of cell 1.

where γ is a propagation path-loss slope[5] determined by the actual terrain environment. In a mobile radio medium, γ usually is assumed to be 4 (see Sec. 1.6). K_I is the number of cochannel interfering cells and is equal to 6 in a fully developed system, as shown in Fig. 2.4. The six cochannel interfering cells in the second tier cause weaker interference than those in the first tier (see Example 2.6 at the end of Sec. 2.4.1).

Therefore, the cochannel interference from the second tier of interfering cells is negligible. Substituting Eq. (2.3-1) into Eq. (2.3-4) yields

$$\frac{C}{I} = \frac{1}{\sum\limits_{k=1}^{K_I} \left(\dfrac{D_k}{R}\right)^{-\gamma}} = \frac{1}{\sum\limits_{k=1}^{K_I} (q_k)^{-\gamma}} \tag{2.3-5}$$

where q_k is the cochannel interference reduction factor with kth cochannel interfering cell

$$q_k = \frac{D_k}{R} \tag{2.3-6}$$

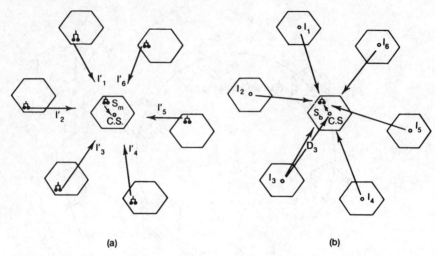

Figure 2.5 Cochannel interference from six interferers. (*a*) Receiving at the cell site; (*b*) receiving at the mobile unit.

2.4 Desired C/I from a Normal Case in an Omnidirectional Antenna System

2.4.1 Analytic solution

There are two cases to be considered: (1) the signal and cochannel interference received by the mobile unit and (2) the signal and cochannel interference received by the cell site. Both cases are shown in Fig. 2.5. N_m and N_b are the local noises at the mobile unit and the cell site, respectively. Usually N_m and N_b are small and can be neglected as compared with the interference level. The effect of the cochannel interference on spectrum efficiency systems will appear in Sec. 13.4. As long as the received carrier-to-interference ratios at both the mobile unit and the cell site are the same, the system is called a *balanced system*. In a balanced system, we can choose either one of the two cases to analyze the system requirement; the results from one case are the same for the others.

Assume that all D_k are the same for simplicity, as shown in Fig. 2.4; then $D = D_k$, and $q = q_k$, and

$$\frac{C}{I} = \frac{R^{-\gamma}}{6D^{-\gamma}} = \frac{q^{\gamma}}{6} \tag{2.4-1}$$

Thus

$$q^{\gamma} = 6\frac{C}{I} \tag{2.4-2}$$

and
$$q = \left(6\,\frac{C}{I}\right)^{1/\gamma} \tag{2.4-3}$$

In Eq. (2.4-3), the value of C/I is based on the required system performance and the specified value of γ is based on the terrain environment. With given values of C/I and γ, the cochannel interference reduction factor q can be determined. Normal cellular practice is to specify C/I to be 18 dB or higher based on subjective tests and the criterion described in Sec. 1.5. Since a C/I of 18 dB is measured by the acceptance of voice quality from present cellular mobile receivers, this acceptance implies that both mobile radio multipath fading and cochannel interference become ineffective at that level. The path-loss slope γ is equal to about 4 in a mobile radio environment.[5]

$$q = D/R = (6 \times 63.1)^{1/4} = 4.41 \tag{2.4-4}$$

The 90th percentile of the total covered area would be achieved by increasing the transmitted power at each cell; increasing the same amount of transmitted power in each cell does not affect the result of Eq. (2.4-4). This is because q is not a function of transmitted power. The computer simulation described in the next section finds the value of $q = 4.6$, which is very close to Eq. (2.4-4). The factor q can be related to the finite set of cells K in a hexagonal-shaped cellular system by

$$q = \overset{\Delta}{=} \sqrt{3K} \tag{2.4-5}$$

Substituting q from Eq. (2.4-4) into Eq. (2.4-5) yields

$$K = 7 \tag{2.4-6}$$

Equation (2.4-6) indicates that a seven-cell reuse pattern* is needed for a C/I of 18 dB. The seven-cell reuse pattern is shown in Fig. 2.3.

Based on $q = D/R$, the determination of D can be reached by choosing a radius R in Eq. (2.4-4). Usually, a value of q greater than that shown in Eq. (2.4-4) would be desirable. The greater the value of q, the lower the cochannel interference. In a real environment, Eq. (2.3-5) is always true, but Eq. (2.4-1) is not. Since Eq. (2.4-4) is derived from Eq. (2.4-1), the value q may not be large enough to maintain a carrier-to-interference ratio of 18 dB. This is particularly true in the worst case, as shown in Chap. 6.

---

* In this seven-cell reuse pattern, the total allocated frequency band is divided into seven subsets. Each particular subset of frequency channels is assigned to one of seven cells.

Example 2.6 Compare interference from the first tier of six interferers with that from twelve interferers (first and second tiers) (see Fig. 2.4).
From the first tier,

$$\frac{C}{I} = \frac{C}{\displaystyle\sum_{i=1}^{6} I_i} = \frac{R_1^{-4}}{6D_1^{-4}} = \frac{a_1^4}{6} \tag{E2.6-1}$$

From the first and second tiers,

$$\frac{C}{I} = \frac{C}{\displaystyle\sum_{i=1}^{6} (I_{1i} + I_{2i})} = \frac{1}{6(a_1^{-4} + a_2^{-4})} \tag{E2.6-2}$$

Since we have found $a_1 = 4.6$, then from the second tier, $a_2 = D_2/R_1 \approx 2D_1/R_1 = 2a_1 = 9.2$. Substituting a_1 and a_2 into Eqs. (E2.6-1) and (E2.6-2), respectively, yields

$$\left(\frac{C}{I}\right)_{\text{1st tier}} = 18.72 \text{ dB}$$

$$\left(\frac{C}{I}\right)_{\text{1st and 2nd tiers}} = 18.46 \text{ dB}$$

We realize that a negligible amount of interference is contributed by the six interferers from the second tier.

2.4.2 Solution obtained from simulation

The required cochannel reduction factor q can be obtained from the simulation also. Let one main cell site and all six possible cochannel interferers be deployed in a pattern, as shown in Fig. 2.4. The distance D from the center cell to the cochannel interferers in the simulation is a variable. $D = 2R$ can be used initially and incremented every $0.5R$ as $D = 2R, 2.5R, 3R$. For every particular value of D, a set of simulation data is generated.

First, the location of each mobile unit in its own cell is randomly generated by a random generator. Then the distance D_k from each of the six interfering mobile units to the center cell site (assuming $K_I = 6$) is obtained. The desired mobile signal as well as six interference levels received at the center cell site would be randomly generated following the mobile radio propagation path-loss rule, which is 40 dB/dec, along with a log-normal standard deviation of 8 dB at its mean value. Summing up all the data from six simulated interferences,

$$I = \sum_{k=1}^{K_I=6} I_k$$

and dividing it by the simulated main carrier, value C becomes C/I.

This C/I is for a particular D, the distance between the center cell site and the cochannel cell sites (cochannel interferers). Repeat this process, say 1000 times, for each particular value of D, based on the criterion stated in Sec. 1.5 (that 75 percent of the users say voice quality is "good" or "excellent" in 90 percent of the total covered area). Then from 75 percent of the users' opinion, $C/I = 18$ dB needs to be achieved[4] with a proper value of D. Assuming that mobile unit locations are chosen randomly and uniformly, then 90 percent of the area corresponds to 900 out of 1000 mobile unit locations.

To find a proper value for D, each mobile unit location associates with its received C/I. Some C/I values are high and some are low. This means that the lowest 100 values of C/I should be discarded. The main C/I value should be derived from the remaining 900 C/I values. This associates a particular C/I for a particular separation D. Repeating this process for different values of D, the corresponding mean C/I values are found. The C/I versus D curve can be plotted, depicting $C/I = 18$ dB as corresponding to $D = 4.6R$, as illustrated in the Bell Lab publication.[4] Then

$$q = \frac{D}{R} = 4.6 \qquad (2.4\text{-}7)$$

Comparing the values of q obtained from an analytic solution shown in Eq. (2.4-4) and q obtained from a simulation solution shown in Eq. (2.4-7), the results are surprisingly close.

Although a simulation (statistical) approach deals with a real-world situation, it does not provide a clear physical picture. The two agreeable solutions illustrated in this section prove that the simple analytic method is implementable in a cellular system based on hexagonal cells.

2.5 Handoff Mechanism

The handoff is the process mentioned in Sec. 1.7. It is a unique feature that allows cellular systems to operate as effectively as demonstrated in actual use. To clearly describe the handoff concept, it is easy to use a one-dimensional illustration as shown in Fig. 2.6, although a real two-dimensional cellular configuration would cover an area with cells. The handoff concept as applied to a one-dimensional case will also apply to two-dimensional cases.

Two cochannel cells using the frequency F_1 separated by a distance D are shown in Fig. 2.6a. The radius R and the distance D are governed by the value of q. Now we have to fill in with other frequency

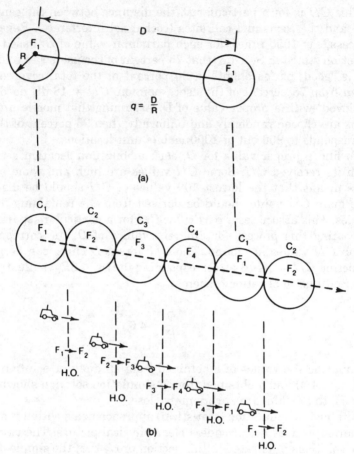

Figure 2.6 Handoff mechanism. (*a*) Cochannel interference reduction ratio *q*. (*b*) Fill-in frequencies.

channels such as F_2, F_3, and F_4 between two cochannel cells in order to provide a communication system in the whole area.

The fill-in frequencies F_2, F_3, and F_4 are also assigned to their corresponding cells C_2, C_3, and C_4 (see Fig. 2.6*b*) according to the same value of *q*.

Suppose a mobile unit is starting a call in cell C_1 and then moves to C_2. The call can be dropped and reinitiated in the frequency channel from F_1 to F_2 while the mobile unit moves from cell C_1 to cell C_2. This process of changing frequencies can be done automatically by the system without the user's intervention. This process of handoff is carried on in the cellular system.

The handoff processing scheme is an important task for any successful mobile system. How does one make any one of the necessary handoffs successful? How does one reduce all unnecessary handoffs in the system? How is the individual cell traffic capacity controlled by altering the handoff algorithm? All these questions will be answered in Chap. 9.

2.6 Cell Splitting

2.6.1 Why splitting?

The motivation behind implementing a cellular mobile system is to improve the utilization of spectrum efficiency.[5,9] The frequency reuse scheme is one concept, and cell splitting is another concept. When traffic density starts to build up and the frequency channels F_i in each cell C_i cannot provide enough mobile calls, the original cell can be split into smaller cells. Usually the new radius is one-half the original radius (see Fig. 2.7). There are two ways of splitting: In Fig. 2.7a, the original cell site is not used, while in Fig. 2.7b, it is.

$$\text{New cell radius} = \frac{\text{old cell radius}}{2} \tag{2.6-1}$$

Then based on Eq. (2.6-1), the following equation is true.

$$\text{New cell area} = \frac{\text{old cell area}}{4} \tag{2.6-2}$$

Let each new cell carry the same maximum traffic load of the old cell; then, in theory,

$$\frac{\text{New traffic load}}{\text{Unit area}} = 4 \times \frac{\text{traffic load}}{\text{unit area}}$$

2.6.2 How splitting?

There are two kinds of cell-splitting techniques:

1. *Permanent splitting*. The installation of every new split cell has to be planned ahead of time; the number of channels, the transmitted power, the assigned frequencies, the choosing of the cell-site selection, and the traffic load consideration should all be considered. When ready, the actual service cut-over should be set at the lowest traffic point, usually at midnight on a weekend. Hopefully, only a few calls will be dropped because of this cut-over, assuming that the downtime of the system is within 2 h.

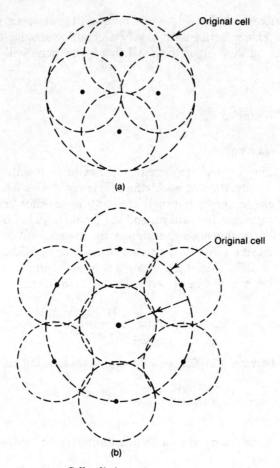

Figure 2.7   Cell splitting.

2. *Dynamic splitting.* This scheme is based on utilizing the allocated spectrum efficiency in real time. The algorithm for dynamically splitting cell sites is a tedious job since we cannot afford to have one single cell unused during cell splitting at heavy traffic hours. Section 10.4.2 will discuss this topic in depth.

2.7   Consideration of the Components of Cellular Systems

The elements of cellular mobile radio system design have been mentioned in the previous sections. Here we must also consider the components of cellular systems, such as mobile radios, antennas, cell-site

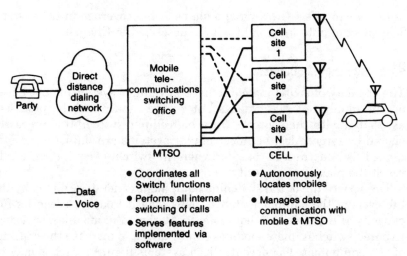

Figure 2.8 A general view of cellular telecommunications systems.

controller, and MTSO. They would affect our system design if we do not choose the right one. The general view of the cellular system is shown in Fig. 2.8. Even though the EIA (Electronic Industries Association) and the FCC have specified standards for radio equipment at the cell sites and the mobile sites, we still need to be concerned about that equipment. The issues affecting choice of antennas, switching equipment, and data links are briefly described here.

2.7.1 Antennas

Antenna pattern, antenna gain, antenna tilting, and antenna height[6] all affect the cellular system design. The antenna pattern can be omnidirectional, directional, or any shape in both the vertical and the horizon planes. Antenna gain compensates for the transmitted power. Different antenna patterns and antenna gains at the cell site and at the mobile units would affect the system performance and so must be considered in the system design.

The antenna patterns seen in cellular systems are different from the patterns seen in free space. If a mobile unit travels around a cell site in areas with many buildings, the omnidirectional antenna will not duplicate the omnipattern. In addition, if the front-to-back ratio of a directional antenna is found to be 20 dB in free space, it will be only 10 dB at the cell site. An explanation for these phenomena is given in Chapter 5.

Antenna tilting can reduce the interference to the neighboring cells and enhance the weak spots in the cell. Also the height of the cell-site antenna can affect the area and shape of the coverage in the system. The effect of antenna height will be described in Chap. 4.

2.7.2 Switching equipment

The capacity of switching equipment in cellular systems is not based on the number of switch ports but on the capacity of the processor associated with the switches. In a big cellular system, this processor should be large. Also, because cellular systems are unlike other systems, it is important to consider when the switching equipment would reach the maximum capacity.

The service life of the switching equipment is not determined by the life cycle of the equipment but by how long it takes to reach its full capacity. If the switching equipment is designed in modules, or as distributed switches, more modules can be added to increase the capacity of the equipment. For decentralized systems digital switches may be more suitable. The future trend seems to be the utilization of system handoff. This means that switching equipment can link to other switching equipment so that a call can be carried from one system to another system without the call being dropped. We will discuss these issues in Chap. 11.

2.7.3 Data links

The data links are shown in Fig. 2.8. Although they are not directly affected by the cellular system, they are important in the system. Each data link can carry multiple channel data (10 kbps data transmitted per channel) from the cell site to the MTSO. This fast-speed data transmission cannot be passed through a regular telephone line. Therefore, data bank devices are needed. They can be multiplexed, many-data channels passing through a wideband T-carrier wire line or going through a microwave radio link where the frequency is much higher than 850 MHz.

Leasing T1-carrier wire lines through telephone companies can be costly. Although the use of microwaves may be a long-term money saver, the availability of the microwave link has to be considered and is described in Chap. 12. The arrangement of data links will be described in Chap. 11.

References

1. Siemens, "Telephone Traffic Theory and Table and Charts, Part 1," Telephone and Switching Division, Siemens, Munich, 1970.
2. V. H. MacDonald, "The Cellular Concept," *Bell System Technical Journal,* Vol. 58, No. 1, January 1979, pp. 15–42.

3. W. C. Y. Lee, "Mobile Communications Design Fundamentals," Howard W. Sams & Co., 1986, p. 141.
4. Bell Laboratories, "High-Capacity Mobile Telephone System Technical Report," December 1971, submitted to FCC.
5. W. C. Y. Lee, "Mobile Cellular System Conserves Frequency Resource," *Microwave Systems News & Communications Technology,* Vol. 15, No. 7, June 1985, pp. 139–150.
6. W. C. Y. Lee, *Mobile Communications Engineering,* McGraw-Hill Book Co., 1982, p. 102.
7. D. Bodson, G. F. McClure, and S. R. McConoughey (eds.). *Land-Mobile Communications Engineering,* IEEE Press, 1984, Part III.
8. F. H. Blecher, "Advanced Mobile Phone Service," *IEEE Transactions on Vehicular Technology,* Vol. VT-29, May 1980, pp. 238–244.
9. S. W. Halpren, "Reuse Petitioning in Cellular Systems," *33rd IEEE Vehicular Technology Conference Record,* Toronto, May 1983.

3

Specifications
of Analog Systems

In this chapter, we concentrate on U.S. analog cellular mobile specifications.[1] Also, we touch on the differences in other foreign analog cellular mobile systems.

3.1 Definitions of Terms and Functions

1. *Home mobile station* (*unit*). A mobile station that is subscribed in its cellular system.

2. *Land station.* A station other than a mobile station, which links to the mobile station.

3. *Control channel.* A channel used for the transmission of digital control information from a land station to a mobile station, or vice versa.

4. *Forward control channel* (FDCC). A control channel used from a land station to a mobile station.

5. *Reverse control channel* (RECC). A control channel used from a mobile station to a land station.

6. *Forward voice channel* (FVC). A voice channel used from a land station to a mobile unit.

7. *Reverse voice channel* (RVC). A voice channel used from a mobile station to a land station.

8. *Set-up channels.* A number of designated control channels.

9. *Access channel.*   A control channel used by a mobile station to access a system and obtain service. The access channel always accesses from the mobile station to the cell site.

10. *Paging channel.*   The act of seeking a mobile station when an incoming call from the land line has been placed to it.

11. *Digital color code* (DC).   A digital signal transmitted by a forward control channel to detect capture of an interfering mobile station. There are four codes (See Sec. 3.2.8.)

12. *Flash request.*   A message sent on a voice channel from a mobile station to a land station indicating a user's desire to invoke special processing, such as an emergency.

13. *Signaling tone.*   A 10-kHz tone transmitted by the mobile station on a voice channel. It serves several functions.

14. *Handoff.* The act of transferring a mobile station from one voice channel to another voice channel. There are two kinds of handoffs:
    a. Interhandoff, from one cell to another cell
    b. Intrahandoff, within a cell

15. *Numeric information.*   Used to describe the operation of the mobile station.

Numeric indicators	
MIN	mobile identification number
MIN1	24 bits that correspond to the seven-digit directory number assigned to the mobile station
MIN2	10 bits that correspond to the three-digit area code
BIS	Identifies whether a mobile station must check an idle-to-busy transition on a reverse control channel when accessing a system. In a forward control-channel busy-idle bit inserts in every 10-bit interval of a transmitted bit stream.
CCLIST	scanned by a mobile station on a list of control channels
CMAX	maximum number of control channels to be scanned by the mobile station (up to 21 channels)
MAXBUSY	maximum number of busy occurrences allowed on a reversed control channel
MAXSZTR	maximum number of seizure attempts allowed on a reversed control channel
NBUSY	number of times a mobile station attempts to seize a reverse control channel and finds it busy
NSZTR	number of times a mobile station attempts to seize a reverse control channel and fails
PL	mobile station RF power level

SCC   a digital number that is stored and used to identify which SAT (see item 20 below) frequency a mobile station should be received on

16. *Paging.* The act of seeking a mobile station when an incoming call from the land station has been placed to it.

17. *Paging channel.* A forward control channel which is used to page mobile stations and send orders.

18. *Registration.* The procedure by which a mobile station identifies itself to a land station as being active.

19. *Roamer.* A mobile station which operates in a cellular system other than the one from which service is subscribed.

20. *Supervisory audio tone* (SAT). One of three tones in the 6-kHz region; there is one SAT frequency for each land station. In certain circumstances, there is one SAT frequency for each sector of each land station.

21. *System identification* (SID). A digital identification uniquely associated with a cellular system.

22. *Electronic serial number* (ESN). Each mobile station has an ESN assigned by the manufacturer.

23. *Group identification.* A subset of the most significant bits of SID that is used to identify a group of cellular systems, such as NYNEX systems, PacTel systems, Southwestern Bell systems.

24. *Channel spacing.* 30 kHz per one-way channel. As an example, channel 1 is 825.030 MHz (mobile transmit) and 870.030 MHz (land transmit). Additional spectrum allocation of 10 MHz for the cellular industry changes the channel numbering order.

3.2 Specification of Mobile Station (Unit) in the United States[1-3]

3.2.1 Power

Let P_0 be the specified power and f_0 be the specified frequency channel. P and f are the operating power and frequency, respectively.

Power level (carrier-off condition) requires $P < -60$ dBm in 2 ms

Power level (carrier-on-condition) within 3 dB of specified power(P_0) within 2 ms

Power level (off-frequency condition), if $|f - f_c| > 1$ kHz, do not transmit; then $P < -60$ dBm

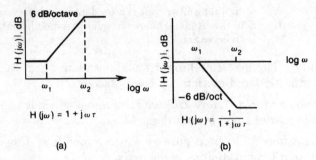

Figure 3.1 Preemphasis/deemphasis response. (*a*) Preemphasis response $\omega_1 = 1/\tau$, $\omega_2 = 2\pi f_2$. (*b*) Deemphasis response $\omega_1 = 1/\tau$, $\omega_2 = 2\pi f_2$.

Power transmitted levels are maximum effective radiated power (ERP) with respect to a half-wave dipole

Mobile stations have three station class marks:

Power class	P, power level = 0	Tolerance
I	6 dBW (4.0 W)	(8 dBW $\leq P \leq$ 2 dBW)
II	2 dBW (1.6 W)	(4 dBW $\leq P \leq$ −2 dBW)
III	−2 dBW (0.6 W)	(0 dBW $\leq P \leq$ −6 dBW)

Each mobile station power class I has eight full power levels (0 to 7), with power level 0 being the highest. Each level has a 4-dB drop. The total power control range for power class I (mobile) is 28 dB. The names CMAC and VMAC indicate the maximum control and maximum voice attenuation codes, respectively. For all three mobile station power classes, power level 7 is −22 dBW (or 8 dBm or 6.3 mW).

3.2.2 Modulation

1. *Compressor / expandor (compander)*.   A 2:1 syllabic compander is used. Every 2-dB change in input level converts (compresses) to 1 dB at output (at the transmitted side). Then reverse the two numbers (expand) at the received side. It serves two purposes:
    a. To confine the energy in the channel bandwidth
    b. To generate a quieting effect during a speech pulse
2. *Preemphasis / deemphasis* (see Fig. 3.1). The preemphasis network and its response are shown in Fig. 3.1. The improvement factor ρ_{FM} is[4]

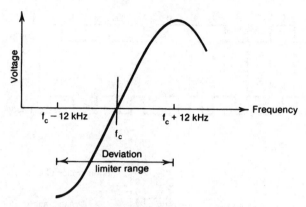

Figure 3.2 Deviation limiter characteristics.

$$\rho_{FM} = \frac{(f_2/f_1)^3}{3[f_2/f_1 - \tan^{-1}(f_2/f_1)]}$$

For $f_2/f_1 < 2$, ρ_{FM} approaches 1.
For wideband, $f_2 \gg f_1$. Thus

$$\rho_{FM} = \frac{(f_2/f_1)^2}{3}$$

For $f_2 = 3$ kHz and $f_1 = 300$ Hz, the improvement factor ρ_{FM} is

$$\rho_{FM} = 10 \log \frac{100}{3} = 15.23 \text{ dB}$$

3. *Deviation limiter.* A mobile station must limit the instantaneous frequency deviation to ± 12 kHz. The deviation limiter of the frequency-to-voltage characteristic is shown in Fig. 3.2.
4. *Wideband data signal.* A NRZ (non-return-to-zero) binary data stream is encoded to a Manchester (biphase) code, as shown in Fig. 3.3. It modulates to a ± 8-kHz binary FSK with a transmission rate of 10 kbps. The advantage of using a Manchester code in a voice channel is that the energy of this code is concentrated at the transmisson rate of 10 kHz. Therefore, a burst of signals transmitted over the voice channel can be detected. The Manchester code is applied to both control channels and voice channels.

3.2.3 Limitation on emission

It is very important that each mobile station have a limit on its emission. An RF signal emitted by any receiver has to be less than some

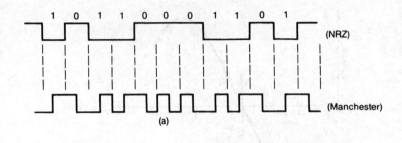

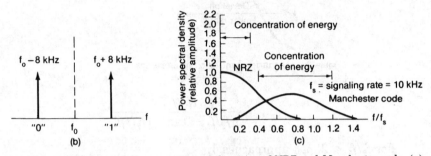

(a)

(b)

(c)

Figure 3.3 Waveforms and power spectral densities of NRZ and Manchester code. (a) Waveforms of NRZ and Manchester code. (b) Use of two frequency deviations to represent two levels. (c) Power spectral densities of NRZ and Manchester code.

value, which in turn depends on different frequency bands. The values are shown in the box as follows.

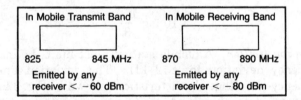

In Mobile Transmit Band	In Mobile Receiving Band
825 845 MHz	870 890 MHz
Emitted by any receiver < -60 dBm	Emitted by any receiver < -80 dBm

3.2.4 Security and identification

1. *Mobile identification number* (MIN). A binary number of 34 bits ($2^{34} \approx 1.7 \times 10^{10}$) derived from a 10-digit directory telephone number.

2. *Electronic serial number* (ESN). A 32-bit binary number that uniquely identifies a mobile unit and must be set by the factory. Attempts to change the serial number circuit should render the mobile unit *inoperative*.

The manufacturer should not overlook this requirement, and each operating system has to check that this requirement is enforced from different manufactured units. This number will be used in the system for security purposes.

An amount of revenue can be lost to the system operation if the manufacturer fails to implement this requirement in the mobile unit and a false ESN can be used in place easily.

3. *First paging channel* (FIRSTCHP).   An 11-bit system which identifies the channel number of the first paging channel when the mobile station is "home." It is stored in the mobile unit.

4. *Home system identification* (SID).   Fifteen-bit system used to identify the home station. The least significant bit is 1 for a Block A system, otherwise 0 for a Block B system.

5. *Preferred system selection.*   Provided as a means for selecting the preferred system as either system A or system B within a mobile station.

3.2.5   Supervision

SAT (supervisory audio tone)

1. *SAT function.*   There are three SAT tones: 5970, 6000, and 6030 Hz. The tolerance of each tone is ± 15 Hz. The features of the SAT tones are

   a. Each land station is assigned to one SAT among the three listed above.

   b. One SAT tone is added to each forward voice channel (FVC) by a land station. The mobile station detects, filters, and modulates on the reversed voice channel (RVC) with this same tone.

   c. SAT is suspended during transmission of wideband data (a burst of signaling of 10 kbps), information, or other control features on the reverse voice channel.

   d. It is not suspended when a signaling tone (10 kHz) is sent.

   e. The received audio transmission must be muted if the measured SAT and SCC do not agree with each other.

2. *SAT transmission.*   The tone modulation index is $\frac{1}{3}$. It is a narrowband FM. The deviation $\Delta F = \pm 2$ kHz is centered around each SAT tone.

3. *Fade timing status of SAT.*   The transmitter is turned off if no valid SAT tone can be detected or the measured SAT does not agree with the SAT color code (SCC) of each cell site transmitted by the control signal, or if no SAT is received when the timer counts to 5 s. SCC is indicated in the set-up channel.

Signaling tone.   Signaling tones must be kept within 10 kHz ± 1 Hz and produce a nominal frequency deviation of ±8 kHz of the carrier frequency. It is used over the voice channel. It serves three functions.

1. Flush for special orders
2. Terminate the cells
3. Order confirmation

Malfunction timer set at 60 s. The transmission will cease when the timer exceeds 60 s. The timer never expires as long as the proper sequence of operations is taking place.

3.2.6 Call processing

Initialization. When the user turns on the mobile station, the initialization work starts.

1. Mobile station must tune to the strongest dedicated control channel (usually one of 21 channels) within 3 s. Then a system parameter message should be received.
2. If it cannot complete this task on the strongest dedicated control channel, it turns to the second strongest dedicated control channel and attempts to complete this task within the next 3-s interval.
3. Check whether it is in Enable or Disable status. Serving system status changed to Enable if the preferred system is System A. Serving system status changed to Disable if the preferred system is System B.

Paging channel selection

1. The mobile station must then tune to the strongest paging channel within 3 s. Usually paging channels are control channels.
2. Receive an overhead message train and update the following: Roam, system identification, local control status.

Idle stage. A mobile station executes each of four tasks at least every 46.3 ms.

1. *Responds to overhead information.* Compare SID_s (stored) with SID_r (received). If $SID_s \neq SID_r$, return to initialization stage.
2. *Page match.* If the roam status is Disable, the mobile station must attempt to match for one-word message or two-word messages. If the roam status is Enable, the mobile station must attempt to match two words.
3. *Order.* After matching the MIN, respond to the order.

4. *Rescan access channels.* Because the vehicle is moving, information must be updated. Rescan at 2- to 5-min intervals, depending on different manufacturers.

Call initiation. Origination indication (system access task).

1. *System access task.* When the system access task is started, an access timer is set for

Origination	max 12 s
Page response	max 6 s
An order response	max 6 s
Registration	max 6 s

2. *Scan access channels.* Choose one or, at most, two channels with the strongest signals. If the service request cannot be completed on the strongest signal, the mobile station can select the second strongest access channel (called the *alternate access channel*).
3. *Seize reverse control channel* (RECC).
   a. BIS = 1 status: The mobile station is ready for sending. The land station may ask the mobile station to check and wait for overhead message (WFOM) bit.

WFOM = 1	The mobile station waits to update overhead information (see item 5).
WFOM = 0	Delay a random time (0 to 92 ms) and send service request. It is an access attempt.

   b. BIS = 0 status: This is the status of a "busy" condition. The mobile station increments NBUSY by 1, and then has to wait a random time interval of 0 to 200 ms to check the BIS status (0 to 1) again. When NBUSY exceeds MAXBUSY the call is terminated.
4. *Access attempt parameters.* Maximum of 10 attempts. There is a random delay interval of 0 to 92 ± 1 ms for each attempt at checking the status of BIS. If BIS = 1, the mobile station just waits for the transmitting power to come up and sends out the service request message. The random time delay is used to avoid two or more mobile stations requesting services at the same time.
5. *Update overhead information.* Update overhead information should be completely received by the mobile station within 1.5 s after a call is initiated. Update overhead information is as follows.
   Overhead information (OHD). The OHD will be sent by the land station and updated by the mobile station.

    a. Overhead control message (whether the system is overloaded or not)

    b. Access type parameter message sets the busy-idle status bits in the BIS field.

    c. Access attempt parameters message provides the following parameters.

        (1) Maximum number of seizure tries allowed

        (2) Maximum number of busy occurrences

After the update overhead information has been completely received, the mobile station waits a random time interval of 0- to 750-ms and enters the seize reverse control channel task stated in item 3.

6. *Delay after failure.* The mobile station must examine the access timer every 1.5 s; if it does not expire it reenters the access task after failure. The three failure conditions are as follows.

    a. Collision with other mobile station messages. If the collision occurs before the first 56 bits, the BIS changes from 1 to 0.

    b. The land station does not receive the signaling bits. the BIS remains 1 after the mobile station has sent 104 bits.

    c. The land station receives all the signaling bits but cannot interpret them and respond.

When these conditions occur, the mobile station must wait a random time before making the next attempt. A random delay should be in the interval of 0 to 200 ms.

7. *Service request message.* A whole package of service request messages must be continuously sent to the land station. The format of each signaling word is shown in Fig. 3.4. There is a maximum length to the message consisting of five words: A, B, C, D, E. After a complete message is sent by the mobile station, an unmodulated carrier follows for 25 ms to indicate the end of the message.

8. *Await message response.* If there is no response after the request is sent for 5 s, the call is terminated, and a 120-impulse-per-minute fast tone is generated to the user. If decoded MIN bits match within 5 s, the mobile station must respond with the following messages.

    a. If access is an *origination* or *page response.*

        (1) Initiate voice channel designation message. Update the parameters as set in the message.

        (2) For a directed-retry message the mobile station must examine the signal strength on each of the retry channels and choose up to two channels with the strongest signals. The mobile station must then tune to the strongest retry access channel.

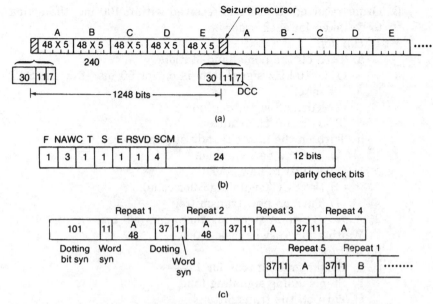

Figure 3.4 Signaling format of both RECC and RVC. (*a*) Reverse control channel (RECC) data stream. (*b*) A message word of RECC. (*c*) Signaling format of RVC.

    b. If the mobile station encounters the start of a new message before it receives the directed-retry message, the call has to be terminated.

3.2.7 Mobile station controls on the voice channel

Loss of radio-link continuity. While the mobile station is tuned to a voice channel, a fade timer must be started when no SAT tone is received. If the fade timer counts to 5 s, the mobile station must turn off.

Confirm initial voice channel. Within 100 ms of the receipt of the initial voice channel designation, the mobile station must determine that the channel number is within the set allocated to the home land station or from the other source.

Alerting

I. Waiting for order
   A. If an order cannot be received in 5 s, terminate the call.

B. Order received. If order is received within 100 ms, the action to be taken for each order is

1. *Handoff*
   a. Turn off the home land station.
      (1) a 10-kHz signal tone is on for 50 ms after the SAT tone.
      (2) Turn off signaling tone.
      (3) Turn off transmitter.
   b. Turn on the new site, adjust power level.
      (1) Turn to new channel.
      (2) Adjust to new SAT.
      (3) Set SCC (signaling color code).
      (4) Turn on new transmitter.

2. *Alert.* Turn on signaling tone, run 500 ms, and enter the Waiting for Answer task.

3. *Release*
   a. Send signaling tone for 1.8 s.
   b. Stop sending signaling tone.
   c. Turn off the transmitter.

4. *Audit.* Send order confirmation message to land station, remain in Waiting for Answer task, and reset the order timer for 5 s.

5. *Maintenance.* Turn on signaling tone, run for 500 ms, and enter the Waiting for Answer task.

6. *Change power*
   a. Adjust the transmitter to new ordered level.
   b. Send order confirmation to the land station.
   c. Local control. If the local control option is enabled in the mobile station, the local control order can be enabled if the group identification matches the SID_p in the mobile station's permanent security memory. A system operator can have a "local control" order for several markets and order under a group identification.

II. Waiting for answer. After requesting orders from the land station, the mobile station is in the Waiting for Answer status. An alert time must be set to 65 s. If no answer comes back in 65 s, the call is terminated. Events occur in the same order as listed in the Waiting for Order section above.

Conversation.   A release-delay timer must be set to 500 ms during the conversation. The task can be used for the following conditions.

1. If the user terminates the call.

2. If the user requests a flush.

3. Within 100 ms of receipt of any orders, action will be taken by the mobile station for each order

3.2.8 Signaling format

Signaling rate.   The signaling rate is 10 kbps ± 1 bps. It is slow enough to not cause the intersymbol interference. The Manchester code waveform is applied so that the energy of this signaling waveform is concentrated at 10 kHz, which can be distinguished from the energy concentrated around the carrier frequency for the baseband voice. (See Fig. 3.3c.)

Signaling format.   The reverse control channel (RECC) data stream is shown in Fig. 3.4a. The first word of 48 bits is called the *seizure precursor*, which consists of 30 synchronization bits, 11 frame bits, and 7 coded DCC (digital color code) bits.

Function	Coding
30 synchronization bits	10101010--
11 frame bits (word synchronization)	11100010010
7-bit coded DCC (00)	0000000
(01)	0011111
(10)	1100011
(11)	1111100

Each information word contains 48 bits. Each word block contains 240 bits, where each word is repeated five times.

The maximum data stream is one seizure precursor plus five word blocks: A, B, C, D, and E. The total number of bits is 1248 bits as shown in Fig. 3.4a.

In each information word, 36 bits are information bits and the other 12 bits are parity check bits, formed by encoding 36 bits into a (48, 36) BCH code that has a Hamming distance of 5 (described in Chap. 13). The format is shown in Fig. 3.4b for the first word. The interpretation of the data field is as follows.

F   First word indication field, 1—first word, 0—subsequent words

NAWC   Number of additional words coming

T   T field, 1—indicates an origination, 0—indicates page response

S   S field, 1—send serial number word, 0—otherwise

E   Extended address field, 1—extended address word sent, 0—not sent

SCM   Station class mark field (see Sec. 2.2.1)

Types of messages. The types of messages to be transmitted over the reverse control channel (RECC) are:

- Page response message. When the mobile station receives a page from the land station, the mobile station responds back.
- Origination message. The mobile station originates the call.
- Order confirmation message. The mobile station responds to the order from the land station.
- Order message. The mobile station orders the tasks which should be performed by the land station and the mobile transmission switching office (MTSO).

Function of each word

Word A	An abbreviated address word. It is always sent to identify the mobile station.
Word B	An extended address word. It will be sent on request from the land station or in a roam situation. In addition, the local control field and the other field are shown in this word.
Word C	A serial number word. Every mobile unit has a unique serial number provided by the manufacturer. It is used to validate the eligible users.
Word D	The first word of the called address.
Word E	The second word of the called address.

Reverse voice channel (RVC). The reverse voice channel (RVC) is also used by a wideband data stream sent from the mobile station to the land station. A 10 kbps ± 1 bps data stream is generated. A word is formed by encoding the 36 content bits into a (48, 36) BCH code, the same as the RECC.

1. *Signaling format.* The first 101 syn bits are used for increasing the possibility of successful syn. The signaling format of RVC is shown in Fig. 3.4c. There are two words: The first word repeats five times, and then the second word repeats five times.
2. *Types of messages.* There are two types of messages:
   a. Order confirmation message (one word) responds to the land station to confirm the order, e.g., handoff confirmation.
   b. Called-address message (two words) establishes a three-party call.

3.3 Specification of Land Station (United States)[1-3]

Most parts in the specification of the land station are the same as the specification of the mobile station, such as the modulation of voice signals (sec. 3.2.2), security and identification (Sec. 3.2.4),and supervision (Sec. 3.2.5). These sections will not be repeated in the specification of the land station.

3.3.1 Power

Maximum effective radiated power (ERP) and antenna height above the average terrain (HAAT) must be coordinated locally on an ongoing basis. Maximum power is 100 W at a HAAT of 500 ft. Normally, the transmitting 20 W at an antenna height of 100 ft above the local terrain is implemented.

3.3.2 Limit on emission

The field strength limit at a distance of 100 ft or more from the receiver is 500 μV/m.

3.3.3 Call processing

Call processing is the land station operation that controls the mobile station.

Overhead functions for mobile station initiation. The overhead message train contains the first part of the system identification (SID1) and the number of paging channels (N).

On control channel

1. Overhead information is sent on the forward control channel and requires all mobile stations to either update or respond with new information during a system access.

    Update the following information
    - First part of the system identification (SID1)
    - Serial number (S). If S = 1, all mobile stations send their serial numbers during a system access; if S = 0, no need to send serial number.
    - Registration (REGH, REGR). The land station is capable of registering the mobile stations.

REGH = 1   Enables registration for home mobile stations

REGH = 0   Otherwise

REGR = 1   Enables registration for roaming mobile stations

REGR = 0   Otherwise

- Extended address (E)

E = 1   Both MIN1 and MIN2 required

E = 0   Otherwise

- Discontinuous transmission (DTX)

DTX = 1   Let mobile stations use discontinuous transmission mode on the voice channel (reducing power consumption for portable units)

DTX = 0   Otherwise

- Number of paging channels (N)
- Read control-filler message (RCF) (see Sec. 3.3.4)

RCF = 1   ask the mobile unit to read control-filler message before accessing a system on a reverse control channel

RCF = 0   Otherwise

- Combined paging/access (CPA)

CPA = 1   Paging channel and access channel are the same

CPA = 0   Paging channel and access channel are not the same

- Number of access channel (CMAX)

Respond with the following information
- Local control.   A system operation for home mobile stations and for the roaming mobile stations that are members.
- New access channels (NEWWACC).   Send NEWACC information along the first access channel.
- Registration increment (REGINCR).   Each time the mobile station increments a fixed value received on FOCC for its updated registration ID if it is equipped for autonomous registration.
- Registration ID (REGID). The last registration number received on FOCC and stored at the mobile station. Every time an increment occurs REGID (new) = REGID (old) + REGINCR, the mobile station identifies itself to a land station.
- Rescan.   The rescan global action message must be sent to require all mobile stations to enter the initialization task.

2. The land station will use a control message (one word or two words) to page a mobile station through its home land station. The roaming mobile station must be paged with a two-word message.

3. Orders must be sent to mobile stations with a two-word control message. The orders can be audit and local control. By sending local orders with the order field set to local control and using system identifications (SID) that have identical group identifications, a home mobile station or a roaming mobile station which is a member of a group can be distinguished.

Land station support of system access

1. *Overhead information.* The following information must be sent on a forward control channel to support system access which is used by mobile stations.

   - *Digital color code* (DCC). The mobile station uses DCC to identify the land station.
   - *Control mobile attenuation code* (CMAC). When a control-filler message is transmitted, the mobile station receiving the code has to adjust its transmitter power level before accessing a system on a reverse control channel.
   - *Wait for overhead message* (WFOM). Set WFOM to 1 in the control-filler message; then the mobile station must wait for WFOM before accessing a system on a reverse control channel.
   - *Overload control* (OLC). The mobile stations that are assigned to one or more of the 16 overload classes (N = 1 to 16) must not access the system for originations on the RECC.
   - *Access-type parameters.* When the access-type parameters' global action message with the BIS field set to 0 is appended to a system parameter overhead message, the mobile stations do not check for an idle-to-busy status.
   - *Access-attempt parameters* are the limit on the number of "busy" occurrences for mobile stations or the default values for the number of seizure attempts.

2. *Reverse control channel seizure by a mobile station.* When this equals 1 all mobile stations must check for an idle-to-busy status when accessing a system. A seizure precursor (48 bits including coded DCC) sent by a mobile station and received by the land station should match its encoded form of DCC.

   It must set the status of the busy-idle bits on the forward control channel between 0.8 and 2.9 ms of receipt of the last bit of 48 bits of the seizure precursor. The busy-idle bits must remain busy until
   - 30 ms after the last word of message has been received

■ (24 N + 55) ms otherwise, where N is the maximum number of words. It will not exceed 175 ms.
3. *Response to mobile station messages.* It is not required that the land station respond to the mobile station message. During periods of system overload or high usage, it may be desirable to permit mobile stations to "time-out" rather than sending release or other orders which use system capacity. The usual time-out period is 5 s. It means that after 5 s, if the mobile station does not receive any response from the land station, the mobile station terminates the transmitted power. The following responses to mobile stations may be sent:
   a. *Origination message.*   Send one of the following orders.
      (1) Initial voice channel designation
      (2) Directed retry—direct to other cell site
      (3) Intercept—priority feature
      (4) Reorder—initiate again
   b. *Page response message.*   Send one of the following orders.
      (1) Initial voice channel designation
      (2) Directed retry
      (3) Release—turn off signaling tone and release the channel
   c. *Order message.*   Send one of the following orders.
      (1) Order confirmation
      (2) Release
   d. *Order confirmation message.*   "No message is sent."

Mobile station control on voice channel.   The change of status of the supervisory audio tone (SAT) and signaling tone (ST) are used to signal the occurrence of certain events during the progress of a call, such as confirming orders, sending a release request, sending a flash request, and loss of radio-link continuity. In addition to the analog signaling (SAT and ST) to and from the mobile station, digital messages (in a burst mode with 10 kbps transmission rate) can be sent to and received from the mobile station. Response to the digital message is either a digital message or a status change of SAT and ST.

We use the notation "(SAT, ST) status" to describe the signaling condition.

SAT		ST		(SAT, ST)	
On	Off	On	Off	status	Conditions
1			0	(1,0)	Mobile off-hook
1		1		(1,1)	Mobile on-hook
	0	1		(0,1)	Mobile in fade
	0		0	(0,0)	Mobile transmitter off

1. *Loss of radio-link continuity.* A designated SAT tone is continuously sent to the mobile station; the same SAT should be sent back on a reverse voice channel. If within 5 s the SAT has not been received, the land station would assume that the mobile station is lost and terminates the call.

2. *Initial voice channel confirmation*
   a. Confirmation will be received by the land station as a change in the SAT, ST status from (0,0) to (1,0).
   b. If the confirmation is not received, the land station must either resend the message or turn off the voice channel transmitter.
   c. If the mobile station was paged, the land station must enter the Wait for Order task or Conversation task.

3. *Alerting*
   a. *Waiting for Order task.* After being paged, the mobile station confirms the initial voice channel designation.
      (1) *Handoff.* The mobile station confirms the order by a change in the (SAT, ST) status from (1,0) to (1,1) for 50 ms. The land station must remain in the Waiting for Order task.
      (2) *Alert.* The mobile station confirms the order by changing (SAT, ST) status from (1,0) to (1,1). The land station must then enter the Waiting for Answer task.
      (3) *Release.* The mobile station confirms the order by a change of (SAT, ST) status from (1,0) to (1,1) and holds the (1,1) status for 1.8 s. The land station must then turn off the transmitter.
      (4) *Audit.* The mobile station confirms the order by a digital message. The land station remains in the Waiting for Order task.
      (5) *Maintenance.* The mobile station confirms the order by a change in (SAT, ST) status from (1,0) to (1,1). The land station remains in the Waiting for Order mode.
      (6) *Change power.* The mobile station confirms the order by a digital message.
      (7) *Local control.* The confirmation and action depends on the message.
   b. *Waiting for Answer task.* When this task is entered, an alert timer must be set for 30 s. The following orders can be sent:
      (1) *Handoff.* The mobile station confirms the order by a change of (SAT, ST) status from (1,1) to (1,0) for 500 ms followed by (1,0) to (1,1) held for 50 ms on the old channel. Then (1,1) status is sent on the new channel.
      (2) *Alert.* If no confirmation is received, the land station must reset the alert timer to 30 s.

(3) *Stop alert.* The mobile station confirms the order by a change of (SAT, ST) status from (1,1) to (1,0).

(4) *Release.* The mobile station confirms the order by changing (SAT, ST) status from (1,1) to (1,0) for 500 ms followed by a change of (SAT, ST) from (1,0) to (1,1), which is then held for 1.8 s. The land station must turn off the transmitter.

(5) *Audit.* The mobile station confirms the order by a digital message.

(6) *Maintenance.* If no confirmation is received, the land station resets the alert timer to 30 s.

(7) *Change power.* The mobile station confirms the order by a digital message (see Sec. 3.2.1).

(8) *Local control.* The confirmation and action depends on the message.

4. *Conversation.* The mobile station signals an answer by a change in the (SAT, ST) status from (1,1) to (1,0). The land station enters the conversation task.

a. *Handoff.* The mobile station confirms the order by a change in the (SAT, ST) from (1,0) to (1,1), which is then held for 50 ms. Then the land station must remain in the Conversation task.

b. *Send called address.* The called mobile station confirms the order by a digital message with the called address information. This feature would save the established link if the called address were in error because of the transmission medium.

c. The functions *alert, release, audit, maintenance,* and *local control.* Same as in the Waiting for Order task.

d. *Change power.* The mobile station confirms the order by a digital message.

e. *Flash request.* The mobile station signals a flash by changing (SAT, ST) from (1,0) to (1,1) then holding (1,1) for 400 ms, then following with a transition to (1,0).

f. *Release request.* The mobile station signals a release by changing the (SAT, ST) status from (1,0) to (1,1), which is then held for 1.8 s. The land station must turn off the transmitter. This would be used for the mobile user who dials a called number and decides to terminate for any reason.

3.3.4 Signaling formats

Forward control channel (FOCC). The FOCC is a continuous wideband 10 kbps ± 0.1 bps data stream sent from the land station to the mobile station. Each forward control channel consists of three discrete information streams (see Fig. 3.5):

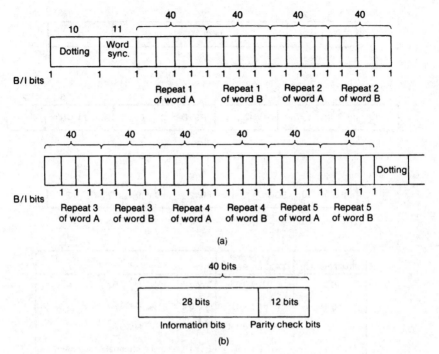

Figure 3.5 Forward control channel message stream (land-to-mobile). (*a*) Signaling format of FOCC. (*b*) A message word of FOCC.

Stream A (least significant bit of MIN = 0)

Stream B (least significant bit of MIN = 1)

Busy-idle stream (busy = 0, idle = 1); it is at a 1 kbps rate, i.e., one busy-idle bit every 10 data bits.

The 10-bit dotting sequence (1010101010) is for bit syn. The 10-bit length is assumed to be sufficient for bit syn because the mobile station is always monitoring the FOCC after initialization (see 3.2.6). The frame syn bits are the Barker sequence (11100010010). A word is formed by coding 28 control bits into a (40, 28, 5) BCH code. The total number of bits is 40, the number of information bits is 28, and the hamming distance is 5. The hamming distance d can be translated to the capability of error correction bits, t, as follows:

$$t = \frac{d - 1}{2} = 2$$

Since this code will detect errors as well as correct them, it reduces to correct one bit in error and assure detection of two bits in error.

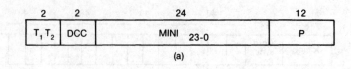

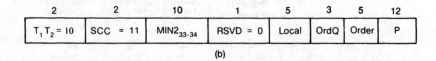

2	2	10	3	1J	12
T_1T_2 = 10	SCC ≠ 11	MIN2$_{33-24}$	VMAC	Chan	P

(c)

DCC (from FOCC)	7 bit coded DCC on RECC	SCC		
0 0	0 0 0 0 0 0 0	SCC ≠ 11 ⎰ 0 0	5970 Hz
0 1	0 0 1 1 1 1 1	0 1	6000 Hz
1 0	1 1 0 0 0 1 1	1 0	6030 Hz
1 1	1 1 1 1 1 0 0	1 1	Not a channel designation

(d)

Figure 3.6 Mobile station control message. (*a*) Word 1—abbreviated address word (the busy-idle stream is not shown); (*b*) word 2—extended address word (SCC = 11); (*c*) word 2—extended address word (SCC ≠ 11); (*d*) DCC and SCC codes.

The code is a shortened version of the primitive (63,51,5) BCH code. It has 12 parity check bits. As long as the 12 parity check bits are retained, the shortened version can be any length. We use (40,28,5), for which a message of 28 bits is suitable (see Fig. 3.5*b*). The transmission rate is 10 kbps. The throughput is 1200 bps (see Fig. 3.5*a*).

Types of messages

1. *Mobile station control message.* Consists of one, two, or four words.
   a. *Word 1.* Abbreviated address word (see Fig. 3.6*a*), T_1T_2 = type field.
      $T_1T_2 = 00$ Only word 1 is sent
      $T_1T_2 = 01$ Multiple words are sent.
      DCC—digital color code field, 2 bits sent on FOCC. Then it is received by the mobile station and translated to a 7-bit coded DCC on RECC (see Fig. 3.6*d*).

Overhead message types

Code	Order
000	registration ID
001	control-filler
010	reserved
011	reserved
100	global action
101	reserved
110	word 1 of system parameter message
111	word 2 of system parameter message

(a)

2	2	20	1	3	12
T_1T_2	DCC	REGID	END	OHD	P

(b)

Figure 3.7 Overhead message types and format.

b. *Words 2 to 4.* Extended address word $T_1T_2 = 10$, set in each additional word (see Fig. 3.6b and c). Let SCC = 11 or SCC $\neq$ 11 (see two bits indicated in Fig. 3.6d).

(1) *Word 2*

(a) By combining "order code" (ORDER) and "order qualification code" (ORDQ) in this word (see Fig. 3.6b), we can describe 11 functions.

Page (or origination)	Registration
Alert	Intercept
Release	Maintenance
Reorder	Send called address
Stop alert	Direct-retry status
Audit	

(b) Also changes the mobile station power levels in eight levels

(2) *Word 3.* First directed-retry word.

(3) *Word 4.* Second directed-retry word.

2. *Overhead message (OHD).* A 3-bit OHD field (see Fig. 3.7a) is used to identify the overhead message types. It locates just before the 12 parity check bits shown in Fig. 3.7b. The overhead types are grouped into the following functional classes.

a. System parameter overhead message. The system parameter overhead message must be sent every 0.8 $\pm$ 0.3 s. It consists of two words: word 1 contains the first part of the system identi-

	BS	WS	Word	BS	WS	Word	BS	WS	Word	• • • • •
FVC	101	11	40	37	11	40	37	11	40	(28, 12)
RVC	100	11	48	37	11	48	37	11	48	(36, 12)

Repeat 11 times for FVC   BS - Bit sync.
Repeat 5 times for RVC   WS - Word sync.

Figure 3.8 Signaling format of FVC and RVC.

fication field, and word 2 contains the number of paging channels and number of access channels.

b. Global action overhead message. There are many global action overhead messages. Each of them consists of one word. The actions are registration increment, new access channel starting point, maximum busy occurrences, maximum seizure tries, etc.

c. Registration ID message. It consists of one word containing the registration ID field.

d. Control-filler message. It consists of one word and is sent whenever there is no other message to be sent on FOCC. It is used to specify a control mobile attenuation code (CMAC) which is used by mobile stations accessing the system, and a Wait for Overhead Message bit (WFOM) indicating whether mobile stations must read an overhead message train before accessing the system.

3. Data restriction

a. The overhead message transmission rate is about once per second.

b. Design the control-filler message to exclude the frame-sync (word sync) sequence.

c. Restrict the use of certain control office codes.

Forward voice channel (FVC). During the call period, FVC is used for signaling. At the beginning, the 101-bit dotting sequence is used for bit sync then for all the repeat dotting sequences; each of them only contains 37 bits, as shown in Fig. 3.8. The word length is 40 bits and repeats 11 times. The reason for repeating 11 times is to be sure that the handoff message would reach the mobile station before the signal dropped below the unacceptable level at the mobile station. The FVC signaling is mainly used for handoff, and the signal level when the handoff occurs is usually very weak. Therefore, the purpose of the FVC is to be sure that the mobile station will get the message and not have a chance to send back a response on receipt because of a weak signal condition.

3.3.5 Additional spectrum radio (ASR) issues

The FCC has allocated an additional 83 voice channels to each system (Band A and Band B). ASR will have 832 channels, half of them, 416 channels (333 channels plus 83 channels), are operated for each system. The 21 control channels still remain the same, but the total number of voice channels becomes 395. The new numbering scheme is shown below.

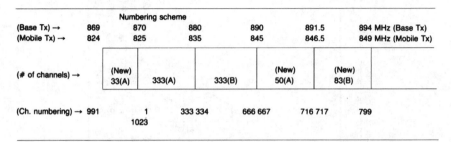

		Numbering scheme				
(Base Tx) →	869	870	880	890	891.5	894 MHz (Base Tx)
(Mobile Tx) →	824	825	835	845	846.5	849 MHz (Mobile Tx)
(# of channels) →	(New) 33(A)	333(A)	333(B)	(New) 50(A)	(New) 83(B)	
(Ch. numbering) →	991	1 1023	333 334	666 667	716 717	799

ASR is identified by using the station class mark field as shown in Fig. 3.4b. The station class mark (SCM) consists of 4 bits and is specified as shown below.

Power class	SCM	Transmission	SCM	Bandwidth	SCM
Class I (4 W)	XX00	Continuous	X0XX	20 Mhz	0XXX
Class II (1.2 W)	XX01	Discontinuous	X1XX	25 MHz (ASR)	1XXX
Class III (0.6 W)	XX10				
Reserved	XX11				

The new frequency management charts for Block A and Block B are shown in Table 3.1.

3.4 Different Specifications of the World's Analog Cellular Systems

In general, cellular systems can be classified by their operating frequencies: 450-MHz or 800-MHz. Also, they can be distinguished by the spacing between their channels (also called the *channel bandwidth*): 30, 25, or 20 kHz. Japanese NTT has deployed a 12.5-kHz channel spacing in their cellular system.

The large-capacity cellular telephones used in the world are listed in Table 3.2. There are five major systems, and their message protection schemes are different. The major differences are

TABLE 3.1 New Frequency Management (Full Spectrum)

Block A

1A	2A	3A	4A	5A	6A	7A	1B	2B	3B	4B	5B	6B	7B	1C	2C	3C	4C	5C	6C	7C
1	2	3	4	5	6	7	8	9	10	11	12	13	14	15	16	17	18	19	20	21
22	23	24	25	26	27	28	29	30	31	32	33	34	35	36	37	38	39	40	41	42
43	44	45	46	47	48	49	50	51	52	53	54	55	56	57	58	59	60	61	62	63
64	65	66	67	68	69	70	71	72	73	74	75	76	77	78	79	80	81	82	83	84
85	86	87	88	89	90												102	103	104	105
106	107	108	109	110	111												123	124	125	126
127	128	129	130	131	132												144	145	146	147
148	149	150	151	152	153												165	166	167	168
169	170	171	172	173	174												186	187	188	189
190	191	192	193	194	195												207	208	209	210
211	212	213	214	215	216												228	229	230	231
232	233	234	235	236	237												249	250	251	252
253	254	255	256	257	258												270	271	272	273
274	275	276	277	278	279												291	292	293	294
295	296	297	298	299	300												312	X	X	X
313*	314	315	316	317	318	319	320	321	322	323	324	325	326	327	328	329	330	331	332	333
667	668	669	670	671	672	673	674	675	676	677	678	679	680	681	682	683	684	685	686	687
688	689	690	691	692	693	694	695	696	697	698	699	700	701	702	703	704	705	706	707	708
709	710	711	712	713	714	715	716	X	991	992	993	994	995	996	997	998	999	1000	1001	1002
1003	1004	1005	1006	1007	1008	1009	1010	1011	1012	1013	1014	1015	1016	1017	1018	1019	1020	1021	1022	1023

Diagram labels: 1C, 1B, 1A, 3C, 3B, 3A, 4A, 4B, 4C, 5A, 5B, 5C, 6C, 6A, 6B, 7C, 7A, 7B, 2C, 2A, 2B

Block B

1A	2A	3A	4A	5A	6A	7A	1B	2B	3B	4B	5B	6B	7B	1C	2C	3C	4C	5C	6C	7C
334	**335**	**336**	**337**	**338**	**339**	**340**	**341**	**342**	**343**	**344**	**345**	**346**	**347**	**348**	**349**	**350**	**351**	**352**	**353**	**354**
355	356	357	358	359	360	361														375
376	377	378	379	380	381	382														396
397	398	399	400	401	402	403														417
418	419	420	421	422	423	424														438
439	440	441	442	443	444	445														459
460	461	462	463	464	465	466														480
481	482	483	484	485	486	487														501
502	503	504	505	506	507	508														522
523	524	525	526	527	528	529														543
544	545	546	547	548	549	550														564
565	566	567	568	569	570	571														585
586	587	588	589	590	591	592	593	594	595	596	597	598	599	600	601	602	603	604	605	606
607	608	609	610	611	612	613	614	615	616	617	618	619	620	621	622	623	624	625	626	627
628	629	630	631	632	633	634	635	636	637	638	639	640	641	642	643	644	645	646	647	648
649	650	651	652	653	654	655	656	657	658	659	660	661	662	663	664	665	666			
																		717	718	719
720	721	722	723	724	725	726	727	728	729	730	731	732	733	734	735	736	737	738	739	740
741	742	743	744	745	746	747	748	749	750	751	752	753	754	755	756	757	758	759	760	761
762	763	764	765	766	767	768	769	770	771	772	773	774	775	776	777	778	779	780	781	782
783	784	785	786	787	788	789	790	791	792	793	794	795	796	797	798	799				

Central hexagonal cell-reuse diagram (cell labels): 3A, 3B, 3C, 4A, 4B, 4C, 5A, 5B, 5C, 6A, 6B, 6C, 7A, 7B, 7C.

*Boldface numbers indicate 21 control channels for Block A and Block B respectively.

TABLE 3.2 Large-Capacity Analog Cellular Telephones Used in the World

	Japan	North America	England	Scandinavia	Germany
System	NTT	AMPS	TACS	NMT	C450
Transmission frequency (MHz):					
Base station	870–885	869–894	917–950	463–467.5	461.3–465.74
Mobile station	925–940	824–849	872–905	453–457.5	451.3–455.74
Spacing between transmission and receiving frequencies (MHz)	55	45	45	10	10
Spacing between channels (kHz)	25, 12.5	30	25	25	20
Number of channels	600	832 (control channel 21 × 2)	1320 (control channel 21 × 2)	180	222

Coverage radius (km)	5 (urban area) 10 (suburbs)	2–20	2–20	1.8–40	5–30
Audio signal: Type of modulation	FM	FM	FM	FM	FM
Frequency deviation (kHz)	±5	±12	±9.5	±5	±4
Control signal: Type of modulation	FSK	FSK	FSK	FSK	FSK
Frequency deviation (kHz)	±4.5	±8	±6.4	±3.5	±2.5
Data transmission rate (kb/s)	0.3	10	8	1.2	5.28
Message protection	Transmited signal is checked when it is sent back to the sender by the receiver.	Principle of majority decision is employed.	Principle of majority decision is employed.	Receiving steps are predetermined according to the content of the message.	Message is sent again when an error is detected.

SOURCE: Report from International Radio Consultative Committee (CCIR).

TABLE 3.3 World's Analog Cellular Systems

NTT	AMPS	TACS	NMT	C450	NEC
Japan	U.S.	England	4 Nordic countries	Germany	Australia
	Canada	Hong Kong	Spain		Singapore
	South Korea	China	The Netherlands		Hong Kong
	Hong Kong		Belgium		Jordan
	Taiwan		Oman		Colombia
	Australia		Saudi Arabia		Mexico
	China		Tunisia		Kuwait
	Mexico		Malaysia		
	Brazil		Australia		
	Argentina		Ireland		
	Chile				

SOURCE: Report from International Radio Consultative Committee (CCIR).

1. The principle of majority decision (PMD)

2. The automatic repeat request (ARQ)

These schemes each have their merits. In a very severe fading environment, PMD is a good candidate. In a fairly light fading environment, ARQ is a good candidate. Table 3.3 lists the system of each country's use. There are 25 countries which have cellular systems.

Today there is no compatibility among the analog cellular systems serving European countries. In 1982, they agreed to set up a pan-European digital cellular standard, called GSM (special mobile group) specification, which was implemented in 1992 and is described in Chap. 15.

References

1. FCC OST, "Cellular System Mobile Station—Land Station Compatibility Specification," *OST Bulletin,* No. 53, July 1983.
2. "Advanced Mobile Phone Service," special issue, *Bell System Technical Journal,* Vol. 58, January 1979.
3. "Code of Federal Regulations," FCC, part 22 1986, pp. 85–190.
4. P. F. Panter, *Modulation, Noise, and Spectral Analysis,* McGraw-Hill Book Co., 1965, p. 447.
5. EIA Interim Standard, "Cellular System Mobile Station-Land Station Compatibility Specification," IS-3-D EIA, March, 1987.

Cell Coverage
for Signal and Traffic

4.1 General Introduction

Cell coverage can be based on signal coverage or on traffic coverage. Signal coverage can be predicted by coverage prediction models and is usually applied to a start-up system. The task is to cover the whole area with a minimum number of cell sites. Since 100 percent cell coverage of an area is not possible, the cell sites must be engineered so that the holes are located in the no-traffic locations. The prediction model is a point-to-point model which is discussed in this chapter. We have to examine the service area as occurring in one of the following environments:

Human-made structures

In an open area

In a suburban area

In an urban area

Natural terrains

Over flat terrain

Over hilly terrain

Over water

Through foliage areas

The results generated from the prediction model will differ depending on which service area is used.

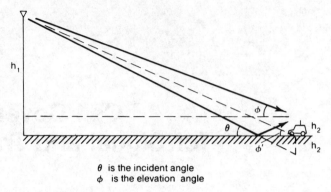

θ is the incident angle
φ is the elevation angle

Figure 4.1 A coordinate sketch in a flat terrain.

There are many field-strength prediction models in the literature.[1-28] They all provide more or less an area-to-area prediction. As long as 68 percent of the predicted values from a model are within 6 to 8 dB (one standard deviation) of their corresponding measured value, the model is considered a good one. However, we cannot use area-to-area prediction models for cellular system design because of the large uncertainty of the prediction.

The model being introduced here is the point-to-point prediction model which would provide a standard deviation from the predicted value of less than 3 dB. An explanation of this model appears in Refs. 23, 24, and 41. Many tools can be developed based upon this model, such as cell-site choosing, interference reduction, and traffic handling.

4.1.1 Ground incident angle and ground elevation angle

The ground incident angle and the ground elevation angle over a communication link are described as follows. The ground incident angle θ is the angle of wave arrival incidently pointing to the ground as shown in Fig. 4.1. The ground elevation angle φ is the angle of wave arrival at the mobile unit as shown in Fig. 4.1.

Example 4.1 In a mobile radio environment, the average cell-site antenna height is about 50 m, the mobile antenna height is about 3 m, and the communication path length is 5 km. The incident angle is (see Fig. 4.1)

$$\theta = \tan^{-1} \frac{50 \text{ m} + 3 \text{ m}}{5 \text{ km}} = 0.61°$$

The elevation angle at the antenna of the mobile unit is

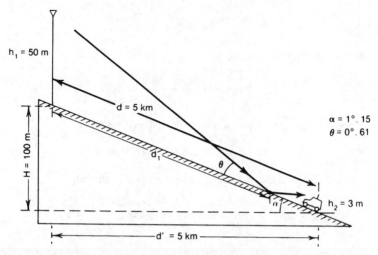

Figure 4.2 A coordinate sketch in a hilly terrain.

$$\phi = \tan^{-1} \frac{50 \text{ m} - 3 \text{ m}}{5 \text{ km}} = 0.54°$$

The elevation angle at the location of the mobile unit is

$$\phi' = \tan^{-1} \frac{50 \text{ m}}{5 \text{ km}} = 0.57°$$

4.1.2 Ground reflection angle and reflection point

Based on Snell's law, the reflection angle and incident angle are the same. Since in graphical display we usually exaggerate the hilly slope and the incident angle by enlarging the vertical scale, as shown in Fig. 4.2, then as long as the actual hilly slope is less than 10°, the reflection point on a hilly slope can be obtained by following the same method as if the reflection point were on flat ground. Be sure that the two antennas (base and mobile) have been placed vertically, not perpendicular to the sloped ground. The reason is that the actual slope of the hill is usually very small and the vertical stands for two antennas are correct. The scale drawing in Fig. 4.2 is somewhat misleading; however, it provides a clear view of the situation.

Example 4.2 Let $h_1 = 50$ m, $h_2 = 3$ m, $d = 5$ km, and $H = 100$ m as shown in Fig. 4.2.

(a) Using the approximate method ($d \approx d' \approx 5$ km), the slope angle α of the hill is

$$\alpha = \tan^{-1} \frac{100 \text{ m}}{5 \text{ km}} = 1.14576°$$

the incident angle is

$$\theta = \tan^{-1} \frac{50 \text{ m} + 3 \text{ m}}{5 \text{ km}} = 0.61$$

and the reflection point location from the cell-site antenna

$$d_1 = 50/\tan \theta = 4.717 \text{ km}.$$

(b) Using the accurate method, the slope angle α of the hill is

$$\alpha = \tan^{-1} \frac{100 \text{ m}}{\sqrt{(5 \text{ km})^2 - (100 \text{ m})^2}} = \tan^{-1} \frac{100}{4999} = 1.14599°$$

The incident angle θ and the reflection point location d_1 are the same as above.

4.2   Obtaining the Mobile Point-to-Point Model (Lee Model)

This mobile point-to-point model is obtained in three steps: (1) generate a standard condition, (2) obtain an area-to-area prediction model, (3) obtain a mobile point-to-point model using the area-to-area model as a base. The philosophy of developing this model is to try to separate two effects, one caused by the natural terrain contour and the other by the human-made structures, in the received signal strength.

4.2.1   A standard condition

To generate a standard condition and provide correction factors, we have used the standard conditions shown on the left side and the correction factors on the right side[10] of Table 4.1. The advantage of using these standard values is to obtain directly a predicted value in decibels above 1 mW expressed in dBM.

4.2.2   Obtain area-to-area prediction curves for human-made structures

The area-to-area prediction curves are different in different areas. In area-to-area prediction, all the areas are considered flat even though the data may be obtained from nonflat areas. The reason is that area-to-area prediction is an average process. The standard deviation of the average value indicates the degree of terrain roughness.

TABLE 4.1 Generating a Standard Condition

Standard condition	Correction factors*
At the Base Station	
Transmitted power P_t = 10 W (40 dBm)	$\alpha_1 = 10 \log \dfrac{P_t'}{10}$
Antenna height h_1 = 100 ft (30 m)	$\alpha_2 = 20 \log \dfrac{h_1'}{h_1}$
Antenna gain g_t = 6 dB/dipole	$\alpha_3 = g_{t2}' - 6$
At the Mobile Unit	
Antenna height, h_2 = 10 ft (3 m)	$\alpha_4 = 10 \log \dfrac{h_2'}{h_2}$
Antenna gain, g_m = 0 dB/dipole	$\alpha_5 = g_m'$

*All the parameters with primes are the new conditions.

Effect of the human-made structures. Since the terrain configuration of each city is different, and the human-made structure of each city is also unique, we have to find a way to separate these two. The way to factor out the effect due to the terrain configuration from the man-made structures is to work out a way to obtain the path loss curve for the area as if the area were flat, even if it is not. The path loss curve obtained on virtually flat ground indicates the effects of the signal loss due to solely human-made structures. This means that the different path loss curves obtained in each city show the different human-made structure in that city. To do this, we may have to measure signal strengths at those high spots and also at the low spots surrounding the cell sites, as shown in Fig. 4.3a. Then the average path loss slope (Fig. 4.3b), which is a combination of measurements from high spots and low spots along different radio paths in a general area, represents the signal received as if it is from a flat area affected only by a different local human-made structured environment. We are using 1-mi intercepts (or, alternatively, 1-km intercepts) as a starting point for obtaining the path loss curves.

Therefore, the differences in area-to-area prediction curves are due to the different man-made structures. We should realize that measurements made in urban areas are different from those made in suburban and open areas. The area-to-area prediction curve is obtained from the mean value of the measured data and used for future predictions in that area. Any area-to-area prediction model[1-28] can be used as a first step toward achieving the point-to-point prediction model.

One area-to-area prediction model which is introduced here[10] can be represented by two parameters. (1) the 1-mi (or 1-km) intercept point

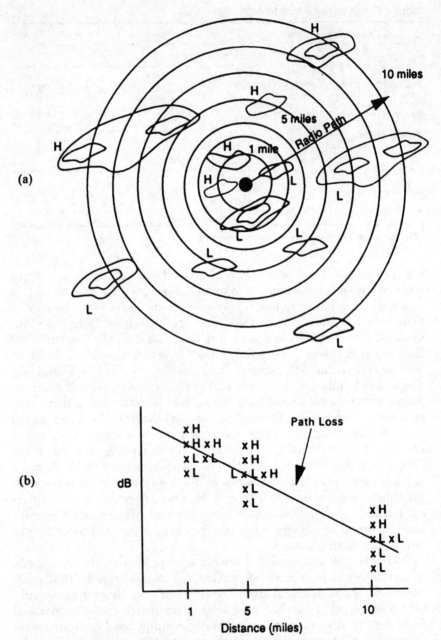

Figure 4.3 Propagation path loss curves for human-made structures. (a) For selecting measurement areas (b) path loss phenomenon.

and (2) the path-loss slope. The 1-mi intercept point is the power received at a distance of 1 mi from the transmitter. There are two general approaches to finding the values of the two parameters experimentally.

1. Compare an area of interest with an area of similar human-made structures which presents a curve such as that shown in Fig. 4.3c. The suburban area curve is a commonly used curve. Since all suburban areas in the United States look alike, we can use this curve for all suburban areas. If the area is not suburban but is similar to the city of Newark, then the curve for Newark should be used.

2. If the human-made structures of a city are different from the cities listed in Fig. 4.3c, a simple measurement should be carried out. Set up a transmitting antenna at the center of a general area. As long as the building height is comparable to the others in the area, the antenna location is not critical. Take six or seven measured data points around the 1-mi intercept and around the 10-mi boundary based on the high and low spots. Then compute the average of the 1 mi data points and of the 10 mi data points. By connecting the two values, the path-loss slope can be obtained. If the area is very hilly, then the data points measured at a given distance from the base station in different locations can be far apart. In this case, we may take more measured data points to obtain the average path-loss slope.

If the terrain of the hilly area is generally sloped, then we have to convert the data points that were measured on the sloped terrain to a fictitiously flat terrain in that area. The conversion is based on the effective antenna-height gain as[29]

$$\Delta G = \text{effective antenna-height gain} = 20 \log \frac{h_e}{h_1} \qquad (4.2\text{-}1)$$

where h_1 is the actual height and h_e is the effective antenna height at either the 1- or 10-mi locations. The method for obtaining h_e is shown in the following section.

3. An explanation of the path-loss phenomenon is as follows. The plotted curves shown in Fig. 4.3c have different 1-mi intercepts and different slopes. The explanation can be seen in Fig. 4.3d. When the base station antenna is located in the city, then the 1-mi intercept could be very low and the slope is flattened out, as shown by Tokey's curve. When the base station is located outside the city, the intercept could be much higher and the slope is deeper, as shown by the Newark curve. When the structures are uniformly distributed, depending on the density (average separation between buildings) s, the 1-mi intercept could be high or low, but the slope may also keep at 40 dB/dec.

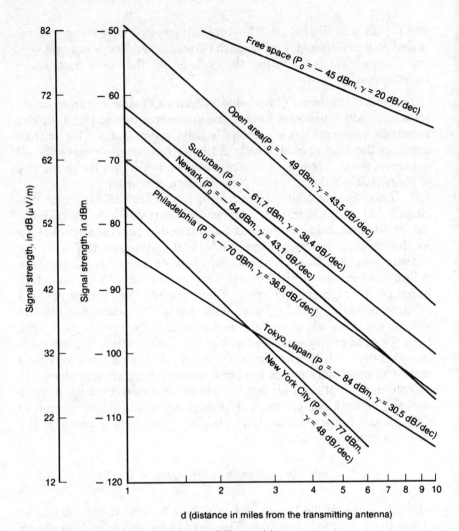

d (distance in miles from the transmitting antenna)

Figure 4.3 (c) Propagation path loss in different cities.

4.2.3 The phase difference between a direct path and a ground-reflected path

Based on a direct path and a ground-reflected path (see Fig. 4.4),

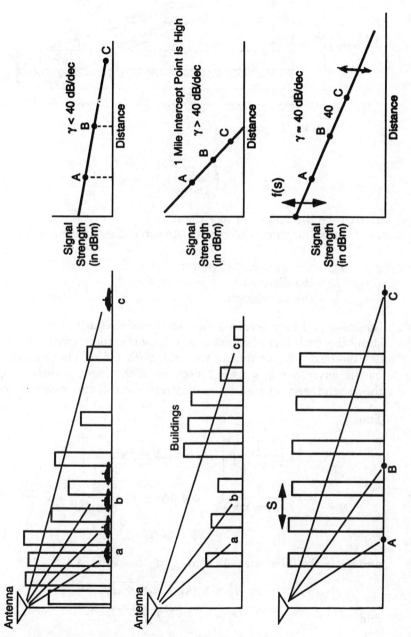

Figure 4.3 (d) Explanation of the path-loss phenomenon.

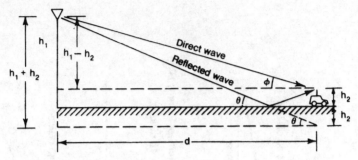

Figure 4.4 A simple model.

$$P_r = P_0 \left(\frac{1}{4\pi d/\lambda}\right)^2 \left|1 + a_v e^{j\Delta\phi}\right|^2 \qquad (4.2\text{-}2)$$

where a_v = the reflection coefficient
$\Delta\phi$ = the phase difference between a direct path and a reflected
        path
P_0 = the transmitted power
d = the distance
λ = the wavelength

Equation (4.2-2) indicates a two-wave model which is used to understand the path-loss phenomenon in a mobile radio environment. It is not the model for analyzing the multipath fading phenomenon. In a mobile environemnt $a_v = -1$ because of the small incident angle of the ground wave caused by a relatively low cell-site antenna height.

Thus

$$P_r = P_0 \left(\frac{1}{4\pi d/\lambda}\right)^2 \left|1 - \cos\Delta\phi - j\sin\Delta\phi\right|^2$$

$$= P_0 \frac{2}{(4\pi d/\lambda)^2}(1 - \cos\Delta\phi) = P_0 \frac{4}{(4\pi d/\lambda)^2}\sin^2\frac{\Delta\phi}{2} \quad (4.2\text{-}3)$$

where $$\Delta\phi = \beta\,\Delta d \qquad (4.2\text{-}4)$$

and Δd is the difference, $\Delta d = d_1 - d_0$, from Fig. 4.4.

$$d_1 = \sqrt{(h_1 + h_2)^2 + d^2} \qquad (4.2\text{-}5)$$

and $$d_2 = \sqrt{(h_1 - h_2)^2 + d^2} \qquad (4.2\text{-}6)$$

Since Δd is much smaller than either d_1 or d_2,

$$\Delta\phi = \beta\ \Delta d \approx \frac{2\pi}{\lambda}\ \frac{2h_1 h_2}{d} \tag{4.2-7}$$

Then the received power of Eq. (4.2-3) becomes

$$P_r = P_0\ \frac{\lambda^2}{(4\pi)^2 d^2}\ \sin^2 \frac{4\pi h_1 h_2}{\lambda d} \tag{4.2-8}$$

If $\Delta\phi$ is less than 0.6 rad, then $\sin(\Delta\phi/2) \approx \Delta\phi/2$, $\cos(\Delta\phi/2) \approx l$ and Eq. (4.2-8) simplifies to

$$P_r = P_0\ \frac{4}{16\pi^2 (d/\lambda)^2}\left(\frac{2\pi h_1 h_2}{\lambda d}\right)^2 = P_0\left(\frac{h_1 h_2}{d^2}\right)^2 \tag{4.2-9}$$

From Eq. (4.2-9), we can deduce two relationships as follows:

$$\Delta P = 40\ \log\frac{d_1}{d_2} \quad \text{(a 40 dB/dec path loss)} \tag{4.2-10a}$$

$$\Delta G = 20\ \log\frac{h'_1}{h_1} \quad \text{(an antenna height gain of 6 dB/oct)} \tag{4.2-10b}$$

where ΔP is the power difference in decibels between two different path lengths and ΔG is the gain (or loss) in decibels obtained from two different antenna heights at the cell site. From these measurements, the gain from a mobile antenna height is only 3 dB/oct, which is different from the 6 dB/oct for h'_1 shown in Eq. (4.2-10b). Then

$$\Delta G' = 10\ \log\frac{h'_2}{h_2} \quad \text{(an antenna-height gain of 3 dB/oct)} \tag{4.2-10c}$$

Example 4.3 The distance $d = 8$ km. The antenna height at the cell site is 30 m, and at the mobile unit it is 3 m. Then the phase difference at 850 MHz is $\beta \cdot \Delta d$, or 0.4 rad, which is less than 0.6 rad. Therefore, Eq. (4.2-9) can be applied.

4.2.4 Why there is a constant standard deviation along a path-loss curve

When plotting signal strengths at any given radio-path distance, the deviation from predicted values is approximately 8 dB.[10,12] This standard deviation of 8 dB is roughly true in many different areas. The explanation is as follows. When a line-of-sight path exists, both the direct wave path and reflected wave path are created and are strong (see Fig. 4.2). When an out-of-sight path exists, both the direct wave path and the reflected wave path are weak. In either case, according to the theoretical model, the 40-dB/dec path-loss slope applies. The difference between these two conditions is the 1-mi intercept (or 1-km intercept) point. It can be seen that in the open area, the 1-mi intercept is high. In the urban area, the 1-mi intercept is low. The standard deviation obtained from the measured data remains the same along the different path-loss curves regardless of environment.

Support for the above argument can also be found from the observation that the standard deviation obtained from the measured data along the predicted path-loss curve is approximately 8 dB. The explanation is that at a distance from the cell site, some mobile unit radio paths are line-of-sight, some are partial line-of-sight, and some are out-of-sight. Thus the received signals are strong, normal, and weak, respectively. At any distance, the above situations prevail. If the standard deviation is 8 dB at one radio-path distance, the same 8 dB will be found at any distance. Therefore a standard deviation of 8 dB is always found along the radio path as shown in Fig. 4.5. The standard deviation of 8 dB from the measured data near the cell site is due mainly to the close-in buildings around the cell site. The same standard deviation from the measured data at a distant location is due to the great variation along different radio paths.

4.2.5 The straight-line path-loss slope with confidence

As we described earlier, the path-loss curves are obtained from many different runs at many different areas. As long as the distances of the radio path from the cell site to the mobile unit are the same in different runs, the signal strength data measured at that distance would be used to calculate the mean value for the path loss at that distance. In the experimental data, the path-loss deviation is 8 dB across the distance from 1.6 to 15 km (1 to 10 mi) where the general terrain contours are not generally flat. Figure 4.5 depicts this. The path-loss curve is γ. The received power can be expressed as

$$P_r = P_0 - \gamma \log \frac{r}{r_0} \qquad (4.2\text{-}11)$$

The slope γ is different in different areas, but it is always a straight line in a log scale. If $\gamma = 20$ is a free-space path loss, $\gamma = 40$ is a mobile path loss.

Confidence level.[31] A confidence level can only be applied to the path-loss curve when the standard deviation σ is known. In American suburban areas, the standard deviation $\sigma = 8$ dB. The values at any given distance over the radio path are concentrated close to the mean and

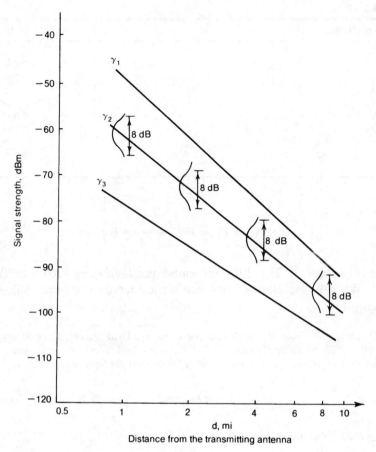

Figure 4.5 An 8-dB local mean spread.

have a bell-shaped (normal) distribution. The probability that 50 percent of the measured data are equal to or below a given level is

$$P(x \geq C) = \int_C^\infty \frac{1}{\sqrt{2\pi}\sigma} e^{-(x-A)^2/2\sigma^2} \, dx = 50\% \qquad (4.2\text{-}12)$$

where A is the mean level obtained along the path-loss slope, which is shown in Eq. (4.2-11) as

$$A = P_0 - \gamma \log \frac{r_1}{r_0}$$

Thus level A corresponds to the distance r_1. If level A increases, the confidence level decreases, as shown in Eq. (4.2-12).

TABLE 4.2

$P(x \le C)$, %	$C = B\sigma + A$
80	$-0.85\sigma + A$
70	$-0.55\sigma + A$
60	$-0.25\sigma + A$
50	A
40	$0.25\sigma + A$
30	$0.55\sigma + A$
20	$0.85\sigma + A$
16	$1\sigma + A$
10	$1.3\ \sigma + A$
2.28	$2\ \sigma + A$

$$P(x \ge C) = P\left(\frac{x - A}{\sigma} \ge B\right) \qquad (4.2\text{-}13)$$

Let $C = B\sigma + A$. The different confidence levels are shown in Table 4.2. We can see how to use confidence levels from the following example.

Example 4.4 From the path-loss curve, we read the expected signal level as -100 dBm at 16 km (10 mi). If the standard deviation $\sigma = 8$ dB, what level would the signal equal or exceed for a 20 percent confidence level?

$$P\left(\frac{x - A}{\sigma} \ge B\right) = 20\% \qquad x \ge B\sigma + A \qquad (\text{E4.4-1})$$

or from Table 4.2 we obtain

$$x \ge 0.85 \times 8 + (-100) = -93.2 \text{ dBm}$$

The log normal curve with a standard deviation of 8 dB is shown in Fig. 4.6.

4.2.6 Determination of confidence interval

The confidence interval is often confused with confidence level. This usually happens when dealing with a particular run in a particular terrain contour. The signal strength of a run is shown in Fig. 1.6. The local mean is the envelope of the received signal, which also follows a log-normal distribution as shown in Fig. 4.6. The standard deviation of the local mean curve is a reflection of how much variation there is in terrain contour. If we know the standard deviation, then we can estimate how often the local mean (average power of the signal) falls within given limits (confidence interval).

The confidence intervals are defined as

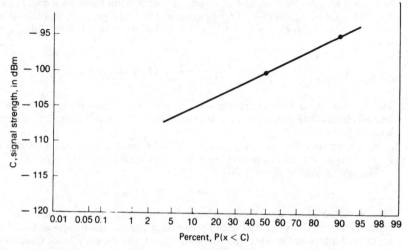

Figure 4.6 A log-normal curve.

$$P(m - \sigma \leq x \leq m + \sigma) = \int_{m-\sigma}^{m+\sigma} P(z)\, dz$$

$$= \int_{m-\sigma}^{m+\sigma} \frac{1}{\sqrt{2\pi}\sigma}\, e^{-(x-m)^2/2\sigma^2}\, dx$$

$$= \int_{-1}^{+1} \frac{1}{\sqrt{2\pi}}\, e^{-y^2} \, dy = 68\% \qquad (4.2\text{-}14)$$

or

$$P(m - 2\sigma \leq x \leq m + 2\sigma)$$

$$= \int_{m-2\sigma}^{m+2\sigma} \frac{1}{\sqrt{2\pi}\sigma}\, e^{-(x-m)^2/2\sigma^2}\, dx = 95.45\% \quad (4.2\text{-}15)$$

where m is the mean of all the data and σ is the standard deviation of all the data.

Equation (4.2-14) indicates that 68 percent of predicted data will fall in the range between $-\sigma$ and $+\sigma$ around this mean value. In other words, we are 68 percent confident that a predicted data point will fall between $m - \sigma$ and $m + \sigma$.

The standard deviation from Fig. 4.6 can be found from Table 4.2 as

$$\sigma = \frac{C - A}{B} \qquad \text{[for a given percentage, } P(x \geq C)] \quad (4.2\text{-}16)$$

Example 4.5   Find the standard deviation of a local mean curve as shown in Fig. 4.6. The confidence level for -95 dBm is found to be 10 percent, and the mean is -110 dBm. Then $C = -95$ dBm, $A = -110$ dBm, $B = 1.3$ (from Table 4.2), and

$$\sigma = \frac{-95 - (-110)}{1.3} = 11.54 \text{ dB}$$

Example 4.6   If we do not have Fig. 4.6 in hand but we know the average power and its standard deviation, we can determine the percentage of signal above any level.

Assuming that the average power is -90 dBm and the standard deviation is 9 dB, what would the signal level be if the confidence level is 30 percent?

The level would be (from Table 4.2)

$$0.55 \times 9 + (-90) = -85.05 \text{ dBm}$$

The confidence level and the confidence interval of a signal strength can be calculated from the predicted data applied to a mobile point-to-point model in an area of interest. However, the confidence level and the confidence interval of a signal strength cannot be found from a simple path-loss slope. In other words, it cannot be obtained from an area-to-area model unless the standard deviation of the model from which the curves were generated is known.

F(50,70) is a common notation to indicate that a signal strength is predicted under a confidence level of 50 percent for time to 70 percent for coverage. A detailed description can be found in Ref. 30.

4.2.7   A general formula for mobile radio propagation

Here we are only interested in a general propagation path-loss formula in a general mobile radio environment, which could be a suburban area. The 1-mi intercept level in a suburban area is -61.7 dBm under the standard conditions listed in Table 4.1. Combining these data with the equation shown in Eq. (4.2-10b) from the theoretical prediction model, and Eqs. (4.2-10c) and (4.2-11) from the measured data, the received power P_r at the suburban area can be expressed as

$$P_r = (P_t - 40) - 61.7 - 38.4 \log \frac{r_1}{1 \text{ mi}} + 20 \log \frac{h_1}{100 \text{ ft}}$$

$$+ 10 \log \frac{h_2}{10 \text{ ft}} + (G_t - 6) + G_m \quad (4.2\text{-}17)$$

Equation (4.2-17) can be simplified as

$$P_r = P_t - 157.7 - 38.4 \log r_1 + 20 \log h_1$$

$$+ 10 \log h_2 + G_t + G_m \quad (4.2\text{-}18)$$

where P_t is in decibels above 1 mW, r_1 is in miles, h_1 and h_2 are in feet, and G_t and G_m are in decibels. Equation (4.2-18) is used for suburban areas. We may like to change Eq. (4.2-18) to a general formula

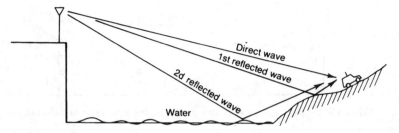

Figure 4.7 A model for propagation over water.

by using P_r at 10 mi as a reference which is -100 dBm, as shown in Fig. 4.3. Also the 40 dB/oct slope used is generous. Then Eq. (4.2-18) changes to

$$P_r = P_t - 156 - 40 \log r_1 + 20 \log h_1$$
$$+ 10 \log h_2 + G_t + G_m \quad (4.2\text{-}19)$$

where the units of P_t, r_1, h_1, h_2, G_t, and G_m are stated below Eq. (4.2-18). Equation (4.2-19) can be used as a general formula in a mobile radio environment.

The most general formula is expressed as follows

$$P_r = P_t - K - \gamma \log r_1 + 20 \log h_1 + 10 \log h_2 + G_t + G_m \quad (4.2\text{-}20)$$

where $P_r = P_t - K$ at $r_1 = 1$ mile, $h_1 = h_2 = 1'$, and $G_t = G_m = 0$. The value of K and γ will be different and need to be measured in different human-made environment.

4.3   Propagation over Water or Flat Open Area

Propagation over water or flat open area is becoming a big concern because it is very easy to interfere with other cells if we do not make the correct arrangements. Interference resulting from propagation over the water can be controlled if we know the cause.

In general, the permittivities ϵ_r of seawater and fresh water are the same, but the conductivities of seawater and fresh water are different. We may calculate the dielectric constants ϵ_c, where $\epsilon_c = \epsilon_r - j60\sigma\lambda$. The wavelength at 850 MHz is 0.35 m. Then ϵ_c (seawater) $= 80 - j84$ and ϵ_c (fresh water) $= 80 - j0.021$.

However, based upon the reflection coefficients formula[32,33] with a small incident angle, both the reflection coefficients for horizontal polarized waves and vertically polarized waves approach 1. Since the 180° phase change occurs at the ground reflection point, the reflection coefficient is -1. Now we can establish a scenario, as shown in Fig. 4.7. Since the two antennas, one at the cell site and the other at the mobile unit, are well above sea level, two reflection points are gener-

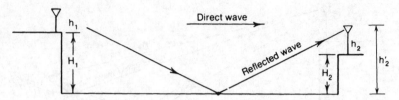

Figure 4.8 Propagation between two fixed stations over water or flat open land.

ated. The one reflected from the ground is close to the mobile unit; the other reflected from the water is away from the mobile unit. We recall that the only reflected wave we considered in the land mobile propagation is the one reflection point which is always very close to the mobile unit. We are now using the formula to find the field strength under the circumstances of a fixed point-to-point transmission and a land-mobile transmission over a water or flat open land condition.

4.3.1 Between fixed stations

The point-to-point transmission between the fixed stations over the water or flat open land can be estimated as follows. The received power P_r can be expressed as (see Fig. 4.8)

$$P_r = P_t \left(\frac{1}{4\pi d/\lambda} \right)^2 \left| 1 + a_v e^{-j\phi_v} \exp\left(j \, \Delta\phi \right) \right|^2 \qquad (4.3\text{-}1)$$

where P_t = transmitted power
d = distance between two stations
λ = wavelength
a_v, ϕ_v = amplitude and phase of a complex reflection coefficient, respectively

$\Delta\phi$ is the phase difference caused by the path difference Δd between the direct wave and the reflected wave, or

$$\Delta\phi = \beta \, \Delta d = \frac{2\pi}{\lambda} \, \Delta d \qquad (4.3\text{-}2)$$

The first part of Eq. (4.3-1) is the free-space loss formula which shows the 20 dB/dec slope; that is, a 20-dB loss will be seen when propagating from 1 to 10 km.

$$P_0 = \frac{P_t}{(4\pi d/\lambda)^2} \qquad (4.3\text{-}3)$$

The $a_v e^{-j\phi_v}$ are the complex reflection coefficients and can be found from the formula[32]

$$a_v e^{-j\phi_v} = \frac{\epsilon_c \sin \theta_1 - (\epsilon_c - \cos^2 \theta_1)^{1/2}}{\epsilon_c \sin \theta_1 + (\epsilon_c - \cos^2 \theta_1)^{1/2}} \qquad (4.3\text{-}4)$$

When the vertical incidence is small, θ is very small and

$$a_v \approx -1 \quad \text{and} \quad \phi_v = 0 \qquad (4.3\text{-}5)$$

can be found from Eq. (4.3-4). ϵ_c is a dielectric constant that is different for different media. However, when $a_v e^{-j\phi_v}$ is independent of ϵ_c, the reflection coefficient remains -1 regardless of whether the wave is propagated over water, dry land, wet land, ice, and so forth. The wave propagating between fixed stations is illustrated in Fig. 4.8.

Equation (4.3-1) then becomes

$$P_r = \frac{P_t}{(4\pi d/\lambda)^2} |1 - \cos \Delta\phi - j \sin \Delta\phi|^2$$

$$= P_0(2 - 2 \cos \Delta\phi) \qquad (4.3\text{-}6)$$

since $\Delta\phi$ is a function of Δd and Δd can be obtained from the following calculation. The effective antenna height at antenna 1 is the height above the sea level.

$$h_1' = h_1 + H_1$$

The effective antenna height at antenna 2 is the height above the sea level.

$$h_2' = h_2 + H_2$$

as shown in Fig. 4.8, where h_1 and h_2 are actual heights and H_1 and H_2 are the heights of hills. In general, both antennas at fixed stations are high, so the reflection point of the wave will be found toward the middle of the radio path. The path difference Δd can be obtained from Fig. 4.8 as

$$\Delta d = \sqrt{(h_1' + h_2')^2 + d^2} - \sqrt{(h_1' - h_2')^2 + d^2} \qquad (4.3\text{-}7)$$

Since $d \gg h_1'$ and h_2', then

$$\Delta d \approx d \left[1 + \frac{(h_1' + h_2')^2}{2d^2} - 1 - \frac{(h_1' - h_2')^2}{2d^2} \right] = \frac{2h_1'h_2'}{d} \qquad (4.3\text{-}8)$$

Then Eq. (4.3-2) becomes

$$\Delta\phi = \frac{2\pi}{\lambda}\frac{2h_1'h_2'}{d} = \frac{4\pi h_1'h_2'}{\lambda d} \tag{4.3-9}$$

Examining Eq. (4.3-6), we can set up five conditions:

1. $P_r < P_0$. The received power is less than the power received in free space; that is,

$$2 - 2 \cos \Delta\phi < 1 \quad \text{or} \quad \Delta\phi < \frac{\pi}{3} \tag{4.3-10}$$

2. $P_r = 0$; that is,

$$2 - 2 \cos \Delta\phi = 0 \quad \text{or} \quad \Delta\phi = \frac{\pi}{2}$$

3. $P_r = P_0$; that is,

$$2 - 2 \cos \Delta\phi = 1 \quad \text{or} \quad \Delta\phi = \pm 60° = \pm\frac{\pi}{3} \tag{4.3-11}$$

4. $P_r > P_0$; that is,

$$2 - 2 \cos \Delta\phi > 1 \quad \text{or} \quad \frac{\pi}{3} < \Delta\phi < \frac{5\pi}{3} \tag{4.3-12}$$

5. $P_r = 4P_0$; that is,

$$2 - 2 \cos \Delta\phi = \text{max} \quad \text{or} \quad \Delta\phi = \pi \tag{4.3-13}$$

The value of $\Delta\phi$ can be found from Eq. (4.3-9). Now we can examine the situations resulting from Eq. (4.3-9) in the following examples.

Example 4.7 Let a distance between two fixed stations be 30 km. The effective antenna height at one end h_1 is 150 m above sea level. Find the h_2 at the other end so that the received power always meets the condition $P_r < P_0$ at 850-MHz transmission ($\lambda = 0.35$ m).

solution

$$\frac{4\pi h_1'h_2'}{\lambda d} \leq \frac{\pi}{3} \tag{E4.7-1}$$

or

$$h_1' \le \frac{d\lambda}{12h_1'} = \frac{30,000 \times 0.35}{12 \times 150} = 6 \text{ m} \qquad \text{(E4.7-2)}$$

Example 4.8 Using the same parameters given in Example 4.7, find the range of h_2 which would keep $P_r > P_0$, and find the maximum received power P_r for $P_r = 4P_0$.

solution

a. $\dfrac{\pi}{3} \le \dfrac{4\pi h_1' h_2'}{\lambda d} \le \dfrac{5\pi}{3}$ the range of h_2 for $P_r > P_0$ $\qquad$ (E4.8-1)

Substituting the values given in Example 4.7, we obtain

$$6 \text{ m} < h_2 < 30 \text{ m} \qquad 42 \text{ m} < h_2 < 66 \text{ m} \qquad \text{(E4.8-2)}$$

b. $\Delta\phi = \pi$ for the maximum received power.

$$h_2 = 18 \text{ m} \qquad h_2 = 54 \text{ m} \qquad h_2 = 6 \text{ m}[3(2n - 1)] \qquad \text{(E4.8-3)}$$

where n is any integer.

4.3.2 Land-to-mobile transmission over water

The propagation model would be different for land-to-mobile transmission over water. As depicted in Fig. 4.7, there are always two equal-strength reflected waves, one from the water and one from the proximity of the mobile unit, in addition to the direct wave. The reflected wave, whose reflected point is on the water is counted because there are no surrounding objects near this point. Therefore the reflected energy is strong. The other reflected wave that has a reflection point proximal to the mobile unit also carries strong reflected energy to it.

Therefore, the reflected power of the two reflected waves can reach the mobile unit without noticeable attenuation. The total received power at the mobile unit would be obtained by summing three components.

$$P_r = \frac{P_t}{(4\pi d/\lambda)^2} \left|1 - e^{j\Delta\phi_1} - e^{j\Delta\phi_2}\right|^2 \qquad (4.3\text{-}14)$$

where $\Delta\phi_1$ and $\Delta\phi_2$ are the path-length difference between the direct wave and two reflected waves, respectively. Since $\Delta\phi_1$ and $\Delta\phi_2$ are very small usually for the land-to-mobile path, then

$$P_r = \frac{P_t}{(4\pi d/\lambda)^2} \left|1 - \cos \Delta\phi_1 - \cos \Delta\phi_2\right.$$

$$\left. - j(\sin \Delta\phi_1 + \sin \Delta\phi_2)\right|^2 \qquad (4.3\text{-}15)$$

Follow the same approximation for the land-to-mobile propagation over water.

$$\cos \Delta\phi_1 \approx \cos \Delta\phi_2 \approx 1 \qquad \sin \Delta\phi_1 \approx \Delta\phi_1 \qquad \sin \Delta\phi_2 \approx \Delta\phi_2$$

Then

$$P_r = \frac{P_t}{(4\pi d/\lambda)^2} |-1 - j(\Delta\phi_2 + \Delta\phi_2)|^2$$

$$= \frac{P_t}{(4\pi d/\lambda)^2} [1 + (\Delta\phi_1 + \Delta\phi_2)^2] \qquad (4.3\text{-}16)$$

In most practical cases, $\Delta\phi_1 + \Delta\phi_2 < 1$; then $(\Delta\phi_1 + \Delta\phi_2)^2 \ll 1$ and Eq. (4.3-16) reduces to

$$P_r \approx \frac{P_t}{(4\pi d/\lambda)^2} \qquad (4.3\text{-}17)$$

Equation (4.3-17) is the same as that expressing the power received from the free-space condition. Therefore, we may conclude that the path loss for land-to-mobile propagation over land, 40 dB/dec, is different for land-to-mobile propagation over water. In the case of propagation over water, the free-space path loss, 20 dB/dec, is applied.

4.4 Foliage Loss

Foliage loss is a very complicated topic that has many parameters and variations. The sizes of leaves, branches, and trunks, the density and distribution of leaves, branches, and trunks, and the height of the trees relative to the antenna heights will all be considered. An illustration of this problem is shown in Fig. 4.9. There are three levels: trunks, branches, and leaves. In each level, there is a distribution of sizes of trunks, branches, and leaves and also of the density and spacing between adjacent trunks, branches, and leaves. The texture and thickness of the leaves also count. This unique problem can become very complicated and is beyond the scope of this book. For a system design, the estimate of the signal reception due to foliage loss does not need any degree of accuracy.

Furthermore, some trees, such as maple or oak, lose their leaves in winter, while others, such as pine, never do. For example, in Atlanta, Georgia, there are oak, maple, and pine trees. In summer the foliage is very heavy, but in winter the leaves of the oak and maple trees fall and the pine leaves stay. In addition, when the length of pine needles reaches approximately 6 in., which is the half wavelength at 800 MHz,

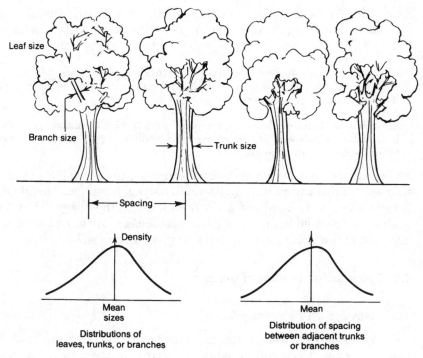

Figure 4.9 A characteristic of foliage environment.

a great deal of energy can be absorbed by the pine trees. In these situations, it is very hard to predict the actual foliage loss.

However, a rough estimate should be sufficient for the purpose of system design. In tropic zones, the sizes of tree leaves are so large and thick that the signal can hardly penetrate. In this case, the signal will propagate from the top of the tree and deflect to the mobile receiver. We will include this calculation also.

Sometime the foliage loss can be treated as a wire-line loss, in decibels per foot or decibels per meter, when the foliage is uniformly heavy and the path lengths are short. When the path length is long and the foliage is nonuniform, then decibels per octaves or decibels per decade is used. Detailed discussion of foliage loss can be found in Refs. 34 to 38. In general, foliage loss occurs with respect to the frequency to the fourth power ($\sim f^{-4}$). Also, at 800 MHz the foliage loss along the radio path is 40 dB/dec, which is 20 dB more than the free-space loss, with the same amount of additional loss for mobile communications. Therefore, if the situation involves both foliage loss and mobile communications, the total loss would be 60 dB/dec (=20 dB/ dec of free-space loss + additional 20 dB due to foliage loss + additional 20 dB due to mobile communiation). This situation would be the case if the foliage would line up along the radio path. A foliage loss in a suburban area of 58.4 dB/dec is shown in Fig. 4.10.

Example 4.9 In a suburban area two places are covered with trees: 2 to 2.5 mi away from the cell site and 3 to 3.5 mi away from the cell site. The additional loss due to foliage is 3 dB, according to Fig. 4.10.

Example 4.10 In a suburban area, one place is 0.3 to 0.5 mi (a distance of 1056 ft) from the cell site with additional trees. The additional path loss is 5 dB due to the foliage, according to Fig. 4.10.

As demonstrated from the above two examples, close-in foliage at the transmitter site always heavily attenuates signal reception. Therefore, the cell site should be placed away from trees. If the heavy foliage is close in at the mobile unit, the additional foliage loss must be calculated using the diffraction loss formula given in Sec. 4.7.3.

4.5 Propagation in Near-in Distance

4.5.1 Why use a 1-mi intercept?

1. Within a 1-mi radius, the antenna beamwidth, especially of a high-gain omnidirectional antenna, is narrow in the vertical plane. Thus the signal reception at a mobile unit less than 1 mi away will be reduced because of the large elevation angle which causes the mobile unit to be in the shadow region (outside the main beam). The larger the elevation angle, the weaker the reception level due to the antenna's vertical pattern, as shown in Fig. 4.11.

2. There are fewer roads within the 1-mi radius around the cell site. The data are insufficient to create a statistical curve. Also the road orientation, in-line and perpendicular, close to the cell site can cause a big difference in signal reception levels (10–20 dB) on those roads.

3. The near-by surroundings of the cell site can bias the reception level either up or down when the mobile unit is within the 1-mi radius. When the mobile unit is 1-mi away from the cell site, the effect due to the near-by surroundings of the cell site becomes negligible.

4. For land-to-mobile propagation, the antenna height at the cell site strongly affects the mobile reception in the field; therefore, mobile reception 1 mi away has to refer to a given base-station antenna height.

4.5.2 Curves for near-in propagation

We usually worry about propagation at the far distance for coverage purposes. Now we also should investigate the near-in distance prop-

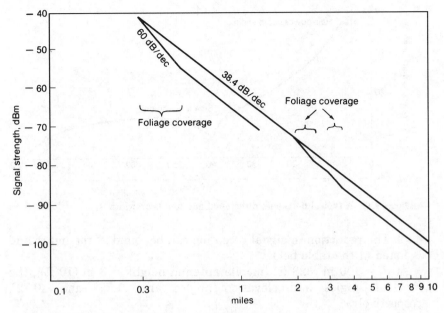

Figure 4.10 Foliage loss calculation in suburban areas.

agation. We may use the suburban area as an example. At the 1-mi intercept the received level is −61.7 dBm based on the reference set of parameters; i.e., the antenna height is 30 m (100 ft). If we increase the antenna height to 60 m (200 ft), a 6-dB gain is obtained. From 60 to 120 m (20 to 400 ft), another 6 dB is obtained. At the 120-m (400-ft) antenna height, the mobile received signal is the same as that received at the free space.

The antenna pattern is not isotropic in the vertical plane. A typical 6-dB omnidirectional antenna vertical beamwidth is shown in Fig.

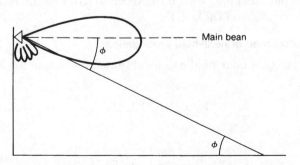

Figure 4.11 Elevation angle of the shadow of the antenna pattern.

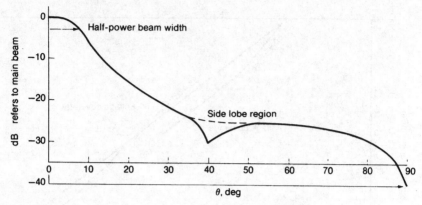

Figure 4.12 A typical 6-dB omnidirectional antenna beamwidth.

4.12. The reduction in signal reception can be found in the figure and is listed in the table below.

At $d = 100$ m (328 ft) [mobile antenna height = 3 m (10 ft)], the incident angles and elevation angles are 11.77° and 10.72°, respectively.

Antenna height h_1, m (ft)	Incident angle θ, degrees	Elevation angle ϕ, degrees	Attentuation α, dB
90 (300)	30.4	29.6	21
60 (200)	21.61	20.75	16
30 (100)	11.77	10.72	6

Since the incident angle becomes larger, the 40-dB/dec slope is no longer valid. If the antenna beam is aimed at the mobile unit, we will observe 24 dB/dec for an antenna height of 100 ft, 22 dB/dec for an antenna height of 200 ft, and 20 dB/dec for an antenna height of 400 ft or higher. The slope of 20 dB/dec is the free-space loss as shown in Fig. 4.13. The power of 11 dBm received at 0.001 mi is obtained from the free-space formula with an ERP of 46 dBm at the cell site as the standard condition (Table 4.1).

4.5.3 Calculation of near-field propagation

The range d_F of near field can be obtained by letting $\Delta\phi$ in Eq (4.2-7) be π.

$$\Delta\phi = \frac{4\pi h_1 h_2}{\lambda d_F} = \pi \tag{4.5-1}$$

and then

$$d_F = \frac{4 h_1 h_2}{\lambda} \tag{4.5-2}$$

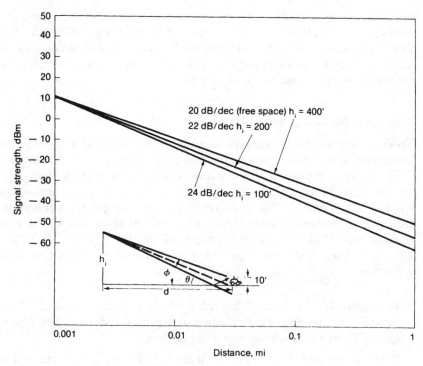

Figure 4.13 Curves for near-in propagation.

The signal received within the nearfield ($d < d_F$) uses the free space loss formula (Eq. (4.3-17)), and the signal received outside the near-field ($d > d_F$) can use the mobile radio path loss formula (Eq. (4.2-18)), for the best approximation.

4.6 Long-Distance Propagation

The advantage of a high cell site is that it covers the signal in a large area, especially in a noise-limited system where usually different frequencies are repeatedly used in different areas. However, we have to be aware of the long-distance propagation phenomenon. A noise-limited system gradually becomes an interference-limited system as the traffic increases. The interference is due to not only the existence of many cochannels and adjacent channels in the system, but the long-distance propagation also affects the interference.

4.6.1 Within an area of 50-mile radius

For a high site, the low-atmospheric phenomenon would cause the ground wave path to propagate in a non-straight-line fashion. The phenomenon is usually more pronounced over seawater because the atmospheric situation over the ocean can be varied based on the different altitudes. The wave path can bend either upward or downward. Then we may have the experience that at one spot the signal may be strong at one time but weak at another.

4.6.2 At a distance of 320 km (200 mi)

Tropospheric wave propagation prevails at 800 MHz for long-distance propagation; sometimes the signal can reach 320 km (200 mi) away.

The wave is received 320 km away because of an abrupt change in the effective dielectric constant of the troposphere (10 km above the surface of the earth). The dielectric constant changes with temperature, which decreases with height at a rate of about 6.5°C/km and reaches −50°C at the upper boundary of the troposphere. In tropospheric propagation, the wave may be divided by refraction and reflection.

Tropospheric refraction.   This refraction is a gradual bending of the rays due to the changing effective dielectric constant of the atmosphere through which the wave is passing.

Tropospheric reflection.   This reflection will occur where there are abrupt changes in the dielectric constant of the atmosphere. The distance of propagation is much greater than the line-of-sight propagation.

Moistness.   Actually water content has much more effect than temperature on the dielectric constant of the atmosphere and on the manner in which the radio waves are affected. The water vapor pressure decreases as the height increases.

If the refraction index decreases with height over a portion of the range of height, the rays will be curved downward, and a condition known as *trapping,* or *duct propagation,* can occur. There are surface ducts and elevated ducts. Elevated ducts are due to large air masses and are common in southern California. They can be found at elevations of 300 to 1500 m (1000 to 5000 ft) and may vary in thickness from a few feet to a thousand feet. Surface ducts appear over the sea and are about 1.5 m (5 ft) thick. Over land areas, surface ducts are produced by the cooling air of the earth.

Tropospheric wave propagation does cause interference and can only be reduced by umbrella antenna beam patterns, a directional antenna pattern, or a low-power-low-antenna-mast approach.

4.7   Obtain Path Loss from a Point-to-Point Prediction Model—A General Approach[24,41]

4.7.1   In nonobstructive condition[42]

In this condition, the direct path from the cell site to the mobile unit is not obstructed by the terrain contour. Here, two general terms should be distinguished. The *nonobstructive direct path* is a path unobstructed by the terrain contour. The *line-of-sight path* is a path which is unobstructed by the terrain contour and by man-made structures. In the former case, the cell-site antenna cannot be seen by the mobile user whereas in the latter case, it can be. Therefore, the signal reception is very strong in the line-of-sight case, which is not the case we are worrying about.

In the mobile environment, we do not often have line-of-sight conditions. Therefore we use direct-path conditions which are unobstructed by the terrain contour. Under these conditions, the antenna-height gain will be calculated for every location in which the mobile unit travels, as illustrated in Fig. 4.14. The method for finding the antenna-height gain is as follows.

Finding the antenna-height gain

1. Find the specular reflection point. Take two values from two conditions stated as follows.
   a. Connect the image antenna of the cell-site antenna to the mobile antenna; the intercept point at the ground level is considered as a potential reflection point.
   b. Connect the image antenna of the mobile antenna to the cell-site antenna; the intercept point at the ground level is also considered as a potential reflection point.
   Between two potential reflection points we choose the point which is close to the mobile unit to be the real one because more energy would be reflected to the mobile unit at that point.
2. Extend the reflected ground plane. The reflected ground plane which the reflection point is on can be generated by drawing a tangent line to the point where the ground curvature is, then extending the reflected ground plane to the location of the cell-site antenna.
3. Measure the effective antenna height. The effective antenna height is measured from the point where the reflected ground plan and the cell-site antenna location meet. Between these two cases shown in Fig. 4.14, h_e equals 40 m in Fig. 4.14a and 200 m in Fig. 4.14b. The actual antenna height h_1 is 100 m.

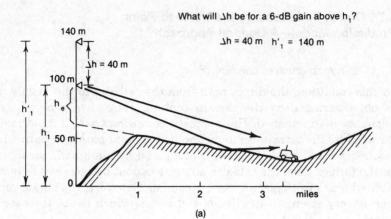

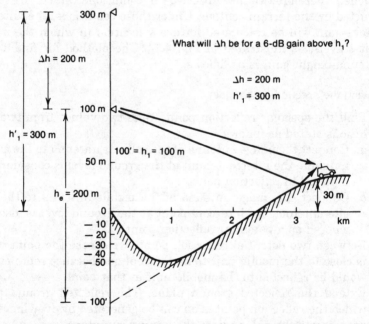

Figure 4.14 Calculation of effective antenna height: (a) case 1; (b) case 2.

4. Calculate the antenna-height gain ΔG. The formula of ΔG is expressed as [see Eq. (4.2-10b)]

$$G = 20 \log \frac{h_e}{h_1} \qquad (4.7\text{-}1)$$

Then the ΔG from Fig. 4.14a is

$$\Delta G = 20 \log \frac{40}{100} = -8 \text{ dB} \qquad \text{(a negative gain in Fig. 4.14}a\text{)}$$

The ΔG from Fig. 4.14b is

$$\Delta G = 20 \log \frac{200}{100} = 6 \text{ dB} \qquad \text{(a positive gain in Fig. 4.14}b\text{)}$$

We have to realize that the antenna-height gain ΔG changes as the mobile unit moves along the road. In other words, the effective antenna height at the cell site changes as the mobile unit moves to a new location, although the actual antenna remains unchanged.

Another physical explanation of effective antenna height. Another physical explanation of effective antenna height is shown in Fig. 4.15. In Fig. 4.15a, we have to ask which height is the actual antenna height h_1, or is the actual antenna height very important in this situation?

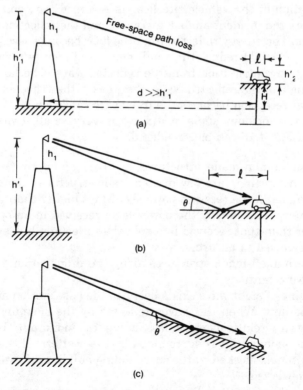

Figure 4.15 Physical explanation of effective antenna heights.

As long as the value of H is much larger than h_2 and the length l of the floor is roughly equal to the length of the vehicle, there is only one direct wave, and the free-space path loss is applied to the situation which provides a strong reception.

In Fig. 4.15b, the situation remains the same, except the length l is longer to allow a reflection point to be generated on the floor. Now two waves are created, one direct wave and one reflected wave. The stronger the reflected wave is, the larger the path loss is. The stronger reflected wave occurs at a very small incident angle θ. This means that a small incident angle corresponds to a large reflection coefficient because of the nature of the reflection mechanism, and the wave reflected from the ground is a 180° phase shift. Therefore, no matter what, the amount of reflected energy always becomes negative. The addition of a strong reflected wave to a direct wave tends to weaken the direct wave.

In Fig. 4.15c, as the incident angle θ approaches zero, the signal reception becomes very weak. The shadow-loss condition starts when both the direct wave and the reflected wave have been blocked. When the direction of the vehicle-site floor is reversed (i.e., going counterclockwise), the incident angle θ increases and the reflection coefficient decreases. The energy reflected from the floor becomes less, and so the direct wave would reduce the small amount of energy resulting from the negative contribution from the reflected wave. The larger the incident angle of the reflected wave, the weaker the reflected wave, and the signal reception becomes the free-space condition.

When the incident angle of a wave is very small, two conditions shown in Fig. 4.16 can be considered.

1. Sparse human-made structures or trees along the propagation path. When there are few human-made structures along the propagation path, the received power is always higher than when there are many. This is why the power level received in an open area is higher than that received in a suburban area and higher still than that received in an urban area.
2. Dense human-made structures along the propagation path. There are two conditions.
   a. A line-of-sight wave exists between the base station and the mobile unit. When the waves reflected by the surrounding buildings are relatively weak, less fading (rician fading)[40] is observed.
   b. The mobile unit is surrounded by the scatters. If the direct reception is blocked by the surrounding buildings, Rayleigh fading is observed.

   In the above two conditions the average received powers are not the same. However, if the reflected waves from surrounding build-

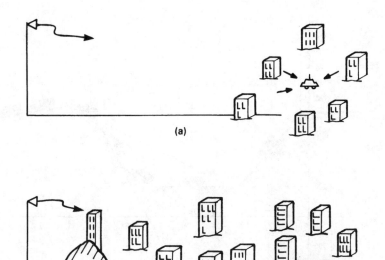

Figure 4.16 Man-made environment. (*a*) Sparse man-made structures. (*b*) Dense man-made structures.

ings are very strong, the average received power from the two different conditions can be very close. It can be seen as an analog to conservation of energy. The total signal received at the mobile unit (or at the cell site) either from a single wave or from many reflected waves tentatively remains a constant. In both conditions, the propagation path loss is 40 dB/dec because both conditions are in a mobile radio environment.

Comments on the contribution of antenna-height gain.   If we do not take into account the changes in antenna-height gain due to the terrain contour between the cell site and the mobile unit the path-loss slope will have a standard deviation of 8 dB. If we do take the antenna-height gain into account, values generally have a standard deviation within 2 to 3 dB.

The effects of terrain roughness are illustrated in Fig. 4.17*a* as changing different effective antenna heights, h_e and h_e' at different positions of the mobile unit. Then the effective antenna gain ΔG can be obtained from Eq. (4.7-1) as

$$\Delta G = 20 \log \frac{h_e}{h_e'}$$

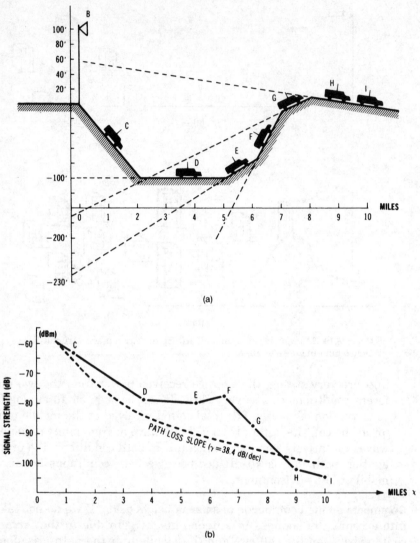

Figure 4.17 Illustration of the terrain effect on the effective antenna gain at each position. (*a*) Hilly terrain contour. (*b*) Point-to-point prediction. (*After Lee, Ref. 40, p. 86.*)

Assume that the mobile unit is traveling in a sururban area, say northern New Jersey. The path-loss slope of this suburban area is shown in Fig. 4.3 and then plotted in Fig. 4.17*b*. Then the antenna-height gains or losses are added or subtracted from the slope at their corresponding points. Now we can visualize the difference between an area-to-area prediction (use a path-loss slope) and a point-to-point prediction (after the antenna-height gain correction). The point-to-point prediction is based on the actual terrain contour along a particular radio path (in this case, the radio path and the mobile path are the same for simplicity), but the area-to-area prediction is not. This is why the area-to-area prediction has a standard deviation of 8 dB but the point-to-point prediction only has a standard deviation of less than 2 to 3 dB (see Sec. 4.8.2).

4.7.2 In obstructive condition

In this condition, the direct path from the cell site to the mobile unit is obstructed by the terrain contour. We would like to treat this condition as follows.

1. *Apply area-to-area prediction.* First, just apply the same steps in the area-to-area prediction as if the obstructive condition did not exist. If the area is in Philadelphia, the Philadelphia path-loss slope applies. All the correction factors would apply to finding the area-to-area prediction for a particular situation.
2. *Obtain the diffraction loss.* The diffraction loss can be found from a single knife-edge or double knife-edge case, as shown in Fig. 4.18.

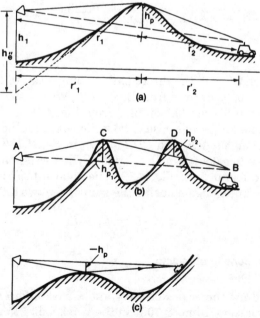

Figure 4.18 Diffraction loss due to obstructive conditions. (*a*) Single knife-edge; (*b*) double knife-edges; (*c*) nonclear path.

a. Find the four parameters for a single knife-edge case. The four parameters, the distances r_1 and r_2 from the knife-edge to the cell site and to the mobile unit, the height of the knife-edge h_p, and the operating wavelength λ, are used to find a new parameter v.

$$v = -h_p \sqrt{\frac{2}{\lambda} \left(\frac{1}{r_1} + \frac{1}{r_2} \right)} \qquad (4.7\text{-}2)$$

h_p is a positive number as shown in Fig. 4.18*a*, and h_p is a negative number as shown in Fig. 4.18*c*. As soon as the value of v is obtained, the diffraction loss L can be found from the curves shown in Fig. 4.19. The approximate formula below can be used with different values of v to represent the curve and be programmed into a computer.

$$1 \leq v \quad L = 0 \text{ dB}$$

$$0 \leq v < 1 \quad L = 20 \log (0.5 + 0.62\, v)$$

$$-1 \leq v < 0 \quad L = 20 \log (0.5 e^{0.95v})$$

$$-2.4 \leq v < -1 \quad L = 20 \log (0.4 - \sqrt{0.1184 - (0.1\, v + 0.38)^2})$$

$$v < -2.4 \quad L = 20 \log \left(-\frac{0.225}{v} \right)$$

$$(4.7\text{-}3)$$

When $h_p = 0$, the direct path is tangential to the knife-edge, and $v = 0$, as derived from Eq. (4.7-2). With $v = 0$, the diffraction loss $l = 6$ dB can be obtained from Fig. 4.19.

b. A double knife-edge case. Two knife edges can be formed by the two triangles *ACB* and *CDB* shown in Fig. 4.18*b*. Each one can be used to calculate v as v_1 and v_2. The corresponding L_1 and L_2 can be found from Fig. 4.19. The total diffraction loss of this double knife-edge model is the sum of the two diffraction losses.

$$L_t = L_1 + L_2$$

4.7.3 Cautions in obtaining defraction loss

We always draw the scales of the y and x axes differently. The same intervals represent 10 m or 10 ft in the y-axis but 1 km or 1 mi in the x-axis. In this way we can depict the elevation change more clearly. Then we have to be aware of the measurement of r_1 and r_2 shown in

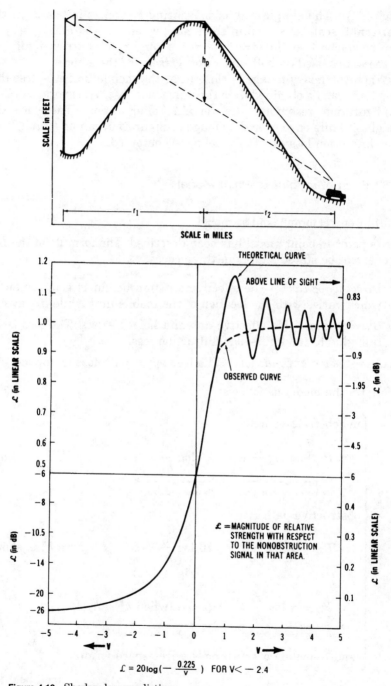

Figure 4.19 Shadow-loss prediction.

Fig. 4.18a. The simple way of measuring r_1 and r_2 is based on the horizontal scale as shown in the figure, where $r_1 = r_1'$ and $r_2 = r_2'$. It can be shown that the errors for $r_1 = r_1'$ and $r_2 = r_2'$ are insignificant if the scales used on both the x and y axes are the same.

When the heavy foliage is close in at the mobile unit, the loss due to foliage can be obtained from the diffraction loss. The average foliage configuration resembles the terrain configuration. Therefore the height of knife edge over the foliage configuration can be found. Then the diffraction loss due to the foliage is obtained.

4.8 Form of a Point-to-Point Model

4.8.1 General formula of Lee model

Lee's point-to-point model has been described. The formula of the Lee model can be stated simply in three cases:

1. *Direct-wave case.* The effective antenna height is a major factor which varies with the location of the mobile unit while it travels.

2. *Shadow case.* No effective antenna height exists. The loss is totally due to the knife-edge diffraction loss.

3. *Over-the-water condition.* The free space path-loss is applied.

We form the model as follows:

$$P_r = \begin{cases} \text{nonobstructive path} \\ \quad = P_{r_0} - \gamma \log \dfrac{r}{r_0} + \quad 20 \log \dfrac{h_e'}{h_1} + \alpha \qquad\qquad (4.8\text{-}1) \\ \qquad\quad \underbrace{\phantom{= P_{r_0} - \gamma \log \tfrac{r}{r_0}}}_{\text{By human-made structure}} \quad \underbrace{\phantom{20 \log \tfrac{h_e'}{h_1} + \alpha}}_{\text{By terrain contour}} \\ \text{obstructive path} \\ \quad = P_{r_0} - \gamma \log \dfrac{r}{r_0} + \quad 20 \log \dfrac{h_e''}{h_1} + L + \alpha \ (\text{where } h_e'' \text{ is shown} \\ \qquad\qquad\qquad\qquad\qquad\qquad \text{in Fig. } 4.18a) \\ \quad = P_{r_0} - \gamma \log \dfrac{r}{r_0} + \quad L + \alpha \ (\text{when } h_e'' \approx h_1) \\ \qquad\quad \underbrace{\phantom{= P_{r_0} - \gamma \log \tfrac{r}{r_0}}}_{\text{By human-made structure}} \quad \underbrace{\phantom{L + \alpha \ (\text{when})}}_{\text{By terrain contour}} \\ \text{land-to-mobile over water} = \text{a free-space formula} \end{cases}$$

(see Sec. 4.3)

Remarks

1. The P_r cannot be higher than that from the free-space path loss.

2. The road's orientation, when it is within 2 mi from the cell site, will affect the received power at the mobile unit. The received power at the mobile unit traveling along an in-line road can be 10 dB higher than that along a perpendicular road.

3. α is the corrected factor (gain or loss) obtained from the condition (see Sec. 4.2.1).

4. The foliage loss (Sec. 4.4) would be added depending on each individual situation. Avoid choosing a cell site in the forest. Be sure that the antenna height at the cell site is higher than the top of the trees.

5. Within one mile (or one kilometer) in a man-made environment, the received signal is affected by the buildings and street orientations. The macrocell prediction formula (Eq. (4.8-1)) can not be applied in such area. A microcell prediction model by Lee is introduced and described in Ref. 45.

4.8.2 The merit of the point-to-point model

The area-to-area model usually only provides an accuracy of prediction with a standard deviation of 8 dB, which means that 68 percent of the actual path-loss data are within the ± 8 dB of the predicted value. The uncertainty range is too large. The point-to-point model reduces the uncertainty range by including the detailed terrain contour information in the path-loss predictions.

The differences between the predicted values and the measured ones for the point-to-point model were determined in many areas. In the following discussion, we compare the differences shown in the Whippany, N.J., area and the Camden-Philadelphia area. First, we plot the points with predicted values at the x-axis and the measured values at the y-axis, shown in Fig. 4.20. The 45° line is the line of prediction without error. The dots are data from the Whippany area, and the crosses are data from the Camden-Philadelphia area. Most of them, except the one at 9 dB, are close to the line of prediction without error. The mean value of all the data is right on the line of prediction without error. The standard deviation of the predicted value of 0.8 dB from the measured one.

In other areas, the differences were slightly larger. However, the standard deviation of the predicted value never exceeds the measured

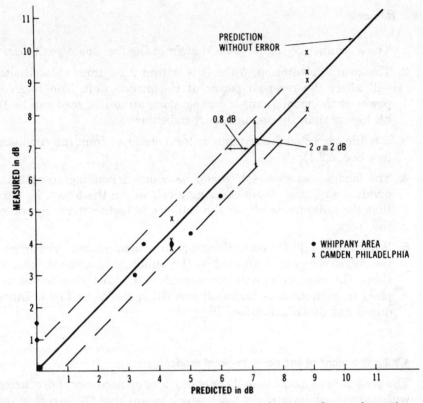

Figure 4.20 Indication of errors in point-to-point predictions under nonobstructive conditions. (*After Lee, Ref. 42.*)

one by more than 3 dB. The standard deviation range is much reduced as compared with the maximum of 8 dB from area-to-area models.

The point-to-point model is very useful for designing a mobile cellular system with a radius for each cell of 10 mi or less. Because the data follow the log-normal distribution, 68 percent of predicted vaues obtained from a point-to-point prediction model are within 2 to 3 dB.

This point-to-point prediction can be used to provide overall coverage of all cell sites and to avoid cochannel interference. Moreover, the occurrence of handoff in the cellular system can be predicted more accurately.

The point-to-point prediction model is a basic tool that is used to generate a signal coverage map, an interference area map, a handoff occurrence map, or an optimum system design configuration, to name a few applications.

4.9   Computer Generation of a Point-to-Point Prediction

The point-to-point prediction described in Sec. 4.8 can easily be used in a computer program. Here we describe the automated prediction in steps.[43]

4.9.1   Terrain elevation data

We may use either a 250,000:1 scale map, called a *quarter-million scale map,* or a 7.5-minute scale map issued by the U.S. Geological Survey. Both maps have the terrain elevation contours. Also, terrain elevation data tapes can be purchased from the DMA (Defense Map Agency). The quarter-million scale map tapes over the whole United States, but the 7.5-minute scale maps are only available for certain areas. Let us discuss the use of these two different scale maps.

1. Use a quarter-million scale map. Each elevation contour increment is 100 ft, which does not provide the fine detail needed for a mobile radio propagation in a hilly area. The quarter-million scale elevation data tapes made by DMA come from the quarter-million scale maps. The elevation data for two adjacent terrain contours is determined by extrapolation. Although the tape gives elevations for every 208 ft (0.01 in on the map), these elevations are not accurate because they are from the quarter-million scale map. However, in most areas, as long as the terrain contour does not change rapidly, the DMA tape can be used as a raw data base. DMA provides two kinds of tapes: a 3-second arc tape and a planar 0.01-in tape. The former has an elevation for every 3-second interval (about 61 to 91.5 m, or 200 to 300 ft, depending on the geographic locations) on the map. The latter has an elevation at intervals of 0.01 in (63.5 m, or 208 ft). Since the arc-second tape provides sample intervals of 3 seconds, a length of 1° has 1200 points. The advantage of using the arc-second tape is the continuity of sample points from one tape to another. However, for the arc-second tape the sample points are not equally far apart on the ground but are in the planar tape. Therefore, if the desired coverage is within one tape's geographic area, the planar tape is used. If the desired area is spread over more than one tape (see Fig. 4.21), the arc-second is used. The structure to a 250,000:1 scale digital elevation format is shown in Fig. 4.22.
2. Use a 7.5-minute map. A 7.5-minute map roughly covers 10 × 13 km², or 6 × 8 mi². The increment of elevation between two contours is 3 or 6 m (10 or 20 ft). The fine resolution of elevation data proves

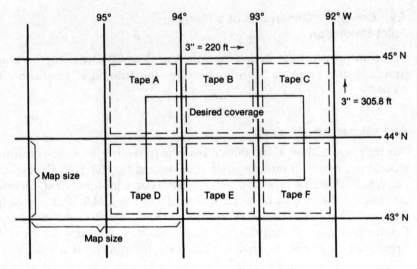

Figure 4.21 A coverage by six tapes.

to be useful for propagation prediction. There are three ways to deal with the 7.5-minute map.

a. Divide a 7.5-minute scale map into either a 30 × 45 grid map, where each grid is 300 × 300 m (1000 × 1000 ft), or into a 15 × 22.5 grid map, where each grid is 600 × 600 m (2000 × 2000 ft). An eyeball estimate of the elevation vaue in each grid is quite adequate.

b. Use DMA tapes in the area if the 7.5-minute tape is available. The 7.5-minute tape can have 150 3-second points. Therefore, a

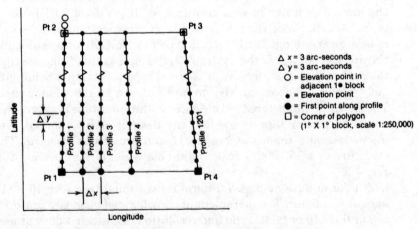

Figure 4.22 Structure of a 250,000:1-scale digital elevation format.

quarter-million scale tape can be replaced by $8 \times 8 = 64$ 7.5 minute maps. The 7.5-minute tapes are used in those areas of the quarter-million map tapes when the terrain contour changes rapidly.

c. The elevation contour line of a 7.5-minute map with the same elevation contour can be digitized in different sample points on the map and stored in a database or on a tape. Then any two points along the terrain elevation contour can be plotted based on the actual contour lines. This is the most accurate method; however, sometimes the accuracy obtained from item a is sufficient for predicting the path loss.

4.9.2   Elevation map

We prefer to use the arc-second tape since the continuity of sample points from tape to tape simplifies the calculation. In the area of N40° latitude, the average elevation of a 2000 × 2000 ft (roughly) grid can be found by taking 7 samples in latitude (a length of 2141 ft) and 10 samples in longitude (a length of 2200 ft).

$$\text{Average elevation} = \frac{\sum\limits_{1}^{70} \text{sample elevations}}{70} \quad \text{(in one grid)}$$

Figure 4.23 shows an elevation map whose grids are approximately 2200 × 2200 ft in area. The average elevation of each grid is given along with a (Y, X) tag.

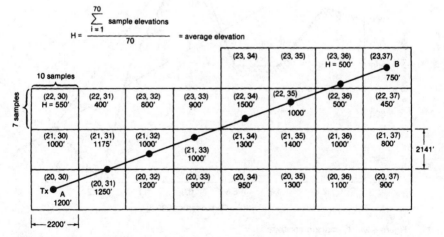

Figure 4.23 Elevation map.

4.9.3 Elevation contour

Assuming that a transmitter is in a grid (20, 30) and the receiver is in a grid (23, 37), then an elevation contour can be plotted with an increment of 2200 ft for every elevation point in its corresponding grid, as shown in Fig. 4.24.

Assume that the antenna height is 100 m (300 ft) at the transmitter and 3 m (10 ft) at the receiver. We can plot a line between two ends and see that shadow-loss condition exists. An example the path loss calculation for this particular path is as follows.

Example 4.11 If the transmitter power is 5 W, the base station antenna gain is 2 dB above dipole, and its height is 300 ft, calculate the path loss from the path shown in Fig. 4.24. The results are given in Table 4.3.

4.10 Cell-Site Antenna Heights and Signal Coverage Cells

4.10.1 Effects of cell-site antenna heights

There are several points which need to be clarified concerning cell-site antenna-height effects.

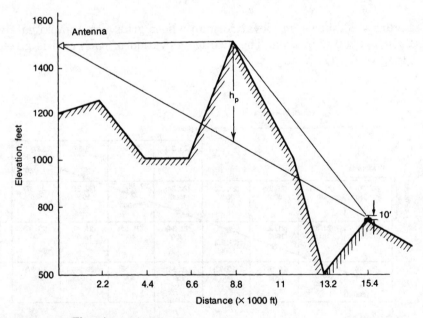

Figure 4.24 Elevation contour.

TABLE 4.3   Work Sheet for Example 4.11

Classification of area: suburban
Distance: 15,400 ft = 2.92 mi
From the curve: (see Fig. 4.3)
P_r = 79.5 dBm

New data:	*Corrections:*
Transmitter power = 5 W	-3 dB
Antenna gain = 2 dB per dipole	-4 dB
Antenna height = 300 ft, $20 \log \dfrac{300}{100} =$	$\dfrac{+9.5 \text{ dB}}{+2.5 \text{ dB}}$

For flat-terrain case:
New path loss $P'_r = -79.5$ dBm + 2.5 dB = -77 dBm
For shadow-region case:
$r_1 = 8800$ ft $r_2 = 6600$ ft $h_p = 450$ ft $f = 850$ MHz
$v = -9.63$
$L = 32.6$ $\left[L = 20 \log \left(-\dfrac{0.225}{v} \right) \right]$
New path loss $P'_r = -77$ dB -33 dB = -110 dBm

Antenna height unchanged. If the power of the cell-site transmitter changes, the whole signal-strength map (obtained from Sec. 4.9) can be linearly updated according to the change in power.

If the transmitted power increases by 3 dB, just add 3 dB to each grid in the signal-strength map. The relative differences in power among the grids remain the same.

Antenna height changed. If the antenna height changes ($\pm \Delta h$), then the whole signal-strength map obtained from the old antenna height cannot be updated with a simple antenna gain formula as

$$\Delta g = 20 \log \frac{h'_1}{h_1} \tag{4.10-1}$$

where h_1 is the old actual antenna height and h'_1 is the new actual antenna height. However, we can still use the same terrain contour data along the radio paths (from the cell-site antenna to each grid) to figure out the difference in gain resulting from the different effective antenna heights in each grid.

$$\Delta g' = 20 \log \frac{h'_e}{h_e} = 20 \log \frac{h_e \pm \Delta h}{h_e} \tag{4.10-2}$$

where h_e is the old effective antenna height and h'_e is the new effective antenna height. The additional gain (increase or decrease) will be added to the signal-strength grid based on the old antenna height.

Example 4.12 If the old cell-site antenna height is 30 m (100 ft) and the new one h_1', is 45 m, the mobile unit 8 km (5 mi) away sees the old cell-site effective antenna height (h_e) being 60 m. The new cell-site effective antenna height h'_e seen from the same mobile spot can be derived.

$$h_e' = h_e + (h_1' - h_1) = h_e + (h_e' - h_e) = h_e + \Delta h = 60 + (45 - 30) = 75 \text{ m}$$

Since the difference between two actual antenna heights is the same as the difference between the two corresponding effective antenna heights seen from each grid, the additional gain (or loss) based on the new change of actual antenna height is

$$\Delta g' = 20 \log \frac{h_e'}{h_e} = 20 \log \left(1 + \frac{h_1' - h_1}{h_e} \right) \qquad \text{(E4.12-1)}$$

Location of the antenna changed. If the location of the antenna changes, the point-to-point program has to start all over again. The old point-to-point terrain contour data are no longer useful. The old effective antenna height seen from a distance will be different when the location of the antenna changes, and there is no relation between the old effective antenna height and the new effective antenna height. Therefore, every time the antenna location changes, the new point-to-point prediction calculation starts again.

Visualization of the effective antenna height. The effective antenna height changes when the location of the mobile unit changes. Therefore, we can visualize the effective antenna height as always changing up or down while the mobile unit is moving. This kind of picture should be kept in mind. In addition, the following facts would be helpful.

Case 1: The mobile unit is driven up a positive slope (up to a high spot). The effective antenna height increases if the mobile unit is driving away from the cell-site antenna, and it decreases if the mobile unit is approaching the cell-site antenna. (See Fig. 4.25a.)

Case 2: The mobile unit is driven down a hill. The effective antenna height decreases if the mobile unit is driving away from the cell-site antenna, and it increases if the mobile unit is approaching the cell-site antenna. (See Fig. 4.25a.)

4.10.2 Visualization of Signal Coverage Cells

A physical cell is usually visualized as a signal-reception region around the cell site. Within the region, there are weak spots called *holes*. This is always true when a cell covers a relative flat terrain.

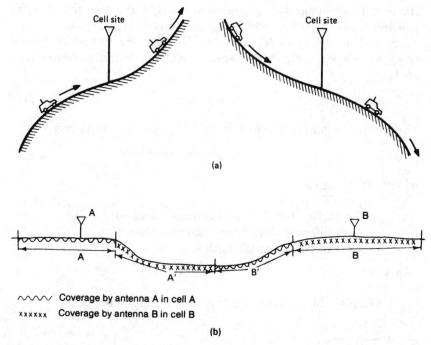

Figure 4.25 Different coverage concept. (a) Signal coverage due to effective antenna heights. (b) Signal coverage served by two cell sites.

However, a cell can contain a hilly area. Then the coverage patterns of the cell will look like those shown in Fig. 4.25b. Here the two cell sites are separated by a river. Because of the shadow loss due to the river bank, cell site A cannot cover area A', but cell site B can. The same situation applies to cell site B in area B'. Now every time the vehicle enters area A', a handoff is requested as if it were in cell B.

Therefore, in most cases, the holes in one cell are covered by the other sites. As long as the processing capacity at the MTSO can handle excessive handoff, this overlapped arrangement for filling the holes is a good approach in a noninterference condition.

4.11   Mobile-to-Mobile Propagation[44]

4.11.1   The transfer function of the propagation channel

In mobile-to-mobile land communication, both the transmitter and the receiver are in motion. The propagation path in this case is usually obstructed by buildings and obstacles between the transmitter and receiver. The propagation channel acts like a filter with a time-varying transfer function $H(f, t)$, which can be found in this section.

The two mobile units M_1 and M_2 with velocities V_1 and V_2, respectively, are shown in Fig. 4.26. Assume that the transmitted signal from M_1 is

$$s(t) = u(t)e^{j\omega t} \tag{4.11-1}$$

The receiver signal at the mobile unit M_2 from an ith path is

$$s_i = r_i u(t - \tau_i)e^{j[(\omega_0 + \omega_{1i} + \omega_{2i})(t-\tau_i) + \phi_i]} \tag{4.11-2}$$

where $u(t)$ = signal
ω_0 = RF carrier
r_i = Rayleigh-distributed random variable
ϕ_i = uniformly distributed random phase
τ_i = time delay on ith path

and

ω_{1i} = Doppler shift of transmitting mobile unit on ith path

$$= \frac{2\pi}{\lambda} V_1 \cos \alpha_{1i} \tag{4.11-3}$$

ω_{2i} = Doppler shift of receiving mobile unit on ith path

$$= \frac{2\pi}{\lambda} V_2 \cos \alpha_{2i} \tag{4.11-4}$$

where α_{1i} and α_{2i} are random angles shown in Fig. 4.26. Now assume that the received signal is the summation of n paths uniformly distributed around the azimuth.

$$s_r = \sum_{i=1}^{n} s_i(t) = \sum_{i=1}^{n} r_i u(t - \tau_i)$$

$$\times \exp\{j[(\omega_0 + \omega_{1i} + \omega_{2i})(t - \tau_i) + \phi_i]\}$$

$$= \sum_{i=1}^{n} Q(\alpha_{i,t}) u(t - \tau_i)e^{j\omega_0(t-\tau_i)} \tag{4.11-5}$$

where

$$Q(\alpha_i, t) = r_i \exp\{j[(\omega_{1i} + \omega_{2i})t + \phi_i']\} \tag{4.11-6}$$

$$\phi_i' = \phi - (\omega_{1i} + \omega_{2i})\tau_i \tag{4.11-7}$$

Equation (4.11-5) can be represented as a statistical model of the channel, as shown in Fig. 4.27.

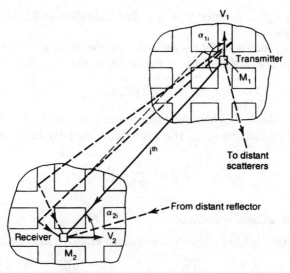

Figure 4.26 Vehicle-to-vehicle transmission.

Let $u(t) = e^{j\omega t}$, then Eq. (4.11-5) becomes

$$e_r(t) = \left[\sum_{i=1}^{n} Q(\alpha_i, t) e^{-j(\omega_0 + \omega)\tau_i} \right] e^{j(\omega_0 + \omega)t} = H(f, t) e^{j(\omega_0 + \omega)t} \quad (4.11\text{-}8)$$

Therefore

$$H(f, t) = \sum_{i=1}^{n} Q(\alpha_i, t) e^{-j(\omega_0 + \omega)\tau_i} \quad (4.11\text{-}9)$$

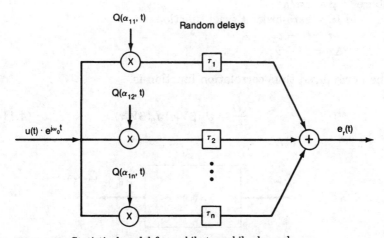

Figure 4.27 Statistical model for mobile-to-mobile channel.

where the signal frequency is $\omega = 2\pi f$. Equation (4.11-9) is expressed in Fig. 4.28.

Let $f = 0$; that is, only a sinusoidal carrier frequency is transmitted. The amplitude of the received signal envelope from Eq. (4.11-8) is

$$r = |H(0, t)| \tag{4.11-10}$$

where r is also a Rayleigh-distributed random variable with its average power of $2\sigma^2$ shown in the probability density function as

$$P(r) = \frac{r}{\sigma^2} e^{-r^2/2\sigma^2} \tag{4.11-11}$$

4.11.2 Spatial time correlation

Let $r_{x_1}(r_1)$ be the received signal envelope at position x_1 at time t_1. Then

$$r_{x_1}(t_1) = \sum_{i=1}^{n} r_i \exp j \left[(\omega_{1i} + \omega_{2i}) + \phi_i + \frac{2\pi}{\lambda} x_1 \cos \alpha_{1i} \right] \tag{4.11-12}$$

The same equation will apply to R_x at position x_2 at time t_2.

The spatial time-correlation function of the envelope is given by

$$R(x_1, x_2, t_1, t_2) = \tfrac{1}{2} \langle r_{x_1}(t_1) r_{x_2}^*(t_2) \rangle \tag{4.11-13}$$

assuming that the random process r is stationary. Then Eq. (4.11-13) can be rewritten as

$$R(\Delta x, \tau) = \sigma^2 J_0(\beta V_1 \tau) J_0(\beta V_2 \tau + \beta \Delta x) \tag{4.11-14}$$

where $\beta = 2\pi/\lambda$
$J_0(\cdot)$ = zero-order Bessel function
$\tau = t_1 - t_2$
$\Delta x = x_1 - x_2$

The normalized time-correlation function is

$$\frac{R(\tau)}{\sigma^2} = J_0(\beta V_1 \tau) J_0(\beta V_2 \tau) \tag{4.11-15}$$

Figure 4.28 A propagation channel model.

Equation (4.11-15) is plotted in Fig. 4.29. The spatial correlation function $R(\Delta x)$ is

$$\frac{R(\Delta x)}{\sigma^2} = J_0(\beta\,\Delta x) \tag{4.11-16}$$

Equation (4.11-16) is the same as for the base-to-mobile channel.

4.11.3 Power spectrum of the complex envelope

The power spectrum $S(f)$ is a Fourier transform of $R(\Delta t)$ from Eq. (4.11-15).

$$S(f) = \int_{-\infty}^{\infty} R(\Delta t)e^{-j2\pi f\Delta t}\,d(\Delta t) \tag{4.11-17}$$

Substituting Eq. (4.11-15) into (4.11-17) yields

$$S(f) = \frac{\sigma^2}{\pi^2 f_1\,\sqrt{a}}\,K\left\{\frac{1+a}{2\sqrt{a}}\,\sqrt{1-\left[\frac{f}{(1+a)f_1}\right]^2}\right\} \tag{4.11-18}$$

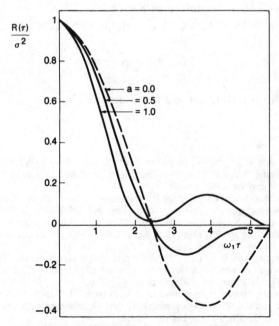

Figure 4.29 Normalized time-correlation function of the complex envelope for different values of $a = V_2/V_1$ versus $\omega_1\,\Delta t$ where $\omega_1 = \beta V_1$. (*After Akki, Ref. 44.*)

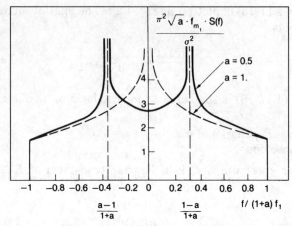

Figure 4.30 Power spectrum of the complex envelope for the case of $a = 0.5$ and $a = 1$ (where $a = V_2/V_1 = f_2/f_1$). (*After Akki, Ref. 44.*)

where $a = f_2/f_1$
$f_1 = V_1/\lambda$
$K(\cdot) = $ complete elliptic integral of the first kind.

Equation (4.11-18) is plotted in Fig. 4.30.
 If $V_2 = 0$ and $a = 0$, Eq. (4.11-18) can be reduced to

$$S(f) = \frac{\sigma^2}{\pi\sqrt{f_1^2 - f^2}} \qquad (4.11\text{-}19)$$

which is the equation for a base-to-mobile channel.

References

1. Y. Okumura et al., "Field Strength and Its Variability in VHF and UHF Land-Mobile Radio Service," *Review of the Electrical Communication Laboratories* Vol. 16, Nos. 9 and 10, 1968, pp. 825–873.
2. M. Hata, "Empirical Formula for Propagation Loss in Land Mobile Radio Services," *IEEE Transactions on Vehicular Technology,* Vol. VT-29, No. 3, 1980, pp. 317–325.
3. A. Akeyama, T. Nagatsu, and Y. Ebine, "Mobile Radio Propagation Characteristics and Radio Zone Design Method in Local Cities," *Review of the Electrical Communication Laboratories* Vol. 30, No. 2, 1982, pp. 308–317.
4. A. G. Longley and P. L. Rice, "Prediction of Tropospheric Radio Transmission Loss Over Irregular Terrain, A Computer Method-1968," ESSA Technical Report ERL 79-ITS 67, NTIS acc. no. 676874, 1968.
5 A. G. Longley, "Radio Propagation in Urban Areas," OT Report 78-144, NTIS, 1978.
6. J. J. Egli, "Radio Propagation Above 40 MC Over Irregular Terrain," *Proceedings of the IRE,* Vol. 45, 1957, pp. 1383–1391.
7. K. Allsebrook and J. D. Parsons, "Mobile Radio Propagation in British Cities at Frequencies in the VHF and UHF Bands," *IEEE Transactions on Vehicular Technology,* Vol. VT-26, No. 4, 1977, pp. 313–323.

7. K. Allsebrook and J. D. Parsons, "Mobile Radio Propagation in British Cities at Frequencies in the VHF and UHF Bands," *IEEE Transactions on Vehicular Technology,* Vol. VT-26, No. 4, 1977, pp. 313–323.
8. K. Bullington, "Radio Propagation for Vehicular Communications," *IEEE Transactions on Vehicular Technology,* Vol. VT-26, No. 4, 1977, pp. 295–308.
9. W. J. Kessler and M. J. Wiggins, "A Simplified Method for Calculating UHF Base-to-Mobile Statistical Coverage Contours over Irregular Terrain," *27th IEEE Vehicular Technology Conference,* 1977, pp. 227–236.
10. W. C. Y. Lee, *Mobile Communications Engineering,* McGraw Hill Book Co., 1982, p. 107.
11. "Bell System Practices Public Land Mobile and UHF Maritime Systems Estimates of Expected Coverage," *Radio Systems General,* July 1963.
12. K. K. Kelly II, "Flat Suburban Area Propagation of 820 MHz," *IEEE Transactions on Vehicular Technology,* Vol. VT-27 November 1978, pp. 198–204.
13. G. D. Ott and A. Plitkins, "Urban Path-Loss Characteristics at 820 MHz," *IEEE Transactions on Vehicular Technology,* Vol. VT-27, November 1978, pp. 189–197.
14. W. R. Young, "Mobile Radio Transmission-Compared at 150 to 3700 MC," *Bell System Technical Journal,* Vol. 31, November 1952, pp. 1068–1085.
15. Robert T. Forrest, "Land Mobile Radio, Propagation Measurements for System Design," *IEEE Transactions on Vehicular Technology,* Vol. VT-24, November 1975, pp. 46–53.
16. G. Hagn, "Radio System Performance Model for Predicting Communications Operational Ranges in Irregular Terrain," *29th IEEE Vehicular Technology Conference Record,* 1970, pp. 322–330.
17. Robert Jensen, "900 MHz Mobile Radio Propagation in the Copenhagen Area," *IEEE Transactions on Vehicular Technology, Vol. VT-26, November 1977.*
18. G. L. Turin, "Simulation of Urban Location Systems," *Proceedings of the 21st IEEE Vehicular Technology Conference,* 1970.
19. D. L. Nielson, "Microwave Propagation Measurements for Mobile Digital Radio Applications," *IEEE Transactions on Vehicular Technology,* Vol. VT-27, August 1978, pp. 117–132.
20. V. Graziano, "Propagation Correlations at 900 MHz," *IEEE Transactions on Vehicular Technology,* Vol. VT-27, November 1978, pp. 182–188.
21. D. O. Reudink, "Properties of Mobile Radio Propagation Above 400 MHz," *IEEE Transactions on Vehicular Technology,* Vol. VT-23, November 1974, pp. 143–160.
22. M. F. Ibrahim and J. D. Parsons, "Urban Mobile Propagation at 900 MHz," *IEEE Electrical Letters,* Vol. 18, No. 3, 1982, pp. 113–115.
23. W. C. Y. Lee, *Mobile Communications Engineering,* McGraw-Hill Book Co., 1982, Chap. 4.
24. W. C. Y. Lee, *Mobile Communications Design Fundamentals,* John Wiley & Sons, 1993, pp. 72–94.
25. N. H. Shepherd, "Radio Wave Loss Deviation and Shadow Loss at 900 MHz," *IEEE Transactions on Vehicular Technology,* Vol. VT-26, November 1977, pp. 309–313.
26. B. Bodson, G. F. McClure, and S. R. McConoughey, *Land-Mobile Communications Engineering,* IEEE Press, 1984.
27. H. F. Schmid, "A Prediction Model for Multipath Propagation of Pulse Signals at VHF and UHF over Irregular Terrain," *IEEE Transactions on Antennas and Propagation,* Vol. AP-18, March 1970, pp. 253–258.
28. H. Suzuki, "A Statistical Model for Urban Radio Propagation," *IEEE Tranactions on Communications,* Vol. Com-25, July 1977.
29. W. C. Y. Lee, *Mobile Communications Engineering,* McGraw-Hill Book Co., 1982, chap. 4.
30. W. C. Y. Lee, *Mobile Communications Engineering,* McGraw-Hill Book Co., 1982, pp. 137–139.
31. J. S. Bendat and A. G. Piersol, *Random Data, Analysis and Measurement Procedures,* Wiley/Interscience, 1971.
32. E. C. Jordan, *Electromagnetic Waves and Radiation Systems,* Prentice-Hall, 1950, p. 141.

33. W. C. Y. Lee, *Mobile Communications Engineering,* McGraw-Hill Book Co., 1982, p. 92.
34. S. Swarup and R. K. Tewari, "Propagation Characteristics of VHF/UHF Signals in Tropical Moist Deciduous Forest," *Journal of the Institution of Electronics and Telecommunication Engineers,* Vol. 21, No. 3, 1975, pp. 123–125.
35. S. Swarup and R. K. Tewari, "Depolarization of Radio Waves in a Jungle Environment," *IEEE Transactions on Antennas and Propagation,* Vol. AP-27, No. 1, January 1979, pp. 113–116.
36. W. R. Vincent and G. H. Hagn, "Comments on the Performance of VHF Vehicular Radio Sets in Tropical Forests," *IEEE Transactions on Vehicular Technology,* Vol. VT-18, No. 2, August 1969, pp. 61–65.
37. T. Tamir, "On Radio-Wave Propagation in Forest Environments," *IEEE Transactions on Antenna and Propagation, Vol. AF-15, November 1967, pp. 806–817.*
38. T. Tamir, "On Radio-Wave Propagation Along Mixed Paths in Forest Environments," *IEEE Transactions on Antennas and Propagation,* Vol. AP-25, July 1971, pp. 471–477.
39. E. C. Jordan (ed.), *Reference Data for Engineers,* 7th ed., Howard W. Sams & Co., 1985, chap. 33.
40. W. C. Y. Lee, *Mobile Communications Design Fundamentals,* John Wiley & Sons, 1993, pp. 30–31.
41. W. C. Y. Lee, "A New Propagation Path-Loss Prediction Model for Military Mobile Access," 1985 IEEE Military Communications Conference, Boston, Conference Record, Vol. 2, pp. 19–21.
42. W. C. Y. Lee, "Studies of Base Station Antenna Height Effect on Mobile Radio," *IEEE Transactions on Vehicular Technology,* Vol. VT-29, May 1980, pp. 252–260.
43. George Washington University, Continuing Engineering Education Program, course 1086, "Mobile Cellular Telecommunications Systems." A lecture note prepared by W. C. Y. Lee.
44. A. S. Akki, F. Haber, "A Statistical Model of Mobile-to-Mobile Land Communication Channels," *IEEE Transactions on Vehicular Technology,* Vol. VT-35, February 1986, p. 2.
45. W. C. Y. Lee, "Mobile Communications Design Fundamentals," John Wiley & Sons, 1993, pp. 88–94.

Cell-Site Antennas
and Mobile Antennas

5.1 Equivalent Circuits of Antennas

The operating conditions of an actual antenna (Fig. 5.1a) can be expressed in an equivalent circuit for both receiving (Fig. 5.1b) and transmitting (Fig. 5.1c). In Fig. 5.1, Z_a is the antenna impedance, Z_L is the load impedance, and Z_T is the impedance at the transmitter terminal.

5.1.1 From the transmitting end
(obtaining free-space path-loss formula)

Power P_t originates at a transmitting antenna and radiates out into space. (Equivalent circuit of a transmitting antenna is shown in Fig. 5.1b.) Assume that an isotropic source P_t is used and that the power in the spherical space will be measured as the power per unit area. This power density, called the *Poynting vector* ρ or the outward flow of electromagnetic energy through a given surface area, is expressed as

$$\rho = \frac{P_t}{4\pi r^2} \quad \text{W/m}^2 \tag{5.1-1}$$

A receiving antenna at a distance r from the transmitting antenna with an aperture A will receive power

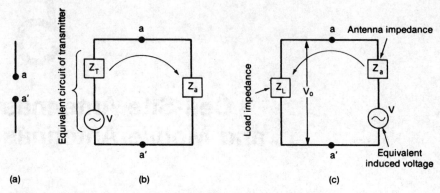

Figure 5.1 An actual antenna and its equivalent circuit. (a) An actual antenna; (b) equivalent circuit of a transmitting antenna; (c) equivalent circuit of a receiving antenna.

$$P_r = \rho A = \frac{P_t A}{4\pi r^2} \quad W \tag{5.1-2}$$

Figure 5.2 is a schematic representation of received power in space.

From Eq. (5.1-2) we can derive the free-space path-loss formula because we know the relationship between the aperture A and the gain G.

$$G = \frac{4\pi A}{\lambda^2} \tag{5.1-3}$$

For a short dipole, $G = 1$. Then

$$A = \frac{\lambda^2}{4\pi} \tag{5.1-4}$$

Substitution of Eq. (5.1-4) into Eq. (5.1-2) yields the free-space formula

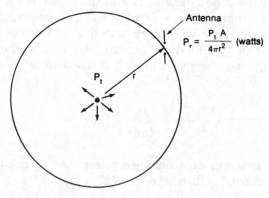

Figure 5.2 Received power in space.

$$P_r = P_t \frac{1}{(4\pi r/\lambda)^2} \tag{5.1-5}$$

5.1.2 At the receiving end dBμV ↔ dBm (decibels above 1 μV ↔ decibels above 1 mW)

We can obtain the received power from Fig. 5.1c

$$P_r = \frac{V^2 Z_L}{(Z_L + Z_a)^2} \tag{5.1-6}$$

where V is the induced voltage in volts. For a maximum power delivery $Z_L = Z_a^*$, where the notation * indicates complex conjugate. Then we obtain $Z_L + Z_a = 2R_L$, where R_L is the real-load resistance. Equation (5.1-6) becomes

$$P_r = \frac{V^2}{4R_L} \tag{5.1-7}$$

Assume that a dipole or a monopole is used as a receiving antenna. The induced voltage V can be related to field strength E as[1]

$$V = \frac{E\lambda}{\pi} \tag{5.1-8}$$

where E is expressed in volts or microvolts per meter. Substitution of Eq. (5.1-7) into Eq. (5.1-8) yields

$$P_r = \frac{E^2 \lambda^2}{4\pi^2 R_L} \tag{5.1-9}$$

If we set $R_L = 50 \ \Omega$, P_r in decibels above 1 mW, and E in decibels (microvolts per meter), Eq. (5.1-9) becomes

$$P_r \text{ (dBm)} = E \text{ (dBμV)} - 113 \text{ dBm} + 10 \log\left(\frac{\lambda}{\pi}\right)^2 \tag{5.1-10}$$

The notation "dBμV" in Eq. (5.1-10) is a simplification of decibels above 1 μV/m, and has been accepted by the Institute of Radio Engineers. We can find the equivalent aperture A of Eq. (5.1-2) because the Poynting vector ρ can be expressed as

$$\rho = \frac{E^2}{Z_0} \tag{5.1-11}$$

where Z_0 is the instrinsic impedance of the space (= 120 π). By substituting Eq. (5.1-11) into Eq. (5.1-2) and comparing it with Eq. (5.1-9) we obtain the equivalent aperture A.

$$A = \frac{\lambda^2 Z_0}{4\pi^2 R_L} \qquad (5.1\text{-}12)$$

Equation (5.1-12) is the same as that given in Eq. 3.11 in Ref. 2.

5.1.3 Practical use of field-strength–received-power conversions

There is always confusion between the following two conversions.

Measuring field strength and converting it to received power. From Eq. (5.1-10), converting field strength in decibels above 1 μV/m to power received in decibels above 1 mW at 850 MHz by a dipole with a 50-Ω load is −132 dB.

$$P_r \quad \text{dBm} = E \quad \text{dB } (\mu V/m) - 132 \text{ dB} \qquad (5.1\text{-}13)$$

$$\text{at 850 MHz}$$

$$39 \text{ dB } (\mu V/m) = -93 \text{ dBm} \qquad (5.1\text{-}14)$$

The notation "39-dBμV contour" is commonly used to mean 39 dB (μV/m) in cellular system design. Equation (5.1-13) is valid only at a given frequency (850 MHz), for a given antenna (monopole or dipole antenna), and for a given antenna load (50 Ω). Otherwise the field strength and the power have to be adjusted accordingly.

Measuring the voltage V_0 at the load terminal (Fig. 5.1c) and converting to received power. Given $P_r = (V_0^2/R_L)$, where $R_L = 50$ Ω, we can obtain a relationship

$$0 \text{ dB}\mu V \langle = \rangle -107 \text{ dBm} \qquad (5.1\text{-}15)$$

For example, if a voltage meter at V_0 is 7 dBμV, then the received power is −100 dBm. Equation (5.1-15) expresses a voltage-to-power ratio which varies with the load impedance but is independent of the frequency and the type of antenna.

5.1.4 Effective radiated power

There is a standard way of specifying the power radiated within a given geographic area. The radiated power that can be transmitted should be converted to the amount of power radiated from an omnidirectional antenna (particularly a dipole antenna). If a high-gain antenna is used, the transmitted power should be reduced. Therefore, if the specification states that the effective radiated power (ERP) or the transmitted power multiplied by the antenna gain is 100 W (50 dBm) and if a 10-dB high-gain antenna is used, the actual transmitted power should be 10 W (40 dBm).

There is also a term called *effective isotropic radiated power* (EIRP). EIRP is referenced to an isotropic point source. The difference between ERP and EIRP is 2 dB.

$$ERP = EIRP - 2 \text{ dB}$$

5.2   The Gain-and-Pattern Relationship

5.2.1   The gain of an antenna or an antenna array

$$G = \frac{4\pi(\text{maximum radiation intensity})}{\text{total power radiated}}$$

$$= \frac{E^2_{\max} (\theta_m, \phi_m)}{\overline{E}^2(\theta, \phi)} \tag{5.2-1}$$

where E = electric field
$E_{\max}$ = maximum of E
$\overline{E}^2$ = average value of E^2 which is related to the radiation intensity
θ, ϕ = angles shown in Fig. 5.3

We can obtain the antenna pattern E from either a measurement or an analytic form. Then the gain G from the pattern E can be calculated.

5.2.2   The pattern and gain of an antenna array

Frequently, it is necessary to calculate the gain of an antenna array. The general field pattern can be expressed as[3]

$$E(\theta, \phi) = \frac{\sin [N\pi(d \cos \phi \cdot \sin \theta + \psi)]}{N \sin [\pi(d \cos \phi \cdot \sin \theta + \psi)]} \times \begin{pmatrix} \text{individual antenna} \\ \text{element pattern} \end{pmatrix}$$

$$\tag{5.2-2}$$

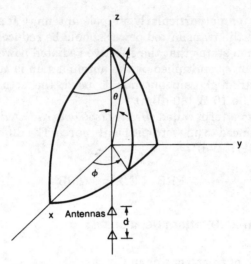

Figure 5.3 A coordinate of antenna arrays.

where N = number of elements in an array

d = spacing between adjacent elements in wavelength

ψ = phase difference between two adjacent elements

ϕ, θ = radiation angles are shown in Fig. 5.3

$\theta = 90°$ = direction perpendicular to the array axis

The gain of the array can be obtained by substituting Eq. (5.2-2) into Eq. (5.2-1). Two cases are listed below.

1. *A broadside-array case* ($\psi = 0$).   The individual elements are in parallel and lie in the *y-z* plane. In this case the gain related to a single element of any type of antenna is shown below.

	Gain G, dB	
N	$d = 0.5\lambda$	$d = 0.9\lambda$
2	4	4.6
3	5.5	6
4	7	8
6	9	10.4
8	10	12
12	12	14
16	13.2	15.4

2. *A collinear-array case* ($\psi = 0$).   The individual elements are collinear in the *z* axis. In this case, the gain related to a single dipole element is as shown in the following table.

	Gain G, dB	.
N	$d = 0.5\lambda$	$d = 0.9\lambda$
2	2.2	3.7
3	3.8	5.7
4	4.9	8.1
6	6.5	9.0
8	8.8	9.4
12	9.4	12.2
16	10.7	13.4

5.2.3 The relationship between gain and beamwidth

A general formula. Assume that the gain G and the directivity D are nearly the same. Then[4]

$$G \approx D = \frac{32,400}{\phi^\circ \theta^\circ} \quad \text{for small } \phi \text{ and } \theta \qquad (5.2\text{-}3)$$

where ϕ° and θ° are the 3-dB beamwidths in two planes. Elliott[4] points out that when ϕ and θ are small ($<40^\circ$) the figure in Eq. (5.2-3) reported by Kraus[5] (41,253) is based on the incorrect assumption that the cross section formed by θ and ϕ is rectangular. It is, in fact, elliptical, so its area is $\pi/4$ that of Kraus' figure, or 32,400. However, for large ($>40^\circ$) ϕ° and θ°, the Kraus equation

$$G \approx D = \frac{41,253}{\phi^\circ \theta^\circ} \quad \text{for large } \phi^\circ \text{ and } \theta^\circ \qquad (5.2\text{-}4)$$

has been shown to be valid. For a high-gain or high-directivity antenna, the gain G resulting from decreased efficiency through the system has only 50 to 70 percent of the directivity D.

For a linear element or collinear array. If a linear element is used, the approximate gain also can be obtained from a vertical 3-dB beamwidth. Assume that the array or the linear element has a doughnut pattern. The relationship between the 3-dB beamwidth θ_0 and the gain G for a linear element or a collinear array can be derived in decibels, above an isotropic source which is a unity gain from Eq. (5.2-3) provided the angle θ_0 is small ($<40^\circ$).[6]

$$G \approx D = \frac{101.5^\circ}{\theta_0} \quad \text{or} \quad D = 10 \log\left(\frac{101.5}{\theta_0}\right) \quad \text{for small } \theta_0$$

$$(5.2\text{-}5)$$

Assume that the gain G and the directivity D are the same. The gain G always refers to an isotropic source which is 2 dB lower than the gain of a dipole. Equation (5.2-5), which is valid for small θ_0, may have an error of $\pm 1°$ as compared with the measured pattern. For large θ_0, the relationship between G and θ_0 can be found from Eq. (5.2-4) as

$$G \approx D = \frac{114.6°}{\theta_0} \tag{5.2-6}$$

Example 5.1 For gain of 9 dB with respect to a dipole antenna or 11 dB with respect to a short dipole, 3-dB beamwidth is

$$\theta_0 = \frac{101.5°}{10^{1.1}} = \frac{101.5°}{12.59} = 8°$$

For a gain of 6 dB with respect to a dipole antenna, the 3-dB beamwidth is

$$\theta_0 = \frac{101.5°}{6.32} = 16°$$

In reality, because of the inefficiency of power delivery, the gain would decrease by 1 dB as a result of a mismatch measured from a (voltage standing-wave ratio) (VSWR)

$$G = 10 \log \frac{101.5}{\theta_0} - 1 \quad \text{dB} \tag{5.2-7}$$

Equation (5.2-7) would be used for an antenna operating within its marginal frequency range.

5.3   Sum-and-Difference Patterns— Engineering Antenna Pattern

After obtaining a predicted field-strength contour (see Sec. 4.8), we can engineer an antenna pattern to conform to uniform coverage. For different antennas pointing in different directions and with different spacings, we can use any of a number of methods. If we know the antenna pattern and the geographic configuration of the antennas, a computer program can help us to find the coverage. Several synthesis methods can be used to generate a desired antenna configuration.

5.3.1   General formula

Many applications of linear arrays are based on sum-and-difference patterns. The mainbeam of the pattern is always known as the sum pattern pointing at an angle θ_0. The difference pattern produces twin mainbeams straddling θ_0. When $2N$ elements are in an array, equispaced by a separation d, the general pattern for both sum and difference is

$$A(\theta) = \sum_{n=1}^{N} I_n \exp\left[j\, \frac{2n-1}{2}\, \beta d(\cos\theta - \cos\theta_0) \right]$$
$$+ I_{-n} \exp\left[-j\, \frac{2n-1}{2}\, \beta d(\cos\theta - \cos\theta_0) \right] \quad (5.3\text{-}1)$$

where β = wavenumber = $2\pi/\lambda$
$\quad I_n$ = normalized current distributions
$\quad N$ = total number of elements

For a sum pattern, all the current amplitudes are the same.

$$I_n = I_{-n} \quad (5.3\text{-}2)$$

For a difference pattern, the current amplitudes of one side (half of the total elements) are positive and the current amplitudes of the other side (half of the total elements) are negative.

$$I_n = -I_{-n} \quad (5.3\text{-}3)$$

Most pattern synthesis problems can be solved by determining the current distribution I_n. A few solutions follow.

5.3.2 Synthesis of sum patterns

Dolph-Chebyshev synthesis of sum patterns. This method can be used to reduce the level of sidelobes; however, one disadvantage of further reduction of sidelobe level is broadening of the mainbeam. The techniques are discussed in Ref. 7.

Taylor synthesis. A continuous line-source distribution or a distribution for discrete arrays can give a desired pattern which contains a single mainbeam of a prescribed beamwidth and pointing direction with a family of sidelobes at a common specified level. The Taylor synthesis is derived from the following equation, where an antenna pattern $F(\theta)$ is determined from an aperture current distribution $g(l)$

$$F(\theta) = \int_{-a}^{a} g(l)e^{j\beta l \, \cos\theta} \, dl \quad (5.3\text{-}4)$$

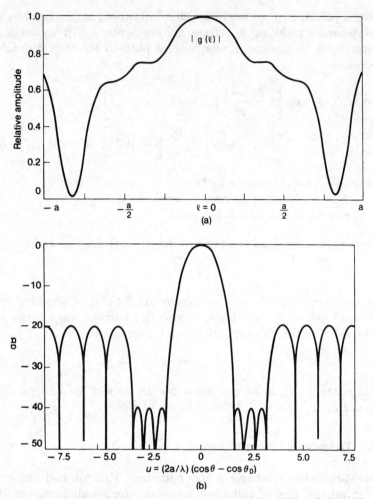

Figure 5.4 A symmetrical sum pattern (*reprinted from Elliot, Ref. 8*). (*a*) The aperture distribution for the two-antenna arrangement shown in Fig. 5.11*b* (© *1976 IEEE; reprinted from IEEE AP Transactions, 1976, pp. 76–83*). (*b*) The evolution of a symmetrical sum pattern with reduced inner sidelobes (© *1976 IEEE; reprinted from IEEE AP Transactions, 1976, pp. 76–83*).

Symmetrical pattern.[8] For production of a symmetrical pattern at the mainbeam, the current-amplitude distribution $|g(l)|$ is the only factor to consider. The phase of the current distribution can remain constant. A typical pattern (Fig. 5.4*a*) would be generated from a current-amplitude distribution (Fig. 5.4*b*).

Asymmetrical pattern. For production of an asymmetrical pattern, both current amplitude $|g(l)|$ and phase $\arg g(l)$ should be considered.

5.3.3 Synthesis of difference patterns
(Bayliss synthesis)[9]

To find a continuous line source that will produce a symmetrical difference pattern, with twin mainbeam patterns and specified sidelobes, we can set

$$D(\theta) = \int_{-a}^{a} g(l)e^{j\beta l \, \cos \, \theta} \, dl \qquad (5.3-5)$$

For a desired difference pattern such as that shown in Fig. 5.5a, the current-amplitude distributions $|g(l)|$ should be designed as shown in Fig. 5.5b and the phase arg $g(l)$ as shown in Fig. 5.5c.

5.3.4 Null-free patterns

In mobile communications applications, field-strength patterns without nulls are preferred for the antennas in a vertical plane. The typical vertical pattern of most antennas is shown in Fig. 5.6a. The field pattern can be represented as

$$F(u) = \sum_{n=0}^{N} K_n \frac{\sin \pi u}{\pi u} \qquad (5.3-6)$$

where $u = (2a/\lambda)(\cos \theta - \cos \theta_n)$. The concept is to add all $(\sin \pi u)/(\pi u)$ patterns at different pointing angles as shown in Fig. 5.6a. K_n is the maximum signal level. The resulting pattern does not contain nulls. The null-free pattern can be applied in the field as shown in Fig. 5.6b.

5.3.5 Practical applications

In designing a collinear array for the high-gain omnidirectional antenna, it is possible to control the current distribution by a microprocessor. This synthesis technique awaits further investigation by antenna research-and-development (R&D) specialists.

5.4 Antennas at Cell Site

5.4.1 For coverage use—
omnidirectional antennas

High-gain antennas. There are standard 6-dB and 9-dB gain omnidirectional antennas. The antenna patterns for 6-dB gain and 9-dB gain are shown in Fig. 5.7.

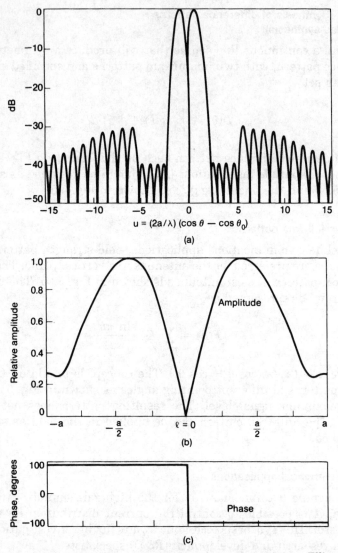

Figure 5.5 A symmetrical difference pattern (*reprinted from Elliot, Ref. 9*). (*a*) A modified Bayliss difference pattern; inner sidelobes symmetrically depressed. (© *IEEE; reprinted from IEEE AP Transactions, 1976, pp. 310–316*). (*b, c*) Aperture distribution for the pattern. (© *IEEE; reprinted from IEEE AP Transactions, 1976, pp. 310–316*).

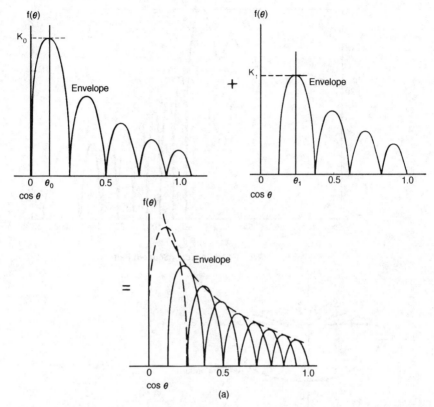

Figure 5.6 Null-free patterns. (*a*) Formation of a null-free pattern; (*b*) application of a null-free pattern (*reprinted from Elliott, Ref. 4, p. 192*).

Start-up system configuration. In a start-up system, an omnicell, in which all the transmitting antennas are omnidirectional, is used. Each transmitting antenna can transmit signals from 16 radio transmitters simultaneously using a 16-channel combiner. Each cell normally can have three transmitting antennas which serve 45 voice radio transmitters* simultaneously. Each sending signal is amplified by its own channel amplifier in each radio transmitter, then 16 channels (radio signals) pass through a 16-channel combiner and transmit signals by means of a transmitting antenna (see Fig. 5.8*a*).

Two receiving antennas commonly can receive all 45 voice radio signals simultaneously. Then in each channel, two identical signals re-

---

* The combiner is designed for combining 16 voice channels. However, the cellular system divides its 312 voice channels into 21 sets; each set consists of only about 15 voice channels. Therefore the dummy loads have to be put on some empty ports of a 16-channel combiner.

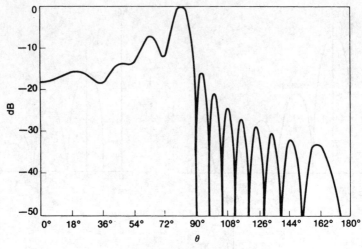

Null-Filled Cosecant Squared Pattern

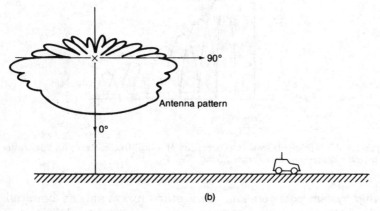

(b)

Figure 5.6 (*Continued*)

ceived by two receiving antennas pass through a diversity receiver of that channel. The receiving antenna configuration on the antenna mast is shown in Fig. 5.8. The separation of antennas for a diversity receiver is discussed in Sec. 5.5.

Abnormal antenna configuration. Usually, the call traffic in each cell increases as the number of customers increases. Some cells require a greater number of radios to handle the increasing traffic. An omnicell site can be equipped with up to 90 voice radios. In such cases six transmitting antennas should be used as shown in Fig. 5.8b. In the meantime, the number of receiving antennas is still two. In order to reduce the number of transmitting antennas, a hybrid ring combiner

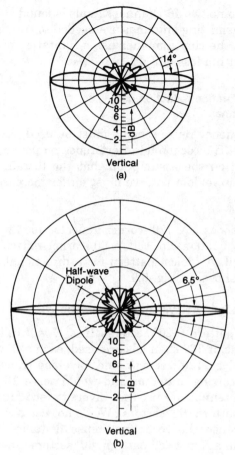

Vertical
(a)

Vertical
(b)

Figure 5.7 High-gain omnidirectional antennas (*reprinted from Kathrein Mobile Communications Catalog*). Gain with reference to dipole: (*a*) 6 dB; (*b*) 9 dB.

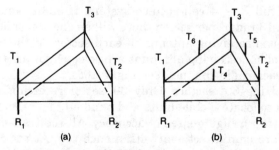

(a)   (b)

Figure 5.8 Cell-site antennas for omnicells: (*a*) for 45 channels; (*b*) for 90 channels.

which can combine two 16-channel signals is found.[10] This means that only three transmitting antennas are needed to transmit 90 radio signals. However, the ring combiner has a limitation of handling power up to 600 W with a loss of 3 dB.

5.4.2 For interference reduction use— directional antennas

When the frequency reuse scheme must be used, cochannel interference will occur. The cochannel interference reduction factor $q = D/R$ = 4.6 is based on the assumption that the terrain is flat. Because actual terrain is seldom flat, we must either increase q or use directional antennas.

Directional antennas. A 120°-corner reflector or 120°-plane reflector can be used in a 120°-sector cell. A 60°-corner reflector can be used in a 60°-sector cell. A typical pattern for a directional antenna of 120° beamwidth is shown in Fig. 5.9.

Normal antenna (mature system) configuration

1. $K = 7$ cell pattern (120° sectors). In a $K = 7$ cell pattern for frequency reuse, if 333 channels are used, each cell would have about 45 radios. Each 120° sector would have one transmitting antenna and two receiving antennas and would serve 16 radios. The two receiving antennas are used for diversity (see Fig. 5.10a).
2. $K = 4$ cell pattern (60° sectors). We do not use $K = 4$ in an omnicell system because the cochannel reuse distance is not adequate. Therefore, in a $K = 4$ cell pattern, 60° sectors are used.[11] There are 24 sectors. In this $K = 4$ cell-pattern system, two approaches are used.
   a. Transmitting-receiving 60° sectors. Each sector has a transmitting antenna carrying its own set of frequency radios and hands off frequencies to other neighboring sectors or other cells. This is a full $K = 4$ cell-pattern system. If 333 channels are used, with 13 radios per sector, there will be one transmitting antenna and one receiving antenna in each sector. At the receiving end, two of six receiving antennas are selected for an angle diversity for each radio channel (see Fig. 5.10b).
   b. Receiving 60° sectors. Only 60°-sector receiving antennas are used to locate mobile units and hand off to a proper neighboring cell with a high degree of accuracy. All the transmitting antennas are omnidirectional within each cell. At the receiving end, the angle diversity for each radio channel is also used in this case.

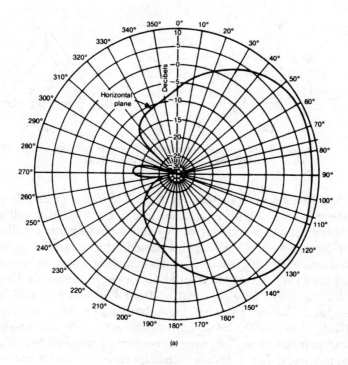

(a)

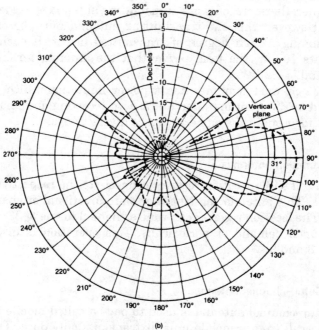

(b)

Figure 5.9 A typical 8-dB directional antenna pattern. (*Reprinted from Bell System Technical Journal, Vol. 58, January 1979, pp. 224–225.*) (*a*) Azimuthal pattern of 8-dB directional antenna. (*b*) Vertical pattern of 8-dB directional antenna.

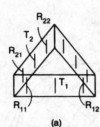 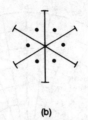 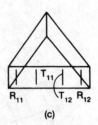

(a)   (b)   (c)

Figure 5.10 Directional antenna arrangement: (a) 120° sector (45 radios); (b) 60° sector; (c) 120° sector (90 radios).

Abnormal antenna configuration. If the call traffic is gradually increasing, there is an economic advantage in using the existing cell systems rather than the new splitting cell system (splitting into smaller cells). In the former, each site is capable of adding more radios. In a $K = 7$ cell pattern with 120° sectors, two transmitting antennas at each sector are used (Fig. 5.10c). Each antenna serves 16 radios if a 16-channel combiner is used. One observation from Fig. 5.10c should be mentioned here. The two transmitting antennas in each sector are placed relatively closer to the receiving antennas than in the single transmitting antenna case. This may cause some degree of desensitization in the receivers. The current technology can combine 32 channels[10] in a combiner; therefore, only one transmitting antenna is needed in each sector. However, this one transmitting antenna must be capable of withstanding a high degree of transmitted power. If each channel transmits 100 W, the total power that the antenna terminal could withstand is 3.2 kW.

The 32-channel combiner has a power limitation which would be specified by different manufacturers. Two receiving antennas in each 120° sector remain the same for space diversity use.

5.4.3 Location antennas

In each cell site a location receiver connects to the respective location antenna. This antenna can be either omnidirectional or shared-directional. The location receiver can tune a channel to one of 333 channels either upon demand or periodically. This operation is discussed in Chaps. 8 and 9.

5.4.4 Setup-channel antennas

The setup-channel antenna is used to page a called mobile unit or to access a call from a mobile unit. It transmits only data. The setup–channel antenna can be an omnidirectional antenna or consist of sev-

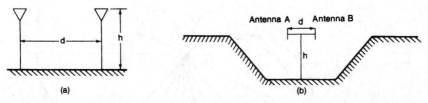

Figure 5.11 Diversity antenna spacing at the cell site: (a) $\eta = h/d$; (b) proper arrangement with two antennas.

eral directional antennas at one cell site. In general, in both omnicell and sector-cell systems, one omnidirectional antenna is used for transmitting signals and another for receiving signals in each cell site. Setup–channel operational procedures are discussed in Chap. 8.

5.4.5 Space-diversity antennas used at cell site

Two-branch space-diversity antennas are used at the cell site to receive the same signal with different fading envelopes, one at each antenna. The degree of correlation between two fading envelopes is determined by the degree of separation between two receiving antennas. When the two fading envelopes are combined, the degree of fading is reduced; this improvement is discussed in Ref. 12. Here the antenna setup is shown in Fig. 5.11a. Equation (5.4-1) is presented as an example for the designer to use.

$$\eta = \frac{h}{D} = 11 \tag{5.4-1}$$

where h is the antenna height and D is the antenna separation. From Eq. (5.4-1), the separation $d \geq 8\lambda$ is needed for an antenna height of 100 ft (30 m) and the separation $d \geq 14\lambda$ is needed for an antenna height of 150 ft (50 m). In any omnicell system, the two space-diversity antennas should be aligned with the terrain, which should have a U shape[13] as shown in Fig. 5.11b.

Space-diversity antennas can separate only horizontally, not vertically; thus, there is no advantage in using a vertical separation in the design.[13] The use of space-diversity antennas at the base station is discussed in detail in Ref. 13.

5.4.6 Umbrella-pattern antennas

In certain situations, umbrella-pattern antennas should be used for the cell-site antennas.

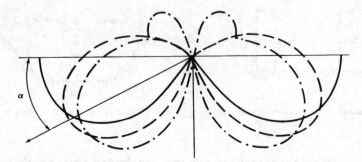

Figure 5.12 Vertical-plane patterns of quarter-wavelength stub antenna on infinite ground plane (solid) and on finite ground planes several wavelengths in diameter (dashed line) and about one wavelength in diameter (dotted line) (*after Kraus, Ref. 14*).

Normal umbrella-pattern antenna.[14] For controlling the energy in a confined area, the umbrella-pattern antenna can be developed by using a monopole with a top disk (top-loading) as shown in Fig. 5.12. The size of the disk determines the tilting angle of the pattern. The smaller the disk, the larger the tilting angle of the umbrella pattern.

Broadband umbrella-pattern antenna. The parameters of a *discone antenna* (a biconical antenna in which one of the cones is extended to 180° to form a disk) are shown in Fig. 5.13*a*. The diameter of the disk, the length of the cone, and the opening of the cone can be adjusted to create an umbrella-pattern antenna as described in Ref. 15.

High-gain broadband umbrella-pattern antenna. A high-gain antenna can be constructed by vertically stacking a number of umbrella-pattern antennas as shown in Fig. 5.13*b*

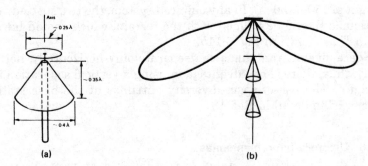

Figure 5.13 Discone antennas. (*a*) Single antenna. (*b*) An array of antennas.

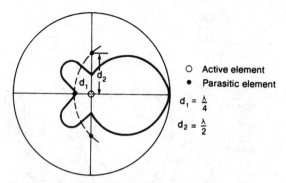

Figure 5.14 Application of parasitic elements (*after Jasik, Ref. 16*).

$$E_0 = \frac{\sin\ [(Nd/2\lambda)\cos\ \phi]}{\sin\ [(d/2\lambda)\ \cos\ \phi]} \cdot \text{(individual umbrella pattern)}$$

where ϕ = direction of wave travel
N = number of elements
d = spacing between two adjacent elements

5.4.7 Interference reduction antenna[16]

A design for an antenna configuration that reduces interference in two critical directions (areas) is shown in Fig. 5.14. The parasitic (insulation) element is about 1.05 times longer than the active element. The separation d and the various values of parasitic element reactance X_{22} were shown by Brown for this application.[17]

5.5 Unique Situations of Cell-Site Antennas

5.5.1 Antenna pattern in free space and in mobile environments

The antenna pattern we normally use is the one measured from an antenna range (open, nonurban area) or an antenna darkroom. However, when the antenna is placed in a surburban or urban environment and the mobile antenna is lower than the heights of the surroundings, the cell-site antenna pattern as a mobile unit received in a circle equidistant around the cell site is quite different from the free-space antenna pattern. Consider the following facts in the mobile radio environment.

1. The strongest reception still coincides with the strongest signal strength of the directional antenna.

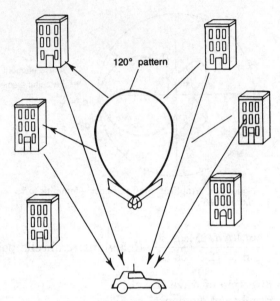

Figure 5.15 Front-to-back ratio of a directional antenna in a mobile radio environment.

2. The pattern is distorted in an urban or suburban environment.

3. For a 120° directional antenna, the backlobe (or front-to-back ratio) is about 10 dB less than the frontlobe, regardless of whether a weak sidelobe pattern or no sidelobe pattern is designed in a free-space condition. This condition exists because the strong signal radiates in front, bouncing back from the surroundings so that the energy can be received from the back of the antenna. The energy-reflection mechanism is illustrated in Fig. 5.15.

4. A design specification of the front-to-back ratio of a directional antenna (from the manufacturer's catalog) is different from the actual front-to-back ratio in the mobile radio environment. Therefore the environment and the antenna beamwidth determine how the antenna will be used in a mobile radio environment. For example, if a 60° directional antenna is used in a mobile radio environment, the actual front-to-back ratio can vary depending on the given environment. If the close-in man-made structures in front of the antenna are highly reflectable to the signal, then the front-to-back ratio of a low-master directional antenna can be as low as 6 dB in some circumstances. In this case, the directional antenna beamwidth pattern has no correlation between it measured in the free space and it measured in the mobile radio environment. If all the buildings are far away from the

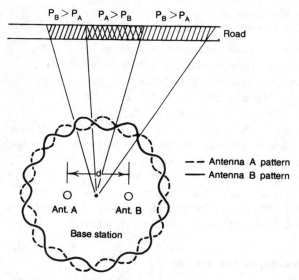

Figure 5.16 Antenna pattern ripple effect.

directional antenna, then the front-to-back ratio measured in the field will be close to the specified antenna pattern, usually 20 dB.

5.5.2 Minimum separation of cell-site receiving antennas

Separation between two transmitting antennas should be minimized to avoid the intermodulation discussed in Chap. 7. The minimum separation between a transmitting antenna and a receiving antenna necessary to avoid receiver desensitization is also described in Chap. 7. Here we are describing a minimum separation between two receiving antennas to reduce the antenna pattern ripple effects.

The two receiving antennas are used for a space-diversity receiver. Because of the near-field disturbance due to the close spacing, ripples will form in the antenna patterns (Fig. 5.16). The difference in power reception between two antennas at different angles of arrival is shown in Fig. 5.16. If the antennas are located closer, the difference in power between two antennas at a given pointing angle increases. Although the power difference is confined to a small sector, it affects a large section of the street as shown in Fig. 5.16. If the power difference is excessive, use of a space diversity will have no effect reducing fading. At 850 MHz, the separation of eight wavelengths between two receiving antennas creates a power difference of ±2 dB, which is tolerable for the advantageous use of a diversity scheme.[18]

5.5.3 Regular check of
the cell-site antennas

Air-pressurized cable is often used in cell-site antennas to prevent moisture from entering the cable and causing excessive attenuation. One method of checking the cell-site antennas is to measure the power delivered to the antenna terminal; however, few systems have this capability. The other method is to measure the VSWR at the bottom of the tower. In this case the loss of reflected power due to the cable under normal conditions should be considered. For a high tower, the VSWR reading may not be accurate.

If each cable connector has 1-dB loss due to energy leakage and two midsection 1-dB loss connectors are used in the transmitted system as shown in Fig. 5.17, the reflected power P_b indicated in the VSWR would be 4 dB less than the real reflected power.

5.5.4 Choosing an antenna site

In antenna site selection we have relied on the point-to-point prediction method (discussed in Chap. 4), which is applicable primarily for coverage patterns under conditions of light call traffic in the system. Reduction of interference is an important factor in antenna site selection.

When a site is chosen on the map, there is a 50 percent chance that the site location cannot be acquired. A written rule states that[24] an antenna location can be found within a quarter of the size of cell $R/4$. If the site is an 8-mi cell, the antenna can be located within a 2-mi radius. This hypothesis is based on the simulation result that the change in site within a 2-mi radius would not affect the coverage pattern at a distance 8 mi away. If the site is a 2-mi cell, the antenna can be located within a 0.5-mi radius.

The quarter-radius rule can be applied only on relatively flat terrain, not in a hilly area. To determine whether this rule can be applied in a general area, one can use the point-to-point prediction method to plot the coverage at different site locations and compare the differences. Usually when the point-to-point prediction method (tool) can be used to design a system, the quarter-radius rule becomes useless.

5.6 Mobile Antennas

The requirement of a mobile (motor-vehicle–mounted) antenna is an omnidirectional antenna which can be located as high as possible from the point of reception. However, the physical limitation of antenna height on the vehicle restricts this requirement. Generally the an-

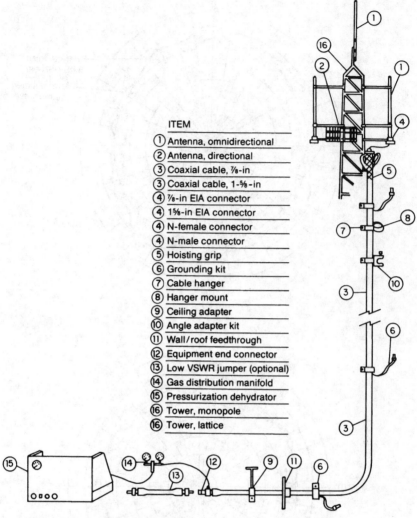

ITEM

①	Antenna, omnidirectional
②	Antenna, directional
③	Coaxial cable, ⅞-in
③	Coaxial cable, 1-⅝-in
④	⅞-in EIA connector
④	1⅝-in EIA connector
④	N-female connector
④	N-male connector
⑤	Hoisting grip
⑥	Grounding kit
⑦	Cable hanger
⑧	Hanger mount
⑨	Ceiling adapter
⑩	Angle adapter kit
⑪	Wall/roof feedthrough
⑫	Equipment end connector
⑬	Low VSWR jumper (optional)
⑭	Gas distribution manifold
⑮	Pressurization dehydrator
⑯	Tower, monopole
⑯	Tower, lattice

Figure 5.17 Antenna system at cell site.

tenna should at least clear the top of the vehicle. Patterns for two types of mobile antenna are shown in Fig. 5.18.

5.6.1 Roof-mounted antenna

The antenna pattern of a roof-mounted antenna is more or less uniformly distributed around the mobile unit when measured at an antenna range in free space as shown in Fig. 5.19. The 3-dB high-gain

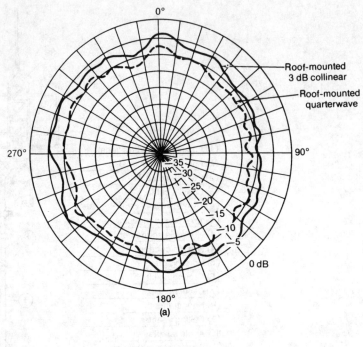

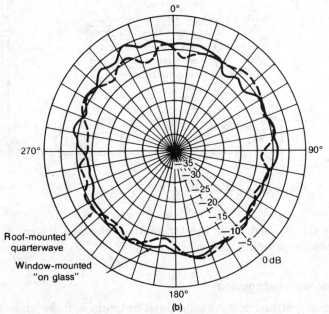

Figure 5.18 Mobile antenna patterns (*from Antenna Specialist Co., Ref. 10*). (*a*) Roof-mounted 3-dB-gain collinear antenna versus roof-mounted quarter-wave antenna. (*b*) Window-mounted "on-glass" gain antenna versus roof-mounted quarter-wave antenna.

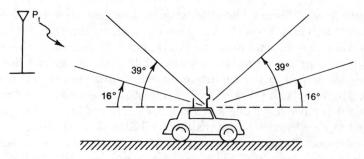

Figure 5.19   Vertical angle of signal arrival.

antenna shows a 3-dB gain over the quarter-wave antenna. However, the gain of the antenna used at the mobile unit must be limited to 3 dB because the cell-site antenna is rarely as high as the broadcasting antenna and out-of-sight conditions often prevail. The mobile antenna with a gain of more than 3 dB can receive only a limited portion of the total multipath signal in the elevation as measured under the out-of-sight condition.[19] This point is discussed in detail in Sec. 5.6.3.

5.6.2   Glass-mounted antennas[10,20]

There are many kinds of glass-mounted antennas. Energy is coupled through the glass; therefore, there is no need to drill a hole. However, some energy is dissipated on passage through the glass. The antenna gain range is 1 to 3 dB depending on the operating frequency.

The position of the glass-mounted antenna is always lower than that of the roof-mounted antenna; generally there is a 3-dB difference between these two types of antenna. Also, glass-mounted antennas cannot be installed on the shaded glass found in some motor vehicles because this type of glass has a high metal content.

5.6.3   Mobile high-gain antennas

A high-gain antenna used on a mobile unit has been studied.[19] This type of high-gain antenna should be distinguished from the directional antenna. In the directional antenna, the antenna beam pattern is suppressed horizontally; in the high-gain antenna, the pattern is suppressed vertically. To apply either a directional antenna or a high-gain antenna for reception in a radio environment, we must know the origin of the signal. If we point the directional antenna opposite to the transmitter site, we would in theory receive nothing.

In a mobile radio environment, the scattered signals arrive at the mobile unit from every direction with equal probability. That is why an omnidirectional antenna must be used. The scattered signals also arrive from different elevation angles. Lee and Brandt[19] used two types of antenna, one $\lambda/4$ whip antenna with an elevation coverage of 39° and one 4-dB-gain antenna (4-dB gain with respect to the gain of a dipole) with an elevation coverage of 16°, and measured the angle of signal arrival in the suburban Keyport-Matawan area of New Jersey. There are two types of test: a line-of-sight condition and an out-of-sight condition. In Lee and Brandt's study the transmitter was located at an elevation of approximately 100 m (300 ft) above sea level. The measured areas were about 12 m (40 ft) above sea level and the path length about 3 mi. The received signal from the 4-dB-gain antenna was 4 dB stronger than that from the whip antenna under line-of-sight conditions. This is what we would expect. However, the received signal from the 4-dB-gain antenna was only about 2 dB stronger than that from the whip antenna under out-of-sight conditions. This is surprising.

The reason for the latter observation is that the scattered signals arriving under out-of-sight conditions are spread over a wide elevation angle. A large portion of the signals outside the elevation angle of 16° cannot be received by the high-gain antenna. We may calculate the portion being received by the high-gain antenna from the measured beamwidth [the beamwidth can be roughly obtained from Eq. (5.2-6)]. For instance, suppose that a 4:1 gain (6 dBi) is expected from the high-gain antenna, but only 2.5:1 is received. Therefore, 63 percent of the signal* is received by the 4-dB-gain antenna (i.e., 6 dBi) and 37 percent is felt in the region between 16 and 39°. Consider the data in the following table.

	Gain, dBi	Linear ratio	$\theta_0/2$, degrees
Whip antenna (2 dB above isotropic)	2	1.58:1	39
High-gain antenna	6	4:1	16
Low-gain antenna	4	2.5:1	24

Therefore, a 2- to 3-dB-gain antenna (4 to 5 dBi) should be adequate for general use. An antenna gain higher than 2 to 3 dB does not serve the purpose of enhancing reception level. Moreover, measurements reveal that the elevation angle for scattered signals received in urban areas is greater than that in suburban areas.

5.6.4   Horizontally oriented space-diversity antennas

A two-branch space-diversity receiver mounted on a motor vehicle has the advantage of reducing fading and thus can operate at a lower

---

* For a Rayleigh fading signal, 63 percent will be below its power level.

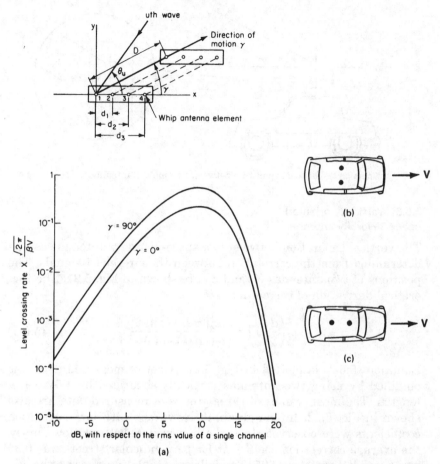

Figure 5.20 Horizontally spaced antennas. (*a*) Maximum difference in lcr of a four-branch equal-gain signal between $\alpha = 0$ and $\alpha = 90°$ with antenna spacing of 0.15λ. (*b*) Not recommended. (*c*) Recommended.

reception level. The advantage of using a space-diversity receiver to reduce interference is discussed in Chap. 7. The discussion here concerns a space-diversity scheme in which two vehicle-mounted antennas separated horizontally by 0.5λ wavelength[21] (15 cm or 6 in) can achieve the advantage of diversity.

We must consider the following factor. The two antennas can be mounted either in line with or perpendicular to the motion of the vehicle. Theoretical analyses and measured data indicate that the in-line arrangement of the two antennas produces fewer level crossings, that is, less fading, than the perpendicular arrangement does. The level crossing rates of two signals received from different horizontally oriented space-diversity antennas are shown in Fig. 5.20.

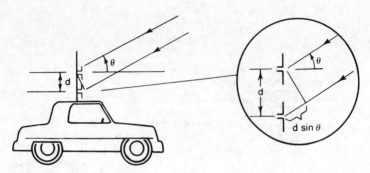

Figure 5.21 Vertical separation between two mobile antennas.

5.6.5 Vertically oriented
space-diversity antennas[22]

The vertical separation between two space-diversity antennas can be determined from the correlation between their received signals. The positions of two antennas X_1 and X_2 are shown in Fig. 5.21. The theoretical derivation of correlation is[23]

$$\rho\left(\frac{d}{\lambda}, \theta\right) = \frac{\sin[(\pi d/\lambda)\sin\theta]}{(\pi d/\lambda)\sin\theta} \qquad (5.6\text{-}1)$$

Equation (5.6-1) is plotted in Fig. 5.22. A set of measured data was obtained by using two antennas vertically separated by 1.5λ wavelengths. The mean values of three groups of measured data are also shown in Fig. 5.22. In one group, in New York City, low correlation coefficients were observed. In two other groups, both in New Jersey, the average correlation coefficient for perpendicular streets was 0.35 and for radial streets, 0.225. The following table summarizes the correlation coefficients in different areas and different street orientations.

	Correlation coefficient	
Area	Average	Standard deviation
New York City	0.1	0.06
Suburban New Jersey		
Radial streets	0.226	0.127
Perpendicular streets	0.35	0.182

From Fig. 5.22 we can also see that the signal arrives at an elevation angle of 29° in the suburban radial streets and 33° in the suburban perpendicular streets. In New York City the angle of arrival approaches 40°.

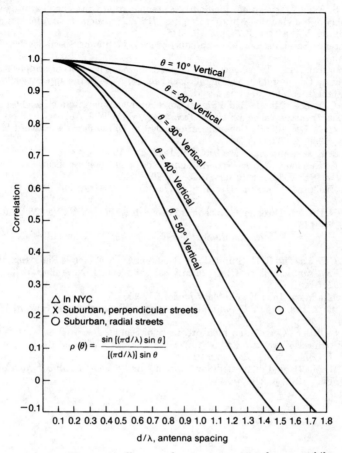

Figure 5.22 Two vertically spaced antennas mounted on a mobile unit.

References

1. E. C. Jordan (ed.), *Reference Data for Engineers: Radio, Electronics, Computer, and Communications,* 7th ed., Howard W. Sams & Co., 1985, pp. 32–34.
2. J. D. Kraus, *Antennas,* McGraw-Hill Book Co., 1950, p. 44.
3. J. D. Kraus, *Antennas,* McGraw-Hill Book Co., 1950, p. 78.
4. R. S. Elliott, *Antenna Theory and Design,* Prentice-Hall, Inc., 1981, p. 206.
5. J. D. Kraus, *Antennas,* McGraw-Hill Book Co., 1950, p. 25.
6. R. S. Elliott, *Antenna Theory and Design,* Prentice-Hall, Inc., 1981, p. 157.
7. C. L. Dolph, "A Current Distribution for Broadside Arrays Which Optimizes the Relationship between Beamwidth and Side-Lobe Level," *Proceedings of the IRE,* Vol. 34, No. 6, June 1946, pp. 335–348.
8. R. S. Elliott, "Design of Live-Source Antennas for Sum Patterns with Sidelobes of Individually Arbitrary Heights," *IEEE Transactions on Antennas and Propagation,* Vol. AP-24, 1976, pp. 76–83.

9. R. S. Elliott, "Design of Line-Source Antennas from Difference Patterns with Side Lobes of Individually Arbitrary Heights," *IEEE Transactions on Antennas and Propagation,* Vol. AP-24, 1976, pp. 310–316.
10. Antenna Specialist Co. catalog product branches, Antenna Specialist Co., Cleveland, Ohio.
11. American Radio Telephone Service (ARTS) development license application to the Federal Communications Commission, Feb. 14, 1977. In the application, a proposed system from Motorola was described.
12. W. C. Y. Lee, "Mobile Radio Signal Correlation versus Antenna Height and Space," *IEEE Transactions on Vehicular Technology,* Vol. VT-25, August 1977, pp. 290–292.
13. W. C. Y. Lee, *Mobile Communications Design Fundamentals,* Howard W. Sams & Co., 1986, p. 202.
14. J. D. Kraus, *Antennas,* McGraw-Hill Book Co., 1950, p. 421.
15. A. G. Kandoian, "Three New Antenna Types and Their Applications," *Proceedings of the IRE,* Vol. 34, February 1946, pp. 70W–75W.
16. H. Jasik (ed.), *Antenna Engineering Handbook,* McGraw-Hill Book Co., 1961, pp. 5–7.
17. G. H. Brown, "Directional Antennas," *Proceedings of the IRE,* Vol. 25, January 1937, pp. 75–145.
18. W. C. Y. Lee, *Mobile Communications Engineering,* McGraw-Hill Book Co., 1982, p. 152.
19. W. C. Y. Lee and R. H. Brandt, "The Elevation Angle of Mobile Radio Signal Arrival," *IEEE Transactions on Communications,* Vol. Com-21, November 1973, pp. 1194–1197.
20. Mobile Mark, Inc., Mobile Mark Model OW-900.
21. W. C. Y. Lee, *Mobile Communications Design Fundamentals,* Howard W. Sams & Co., 1986, p. 222.
22. J. S. Bitler, "Correlation Measurements of Signals Received on Vertically Spaced Antennas," Microwave Radio Symposium, Boulder, Colorado, 1972.
23. M. J. Gans, private communications.
24. V. H. MacDonald, "The Cellular Concept," *Bell System Technical Journal,* Vol. 58, January 1979, p. 27.

Cochannel Interference Reduction

6.1 Cochannel Interference

The frequency-reuse method is useful for increasing the efficiency of spectrum usage but results in cochannel interference because the same frequency channel is used repeatedly in different cochannel cells. Application of the cochannel interference reduction factor $q = D/R = 4.6$ for a seven-cell reuse pattern ($K = 7$) is described in Sec. 2.4.[1]

In most mobile radio environments, use of a seven-cell reuse pattern is not sufficient to avoid cochannel interference. Increasing $K > 7$ would reduce the number of channels per cell, and that would also reduce spectrum efficiency. Therefore, it might be advisable to retain the same number of radios as the seven-cell system but to sector the cell radially, as if slicing a pie. This technique would reduce cochannel interference and use channel sharing and channel borrowing schemes to increase spectrum efficiency.

6.2 Exploring Cochannel Interference Areas in a System

Problems in mobile telephone coverage (service), particularly holes (weak signal strength*) which result in call drops during the customer's conversation, have been partially solved by applying the propagation (wave motion) studies discussed in Chap. 4 for the case where no cochannel interference exists.

---

* Signal strength is measured in dBm, and field strength is measured in dB(μV/m). The conversion between these units can be found in Sec. 5.1.3.

When customer demand increases, the channels, which are limited in number, have to be repeatedly reused in different areas, which provides many cochannel cells, which increases the system's capacity. But cochannel interference may be the result. In this situation, the received voice quality is affected by both the grade of coverage and the amount of cochannel interference. For detection of serious channel interference areas in a cellular system, two tests are suggested.

Test 1—find the cochannel interference area from a mobile receiver. Cochannel interference which occurs in one channel will occur equally in all the other channels in a given area. We can then measure cochannel interference by selecting any one channel (as one channel represents all the channels) and transmitting on that channel at all cochannel sites at night while the mobile receiver is traveling in one of the cochannel cells.

While performing this test we watch for any change detected by a field-strength recorder in the mobile unit and compare the data with the condition of no cochannel sites being transmitted. This test must be repeated as the mobile unit travels in every cochannel cell. To facilitate this test, we can install a channel scanning receiver in one car.

One channel (f_1) records the signal level (no-cochannel condition), another channel (f_2) records the interference level (six-cochannel condition is the maximum), while the third channel receives f_3, which is not in use. Therefore, the noise level is recorded only in f_3 (see Fig. 6.1).

We can obtain, in decibels, the carrier-to-interference ratio C/I by subtracting the result obtained from f_2 from the result obtained from f_1 (carrier minus interference $C - I$) and the carrier-to-noise ratio C/N by subtracting the result obtained from f_3 from the result obtained from f_2 (carrier minus noise $C - N$). Four conditions should be used to compare the results.

1. If the carrier-to-interference ratio C/I is greater than 18 dB throughout most of the cell, the system is properly designed.

2. If C/I is less than 18 dB and C/N is greater than 18 dB in some areas, there is cochannel interference.

3. If both C/N and C/I are less than 18 dB and $C/N \approx C/I$ in a given area, there is a coverage problem.

4. If both C/N and C/I are less than 18 dB and $C/N > C/I$ in a given area, there is a coverage problem *and* cochannel interference.

Test 2—find the cochannel interference area which affects a cell site. The reciprocity theorem can be applied for the coverage problem but not for cochannel interference. Therefore, we cannot assume that the first

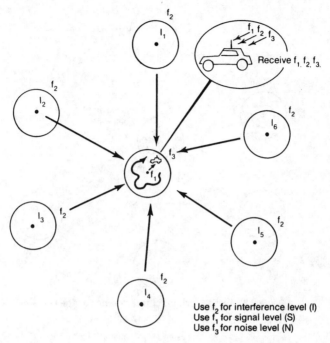

Figure 6.1 Test 1: cochannel interference at the mobile unit.

test result will apply to the second test condition. We must perform the second test as well.

Because it is difficult to use seven cars simultaneously, with each car traveling in each cochannel cell for this test, an alternative approach may be to record the signal strength at every cochannel cell site while a mobile unit is traveling either in its own cell or in one of the cochannel cells shown in Fig. 6.2.

First we find the areas in an interfering cell in which the top 10 percent level of the signal transmitted from the mobile unit in those areas is received at the desired site (*J*th cell in Fig. 6.1). This top 10 percent level can be distributed in different areas in a cell. The average value of the top 10 percent level signal strength is used as the interference level from that particular interfering cell. The mobile unit also travels in different interfering cells. Up to six interference levels are obtained from a mobile unit running in six interfering cells. We then calculate the average of the bottom 10 percent level of the signal strength which is transmitted from a mobile unit in the desired cell (*J*th cell) and received at the desired cell site as a carrier reception level.

Then we can reestablish the carrier-to-interference ratio received at a desired cell, say, the *J*th cell site as follows.

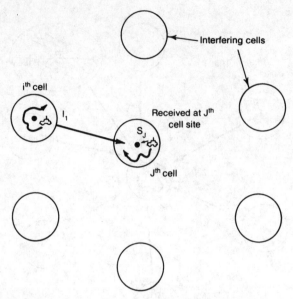

Figure 6.2 Test 2: cochannel interference at the cell site.

$$\frac{C_J}{I} = \frac{C_J}{\sum\limits_{\substack{i=1 \\ i \neq J}}^{6} I_i}$$

The number of cochannel cells in the system can be less than six. We must be aware that all C_J and I_i were read in decibels. Therefore, a translation from decibels to linear is needed before summing all the interfering sources. The test can be carried out repeatedly for any given cell. We then compare

$$\frac{C_J}{I} \quad \text{and} \quad \frac{C_J}{N_J}$$

and determine the cochannel interference condition, which will be the same as that in test 1. N_J is the noise level in the Jth cell assuming no interference exists.

6.3 Real-Time Cochannel Interference Measurement at Mobile Radio Transceivers

When the carriers are angularly modulated by the voice signal and the RF frequency difference between them is much higher than the fading frequency, measurement of the signal carrier-to-interference ratio C / I reveals that the signal is

$$e_1 = S(t) \sin(\omega t + \phi_1) \qquad (6.3\text{-}1)$$

and the interference is

$$e_2 = I(t) \sin(\omega t + \phi_2) \qquad (6.3\text{-}2)$$

The received signal is

$$e(t) = e_1(t) + e_2(t) = R \sin(\omega t + \psi) \qquad (6.3\text{-}3)$$

where

$$R = \sqrt{[S(t) \cos \phi_1 + I(t) \cos \phi_2]^2 + [S(t) \sin \phi_1 + I(t) \sin \phi_2]^2}$$

$$(6.3\text{-}4)$$

and
$$\psi = \tan^{-1} \frac{S(t) \sin \phi_1 + I(t) \sin \phi_2}{S(t) \cos \phi_1 + I(t) \cos \phi_2} \qquad (6.3\text{-}5)$$

The envelope R can be simplified in Eq. (6.3-4), and R^2 becomes

$$R^2 = \{S^2(t) + I^2(t) + 2S(t)I(t) \cos(\phi_1 - \phi_2)\} \qquad (6.3\text{-}6)$$

Following Kozono and Sakamoto's[2] analysis of Eq. (6.3-6) the term $S^2(t) + I^2(t)$ fluctuates close to the fading frequency V/λ and the term $2S(t)I(t) \cos(\phi_1 - \phi_1)$ fluctuates to a frequency close to $d / dt(\phi_1 - \phi_2)$, which is much higher than the fading frequency. Then the two parts of the squared envelope can be separated as

$$X = S^2(t) + I^2(t) \qquad (6.3\text{-}7)$$

$$Y = 2S(t)I(t) \cos(\phi_1 - \phi_2) \qquad (6.3\text{-}8)$$

Assume that the random variables $S(t)$, $I(t)$, ϕ_1, and ϕ_2 are independent; then the average processes on X and Y are

$$\overline{X} = \overline{S^2(t)} + \overline{I^2(t)} \qquad (6.3\text{-}9)$$

$$\overline{Y^2} = 4\overline{S^2(t)}\overline{I^2(t)}(\tfrac{1}{2}) = 2\overline{S^2(t)}\overline{I^2(t)} \qquad (6.3\text{-}10)$$

The signal-to-interference ratio Γ becomes

$$\Gamma = \frac{\overline{S^2(t)}}{\overline{I^2(t)}} = k + \sqrt{k^2 - 1} \qquad (6.3\text{-}11)$$

where
$$k = \frac{\overline{X}^2}{\overline{Y}^2} - 1 \qquad (6.3\text{-}12)$$

Since X and Y can be separated in Eq. (6.3-6), the preceding computation of Γ in Eq. (6.3-11) could have been accomplished by means of an envelope detector, and analog-to-digital converter, and a microcomputer. The sampling delay time Δt should be small enough to satisfy

$$S(t) \approx S(t + \Delta t), \qquad I(t) \approx I(t + \Delta t) \qquad (6.3\text{-}13)$$

and
$$\overline{\cos\,[\phi_1(t) - \phi_2(t)]\,\cos\,[\phi_1(t + \Delta t) - \phi_2(t + \Delta t)]} \approx 0 \qquad (6.3\text{-}14)$$

Determining the delay time Δt to meet the requirement of Eq. (6.3-13) for this calculation is difficult and is a drawback to this measurement technique. Therefore, real-time cochannel interference measurement is difficult to achieve in practice.

6.4   Design of an Omnidirectional Antenna System in the Worst Case

In Sec. 2.4 we proved that the value of $q = 4.6$ is valid for a normal interference case in a $K = 7$ cell pattern.[3] In this section we would like to prove that a $K = 7$ cell pattern does not provide a sufficient frequency-reuse distance separation even when an ideal condition of flat terrain is assumed. The worst case is at the location where the mobile unit would receive the weakest signal from its own cell site but strong interferences from all interfering cell sites.

In the worst case the mobile unit is at the cell boundary R, as shown in Fig. 6.3. The distances from all six cochannel interfering sites are also shown in the figure: two distances of $D - R$, two distances of D, and two distances of $D + R$.

Following the mobile radio propagation rule of 40 dB/dec shown in Chap. 4, we obtain

$$C \propto R^{-4} \qquad I \propto D^{-4}$$

Then the carrier-to-interference ratio is

$$\frac{C}{I} = \frac{R^{-4}}{2(D - R)^{-4} + 2(D)^{-4} + 2(D + R)^{-4}}$$

$$= \frac{1}{2(q - 1)^{-4} + 2(q)^{-4} + 2(q + 1)^{-4}} \qquad (6.4\text{-}1a)$$

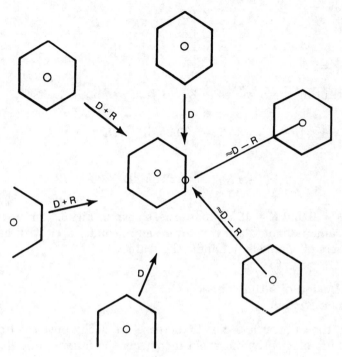

Figure 6.3 Cochannel interference (a worst case).

where $q = 4.6$ is derived from the normal case shown in Eq. (2.4-7). Substituting $q = 4.6$ into Eq. (6.4-1a), we obtain $C/I = 54$ or 17 dB, which is lower than 18 dB. To be conservative, we may use the short-est distance $D - R$ for all six interferers as a worst case; then Eq. (6.4-1a) is replaced by

$$\frac{C}{I} = \frac{R^{-4}}{6(D - R)^{-4}} = \frac{1}{6(q - 1)^{-4}} = 28 = 14.47 \text{ dB} \qquad (6.4\text{-}1b)$$

In reality, because of the imperfect site locations and the rolling nature of the terrain configuration, the C/I received is always worse than 17 dB and could be 14 dB and lower. Such an instance can easily occur in a heavy traffic situation; therefore, the system must be de-signed around the C/I of the worst case. In that case, a cochannel interference reduction factor of $q = 4.6$ is insufficient.

Therefore, in an omnidirectional-cell system, $K = 9$ or $K = 12$ would be a correct choice. Then the values of q are

$$q = \begin{cases} \dfrac{D}{R} = \sqrt{3K} \\ 5.2 \qquad K = 9 \\ 6 \qquad\; K = 12 \end{cases} \qquad (6.4\text{-}2)$$

Substituting these values in Eq. (6.4-1), we obtain

$$\frac{C}{I} = 84.5 \;(=)\; 19.25 \text{ dB} \qquad K = 9 \qquad (6.4\text{-}3)$$

$$\frac{C}{I} = 179.33 \;(=)\; 22.54 \text{ dB} \qquad K = 12 \qquad (6.4\text{-}4)$$

The $K = 9$ and $K = 12$ cell patterns, shown in Fig. 6.4, are used when the traffic is light. Each cell covers an adequate area with adequate numbers of channels to handle the traffic.

6.5   Design of a Directional Antenna System

When the call traffic begins to increase, we need to use the frequency spectrum efficiently and avoid increasing the number of cells K in a seven-cell frequency-reuse pattern. When K increases, the number of frequency channels assigned in a cell must become smaller (assuming a total allocated channel divided by K) and the efficiency of applying the frequency-reuse scheme decreases.

Instead of increasing the number K in a set of cells, let us keep $K = 7$ and introduce a directional-antenna arrangement. The cochannel interference can be reduced by using directional antennas. This means that each cell is divided into three or six sectors and uses three or six directional antennas at a base station. Each sector is assigned a set of frequencies (channels). The interference between two cochannel cells decreases as shown Fig. 6.5.

6.5.1   Directional antennas in $K = 7$ cell patterns

Three-sector case. The three-sector case is shown in Fig. 6.5. To illustrate the worst-case situation, two cochannel cells are shown in Fig. 6.6a. The mobile unit at position E will experience greater interference in the lower shaded cell sector than in the upper shaded cell-sector site. This is because the mobile receiver receives the weakest signal from its own cell but fairly strong interference from the interfering

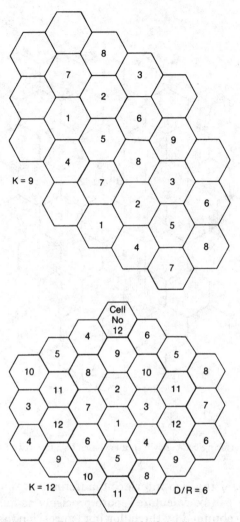

Figure 6.4 Interference with frequency-reuse patterns $K = 9$ and $K = 12$.

cell. In a three-sector case, the interference is effective in only one direction because the front-to-back ratio of a cell-site directional antenna is at least 10 dB or more in a mobile radio environment. The worst-case cochannel interference in the directional-antenna sectors in which interference occurs may be calculated. Because of the use of directional antennas, the number of principal interferers is reduced from six to two (Fig. 6.5). The worst case of C/I occurs when the mobile unit is at position E, at which point the distance between the

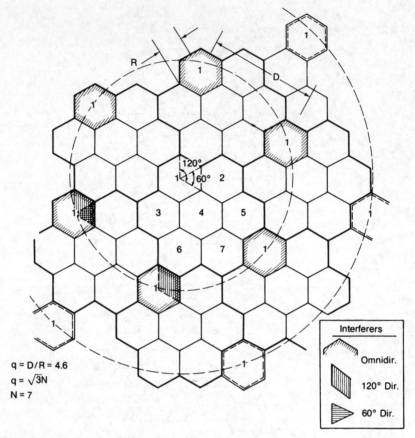

Figure 6.5 Interfering cells shown in a seven-cell system (two-tiers).

mobile unit and the two interfering antennas is roughly $D + (R/2)$; however, C/I can be calculated more precisely as follows.* The value of C/I can be obtained by the following expression (assuming that the worst case is at position E at which the distances from two interferers are $D + 0.7$ and D).

$$\frac{C}{I}\ (\text{worst case}) = \frac{R^{-4}}{(D + 0.7R)^{-4} + D^{-4}}$$

$$= \frac{1}{(q + 0.7)^{-4} + q^{-4}} \qquad (6.5\text{-}1)$$

---

* The difference in results between using a closed form and an approximate calculation is small.

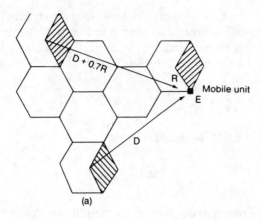

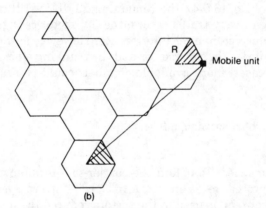

Figure 6.6 Determination of carrier-to-interference ratio C / I in a directional antenna system. (a) Worst case in a 120° directional antenna system ($N = 7$); (b) worst case in a 60° directional antenna system ($N = 7$).

Let $q = 4.6$; then Eq. (6.5-1) becomes

$$\frac{C}{I} \text{ (worst case)} = 285 \; (=) \; 24.5 \text{ dB} \qquad (6.5\text{-}2)$$

The C / I received by a mobile unit from the 120° directional antenna sector system expressed in Eq. (6.5-2) greatly exceeds 18 dB in a worst case. Equation (6.5-2) shows that using directional antenna sectors can improve the signal-to-interference ratio, that is, reduce the cochannel interference. However, in reality, the C / I could be 6 dB weaker than in Eq. (6.5-2) in a heavy traffic area as a result of irregular terrain contour and imperfect site locations. The remaining 18.5 dB is still adequate.

Six-sector case.   We may also divide a cell into six sectors by using six 60°-beam directional antennas as shown in Fig. 6.6b. In this case, only one instance of interference can occur in each sector as shown in Fig. 6.5. Therefore, the carrier-to-interference ratio in this case is

$$\frac{C}{I} = \frac{R^{-4}}{(D + 0.7R)^{-4}} = (q + 0.7)^4 \qquad (6.5\text{-}3)$$

For $q = 4.6$, Eq. (6.5-3) becomes

$$\frac{C}{I} = 794 \ (=) \ 29 \text{ dB} \qquad (6.5\text{-}4)$$

which shows a further reduction of cochannel interference. If we use the same argument as we did for Eq. (6.5-2) and subtract 6 dB from the result of Eq. (6.5-4), the remaining 23 dB is still more than adequate. When heavy traffic occurs, the 60°-sector configuration can be used to reduce cochannel interference. However, fewer channels are generally allowed in a 60° sector and the trunking efficiency decreases. In certain cases, more available channels could be assigned in a 60° sector.

6.5.2   Directional antenna in $K = 4$ cell pattern

Three-sector case.   To obtain the carrier-to-interference ratio, we use the same procedure as in the $K = 7$ cell-pattern system. The 120° directional antennas used in the sectors reduced the interferers to two as in $K = 7$ systems, as shown in Fig. 6.7. We can apply Eq. (6.5-1) here. For $K = 4$, the value of $q = \sqrt{3K} = 3.46$; therefore, Eq. (6.5-1) becomes

$$\frac{C}{I} \text{ (worst case)} = \frac{1}{(q + 0.7)^{-4} + q^{-4}} = 97 = 20 \text{ dB} \qquad (6.5\text{-}5)$$

If, using the same reasoning used with Eq. (6.5-4), 6 dB is subtracted from the result of Eq. (6.5-5), the remaining 14 dB is unacceptable.

Six-sector case.   There is only one interferer at a distance of $D + R$ shown in Fig. 6.7. With $q = 3.46$, we can obtain

$$\frac{C}{I} \text{ (worst case)} = \frac{R^{-4}}{(D + R)^{-4}} = \frac{1}{(q + 1)^{-4}} = 355 = 26 \text{ dB} \qquad (6.5\text{-}6)$$

If 6 dB is subtracted from the result of Eq. (6.5-6), the remaining 21

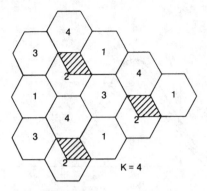

Figure 6.7 Interference with frequency-reuse pattern $K = 4$.

dB is adequate. Under heavy traffic conditions, there is still a great deal of concern over using a $K = 4$ cell pattern in a 60° sector. An explanation of this point is given in the next section.

6.5.3 Comparing $K = 7$ and $K = 4$ systems

A $K = 7$ cell-pattern system is a logical way to begin an omnicell system. The cochannel reuse distance is more or less adequate, according to the designed criterion. When the traffic increases, a three-sector system should be implemented, that is, with three 120° directional antennas in place. In certain hot spots, 60° sectors can be used locally to increase the channel utilization.

If a given area is covered by both $K = 7$ and $K = 4$ cell patterns and both patterns have a six-sector configuration, then the $K = 7$ system has a total of 42 sectors, but the $K = 4$ system has a total of only 26 sectors and, of course, the system of $K = 7$ and six sectors has less cochannel interference.

One advantage of 60° sectors with $K = 4$ is that they require fewer cell sites than 120° sectors with $K = 7$. Two disadvantages of 60° sectors are that (1) they require more antennas to be mounted on the antenna mast and (2) they often require more frequent handoffs because of the increased chance that the mobile units will travel across the six sectors of the cell. Furthermore, assigning the proper frequency channel to the mobile unit in each sector is more difficult unless the antenna height at the cell site is increased so that the mobile unit can be located more precisely. In reality the terrain is not flat, and coverage is never uniformly distributed; in addition, the directional antenna front-to-back power ratio in the field is very difficult to predict (see Sec. 5.4.2). In small cells, interference could become uncontrol-

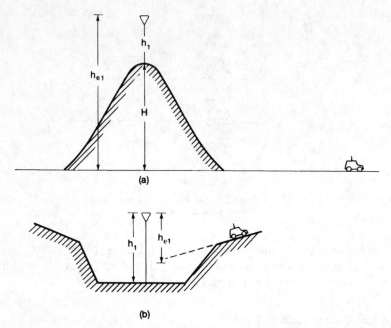

Figure 6.8 Lowering the antenna height (*a*) on a high hill and (*b*) in a valley.

lable; thus the use of a $K = 4$ pattern with 60° sectors in small cells needs to be considered only for special implementations such as portable cellular systems (Sec. 13.5) or narrowbeam applications (Sec. 10.6). For small cells, a better alternative scheme is to use a $K = 7$ pattern with 120° sectors plus the underlay-overlay configuration described in Sec. 11.4.

6.6   Lowering the Antenna Height

Lowering the antenna height does not always reduce the cochannel interference. In some circumstances, such as on fairly flat ground or in a valley situation, lowering the antenna height will be very effective for reducing the cochannel and adjacent-channel interference. However, there are three cases where lowering the antenna height may or may not effectively help reduce the interference.

On a high hill or a high spot.   The effective antenna height, rather than the actual height, is always considered in the system design. Therefore, the effective antenna height varies according to the location of the mobile unit, as described in Chap. 4. When the antenna site is on a hill, as shown in Fig. 6.8*a*, the effective antenna height is $h_1 + H$.

If we reduce the actual antenna height to $0.5h_1$, the effective antenna height becomes $0.5h_1 + H$. The reduction in gain resulting from the height reduction is

$$G = \text{gain reduction} = 20 \log_{10} \frac{0.5h_1 + H}{h_1 + H}$$

$$= 20 \log_{10} \left(1 - \frac{0.5h_1}{h_1 + H}\right) \qquad (6.6\text{-}1)$$

If $h_1 \ll H$, then Eq. (6.6-1) becomes

$$G \approx 20 \log_{10} 1 = 0 \text{ dB}$$

This simply proves that lowering antenna height on the hill does not reduce the received power at either the cell site or the mobile unit.

In a valley. The effective antenna height as seen from the mobile unit shown in Fig. 6.8*b* is h_{e1}, which is less than the actual antenna height h_1. If $h_{e1} = \frac{2}{3}h_1$ and the antenna is lowered to $\frac{1}{2}h_1$, then the new effective antenna height, determined from Chap. 4, is

$$h_{e1} = \tfrac{1}{2}h_1 - (h_1 - \tfrac{2}{3}h_1) = \tfrac{1}{6}h_1$$

Then the antenna gain is reduced by

$$G = 20 \log \frac{\frac{1}{6}h_1}{\frac{2}{3}h_1} = -12 \text{ dB}$$

This simply proves that the lowered antenna height in a valley is very effective in reducing the radiated power in a distant high elevation area. However, in the area adjacent to the cell-site antenna, the effective antenna height is the same as the actual antenna height. The power reduction caused by decreasing antenna height by half is only

$$20 \log \frac{\frac{1}{2}h_1}{h_1} = -6 \text{ dB}$$

In a forested area. In a forested area, the antenna should clear the tops of any trees in the vicinity, especially when they are very close to the antenna. In this case decreasing the height of the antenna would not be the proper procedure for reducing cochannel interference because excessive attenuation of the desired signal would occur in the vicinity of the antenna and in its cell boundary if the antenna were below the treetop level. This phenomenon is described in Sec. 4.4.

6.7   Reduction of Cochannel Interference by Means of a Notch in the Tilted Antenna Pattern

6.7.1   Introduction

Reduction of cochannel interference in a cellular mobile system is always a challenging problem. A number of methods can be considered, such as (1) increasing the separation between two cochannel cells, (2) using directional antennas at the base station, or (3) lowering the antenna heights at the base station. Method 1 is not advisable because as the number of frequency-reuse cells increases, the system efficiency, which is directly proportional to the number of channels per cell, decreases. Method 3 is not recommended because such an arrangement also weakens the reception level at the mobile unit. However, method 2 is a good approach, especially when the number of frequency-reuse cells is fixed. The use of directional antennas in each cell can serve two purposes: (1) further reduction of cochannel interference if the interference cannot be eliminated by a fixed separation of cochannel cells and (2) increasing the channel capacity when the traffic increases. In this chapter we try to further reduce the cochannel interference by intelligently setting up the directional antenna.

6.7.2   Theoretical analysis

Under normal circumstances radiation from a cochannel serving site can easily interfere with another cochannel cell as shown in Fig. 6.8. Installation of a 120° directional antenna can reduce the interference in the system by eliminating the radiation to the rest of its 240° sector. However, cochannel interference can exist even when a directional antenna is used, as the serving site can interfere with the cochannel cell that is directly ahead. Let us assume that a seven-cell cellular system ($K = 7$) is used. The cochannel interference reduction factor q becomes

$$q = \sqrt{3N} = 4.6 \qquad (6.7\text{-}1)$$

and the cochannel cell separation D can be found if the cell radius is known.

$$D = qR = 4.6R \qquad (6.7\text{-}2)$$

With a separation of $4.6R$, the area of interference at the interference-receiving cell is illuminated by the central 19° sector of the entire (120°) transmitting antenna pattern at the serving cell (see Fig. 6.9). If three identical directional antennas are implemented in every cell,

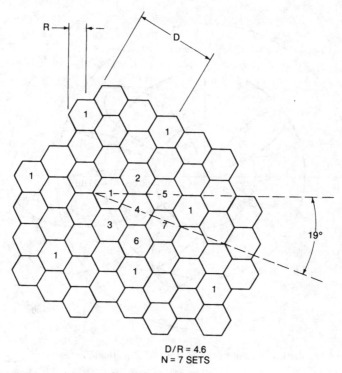

D / R = 4.6
N = 7 SETS

Figure 6.9 A seven-cell cellular configuration.

with each antenna covering a 120° sector, then every sector receives interference in the central 19° sector of the entire 120° angle at the interfering cell. Therefore, attempts should be made to reduce the signal strength of the interference in this 19° sector.

There are two ways to tilt down the antenna patterns; electronically and mechanically. The electronic downtilting is to change the phases among the elements of a colinear array antenna. The mechanical downtilting is to downtilt the antenna physically.

To achieve a significant gain of C/I in the interference-receiving cell, we should consider using a notch in the center of the antenna pattern at the interfering cell. An antenna pattern with a notch in the center can be obtained in a number of ways. One relatively simple way is to tilt the high-gain directional antenna mechanical downward.[4] A discussion of this method follows.

6.7.3 The effect of mechanically downtilting antenna on the coverage pattern

Because the shape of the antenna pattern at the base station relates directly to the reception level of signal strength at the mobile unit,

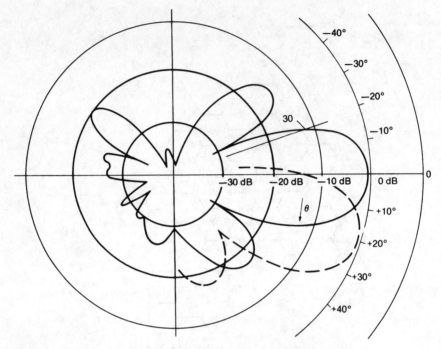

Figure 6.10 Vertical antenna pattern of a 120° directional antenna.

the following antenna pattern effect must be analyzed.

When a high-gain directional antenna (the pattern in the horizontal x-y plane is shown in Fig. 6.10 and in the vertical x-z plane, in Fig. 6.11) is physically (mechanically) tilted at an angle θ in the x-y plane shown in Fig. 6.11, how does the pattern in the x-y plane change? The antenna pattern obtained in the x-y plane after tilting the antenna is shown in Fig. 6.11. When the center beam is tilted downward by an angle θ, the off-center beam is tilted downward by only an angle ψ as shown in Fig. 6.12. The pattern in the x-y plane can be plotted by varying the angle ϕ. From the diagram in Fig. 6.12, we can obtain a derivation which provides the relationship among the angles ψ, θ, and ϕ as

$$\sin \frac{\theta}{2} = \frac{d}{l} \tag{6.7-3}$$

$$\frac{\overline{DB}}{\sin \phi} = \frac{l}{\sin(135° - \phi)} \tag{6.7-4}$$

$$\overline{CD} = l \, \frac{\sin 45°}{\sin(135 - \phi)} \tag{6.7-5}$$

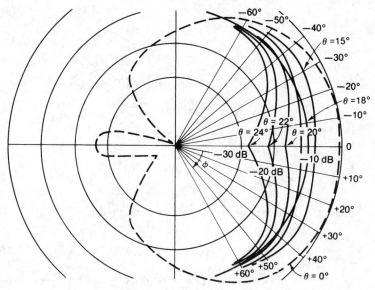

Figure 6.11 Notch appearing in tilted antenna pattern. (*Reprinted after Lee, Ref. 4.*)

$$\frac{\overline{AD}}{\overline{DF}} = \frac{\overline{AB}}{2d} = \frac{\sqrt{2}l}{2d} \qquad (6.7\text{-}6)$$

$$\overline{AD} = \overline{AB} - \overline{DB} \qquad (6.7\text{-}7)$$

$$\cos \psi = \frac{2\overline{CD} - \overline{DF}}{2\overline{CD}} = 1 - \frac{\overline{DF}}{2\overline{CD}} \qquad (6.7\text{-}8)$$

Substituting Eqs. (6.7-3) to (6.7-7) into Eq. (6.7-8), we obtain

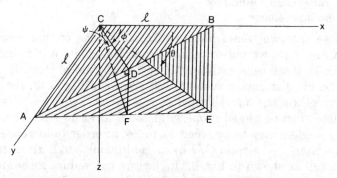

Figure 6.12 Coordinate of the tilting antenna pattern.

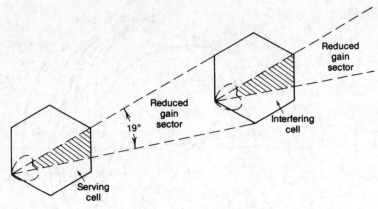

Figure 6.13 Reduced-gain sector of two cochannel cells.

$$\cos \psi = 1 - \cos^2 \phi \ (1 - \cos \theta) \qquad (6.7\text{-}9)$$

or $\qquad \psi = \cos^{-1} [1 - \cos^2 \phi \ (1 - \cos \theta)] \qquad (6.7\text{-}10)$

If the physically tilted angle is $\theta = 18°$, then the off-center beam ψ is tilted downward.

$$\phi = \begin{cases} 0° \\ 45° \\ 90° \end{cases} \qquad \psi = \begin{cases} 18° = \theta \\ 12.7° \\ 0° \end{cases}$$

This list tells us that the physically tilted angle ϕ and the angle ψ are not linearly related, and that when $\phi = 90°$, then $\psi = 0°$. When the angle θ increases beyond $18°$, the notch effect of the pattern in the x-y plane becomes evident, as indicated in Fig. 6.11.

6.7.4 Suggested method for reducing interference

Suppose that we would like to take advantage of this notch effect. From Fig. 6.13, we notice that the interfering site could cause interference at those cells within a $19°$ sector in front of the cell.

In an ideal situation such as that shown in Fig. 6.13, the antenna pattern of the serving cell must be rotated clockwise by $10°$ such that the notch can be aimed properly at the interfering cell. The antenna tilting angle θ may be between 22 to $24°$ in order to increase the carrier-to-interference ratio C/I by an additional 7 to 8 dB in the interfering cell as shown in Fig. 6.11. Now we can reduce cochannel interference by an additional 7 to 8 dB because of the notch in the mechanically tilted-antenna pattern. Although signal coverage is rather weak in a small shaded area in the serving cell, as shown in Fig. 6.13, the use of sufficient transmitting power should correct this situation.

6.7.5 Cautions in tilting antennas

When a base-station antenna is tilted down mechanically or electronically by 10°, the strength of the received signal in the horizontal direction, as shown in Fig. 6.10, is decreased by 4 dB. But the strength of the received signal 1° below the horizontal is decreased by 3.5 dB—only 0.5 dB stronger than in the 0° case. This is a very important observation. For example, the elevation angle at the boundary of a 2-mi serving cell with a 100-ft antenna mast is about 0.5°. This means that the serving cell and the interfering cell are separated by only 0.5° at most. Then by tilting the antenna down by 10°, the interference by the interfering cell is reduced by an additional 0.25 dB. This is an insignificant improvement, yet the total power received is 4 dB less than in the no-tilt case. If the tilt is increased to 20°, the received power drops by 16 dB and the reduction in interference due to tilting the antenna is only 1 dB at the interfering cell (see Fig. 6.10). The justification for implementing the tilting antenna is that the new carrier-to-interference ratio ($\alpha C / \beta I$) after tilting is significantly higher than C / I before tilting, where α and β are those constants which can be expressed if the following expression holds.

$$\frac{\alpha C}{\beta I} \text{ (linear scale)} \Rightarrow \frac{C}{I} + (\alpha - \beta) \text{ (dB scale)}$$

In the above example, at a 10° tilt, $\alpha = 3.75$ dB and $\beta = 4$ dB, and the improved new carrier-to-interference ratio is $(C / I) + 0.25$ dB, which is an insignificant improvement. Therefore, the antenna vertical pattern and the antenna height play a major role in justifying antenna tilting. Some calculations are shown in Sec. 6.8.2. Sometimes, tilting the antenna upward may increase signal coverage if interference is not a problem.

6.8 Umbrella-Pattern Effect

The umbrella pattern can be achieved by use of a staggered discone antenna as discussed in Sec. 5.4.6. The umbrella pattern can be applied to reduce cochannel interference just as the downward tilted directional antenna pattern is. The umbrella pattern can be used for an omnidirectional pattern, but not for a directional antenna pattern. The tilted directional antenna pattern can create a notch after tilting 20° or more in front of the beam, but the umbrella pattern cannot.

Of most concern for future cellular systems is the long-distance interference due to tropospheric propagation as mentioned in Sec. 4.6. In the future, one system may experience long-distance interference resulting from other systems located approximately 320 km (200 mi) away. Cochannel interference, especially cross talk, could be a severe problem. Therefore, the umbrella pattern might be recommended for every cell site where interference prevails.

6.8.1 Elevation angle of long-distance propagation

The elevation of the tropospheric layer is 16 km (10 mi)[5] and the propagation distance is about 320 km (200 mi); thus, the angle of the wave propagating through the tropospheric layers is (see Fig. 6.14) roughly

$$\theta = \tan^{-1} \frac{10 \text{ mi}}{100 \text{ mi}} = 5.7°$$

It indicates that no strong power should be transmitted upward by 5° or more in order to avoid long-distance propagation.

6.8.2 Benefit of the umbrella pattern

The umbrella pattern, in which energy is confined to the immediate area of the antenna, is effective in reducing both cochannel and long-distance interference. Also, in hilly terrain areas there are many holes (weak signal spots). With a normal antenna pattern, we cannot raise the antenna high enough to cover these holes and decrease cochannel interference at the same time. However, the advantage of the umbrella pattern is that we can increase the antenna height and still decrease cochannel interference.

The frequency-reuse distance can be shortened by use of the umbrella pattern. To demonstrate this fact, we first calculate the two angles, one from the cell-site antenna to the cell boundary and the other from the cell-site antenna to the cochannel cell (the two angles are shown in Fig. 6.14).

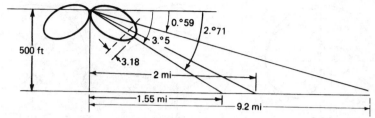

Figure 6.14 Coverage with the tilted-beam pattern.

Antenna height, ft	Desired maximum beam angle at boundary of a 2-mi cell, degrees	Angle toward cochannel cell at a distance of 4.6R (9.2 mi), degrees
100	0.54	0.12
300	1.63	0.35
500	2.71	0.59

Suppose that we are using an umbrella-pattern antenna with 11-dB gain,* and that the half-power beamwidth is above 5°. A tower of 500 ft is also used to cover a 2-mi cell. Then an approximate 3-dB difference due to the antenna pattern shown in Fig. 6.14 is obtained between the area at the maximum beam angle and the area at the angle reaching the cochannel cell.

$$\frac{R^{-4}}{6D^{-4}} = 18 \text{ dB} - 3 \text{ dB} = 15 \text{ dB}$$

$$q^4 = 6 \times 31.6 = 189.74$$

$$q = 3.7$$

where beam strengths in two regions are different by 3 dB. This demonstrates that the required frequency-reuse distance can be reduced. In other words, more protection against cochannel interference is possible with the use of an umbrella pattern than with an omnidirectional beam pattern.

6.9 Use of Parasitic Elements

Interference at the cell site can sometimes be reduced by using parasitic elements, creating a desired pattern in a certain direction. In

---

* Normally antenna gain is measured with respect to a dipole.

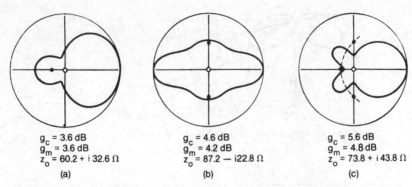

g_c = 3.6 dB
g_m = 3.6 dB
z_0 = 60.2 + i 32.6 Ω

(a)

g_c = 4.6 dB
g_m = 4.2 dB
z_0 = 87.2 − i22.8 Ω

(b)

g_c = 5.6 dB
g_m = 4.8 dB
z_0 = 73.8 + i 43.8 Ω

(c)

Figure 6.15 Parasitic elements with effective interference reduction. (a) One-quarter wavelength spacing; (b) one-half wavelength spacing; (c) combination of a and b. (*Reprint after Jasik, Ref. 6.*)

such instances, the currents appearing in several parasitic antennas are caused by radiation from a nearby drive antenna. A driven antenna and a single parasite can be combined in several ways.

1. *Normal spacing.*[6] We may first generate two separate patterns as shown in Fig. 6.15a and b. A single parasite spaced approximately one-quarter wavelength from the driven element is shown in Fig. 6.15a. Because the current flowing in the parasite is much weaker than that in the driven antenna, the front-to-back ratio is usually high. The two parasites spaced one-half wavelength from the driven element are shown in Fig. 6.15b. A combination of Fig. 6.15a and b forming a pattern very similar to that of a parabola dish is shown in Fig. 6.15c. This is an effective arrangement for cell-site directional antennas with a non-wind-resistant structure: a four-element structure that has only one active element.

2. *Relatively close spacing.*[7] In relatively close spacing two elements are placed as close as 0.04λ. Three cases can be described here.

   a. *The lengths of two elements are identical.* Two elements, one active and one parasitic, are separated by only 0.04λ. At this close spacing, the current flowing in the parasite is very strong. The two elements form a null along the y axis in the horizontal plane and along the z axis in the vertical plane. There is a directive gain of 3 dB relative to a single element. The horizontal pattern and the vertical pattern of the closely spaced arrangement are shown in Fig. 6.16a.

   b. *The length of the parasite is 5 percent longer than that of the active one.* In this case, the parasite acts as a reflector. The pat-

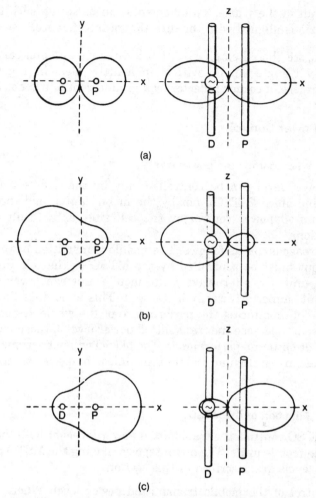

Figure 6.16 A close-in parasitic element with effective interference reduction (D = length of dipole; P = length of parasite). (a) $D = P$; (b) $D < P$; (c) $D > P$. (*Reprint after Jasik, Ref. 6.*)

terns are shown in Fig. 6.16b in both the horizontal and vertical planes. A directive gain of 6 dB is obtained.

c. *The length of the parasite is shorter than that of the active one.* In this case, the parasite acts as a director. The patterns are shown in Fig. 6.16c in the vertical and horizontal planes. A gain of 8 dB is obtained.

The pattern shown in Fig. 6.16a can be used for eliminating the interference to or from a given direction as shown by the null in the y axis. Two elements can be set up such that the y axis is aligned with the direction of interference.

Besides, we should emphasize that a directive antenna can be structured by a single parasite with a single active element (Fig. 6.16b and c). Therefore, a corner reflector or a ground reflector is not needed.

6.10 Power Control

6.10.1 Who controls the power level

The power level can be controlled only by the mobile transmitting switching office (MTSO), not by the mobile units, and there can be only limited power control by the cell sites as a result of system limitations.

The reasons are as follows. The mobile transmitted power level assignment must be controlled by the MTSO or the cell site, not the mobile unit. Or, alternatively, the mobile unit can lower the power level but cannot arbitrarily increase it. This is because the MTSO is capable of monitoring the performance of the whole system and can increase or decrease the transmitted power level of those mobile units to render optimum performance. The MTSO will not optimize performance for any particular mobile unit unless a special arrangement is made.

6.10.2 Function of the MTSO

The MTSO controls the transmitted power levels at both the cell sites and the mobile units. The advantages of having the MSTO control the power levels are described in this section.

1. Control of the mobile transmitted power level. When the mobile unit is approaching the cell site, the mobile unit power level should be reduced for the following reasons.
   a. Reducing the chance of generating intermodulation products from a saturated receiving amplifier. This point is discussed in Chap. 7.
   b. Lowering the power level is equivalent to reducing the chance of interfering with other cochannel cell sites.
   c. Reducing the near-end–far-end interference ratio (see Sec. 7.3.1).
   Reducing the power level if possible is always the best strategy.
2. Control of the cell-site transmitted power level. When the signal received from the mobile unit at the cell site is very strong, the

MTSO should reduce the transmitted power level of that particular radio at the cell site and, at the same time, lower the transmitted power level at the mobile unit. The advantages are as follows.

a. For a particular radio channel, the cell size decreases significantly, the cochannel reuse distance increases, and the cochannel interference reduces further. In other words, cell size and cochannel interference are inversely proportional to cochannel reuse distance.

b. The adjacent channel interference in the system is also reduced.

However, in most cellular systems, it is not possible to reduce only one or a few channel power levels at the cell site because of the design limitation of the combiner. The channel isolation in the combiner is 18 dB. If the transmitted power level of one channel is lower, the channels having high transmitted power levels will interfere with this low-power channel. (The channel combiner is described in Chap. 7.) The manufacturer should design an unequal-power combiner for the system operator so that the power level of each channel can be controlled at the cell site.

3. The power transmitted from a small cell is always reduced, and so is that from a mobile unit. The MTSO can facilitate adjustment of the transmitted power of the mobile units as soon as they enter the cell boundary.

6.11   Diversity Receiver

The diversity scheme applied at the receiving end of the antenna is an effective technique for reducing interference because any measures taken at the receiving end to improve signal performance will not cause additional interference.

The diversity scheme is one of these approaches. We may use a selective combiner to combine two correlated signals as shown in Fig. 6.17. The performance of other kinds of combiners can be at most 2 dB better than that of selective combiners. However, the selective combining technique is the easiest scheme to use.[8]

Figure 6.17 shows a family of curves representing this selective combination. Each curve has an associated correlation coefficient ρ; when using the diversity scheme, the optimum result is obtained when $\rho = 0$.

We have found that at the cell site the correlation coefficient $\rho \leq 0.7$ should be used[9] for a two-branch space diversity; with this coefficient the separation of two antennas at the cell site meets the requirement of $h/d = 11$, where h is the antenna height and d is the antenna separation.

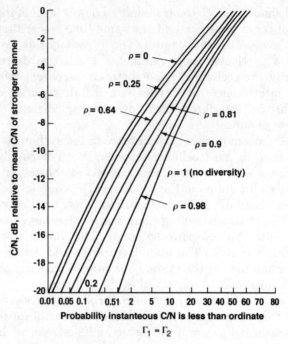

Figure 6.17 Selective combining of two correlated signals.

At the mobile unit we can use $\rho = 0$, which implies that the two roof-mounted antennas of the mobile unit are 0.5λ or more apart. This is verified by the measured data shown in Fig. 6.18.[10]

Now we may estimate the advantage of using diversity. First, let us assume a threshold level of 10 dB below the average power level. Then

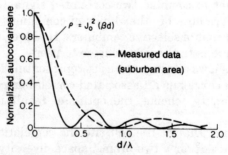

Figure 6.18 Autocorrelation coefficient versus spacing for uniform angular distribution (applied to diversity receiver). (*Reprint after Lee, Ref. 10.*)

we compare the percent of signal below the threshold level both with
, and without a diversity scheme.

1. *At the mobile unit.* The comparison is between curves $\rho = 0$ and
the $\rho = 1$. The signal below the threshold level is 10 percent for no
diversity and 1 percent for diversity. If the signal without diversity
were 1 percent below the threshold, the power would be increased by
10 dB (see Fig. 6.17). In other words, if the diversity scheme is used,
the power can be reduced by 10 dB and the same performance can be
obtained as in the nondiversity scheme. With 10 dB less power transmitted at the cell site, cochannel interference can be drastically
reduced.

2. *At the cell site.* The comparison is between curves of $\rho = 0.7$ and
$\rho = 1$. We use curve $\rho = 0.64$ for a close approximation as shown in
Fig. 6.17. The difference is 10 percent of the signal is below threshold
level when a nondiversity scheme is used versus 2 percent signal below threshold level when a diversity scheme is used. If the nondiversity signal were 2 percent below the threshold, the power would have
to increase by 7 dB (see Fig. 6.17). Therefore, the mobile transmitter
(for a cell-site diversity receiver) could undergo a 7-dB reduction in
power and attain the same performance as a nondiversity receiver at
the cell site. Thus, interference from the mobile transmitters to the
cell-site receivers can be drastically reduced.

6.12   Designing a System to Serve a Predefined Area that Experiences Cochannel Interference

A system for a service area wthout cochannel interference can be designed by using the propagation prediction model described in Chap.
4. When cochannel interference does exist, the service in the area will
deteriorate to one degree or another, depending on the location of the
interference (or interferers). First, let us assume that the ground is
flat; then two theoretical equations for designing a system in a given
service area can be derived for two different interference cases. Then
the same approach can be used to design the systems for a service
area where the ground is not flat.

Flat ground

One-interferer case. An interferer (cochannel site) is a distance d away
from the serving cell site, and the mobile unit is traveling along the
boundary of the serving-cell coverage. When the interferer is inactive,
the coverage boundary is at a distance R from the serving site. When

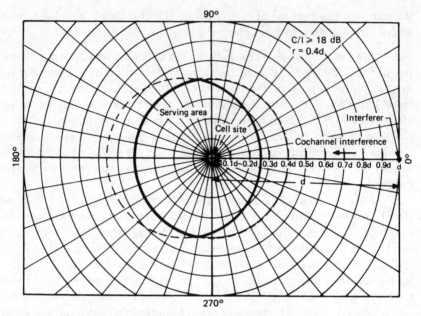

Figure 6.19 Serving area under cochannel interference.

the interferer becomes active, the distance from the serving site to the effective coverage boundary is r, which can be less than R if the interference is either strong or close to the serving site or both. If we use polar coordinates, the serving site is located at $(0, 0)$ with a transmitted power P_0, and the interferer is located at (d, θ_1) with a transmitted power P_1.

The mobile unit is located at $(r, 0)$, where r can be equal to or less than R. Assume the carrier-to-interference ratio requirement is C/I, then

$$\frac{C}{I} \leq \frac{P_0 r^{-4}}{P_1[\sqrt{r^2 + d^2 - 2rd \cos(\theta - \theta_1)}]^{-4}} \qquad r \leq R \qquad (6.12\text{-}1)$$

Let $\theta_1 = 0$ without loss of generality, as shown in Fig. 6.19 as "interferer," and $P_0 = P_1$, then

$$\frac{C}{I} \leq \left(1 + \frac{d^2}{r^2} - \frac{2d}{r} \cos \theta\right)^2 \qquad (6.12\text{-}2)$$

When $r = R$, there is no interference. Then Eq. (6.12-2) becomes

$$\frac{C}{I} \leq \left(1 + \frac{d^2}{R^2} - \frac{2d}{R} \cos \theta\right)^2 \qquad (6.12\text{-}3)$$

Let $\theta = 0$, the strongest interference condition and $C/I = 18$ dB; then Eq. (6.12-3) becomes

$$63 \leq \left(1 - \frac{d}{R}\right)^4$$

or
$$d \geq 3.82R \qquad\qquad (6.12\text{-}4)$$

When $r < d/3.82$, the serving area starts to decrease. Equation (6.12-2) can be converted to rectangular coordinates.

$$\frac{C}{I} \leq \left(1 + \frac{d^2}{x^2 + y^2} - \frac{2dx}{x^2 + y^2}\right)^2$$

or

$$\left(x + \frac{d}{\sqrt{C/I} - 1}\right)^2 + y^2 = d^2 \left(\frac{1}{\sqrt{C/I} - 1} + \frac{1}{(\sqrt{C/I} - 1)^2}\right) \quad (6.12\text{-}5)$$

where $x^2 + y^2 \leq R^2$, the serving area has its center at $(-d/\sqrt{C/I} - 1, 0)$, and the radius is $d(C/I)^{1/4}/(\sqrt{C/I} - 1)$. An illustration of Eq. (6.12-5) is shown in Fig. 6.19 as the boundary area.

Multiple-interference case. For K_I interferers, Eq. (6.12-1) can be modified as

$$\frac{C}{I} = \frac{P_0 r^{-4}}{\sum\limits_{i=1}^{K_I} P_i \left[\sqrt{r^2 + d^2 - 2rd \cos(\theta - \theta_1)}\right]^{-4}}$$

$$K_I \leq 6, r \leq R \quad (6.12\text{-}6)$$

Nonflat ground

The same approach is applied to the propagation model described in Chap. 4 in a real environment for predicting the serving areas.

References

1. W. C. Y. Lee, "Elements of Cellular Mobile Radio Systems," *IEEE Transactions on Vehicular Technology*, Vol. 35, May 1986, pp. 48–56.
2. S. Kozono and M. Sakamoto, "Channel Interference Measurement in Mobile Radio Systems," *Proceedings of the 35th IEEE Vehicular Technology Conference*, Boulder, Colorado, May 21–23, 1985, pp. 60–66.
3. W. C. Y. Lee, *Mobile Communications Design Fundamentals*, Howard W. Sams & Co., 1986, Chap. 4.

4. W. C. Y. Lee, "Cellular Mobile Radiotelephone System Using Tilted Antenna Radiation Pattern," U.S. Patent 4,249,181, February 3, 1981.
5. K. Bullington, "Radio Propagation Fundamentals," *Bell System Technical Journal,* Vol. 36, 1957, pp. 593–626.
6. H. Jasik (Ed.) *Antenna Engineering Handbook,* McGraw-Hill Book Co., 1961, pp. 5–7.
7. A. B. Bailey, *TV and Other Receiving Antenna,* John Francis, Inc., New York, 1950.
8. D. G. Brennan, "Linear Diversity Combining Techniques," *Proceedings of the IRE,* Vol. 47, June 1959, pp. 1075–1102.
9. W. C. Y. Lee, "Mobile Radio Signal Correlation Versus Antenna Height and Spacing," *IEEE Transactions on Vehicular Technology,* Vol. 25, August 1977, pp. 290–292.
10. W. C. Y. Lee, *Mobile Communications Design Fundamentals,* Howard W. Sams & Co., 1986, p. 222.

Types of Noncochannel Interference

7.1 Subjective Test versus Objective Test

Voice quality often cannot be measured by objective testing using parameters such as the carrier-to-noise ratio C/N, the carrier-to-interference ratio C/I, the baseband signal-to-noise S/N, and the signal to noise and distortion ratio (SINAD). In a mobile radio environment, multipath fading plus variable vehicular speed are the major factors causing deterioration of voice quality.

Only the following methods can help to correct this imbalance.

1. Let the received carrier level be high to increase the signal level.

2. Let the receiver sensitivity be high to lower the noise level.

3. Maintain a low distortion level in the receiver to increase SINAD.

4. Use a diversity receiver to reduce the fading.

5. Use a good system design in a mobile radio environment and a good adjacent-channel rejection to reduce the interference.

However, when a transceiver is deployed in a mobile radio environment, a subjective test is still the only way to test this receiver, using different types of modulation, such as single-sideband, double-sideband, amplitude, and frequency modulation (SSB, DSB, AM, FM).

7.1.1 The subjective test

A subjective test can be set up according to the criterion that 75 percent of the customers perceive the voice quality at a given C/N as being "good" or "excellent," the top two levels among the five circuit-merit (CM) grades.[1] The simulator of this test must be adjusted for different mobile speeds. The customers can hear different S/N levels at the baseband on the basis of the carrier-to-noise ratio C/N being changed at the RF transmitter. One typical set of curves from the customers' perception at a mobile speed at 25 km/h (or 16 mi/h) and one at 56 km/h (or 35 mi/h) are shown[2] in Fig. 7.1. Average all the test records for different vehicle speeds and determine a C/N which can satisfy the criterion we have established.

7.1.2 The objective test

There are many objective tests at the baseband for both voice and data. The characterization of voice quality is very difficult, as mentioned previously, but evaluation of data transmission is easy. There are two major terms, bit-error rates and word error rates. The bit-error rate (BER) is the first-order statistic (independent of time or vehicle speed), and the word-error rate (WER) is the second-order statistic which is affected by the vehicle speed. These rates are discussed in Chap. 12.

7.1.3 Measurement of SINAD

SINAD has been used as a measurement of communication signal quality at the baseband or in the cellular mobile receiver to measure the effective FM receiver sensitivity.[3] Some telephone industries use a "notched noise" measurement, in which a 1000-Hz tone is sent down the telephone line. The line noise is added onto the tone when it is received. By notching out the tone frequency, we can determine the remaining noise. This is a type of SINAD measurement.

1. The SINAD of the baseband output signal is defined as the ratio of the total output power to the power of the noise plus distortion only.

$$\text{SINAD} = \frac{\text{total output power}}{\text{nonsignal portion}}$$

$$= \frac{\text{signal} + \text{noise} + \text{distortion}}{\text{noise} + \text{distortion}} \quad (7.1\text{-}1)$$

The output power can be obtained by measuring the output from a voltmeter and then squaring the voltage, or directly from a power

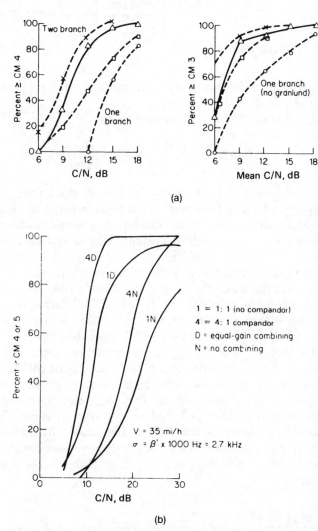

Figure 7.1 Results from subjective tests. (*Reprinted from W. C. Y. Lee, Mobile Communications Engineering, McGraw-Hill Book Co., 1982, pp. 428–429.*) (*a*) System-versus-performance comparison based on circuit merit CM4 vs. CM3. (*b*) System-versus-performance comparisons based on circuit merit CM4 and CM5.

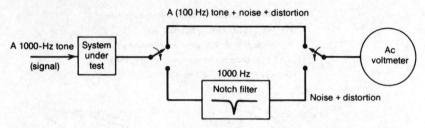

Figure 7.2 A SINAD meter.

meter. In cellular radio equipment, an input of −116 dBm is equivalent to a SINAD of 12 dB.

2. A high signal level can be measured by

$$\text{SINAD} = \frac{\text{signal} + \text{noise}}{\text{noise}} \approx \frac{\text{signal}}{\text{noise}}$$

The SINAD shown in Fig. 7.2 can be obtained by measuring the signal at the upper position and measuring the noise reading received at the lower position, assuming that the distortion is insignificant.

3. Receiver sensitivity can be measured by modulating with a 1-kHz tone at 3-kHz peak modulation deviation as shown in Fig. 7.3. The signal-generated attenuator should be adjusted until the SINAD meter shows 12 dB. Then the microvolt output is read from the attenuator dial, which reveals the "12 dB" of SINAD "sensitivity" of the receiver. This means that the signal input must be of a certain level for the signal at the output to be 12 dB higher than noise plus distortion. If the receiver noise is higher, the minimum input signal level should also be higher in order to maintain the 12-dB SINAD.

4. Noise voltage can be measured from a c-message weighting filter on any kind of telephone circuit. The frequency response of this c-message weighting filter is based on the human voice. The noise measured at the output of the filter is the noise withholding in the speech frequency spectrum. Therefore telephone line performance is measured by the amount of noise voltage through the c-message-weight filter.

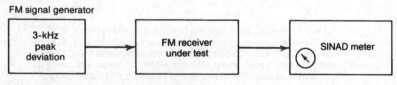

Figure 7.3 Measuring receiver sensitivity.

5. The SINAD meter also can be used as a distortion meter if the noise is very low in comparison to the distortion. The SINAD meter can be used to check the maximum distortion figures of the receiver. The input signal level is increased until no thermal noise can be heard; the receiver volume meter reads the audio power, and the SINAD meter reads the distortion.

7.2 Adjacent-Channel Interference

The scheme discussed in Chap. 6 for reduction of cochannel interference can be used to reduce adjacent-channel interference. However, the reverse argument is not valid here. In addition, adjacent-channel interference can be eliminated on the basis of the channel assignment, the filter characteristics, and the reduction of near-end–far-end (ratio) interference. "Adjacent-channel interference" is a broad term. It includes next-channel (the channel next to the operating channel) interference and neighboring-channel (more than one channel away from the operating channel) interference. Adjacent-channel interference can be reduced by the frequency assignment.

7.2.1 Next-channel interference

Next-channel interference affecting a particular mobile unit cannot be caused by transmitters in the common cell site, but must originate at several other cell sites. This is because any channel combiner at the cell site must combine the selected channels, normally 21 channels (630 kHz) away, or at least 8 or 10 channels away from the desired one. Therefore, next-channel interference will arrive at the mobile unit from other cell sites if the system is not designed properly. Also, a mobile unit initiating a call on a control channel in a cell may cause interference with the next control channel at another cell site. The methods for reducing this next-channel interference use the receiving end. The channel filter characteristics[4] are a 6 dB/oct slope in the voice band and a 24 dB/oct falloff outside the voice-band region (see Fig. 7.4). If the next-channel signal is stronger than 24 dB, it will interfere with the desired signal. The filter with a sharp falloff slope can help to reduce all the adjacent-channel interference, including the next-channel interference.

7.2.2 Neighboring-channel interference

The channels which are several channels away from the next channel will cause interference with the desired signal. Usually, a fixed set of serving channels is assigned to each cell site. If all the channels are

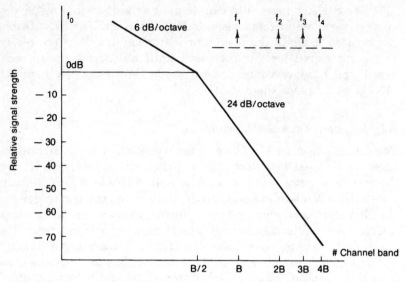

Figure 7.4 Characteristics of channel-band filter.

simultaneously transmitted at one cell-site antenna, a sufficient amount of band isolation between channels is required for a multi-channel combiner (see Sec. 7.7.1) to reduce intermodulation products. This requirement is no different from other nonmobile radio systems. Assume that band separation requirements can be resolved, for example, by using multiple antennas instead of one antenna at the cell site. What channel separation would be needed to avoid adjacent-channel interference? (See Sec. 8.1.)

Another type of adjacent-channel interference is unique to the mobile radio system. In the mobile radio system, most mobile units are in motion simultaneously. Their relative positions change from time to time. In principle, the optimum channel assignments that avoid adjacent-channel interference must also change from time to time. One unique station that causes adjacent-channel interference in mobile radio systems is described in the next section.

7.3 Near-End–Far-End Interference

7.3.1 In one cell

Because motor vehicles in a given cell are usually moving, some mobile units are close to the cell site and some are not. The close-in mobile unit has a strong signal which causes adjacent-channel inter-

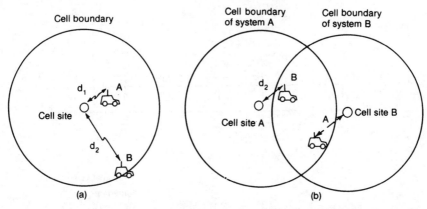

Figure 7.5 Near-end–far-end (ratio) interference. (*a*) In one cell; (*b*) in two-system cells.

ference (see Fig. 7.5*a*). In this situation, near-end–far-end interference can occur only at the reception point in the cell site.

If a separation of $5B$ (five channel bandwidths) is needed for two adjacent channels in a cell in order to avoid the near-end–far-end interference, it is then implied that a minimum separation of $5B$ is required between each adjacent channel used with one cell.

Because the total frequency channels are distributed in a set of N cells, each cell only has $1/N$ of the total frequency channels. We denote $\{F_1\}$, $\{F_2\}$, $\{F_3\}$, $\{F_4\}$ for the sets of frequency channels assigned in their corresponding cells C_1, C_2, C_3, C_4.

The issue here is how can we construct a good frequency management chart to assign the N sets of frequency channels properly and thus avoid the problems indicated above. The following section addresses how cellular system engineers solve this problem in two different systems.

7.3.2 In cells of two systems

Adjacent-channel interference can occur between two systems in a duopoly-market system. In this situation, adjacent-channel interference can occur at both the cell site and the mobile unit.

For instance, mobile unit A can be located at the boundary of its own home cell A in system A but very close to cell B of system B as shown in Fig. 7.5*b*. The other situation would occur if mobile unit B were at the boundary of cell B of system B but very close to cell A of system A. Following the definition of near-end–far-end interference given in Sec. 7.3.1, the solid arrow indicates that interference may occur at cell site A and the dotted arrow indicates that interference

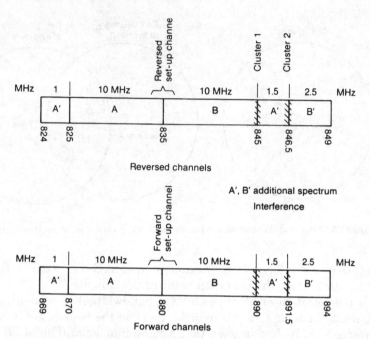

Figure 7.6 Spectrum allocation with new additional spectrum.

may occur at mobile unit A. Of course, the same interference will be introduced at cell site B and mobile unit B.

Thus, the frequency channels of both cells of the two systems must be coordinated in the neighborhood of the two-system frequency bands. This phenomenon will be of greater concern in the future, as indicated in the additional frequency-spectrum allocation charts in Fig. 7.6.

The two causes of near-end–far-end interference of concern here are

1. *Interference caused on the set-up channels.* Two systems try to avoid using the neighborhood of the set-up channels as shown in Fig. 7.6.
2. *Interference caused on the voice channels.* There are two clusters of frequency sets as shown in Fig. 7.6 which may cause adjacent-channel interference and should be avoided. The cluster can consist of 4 to 5 channels on each side of each system, that is, 8 to 10 channels in each cluster. The channel separation can be based on two assumptions.
   a. *Received interference at the mobile unit.* The mobile unit is located away from its own cell site but only 0.25 mi away from the cell site of another system.

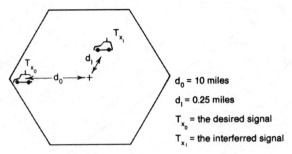

d_0 = 10 miles

d_I = 0.25 miles

T_{x_0} = the desired signal

T_{x_I} = the interferred signal

Figure 7.7   Near-end–far-end ratio interference.

b. Received interference at the cell site. The cell site is located 10 mi away from its own mobile unit but only 0.25 mi from the mobile unit of another system.

These assumptions are discussed in the next section. If the two system operators do not agree to coordinate their use of frequency channels and some of the cell sites of system B are at the coverage boundaries of the cells of system A, then the two groups of frequencies shown in Fig. 7.6 must not be used if interference has to be avoided. Of course, if the two systems do coordinate their use of frequency channels, adjacent channels in the two clusters can be used with no interference.

These observations regarding adjacent-channel interference lead the author to conclude that the existence of two systems having all colocation cell sites in a city is desirable since near-end–far-end ratio interference might be easy to control or might not occur if frequency channel use is coordinated.

7.4   Effect on Near-End Mobile Units

7.4.1   Avoidance of near-end–far-end interference

The near-end mobile units are the mobile units which are located very close to the cell site. These mobile units transmit with the same power as the mobile units which are far away from the cell site. The situation described below is illustrated in Fig. 7.7. The distance d_0 between a calling mobile transmitter and a base-station receiver is much larger than the distance d_I between a mobile transmitter causing interference and the same base-station receiver. Therefore, the transmitter of the mobile unit causing interference is close enough to override the desired base-station signal.[5] This interference, which is based on the distance ratio, can be expressed as

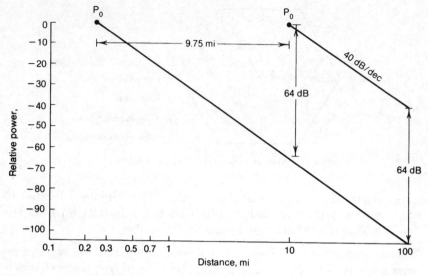

Figure 7.8 Using spacing for cochannel isolation.

$$\frac{C}{I} = \left(\frac{d_0}{d_I}\right)^{-\gamma} \tag{7.4-1}$$

where γ is the path-loss slope. The ratio d_I / d_0 is the near-end–far-end ratio. From Eq. (7.4-1) the effect of the near-end–far-end ratio on the carrier–adjacent-channel interference ratio is dependent on the relative positions of the moving mobile units.

For example, if the calling mobile unit is 10 mi away from the base-station receiver and the mobile unit causing the interference is 0.25 mi away from the base-station receiver, then the carrier-to-interference ratio for interference received at the base-station receiver with $\gamma = 4$ is

$$\frac{C}{I} = \left(\frac{d_0}{d_I}\right)^{-4} = (40)^{-4} = -64 \text{ dB} \tag{7.4-2}$$

This means that the interference is stronger than the desired signal by 64 dB (see Fig. 7.8).

This kind of interference can be reduced only by frequency separation with narrow filter characteristics. Assume that a filter of channel B has a 24 dB/oct slope;[4] then a 24-dB loss begins at the edge of the channel $B/2$. The increase from $B/2$ to B results in 24-dB loss, the increase from B to $2B$ results in another 24-dB loss, and so forth.

In order to achieve a loss of 64 dB, we may have to double the frequency band more than two times as

$$\frac{64}{L} = \frac{64}{24} = 2.67$$

where L is the filter characteristic. The frequency band separation for 64-dB isolation is

$$2^{-(C/I)/L} \left(\frac{B}{2}\right) = 2^{2.67} \left(\frac{B}{2}\right) = 3.18B \qquad (7.4\text{-}3)$$

Therefore, a minimum separation of four channels is needed to satisfy the isolation criterion of 64 dB. The general formula for the required channel separation is based on the filter characteristic L, which is expressed as follows.[5]

$$\text{Frequency band separation} = 2^{G-1}B \qquad (7.4\text{-}4)$$

where

$$G = \frac{\gamma \log_{10}\left(\dfrac{d_0}{d_I}\right)}{L} \qquad (7.4\text{-}5)$$

7.4.2 Nonlinear amplification

When the near-end mobile unit is close to the cell site, its transmitted power is too strong and saturates the IF log amplifier if the received signal at the cell site exceeds -55 dBm. A typical log IF amplifier characteristic is shown in Fig. 7.9. Assume that the mobile unit transmitted power is 36 dBm and the antenna gain is 2 dBi. The power plus the gain is 38 dBm. The receiver power is -55 dBm at the cell cite.

The propagation loss $L = 38$ dBm $- (-55$ dBm$) = 93$ dB. We may calculate the free-space path loss, which is the maximum distance within which the saturation of the IF amplifier will occur. The calculation of free-space loss versus distance at 850 MHz is as follows.

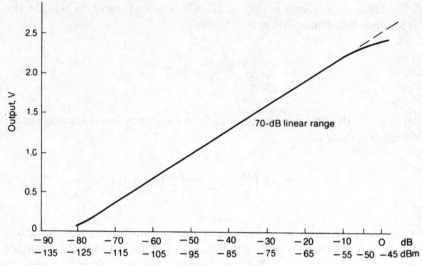

Figure 7.9 A typical intermediate-frequency log amplifier.

$$-55 \text{ dBm} = 10 \log \frac{P}{(4\pi)^2 (d/\lambda)^2}$$

$$= 38 \text{ dBm} - 20 \log 4\pi - 20 \log \left(\frac{d}{\lambda}\right)$$

$$20 \log_{10} \left(\frac{d}{\lambda}\right) = 55 + 38 - 22 = 71$$

(7.4-6)

$$\frac{d}{\lambda} = 10^{71/20} = 3548$$

$$d = 3548\lambda = 4115 \text{ ft}$$

$$= 1241 \text{ m} = 1.24 \text{ km}$$

This means that when the mobile unit is within 1.24 km of the cell-site boundary, it is possible to saturate the IF amplifier, and it is likely that intermodulation will be generated because of the nonlinear portion of the characteristics. If the intermodulation (IM) product matches the frequency channel of another mobile unit far away from the cell site where reception is weak, then the IM can interfere with the other frequency received at the cell site.

Therefore, the near-end mobile unit can cause interference at the cell site with the far-end mobile unit by generating IM at the cell-site amplifier and by leaking into the signal of the far-end mobile unit received at the cell site.

7.5 Cross Talk—A Unique Characteristic of Voice Channels

When the cellular radio system was designed, the system was intended to function like a telephone wire line. A wire pair serves both directions of traffic at the line transmission. In a mobile cellular system there is a pair of frequencies, occupying a bandwidth of 60 kHz, which we simply call a "channel." A frequency of 30 kHz serves a received path, and the other 30 kHz accommodates a transmitted path.

Because of paired-frequency (as a wire pair) coupling through the two-wire–four-wire hybrid circuitry at the telephone central office, it is possible to hear voices in both frequencies (in the frequency pair) simultaneously while scanning on only one frequency in the air. Therefore, just as with a wire telephone line, the full conversation can be heard on a single frequency (either one of the two). This phenomenon does not annoy cellular mobile users; when they talk they also listen to themselves through the phone receiver. They are not even aware that they are listening to their own voices.

This unnoticeable cross-talk phenomenon in frequency pairs has no major impact on both wire telephone line and cellular mobile performance. But when real cross talk occurs it has a larger impact on the cellular mobile system than on the telephone line, because the amount of cross talk could potentially be doubled since cross talk occurring on one frequency will be heard on the other (paired) frequency. Cross talk occurring on the reverse voice channel can be heard on the forward voice channel, and cross talk occurring on the forward voice channel can be heard on the reverse channel. Therefore, the cross-talk effect is twofold. A number of situations are conducive to cross talk.

Near-end mobile unit.   Cross talk can occur when one mobile unit (unit A) is very close to the cell site and the other (unit B) is far from the cell site. Both units are calling to their land-line parties as shown in Fig. 7.10. The near-end mobile unit has a strong signal such that the demultiplexer cannot have an isolation (separation) of more than 30 dB. Then the strong signal can generate strong cross talk while the received signal from mobile unit B is 30 dB weaker than signal A.

Near-end mobile units can belong to one system or to another (foreign) system. If the foreign system units are operating in the new allocated spectrum channels, cross talk can occur. When the mobile

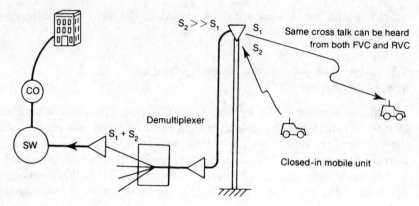

Figure 7.10 Cross-talk phenomenon.

unit is close to the cell site and the cell site is capable of reducing the power of the mobile unit, the near-end mobile interference can be reduced.

If the operating frequencies of both home system units and foreign system units are in the new allocated spectrum channels and the isolation of the multicoupler (demultiplexer) could be only 30 dB, cross talk would occur in the two interfering clusters of channels (Fig. 7.6) and could not be controlled by the system operator.

Close-in mobile units.   When a mobile unit is very close to the cell site and if the reception at the cell site is greater than -55 dBm, the channel preamplifier at the cell site can become saturated and produce IM as a result of the nonlinear portion of the amplification. These IM products are the spurious (unwanted frequency) signal which leaks into the desired signal and produces cross talk. Also, as mentioned previously, the same cross talk can be heard from both the forward and reverse voice channels.

Cochannel cross talk.   The cochannel interference reduction ratio q should be as large as possible to compensate for the cost of site construction and the limitation of available channels at each cellular site. There are other ways to increase q, as mentioned in Chap. 6. An adequate system design will help to reduce the cochannel cross talk.

The channel combiner.   The signal isolation among the forward voice channels in a channel combiner is 17 dB.[4] The loss resulting from inserting the signal into the combiner is about 3 dB. The requirement of IM product suppression is about 55 dB. If one outlet is not matched well, the signal isolation is less than 17 dB. Therefore, for each channel an isolator is installed to provide an additional 30-dB of isolation with a 0.5-dB insertion loss. This isolator prevents any signal from leaking back to the power amplifier (see Sec. 7.7.1). Spurious signals can be cross-coupled to this weak channel while transmitting. This kind of cross-coupled interference can be eliminated by routinely checking impedance matching at the combiner.

Telephone-line cross talk.   Sometimes cross talk can result from cable imbalance or switching error at the central office and be conveyed to the customer through the telephone line. Minimizing this type of cross talk should be given the same priority as reducing the number of call drops, discussed earlier (Chap. 4 and Sec. 6.2).

7.6 Effects on Coverage and Interference by Applying Power Decrease, Antenna Height Decrease, Beam Tilting

Communications engineers sometimes encounter situations where coverage must be reduced to compensate for interference. There are several ways of doing this. Reorienting the directional-antenna patterns, changing the antenna beamwidth, or synthesizing the antenna pattern were discussed in Chap. 6. There are two additional methods, decreasing the power and decreasing the antenna height. Both methods are effective, and engineers often have difficulty choosing between them. Which one is better? The answer is dependent on the situation.

7.6.1 Choosing a proper cell site

Given a fixed transmitted power and a cell-site antenna height, the coverage contours of a cell site for different signal reception levels can be obtained from either the measurement or from the prediction model described in Chap. 4. A typical contour is shown in Fig. 7.11. Because of the irregular terrain contours, contours between different reception levels are not equally spaced.

When a cell site is selected, we must determine whether an ultra-high-frequency (UHF) TV station is nearby (see Sec. 7.9) and whether any future nearby ongoing construction would affect signal coverage from the cell site later. We must check the local noise level and be sure that no spurious signals fall in the cellular frequency band.

Finally, if we are using an existing multiantenna tower, we must ensure that the grounding and shielding are adequate. Otherwise the interference level could become very high and weaken cell-site operation. Sometimes a special isolator may be provided if an AM broadcasting antenna is colocated on the same tower.

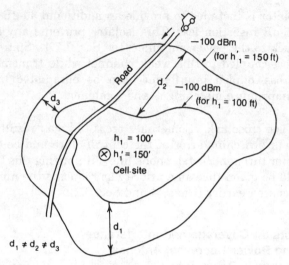

Figure 7.11 Signal-strength contour shape changing as the antenna height changes.

7.6.2 Power decrease

As long as the setup of the antenna configuration at the cell site remains the same, and if the cell-site transmitted power is decreased by 3 dB, then the reception at the mobile unit is also decreased by 3 dB. This is a one-on-one (i.e., linear) correspondence and thus is easy to control.

7.6.3 Antenna height decrease

When antenna height is decreased, the reception power is also decreased. However, the formula [see Eq. (4.10-2)]

$$\text{Antenna height gain (or loss)} = 20 \log \frac{h'_{e1}}{h_{e1}}$$

is based on the difference between the old and new effective antenna heights and not on the actual antenna heights. Therefore, the effective antenna height is the same as the actual antenna height only when the mobile unit is traveling on flat ground. It is easy to decrease antenna height to control coverage in a flat-terrain area. For decreasing antenna height in a hilly area, the signal-strength contour shown in Fig. 7.12a is different from the situation of power decrease shown in Fig. 7.12b. Therefore a decrease in antenna height would affect the coverage; thus antenna height becomes very difficult to control in an

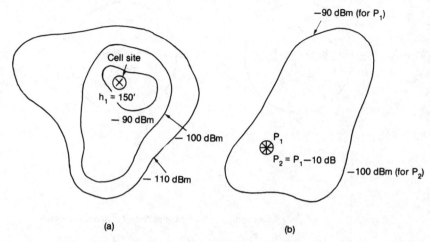

Figure 7.12 The signal-strength effect as measured by different parameters. (*a*) Different signal-strength contours. (*b*) Signal-strength changes with power changes.

overall plan. Some area within the cell may have a high attenuation while another may not.

7.6.4 Antenna patterns

The design of different antenna patterns is discussed and illustrated in Chap. 5. Here we would like to emphasize that the design of the antenna pattern should be based on the terrain contour, the population and building density, and other conditions within a given area. Of course, this is often difficult to do. For instance, implementation of antenna tilting or use of an umbrella pattern might be necessary in certain areas in order to reduce interference.

Sidelobe control (i.e., control of secondary lobe formation in an antenna radiation pattern) is also very critical in the implementation of a directional antenna. Coverage can be controlled by means of the following methods.

Using multiple antennas. In a multiple directional antenna pattern, the antennas can have different power outputs and each antenna can form a desired pattern. Two configurations can be mentioned.

1. All the antennas are facing outward (see Fig. 7.13*a*). The resultant pattern is always difficult to control because ripples and deep nulls frequently form.

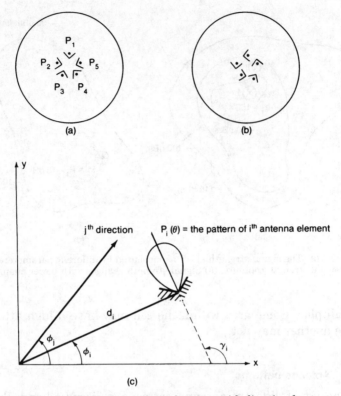

Figure 7.13 Engineering a desired pattern with directional antennas. (a) Five directional antennas facing outward; (b) a skewed configuration of five directional antennas; (c) the coordinate.

2. With skewed directional antennas[6] (see Fig. 7.13b), the resultant pattern becomes smoother. Therefore, this configuration is more attractive.

Using a synthesis of power pattern. The use of steepest descent techniques for searching the antenna parameters by giving an actual pattern and a desired pattern is introduced here. The signal strength contour obtained from Chap. 4 will be used. The difference between the two patterns, actual and desired, or error ϵ, can be expressed as

$$\epsilon(\phi, d, I, \alpha, \gamma) = \sum_{j=1}^{M} W_j (P_j - Q_j)^2 \qquad (7.6\text{-}1)$$

The parameters ϕ, d, and γ are shown in Fig. 7.13c, where I_i and α_i are the amplitude and phase of ith element, respectively. P_j is the desired field strength at the jth direction, and Q_j is the given (measured) field strength at the jth direction. All cells may be divided into M small angles, and the jth direction is one of these angles. In Eq. (7.6-1), W_j is a weighting function. When a nonuniform pattern is to be synthesized $W_j \neq 1$. The steepest descent technique can be applied to find the five parameters associated with pattern P_j which will yield the minimum ϵ in Eq. (7.6-1).

If we are using L elements, then P_j in Eq. (7.6-1) is the desired radiation field strength.

$$P_j = \sum_{i=1}^{L} P_i(\phi_j - \gamma_i)I_i \times \exp\left\{-j\left[\frac{2\pi d_i}{\lambda}\cos(\phi_j - \phi_i) - \alpha_i\right]\right\} \quad (7.6\text{-}2)$$

where $P_i(\phi)$ is the individual pattern of ith element. The magnitude and phase of the ith-element excitation are I_i and α_i, respectively. The remaining variables of Eq. (7.6-2) as shown in Fig. 7.13c. Since ϵ is a function of five parameters are indicated in Eq. (7.6-1), we start with an initial guess for the parameters $(\phi_0, d_0, I_0, \alpha_0, \gamma_0)$, and then apply the iterative equation

$$\beta_{n+1} = \beta_n - k_{\beta_i}\nabla_{\beta_i}\epsilon_n \quad (7.6\text{-}3)$$

where β = one of five parameters

$\nabla_{\beta_i}\epsilon_n$ = component of $\nabla\epsilon$ corresponding to the variable β evaluated at a given point, say, $\beta_n = \phi_n(\beta_n, d_n, I_n, \alpha_n, \gamma_n)$

k_{β_i} = gain constant for the parameter β_i

The value k_{β_i} cannot be small; otherwise the convergent process would be very slow. The iterative process is repeated until $n = N$ is reached, that is, $\nabla_{\beta_i}\epsilon_N = 0$. Then from Eq. (7.6-3), $\beta_{n+1} = \beta_n = \beta_i$ for any one of five parameters for the ith antenna element.

The same procedures apply for all elements, and all calculations can be performed by computer.

Caution: Because the terrain is not flat, the signal strengths in all directions are not uniformly attenuated at equal distances; thus, we must first obtain an antenna pattern (not desired) corresponding to a cell boundary in the actual field from a set of predetermined parameters (assume that the current distributions of all antenna elements are the same) and then convert the undesired pattern through the use of an iteration process to a desirable pattern that can be used in the field. The propagation model described in Chap. 4 will serve this purpose. Thus we can apply this iterative process to practical problems.

7.6.5   Transmitting and receiving antennas at the cell site

At the base station, the transmitted power of 100 W (+50 dBm) plus an antenna gain of 9 dBi is assumed at one transmitting antenna. The receiving antenna, located at the same site, also has a gain of 9 dBi and receives a mobile signal of −100 dBm. The difference in signal strength is

$$(50 + 9 + 9) \text{ dBm} - (-100 \text{ dBm}) = +168 \text{ dB}$$

If the space separation between a transmitting antenna and a receiving antenna is 15 m (50 ft) horizontally, the signal isolation obtained from the free-space formula is 56 dB.

The 45-MHz bandpass filter followed by the receiving antenna has at least a 55-dB rejection for signals arriving from the 870- to 890-MHz transmission band. However, the two numbers added together is 111 dB, which is still not sufficient (57 dB short). That is why the transmitting antenna and receiving antenna are not mounted in the same horizontal plane, but rather on the same vertical pole, if they are omnidirectional. This restriction can be moderated for directional antennas because of the directive patterns.

7.6.6   A 39-dBμ and a 32-dBμ boundary

The Federal Communications Commission (FCC) has used a specified received signal strength[9] for the coverage boundary, which is 39 dBμ (dB in μV/m). This value converts to a received power of −93 dBm for dipole or monopole matching on a 50-Ω load at 850 MHz (see Secs. 5.1.3 or 13.3.3). The value of 39 dBμ (i.e., −93 dBm) should be tested to determine if it is too high for use at the cell boundary in the cellular system.

We can calculate an acceptable level as follows. As we know, the accepted carrier-to-noise ratio for good quality (agreed on by most system operators) is 18 dB. The thermal noise level kTB with a bandwidth of 30 kHz and a temperature of 17°C is −129 dBm.

The receiver front-end noise N_f of an average-quality receiver is 9 dB. The noise figure NF usually would add the front-end noise N_f of the receiver and the noise N_{cm} introduced from the cellular mobile environment.

$$\text{NF} = \sqrt{N_f^2 + N_{cm}^2} \qquad \text{dB}$$

N_{cm} can either increase or decrease, depending on the system design. The earlier data indicate that N_{cm} can be neglected for 900-MHz curves.[7,8] If we now introduce a safety factor and let $N_{cm} = 6$ dB then

$$NF = \sqrt{(9)^2 + (6)^2} = 11 \text{ dB}$$

The total noise level is $N = kTB + NF = -118$ dBm. Because the required C/N is 18 dB, the lowest acceptable signal level is -100 dBm (-32 dBμ), which is 7 dB lower than -93 dBm (39 dBμ). In reality, the cell boundary or the handoff is based on the voice quality, that is, $C/N = 18$ dB or a level of -100 dBm; therefore, the FCC cell boundary of 39 dBμ or -93 dBm is 7 dB higher than the level provided by the system. Thus a cell boundary of 32 dBμ or -100 dBm proved to be sufficient for cellular coverage.

The two main advantages of using a 32-dBμ level (see Fig. 7.14) are that (1) fewer cell sites would be needed to cover a growth area and (2) less interference would be effected at the boundaries. A 32-dB boundary for cells in either boundary of a metropolitan statistical area (MSA) or a rural service area (RSA) is a proper operation, as opposed to a 39-dBμ boundary which is an artificial value.

In September 1991, the FCC was modifying rules pertaining to measurement of coverage. The idea was based on the reason which the author mentioned in the previous edition of this book. The FCC proposes the following formula to define a cellular geographic service area (CGSA):

$$d = 1.05 \times H^{0.34} \times P^{0.17} \qquad \text{(FCC)} \qquad (7.6\text{-}4)$$

where d is the distance from the cell site antenna to the reliable service area boundary in miles, H is the antenna height above average terrain in feet, and P is the effective radiated power (ERP) in watts. This formula approximates this distance to the 32-dBμ contour predicted by Carey.

The prediction based on the Lee model also can be derived from Eq. (4.2-18) as follows

$$d = 0.348 \times H^{0.52} \times P^{0.26} \qquad \text{(Lee)} \qquad (7.6\text{-}5)$$

7.7 Effects of Cell-Site Components

7.7.1 Channel combiner

A fixed-tuned channel combiner at the transmitting side. A channel combiner is installed at each cell site. Then all the transmitted channels can be combined with minimum insertion loss and maximum signal isolation between channels. Of course, we can eliminate the channel combiner by letting each channel feed to its own antenna. Then a 16-

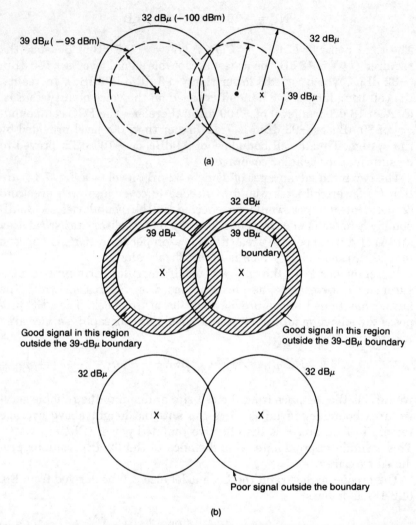

Figure 7.14 (*a*) Using a 32-dB boundary needs fewer cells to cover the area. (*b*) A signal outside its boundary generates noise.

channel site will have 16 antennas for operation. It is an economical and a physical constraint.

A conventional combiner has a 16-channel combined capacity based on the frequency subset of 16 channels, and it causes each channel to lose 3 dB from inserting the signal through the combiner. The signal isolation is 17 dB because each channel is 630 kHz or 21 channels apart from neighboring channels (Fig. 7.15*a*). The intermodulation at the multiplexer is controlled by ferrite isolators, which provide a 30-

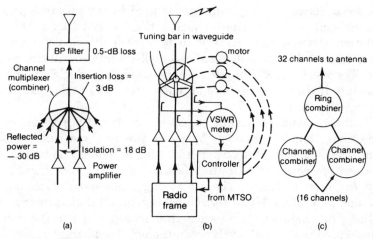

Figure 7.15 Different kinds of channel combiners. (*a*) Fixed-tuned combiner, (*b*) tunable combiner, (*c*) ring combiner.

dB reverse loss. The intermodulation (IM) products are at least 55 dB down from the desired signals. Therefore, the IM will not affect channels within the transmitted band design from this.

Each cable fed into a combiner must be properly shielded. Because it is a nonlinear device, undesired signal leakage into another channel would occur before the combiner can produce the IM products, which would in turn, produce cross-coupled interference. Therefore, proper shielding and impedance match are very important. Fixed-tuned combiners are tuned to match the impedances of a set of fixed frequencies which are assigned to a combiner.

A frequency-agile combiner.[11]   This combiner is capable of returning to any frequency by remote control in real time. The remote control device is a microprocessor. The combiner is a waveguide-resonator combiner with a tuning bar in each input waveguide as shown in Fig. 7.15*b*. The bar is mechanically rotated by a motor, and the voltage standing-wave ratio (VSWR) can be measured when the motor starts to turn. The controller receives an optimum reading after a full turn and is stopped at that position by the controller. The controller also has a self-adjusting potential. This combiner can be used when a dynamic frequency assignment is applied. In many cases, it is preferable to redistribute the frequency channels to avoid prominent interference in certain areas. To use this kind of combiner, cell-site transceivers should also be able to change their operating frequencies, which are controlled by the MTSO, accordingly. This kind of combiner can also be designed to be tuned electronically.

A ring combiner.[12]   A ring combiner is used to combine two groups of channels into a single output. The insertion loss is 3 dB, and the signal isolation between channels is 35 to 40 dB. The function of a ring combiner is to combine two 16-channel combiners into one 32-channel output. Therefore, all 32 channels can be used by a single transmitting antenna. If a cell site has two antennas, up to 64 radio channels can be installed in it.

If all the channel-transmitted powers are low, it is possible to combine more than 32 channels by using two or three ring combiners before feeding them into one transmitting antenna. The total allowed transmitted power is a limiting factor. Some ring combiners have a 600-W power limitation. The use of ring combiners reduces adjacent channel separation. If two 16-channel regular combiners are combined with a ring combiner, the adjacent-channel separation at the ring combiner output can be 315 kHz, even though the adjacent-channel separation of each regular combiner is 630 kHz. It is simply a frequency offset of 315 kHz between two regular combiners.

7.7.2   Demultiplexer at the receiving end

A demultiplexer is used to receive 16 channels from one antenna. The demultiplexer is a filter bank as shown in Fig. 7.16. Then each receiving antenna output passes through a 25-dB-gain amplifier to a demultiplexer. The demultiplexer output has a 12-dB loss from the split of 16 channels.

$$\text{Split loss} = 10 \log 16 = 12 \text{ dB}$$

and the IM product at the output of the demultiplexer should be 65 dB down.[4] The two space-diversity antennas each connect to an umbrella filter (block A or B band filter) and have a 55-dB rejection from the other system band. If the undesired mobile unit is close to the cell site, then the preamplifier becomes saturated and generates IM at the output of the amplifier; these IM products (frequencies) could be felt in one of the weak incoming signals. This situation can lead to cross talk (see Sec. 7.4) which can be heard from both ends of the link because of a unique characteristic of cellular channels (see Sec. 7.5).

7.7.3   SAT tone

General description.   The major function of a supervisory audio tone (SAT) is to ensure that a SAT tone is sent out at the cell site, is received by the mobile unit on a forward voice channel, is converted on a corresponding reverse voice channel, and is then sent back to the

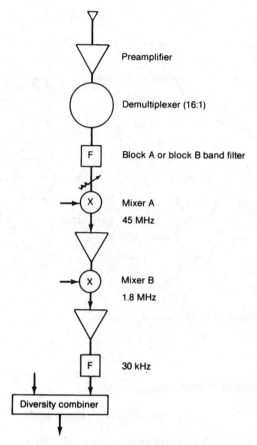

Figure 7.16 A typical cell-site channel receiver.

cell site within 5 s. If the time out is more than 5 s, the cell site will terminate the call.

Every cell site has been assigned to one of three SAT tones. The assignment of three SAT tones in a system is shown in Fig. 7.17. The cells have the same SAT tones, and the same channels are separated by $\sqrt{3}D$, which is farther than the cochannel distance D. Therefore, a receiver located at either the cell site or at the mobile unit and receiving the same frequency with different SAT tones will terminate the call.

Characteristics of SAT. There are three SAT tones, 5970 H, 6000 H, and 6030 Hz, spaced 30 Hz apart. They are narrowband frequency-modulated (FM) with a deviation of $f_\Delta = 2$ kHz. The modulation index is $\beta = \frac{1}{3}$. Let the SAT tone signal be

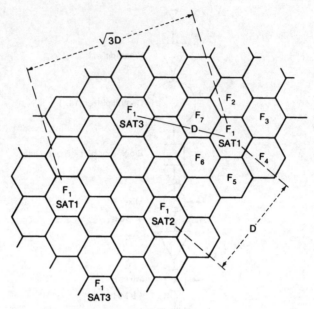

Figure 7.17 SAT spatial allocation.

$$x(t) = A_m \cos \omega_m t \qquad (7.7\text{-}1)$$

and the modulated carrier is

$$x_c(t) = A_c \cos(\omega_c t + \beta \sin \omega_m t) \qquad (7.7\text{-}2)$$

where $\beta = (A_m f_\Delta / f_m)$. Let the amplitude modulation $A_m = 1$; thus, since β is small, Eq. (7.7-2) becomes

$$x_c(t) \approx A_c \cos(\omega_c t) + \frac{A_c \beta}{2}$$

$$\times \cos[2\pi(f_c + f_m)t] - \frac{A_c \beta}{2} \cos[2\pi(f_c - f_m)t]$$

$$= R(t) \cos[\omega_c t + \phi(t)] \qquad (7.7\text{-}3)$$

where[13]

$$R(t) \simeq \sqrt{A_c^2 + \left(2\frac{\beta}{2} A_c \sin \omega_m t\right)^2}$$

$$\approx A_c \left[1 + \frac{\beta^2}{4} - \frac{\beta^2}{4} \cos 2\omega_m t\right] \qquad (7.7\text{-}4)$$

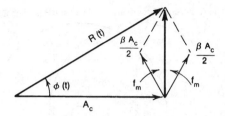

Figure 7.18   Narrowband FM for SAT.

$$\phi(t) \simeq \arctan \left[\frac{2(\beta/2)A_c \sin \omega_m t}{A_c} \right] \simeq \beta \sin \omega_m t \qquad (7.7\text{-}5)$$

The FM phasor diagram for $\beta \ll 1$ is shown in Fig. 7.18. Equation (7.7-4) represents an FM condition in which the amplitude of the carrier always remains constant. This means that the amplitude has no information content. This is a very common consideration in the mobile radio environment because of the severe fading which distorts the constant amplitude.

The SAT generator cannot deviate by more than ± 15 Hz while receiving the signal. The SAT detector uses this criterion to continuously accept or reject a returned SAT. It has been observed that two SATS with two different audio tone amplitudes can arrive at one cell. If the desired SAT tone is weaker than the undesired one by a certain ratio, then the SAT tone will deviate by ± 15 Hz. These conditions are discussed in Sec. 13.1.2. The filter bandwidth of the SAT tone detector relates to call-drop timing, which should be based on the unacceptable voice quality level. In theory, this level is different in different environment. Usually the smaller the filter bandwidth, the lower the call-drop rates. But the voice quality may be very poor before dropping the calls.

7.8   Interference between Systems

7.8.1   In one city

Let us assume that there are two systems operating in one city or one MSA. If a mobile unit of system A is closer to a cell site of system B while a call is being initiated through system A, adjacent channel interference or IM can be produced if the transmitted frequency of mobile unit A is close to the covered band of the received preamplifier at cell site B (see Fig. 7.19a). These IM products will then leak into the receiving channel of system B and cross talk will occur. This cross talk

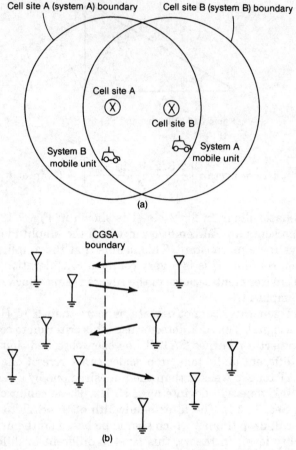

Figure 7.19 Intersystem interference. (*a*) System A cell sites in system B cell coverage; (*b*) interference between two cellular geographic service area (CGSA) systems.

can be heard not only at the land-line side but also at the mobile unit because of the unique characteristics described in Sec. 7.5.

This cross-talk situation can be reduced by any of the following measures.

1. All cell sites in the two systems can be located together (*colocated*).

2. Adjacent channels (four or five channels) at each interface (see Fig. 7.6) of the new allocated voice channels between two of the systems should not be used.

3. To prevent a strong mobile signal from saturating the preamplifier at the cell site, a foreign-system signal should be −55 dBm down

from the cell-site reception point. Otherwise IM products can be produced and mixed with the desired system by passage through the system (band) block filter (see Fig. 7.16).

For instance, IM may occur in either of the following cases.

$$(2 \times 838 - 832) \text{ MHz} = 844 \text{ MHz (system B at the cell site)}$$

Either signal (838 or 832 MHz) is strong; the IM will leak into the 844-MHz channel.

$$(2 \times 834 - 836 \text{ MHz} = 832 \text{ MHz (system A at the cell site)}$$

Either signal (834 or 836 MHz) is strong; the IM will leak into the 832-MHz channel.

7.8.2   In adjacent cities

Two systems operating at the same frequency band and in two adjacent cities or areas may interfere with each other if they do not coordinate their frequency channel use. Most cases of interference are due to cell sites at high altitudes (see Fig. 7.19b). In any start-up system, a high-altitude cell site is always attractive to the designer. Such a system can cover a larger area, and, in turn, fewer cell sites are needed. However, if the neighboring city also uses the same system block, then the result is strong interference, which can be avoided by the following methods.

1. The operating frequencies should be coordinated between two cities. The frequencies used in one city should not be used in the adjacent city. This arrangement is useful only for two low-capacity systems.

2. If both systems are high capacity, then decreasing the antenna heights will result in reduction of the interference not only within each system but also between the two systems.

3. Directional antennas may be used. For example, if one system is high capacity and the other is low capacity, the low-capacity system can use directional antennas but still retain the high tower. In this situation frequency coordination between the two systems has to be worked out at the common boundary because all the allocated frequencies must be used by the high-capacity system in its service area but only some frequencies are used by the low-capacity system.

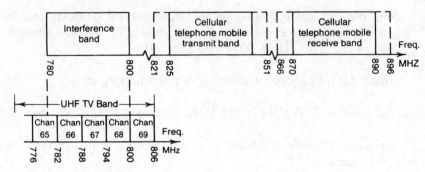

Figure 7.20   Cellular telephone frequency plan.

7.9 UHF TV Interference

Two types of interference can occur between UHF television and 850-MHz cellular mobile phones.

7.9.1 Interference to UHF TV receivers from cellular mobile transmitters

Because of the wide frequency separation between cellular phone systems and the media broadcast services (TV and radio) and the significantly high power levels used by the UHF TV broadcast transmitters, the likelihood of interference from cellular phone transmissions affecting broadcasting is very small.[14,15] There is a slight probability that when the cell-site transmission is 90 MHz above that of a TV channel, it can interfere with the image-response frequency of typical home TV receivers. Interference between TV and cellular mobile channels is illustrated in Fig. 7.20.

Some UHF TV channels overlap cellular mobile channels. These two types of service can interfere with each other only under the following conditions.

1. *Band region with overlapping frequencies.*   Two services have been authorized to operate within the same frequency band region.

2. *Image interference region.*   This is explained as follows. The TV receiver or the cellular receiver (mobile unit or cell site) can receive two transmitted signals, for instance, one from a TV channel and one from a cellular system, and produce a third-order intermodulation product which falls within the TV or the mobile receive band.

Let

$$f_{Tm} = \text{mobile transmit frequency}$$

$$= f_{Rc} = \text{cell-site receive frequency} = f_{Tc} - 45 \text{ MHz}$$

$$f_{Rm} = \text{mobile receive frequency}$$

$$= f_{Tm} + 45 \text{ MHz} = f_{Tc} = \text{cell-site transmit frequency}$$

$$f_{T,TV} = \text{TV transmit frequency}$$

$$f_{R,TV} = \text{TV receive frequency}$$

Third-order intermodulation gives the following results in two cases of interfering UHF TV receivers.

Case 1. Let

$$2f_{Tm} - f_{T,TV} = f_{Rm} \qquad (7.9\text{-}1)$$

$$f_{Tm} = f_{Rm} - 45 \qquad (7.9\text{-}2)$$

then
$$f_{Tm} = f_{T,TV} + 45 \qquad (7.9\text{-}3)$$

Since the mobile transmit frequency f_{Tm} lies in the 825- to 845-MHz band, and the TV transmit frequency $f_{T,TV}$ lies in the 780- to 800-MHz band, f_{Tm} will interfere with the TV receiver as seen from Eq. (7.9-3). This interference region is called the *image interference region.*

Case 2. Let

$$2f_{Rc} - f_{T,TV} = f_{Tc} \qquad (7.9\text{-}4)$$

then
$$f_{Rc} = f_{Tc} - 45 \qquad (7.9\text{-}5)$$

and
$$f_{Tc} = f_{T,TV} + 90 \qquad (7.9\text{-}6)$$

Because the cell-site transmit frequency f_{Tc} lies in the 870- to 890-MHz band, and $f_{T,TV}$ lies in the 780- to 800-MHz band, f_{Tc} will interfere with the TV receiver, as shown in Eq. (7.9-6). This interference region is called the image interference region.

In these two cases an image-interference rejection range of 40 to 50 dB isolation across the UHF TV band is required to prevent this interference. The results from the two cases are as follows.

Case 1: When the mobile transmitter is located near a TV receiver (Eq. 7.9-3). The minimum grade B television service contour of an accepted TV receiver level is −63 dBm with a receiver antenna gain of 6 dB referring to dipole gain. Roughly, this kind of TV station has a coverage of a 56-km (35-mi) radius. Since the cellular telephone mobile unit has an effective radiated power (ERP) of about 37 dBm, the path loss between the TV receiver and the mobile unit must exceed 100 dB (= 63 + 37). The TV antenna height at each residence normally is about $h_2 = 10$ m. The mobile antenna height is about $h_1 = 2$ m. Assume that the cross-modulation loss between two frequency bands is 80 dB and the polarization coupling loss between the bands is 10 dB. Using the formula derived in Eq. (4.2-19), we obtain

$$-63 = 37 - 156 - 40 \log \frac{d_1}{d_0} + 10 \log h_1$$
$$+ 20 \log h_2 + 6 - (80 + 10) \text{ dB} \quad (7.9\text{-}7)$$

Substitution of $h_1 = 2$ m (6 ft) and $h_2 = 10$ m (30 ft) into Eq. (7.9-7) yields

$$140 = -40 \log d_1 + 7.78 + 29.54$$

We can solve d_1 as

$$d_1 = 10^{-2.57} = 0.00239 \text{ mi} = 14 \text{ ft}$$

We find that the required distance from a transmitting cellular mobile unit to a TV receiver is only 14 ft. Besides, a mobile unit is always moving while the TV receivers usually are off; thus, the chance of mobile unit interference occurring within 14 ft of the receiver while TV receivers are operative is very slim. In addition, the chances are that thè mobile unit would remain in the area of interference for only 5 to 10 s.

Case 2: When the cell site transmitter is located near a TV receiver (Eq. 7.9-6). Usually cell-site antennas are located on high towers, and the vertical antenna pattern usually produces a null under the antenna tower. Therefore, even though Eq. (7.9-6) indicates the possibility of cell-site interference, the TV receivers near the cell site will not be in the area of the main antenna beam and, clearly, the horizontally polarized TV wave will not be distorted by the cellular vertically polarized waves when it reaches the TV receiving antenna on the roof of the house. Because of these differences between antenna beam pattern and wave polarization, no strong interference can be seen in this case. We find that the required distance could be less than 200 m (700 ft). We should also consider the following key points.

1. The polarization coupling loss from vertical (cellular) to horizontal (TV) waves can be 10 dB, according to Lee and Yeh's data.[16]

2. The percentage of active mobile units in that area is small.

3. In the UHF TV fringe area, cable TV (CATV) usually provides the service.

4. Only four TV channels (Channels 65 to 68) can experience interference. The chance of one TV set tuning to one of these four "interference channels" and the active mobile unit happening to be in that area at the same time is slim.

5. Even if transmission from the mobile unit does interfere with TV reception, the interference time is very short (<15 s). Therefore, no interference should be encountered.

7.9.2 Interference of cellular mobile receivers by UHF TV transmitters

This type of image interference can occur in the following four cases. Here the image-interference region will be the same as that described in Sec. 7.9.1 but in the reversed direction.

Case 1. Let

$$2f_{Tm} - f_{T,TV} = f_{Rm} \qquad (7.9\text{-}8)$$

Then
$$2f_{Tm} = 2(f_{Rm} - 45) \qquad (7.9\text{-}9)$$

and
$$F_{T,TV} = 2f_{Tm} - f_{Rm} = f_{Rm} - 90 \text{ MHz} \qquad (7.9\text{-}10)$$

Because the mobile unit receiver frequency f_{Rm} lies in the 870- to 890-MHz band, $f_{T,TV}$, which lies in the 780- to 800-MHz band, will interfere with the mobile unit receiver, as shown in Eq. (7.9-10).

Case 2. Let

$$2f_{Rc} - f_{T,TV} = f_{Tc} \qquad (7.9\text{-}11)$$

Then
$$f_{Rc} = f_{Tc} - 45 \qquad (7.9\text{-}12)$$

and
$$f_{Rc} = 2f_{Rc} - f_{T,TV} - 45 = f_{T,TV} + 45 \qquad (7.9\text{-}13)$$

Since the cell-site receiver frequency f_{Rc} lies in the 825- to 845-MHz band, $f_{T,TV}$, which lies in the 780- to 800-MHz band, will interfere with the cell-site receiver as shown in Eq. (7.9-13). There are two additional, but less important, cases.

Case 3. When a mobile receiver approaches a TV transmitter, it is easy to find that transmission from the TV station will not interfere with the reception at the mobile receiver by following the same analysis shown in Sec. 7.9.1, case 2.

Case 4. When the cell-site receiver is only 1 mi or less away from the TV station, interference may result. However when the cell site is very close to the TV station, the interference decreases as a result of the two vertical narrow beams pointing at different elevation levels. For this reason it is advisable to mount a cell-site antenna in the same vicinity as the TV station antenna if the problems of shielding and grounding can be controlled.

7.10 Long-Distance Interference

7.10.1 Overwater path

The phenomenon is mentioned in several reports.[17,18]

1. A 41-mi overwater path operating at 1.5 GHz in Massachusetts Bay[17]
   a. Low ducts (<50 ft thick); steady signal well above normal level is received
   b. High ducts (≥100 ft thick); a high signal level generally on the average is received but with deep fading
2. A 275-mi overwater path operating at 812 and 857 MHz between Charleston, South Carolina and Daytona Beach, Florida
   a. Charleston—antenna height 500 ft above average terrain antenna pattern, omnidirectional ERP 220 W; receiving sensitivity less than 0.5 μV = −113 dBm (1 μV = −107 dBm) with a 50-Ω terminal
   b. Daytona Beach—antenna height 920 ft above average terrain antenna pattern, omnidirectional ERP 440 W; receiving sensitivity 0.7 μV = −110 dBm

Federal Express engineers have discovered the following phenomenon through study of their system.[18] The mobile units in Charleston within 1 to 2 mi of shoreline are capable of clear communication with a repeater station in Daytona Beach. The same situation applies when the mobile unit is in Daytona Beach. These clear path communications occur regardless of weather, time of day, or season. This is a tropospherical propagation, and we should eliminate it in cellular systems to avoid interference among systems in North America. One way of doing this is by use of umbrella antenna patterns.

7.10.2 Overland path

Tropospheric scattering over a land path is not as persistent as that over water and can be varied from time to time. Usually tropospheric propagation is more pronounced in the morning. The distance can be about 200 mi. Federal Express engineers have observed this long-distance propagation throughout their nationwide system.

References

1. V. H. MacDonald, "The Cellular Concept," *Bell System Technical Journal*, Vol. 58, January 1979, pp. 15–42.
2. S. W. Halpern, "Techniques for Estimating Subjective Opinion in High-Capacity Mobile Radio," *Microwave Mobile Symposium*, Boulder, Colorado, 1976.
3. F. E. Terman and J. M. Pettit, *Electronic Measurements*, McGraw-Hill Book Co., 1952.
4. N. Ehrlich, R. E. Fisher, and T. K. Wingard, "Cell-Site Hardware," *Bell System Technical Journal*, Vol. 58, January 1979, pp. 153–199.
5. W. C. Y. Lee, "Elements of Cellular Mobile Radio Systems," *IEEE Transactions on Vehicular Technology*, Vol. VT-35, May 1986, pp. 48–56.
6. J. Pecini and M. H. Idselis, "Radiation Pattern Synthesis for Broadcast Antennas," *IEEE Transactions on Broadcasting*," Vol. BC-18, September 1972, pp. 53–62.
7. *ITT Reference Data for Radio Engineers*, Howard W. Sams & Co., Indianapolis, 1968, pp. 2–27.
8. A. D. Spaulding and R. T. Disney, "Man-Made Radio Noise, Part I: Estimate for Business, Residential and Rural Areas," U.S. Department of Commerce, Office of Technical Services Report 74-38, June 1974.
9. E. N. Skomal, *Man-Made Radio Noise*, Van Nostrand Reinhold, 1978, Chap. 2.
10. FCC Application for Cellular Operation License Requirement.
11. Antenna Specialist Co., A Frequency-Agile Combiner.
12. Antenna Specialist Co., A Ring Combiner.
13. P. F. Panter, *Modulation, Noise, and Spectral Analysis*, McGraw Hill Book Co., 1965, p. 248.
14. S. N. Ahmed and P. C. Constantinon, "A Mobile Interference Model into UHF Television Receivers," *IEEE Transactions on Vehicular Technology*, vol. VT-32, May 1983, pp. 206–208.
15. R. E. Fisher, "UHF Television Interference Associated with Cellular Mobile Telephone Systems," *IEEE Transactions on Vehicular Technology*, Vol. VT-33, August 1984, pp. 244–249.
16. W. C. Y. Lee and Y. S. Yeh, "Polarization Diversity System for Mobile Radio," *IEEE Transactions on Communications*, Vol. COM-20, October 1972, pp. 912–923.
17. S. S. Attwood (Ed.), *Radio Wave Propagation Experiments*, Vol. 2, Columbia University Press, 1946, Part I, pp. 3–47 (Chaps. 1 and 2).
18. G. W. Moor and S. Walton, Federal Express Inc., private communication.

Frequency Management and Channel Assignment

8.1 Frequency Management

The function of frequency management is to divide the total number of available channels into subsets which can be assigned to each cell either in a fixed fashion or dynamically (i.e., in response to any channel among the total available channels).

The terms "frequency management" and "channel assignment" often crate some confusion. *Frequency management* refers to designating set-up channels and voice channels (done by the FCC), numbering the channels (done by the FCC), and grouping the voice channels into subsets (done by each system according to its preference). *Channel assignment* refers to the allocation of specific channels to cell sites and mobile units. A fixed channel set consisting of one or more subsets (see Sec. 8.1.2) is assigned to a cell site on a long-term basis. During a call, a particular channel is assigned to a mobile unit on a short-term basis. For a short-term assignment, one channel assignment per call is handled by the mobile telephone switching office (MTSO). Ideally channel assignment should be based on causing the least interference in the system. However, most cellular systems cannot perform this way.

8.1.1 Numbering the channels

The total number of channels at present (January 1988) is 832. But most mobile units and systems are still operating on 666 channels. Therefore we describe the 666 channel numbering first. A channel consists of two frequency channel bandwidths, one in the low band

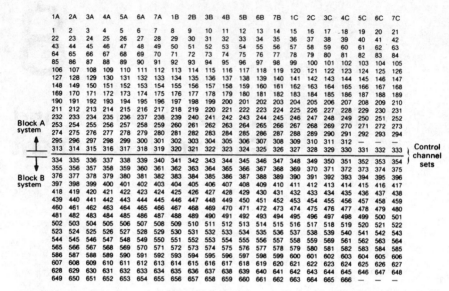

Figure 8.1 Frequency-management chart.

and one in the high band. Two frequencies in channel 1 are 825.030 MHz (mobile transmit) and 870.030 MHz (cell-site transmit). The two frequencies in channel 666 are 844.98 MHz (mobile transmit) and 889.98 MHz (cell-site transmit). The 666 channels are divided into two groups: block A system and block B system. Each market (i.e., each city) has two systems for a duopoly market policy (see Chap. 1). Each block has 333 channels, as shown in Fig. 8.1.

The 42 set-up channels are assigned as follows.

Channels 313–333   block A
Channels 334–354   block B

The voice channels are assigned as follows.

Channels   1–312 (312 voice channels)   block A
Channels 355–666 (312 voice channels)   block B

These 42 set-up channels are assigned in the middle of all the assigned channels to facilitate scanning of those channels by frequency synthesizers (see Fig. 8.1). In the new additional spectrum allocation of 10 MHz (see Fig. 8.2), an additional 166 channels are assigned. Since a 1 MHz is assigned below 825 MHz (or 870 MHz), in the future, additional channels will be numbered up to 849 MHz (or 894 MHz) and will then circle back. The last channel number is 1023 ($=2^{10}$). There are no channels between channels 799 and 991.

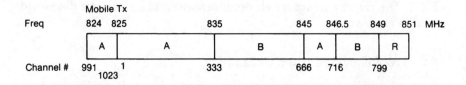

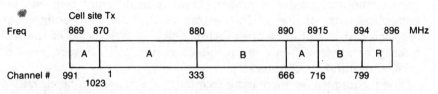

Figure 8.2 New additional spectrum allocation.

8.1.2 Grouping into subsets

The number of voice channels for each system* is 312. We can group these into any number of subsets. Since there are 21 set-up channels for each system, it is logical to group the 312 channels into 21 subsets. Each subset then consists of 16 channels. In each set, the closest adjacent channel is 21 channels away, as shown in Fig. 8.1. The 16 channels in each subset can be mounted on a frame and connected to a channel combiner. Wide separation between adjacent channels is required for meeting the requirement of minimum isolation. Each 16-channel subset is idealized for each 16-channel combiner. In a seven-cell frequency-reuse cell system each cell contains three subsets, $iA + iB + iC$, where i is an integer from 1 to 7. The total number of voice channels in a cell is about 45. The minimum separation between three subsets is 7 channels. If six subsets are equipped in an omnicell site, the minimum separation between two adjacent channels can be only three $(21/6 > 3)$ physical channel bandwidths.

For example,

$$1A + 1B + 1C + 4A + 4B + 4C$$

$$\text{or} \quad 1A + 1B + 1C + 5A + 5B + 5C$$

The antenna arrangement for 90 voice channels was described in Sec.

---

*Not including the new 83 voice channels.

5.4.1. The requirements for channel separation in a cell are discussed in this chapter.

8.2   Frequency-Spectrum Utilization

Since the radio-frequency spectrum is finite in mobile radio systems, the most significant challenge is to use the radio-frequency spectrum as efficiently as possible. Geographic location is an important factor in the application of the frequency-reuse concept in mobile cellular technology to increase spectrum efficiency. Frequency management involving the assignment of proper channels in different cells can increase spectrum efficiency. Thus, within a cell, the channel assignment for each call is studied. Other factors, such as narrowing of the frequency band, off-air call setup, queuing, and call redirect, are described in different chapters.

The techniques for increasing frequency spectrum can be classified as

1. Increasing the number of radio channels using narrow banding, spread spectrum, or time division (Chap. 13)
2. Improving spatial frequency-spectrum reuse (Chaps. 2, 6, and 7)
3. Frequency management and channel assignment (Chap. 8)
4. Improving spectrum efficiency in time (Chap. 14)
5. Reducing the load of invalid calls (Chap. 11)
   a. Off-air call setup—reducing the load of set-up channels
   b. Voice storage service for No-Answer calls
   c. Call forwarding
   d. Reducing the customers' Keep-Dialing cases
   e. Call waiting for Busy-Call situations
   f. Queuing

In this chapter we concentrate on frequency management and channel assignment (item 3).

8.3   Set-up Channels

Set-up channels, also called *control channels,* are the channels designated to set up calls. We should not be confused by the fact that a call always needs a set-up channel. A system can be operated without set-up channels. If we are choosing such a system, then all 333 channels in each cellular system (block A or block B) can be voice channels; however, each mobile unit must then scan 333 channels continuously and detect the signaling for its call. A customer who wants to initiate a call must scan all the channels and find an idle (unoccupied) one to use.

In a cellular system, we are implementing frequency-reuse concepts. In this case the set-up channels are acting as control channels. The 21 set-up channels are taken out from the total number of channels. The number 21 is derived from a seven-cell frequency-reuse pattern with three 120° sectors per cell, or a total of 21 sectors, which require 21 set-up channels. However, now only a few of the 21 set-up channels are being used in each system. Theoretically, when cell size decreases, the use of set-up channels should increase.

Set-up channels can be classified by usage into two types: *access channels* and *paging channels*. An access channel is used for the mobile-originating calls and paging channels for the land-originating calls. In a low-traffic system, access channels and paging channels are the same. For this reason, a set-up channel is sometimes called an "access channel" and sometimes called a "paging channel." Every two-way channel contains two 30-kHz bandwidths. Normally one set-up channel is also specified by two operations as a forward set-up channel (using the upper band) and a reverse set-up channel (using the lower band). In the most common types of cellular systems, one set-up channel is used for both paging and access. The forward set-up channel functions as the paging channel for responding to the mobile-originating calls. The reverse set-up channel functions as the access channel for the responder to the paging call. The forward set-up channel is transmitted at the cell site, and the reverse set-up channel is transmitted at the mobile unit. All set-up channels carry data information only.

8.3.1 Access channels

In mobile-originating calls, the mobile unit scans its 21 set-up channels and chooses the strongest one. Because each set-up channel is associated with one cell, the strongest set-up channel indicates which cell is to serve the mobile-originating calls. The mobile unit detects the system information transmitted from the cell site. Also, the mobile unit monitors the Busy/Idle status bits over the desired forward set-up channel. When the Idle bits are received, the mobile unit can use the corresponding reverse set-up channel to initiate a call.

Frequently only one system operates in a given city; for instance, block B system might be operating and the mobile unit could be set to "preferable A system." When the mobile unit first scans the 21 set-up channels in block A, two conditions can occur.

1. If no set-up channels of block A are operational, the mobile unit automatically switches to block B.

2. If a strong set-up signal strength is received but no message can be detected, then the scanner chooses the second strongest set-up channel. If the message still cannot be detected, the mobile unit switches to block B and scans to block B set-up channels.

The operational functions are described as follows.

1. *Power of a forward set-up channel [or forward control channel (FOCC)].* The power of the set-up channel can be varied in order to control the number of incoming calls served by the cell. The number of mobile-originating calls is limited by the number of voice channels in each cell site. When the traffic is heavy, most voice channels are occupied and the power of the set-up channel should be reduced in order to reduce the coverage of the cell for the incoming calls originating from the mobile unit. This will force the mobile units to originate calls from other cell sites, assuming that all cells are adequately overlapped.
2. *The set-up channel received level.* The set-up channel threshold level is determined in order to control the reception at the reverse control channel (RECC). If the received power level is greater than the given set-up threshold level, the call request will be taken.
3. *Change power at the mobile unit.* When the mobile unit monitors the strongest signal strength from all set-up channels and selects that channel to receive the messages, there are three types of message.
   a. *Mobile station control message.* This message is used for paging and consists of one, two, or four words—DCC, MIN, SCC, and VMAX (see Chap. 3).
   b. *System parameter overhead message.* This message contains two words, including DCC, SID, CMAX, or CPA (see Chap. 3).
   c. *Control-filler message.* This message may be sent with a system parameter overhead message, CMAC—a control mobile attenuation code (seven levels).
4. *Direct call retry.* When a cell site has no available voice channels, it can send a direct call-retry message through the set-up channel. The mobile unit will initiate the call from a neighboring cell which is on the list of neighboring cells in the direct call-retry message.

8.3.2 Paging channels

Each cell site has been allocated its own set-up channel (control channel). The assigned forward set-up channel (FOCC) of each cell site is used to page the mobile unit with the same mobile station control message (discussed in Chap. 3 and Sec. 8.3.1).

Because the same message is transmitted by the different set-up channels, no simulcast interference occurs in the system. The algorithm for paging a mobile unit can be performed in different ways. The simplest way is to page from all the cell sites. This can occupy a large amount of the traffic load. The other way is to page in an area corresponding to the mobile unit phone number. If there is no answer, the system tries to page in other areas. The drawback is that response time is sometimes too long.

When the mobile unit responds to the page on the reverse set-up channel, the cell site which receives the response checks the signal reception level and makes a decision regarding the voice channel assignment based on least interference in the selected sector or underlay-overlay region.

8.3.3 Self-location scheme at the mobile unit

In the cellular system, 80 percent of calls originate from the mobile unit but only 20 percent originate from the land line. Thus, it is necessary to keep the reverse set-up channels as open as possible. For this reason, the self-location scheme at the mobile unit is adapted. The mobile unit selects a set-up channel of one cell site and makes a mobile-originating call. It is called a *self-location scheme*.

However, the self-location scheme at the mobile unit prevents the mobile unit from sending the necessary information regarding its location to the cell site. Therefore, the MTSO does not know where the mobile is. When a land-line call is originated, the MTSO must page all the cell sites in order to search for the mobile unit. Fortunately, land-line calls constitute only 20 percent of land-line originating calls, so the cellular system has no problem in handling them. Besides, more than 50 percent of land-line originating calls are no response.

8.3.4 Autonomous registration

If a mobile station is equipped for autonomous registration, then the mobile station stores the value of the last registration number (REGID) received on a forward control channel. Also, a REGINCR (the increment in time between registrations) is received by the mobile station. The next registration ID should be (see Chap. 3)

$$NXTREG = REGID + REGINCR$$

This tells the mobile unit how long the registration should be repeatedly sent to the cell site, so that the MTSO can track the location of the mobile. This feature is not used in cellular systems at present. However, when the volume of land-line calls begins to increase or the number of cell sites increases, this feature would facilitate paging of the mobile units with less occupancy time on all set-up channels. The trade-off between the self-location scheme and autonomous registration is shown in the following two examples.

Example 8.1 The time spent in the set-up channels for two schemes are compared.

1. Evaluation of a self-location scheme on a land-originating call. Assume that a system has 100 cell sites and a call paging has to reach all 100 cell sites. If every page takes 100 ms and there are 2000 land-originating calls per hour during a busy hour, then the air time spent for the paging during the busy hour is

$$100 \times (100 \text{ ms}) \times 2000 = 20,000 \text{ s} = 333 \text{ min/h}$$

   This is the time spent on all set-up channels.

2. Evaluation of a registration scheme used on an idle stage for locating mobile units. Assume that the registration for each mobile unit is five times per hour. Each registration takes 100 ms. If 20,000 mobile units are on the road, then

$$(5 \times 100 \text{ ms}) \times 20,000 = 1000 \text{ s} = 166.7 \text{ min/h}$$

   This is the time spent on all set-up channels.

In Example 8.1, the time spent on the set-up channels for a self-location scheme is twice as much as that for a registration scheme. In this particular case, the registration scheme is preferable to the self-location scheme.

Example 8.2 Assume that the reverse set-up channels also take the mobile-originating calls, which make up 80 percent of the total number of calls. Assume that 2500 land-originating calls constitute 20 percent of the total number of calls; then the mobile-originating calls represent 10,000 calls per hour handled by the MTSO. Each call initiation takes about 300 ms. Then

$$10,000 \times 300 = 3000 \text{ s} = 50 \text{ min}$$

The 50 min is occupied in both schemes. This is because for a mobile-originating call the self-location scheme provides a negligible time for selecting a desired cell site on a reverse set-up channel. The same negligible time is provided by using the registration scheme for selecting the desired cell site.

In a busy (rush) hour, the attempted call originating at a mobile unit is searching for an idle bit sent from the cell site. If an idle bit cannot be received at the mobile unit after 10 attempts, then a busy tone is heard at the mobile unit.

Therefore, the 50 min calculated above assumes that all 10,000 calls are not blocked. In reality, there is always a certain amount of call blocking during a rush hour. Therefore, even though the MTSO will spend 50 min in a system to process 10,000 calls per hour, the actual attempt calls can be much higher.

8.3.5 Traffic load on a set-up channel and on *N* voice channels

When the traffic of a cell is increasing, more radios will be installed. When a cell has 90 voice channels (radios), one set-up channel must coordinate them in order to set up the calls. On the average, the cell site takes a mobile-originating call on a reverse set-up channel for 100 ms, and the interval between calls is 25 ms (including calls colliding in the air). Thus, in 1 h, if a queuing scheme is applied, the maximum number of calls that a set-up channel can accommodate is

$$\frac{3600 \times 1000}{125 \text{ ms}} = 28{,}800 \text{ calls/h}$$

This equation is based on the assumption that the incoming calls from the mobile units are waiting for the idle bits showing on the forward set-up channel before sending the requests. This is equivalent to a queuing scheme. In general, the waiting period is 1 to 2 s. If the set-up channel is busy during this period, the mobile unit will periodically continue to search for idle channels about every 100 ms. Then the initiating call will be blocked after 10 attempts. An estimate of call blocking can be obtained by using a queuing model. Without queuing schemes, the maximum numer of initiating calls that the set-up channel can take during a busy hour, assuming five attempts per call, is 5760 calls per hour.

To calculate the traffic load on 90 voice channels, let us assume a blocking probability of 0.02 and a holding time of 100 s. Now we can check the offered load a from Table 1.1.

$$a = 78.3$$

The number of calls is

$$M = \frac{78.3 \times 3600}{100} = 2818 \text{ calls/h}$$

The carried load of a set-up channel is always greater than the carried load of the 90 voice channels. A load of more than 90 radios in a cell is not unusual. However, the number of voice channels in a cell rarely exceeds 120. Therefore one set-up channel is used in a cell.

8.3.6 Separation between access and paging

All 21 set-up channels are actually paging channels. The access channel can be assigned by the MTSO as a channel other than the 21 set-up channels in a cell. The mobile unit receives the access channel information from the forward paging channels. In certain cases, as land-originating calls increase, one set-up channel cannot handle all set-up traffic in a cell. In such cases another channel in a group of voice channels is used as an access channel. Now the land-originating calls are using paging channels and the mobile-originating calls are using access channels.

8.3.7 Selecting a voice channel

Assume that a mobile unit calls or responds to a call through a reverse set-up channel which is received from an omnidirectional antenna and the voice channels are assigned from a forward set-up channel at one of three 120°-sector directional antennas.

For mobile-originating calls. The mobile unit selects a cell site based on its received signal-strength indicator (RSSI) reading. When a call of a mobile unit is received by the cell site, the set-up channel receives it through an omnidirectional antenna. The cell-site RSSI scans the incoming signals through three directional antennas and determines which sector is the strongest one. The MTSO then assigns a channel from among those channels designated in that sector. In some systems, a set-up channel is assigned to each sector of a cell.

For paging calls. When any call responds to the cell site, the cell-site RSSI will measure the incoming signal from the three directional antennas and find the strongest sector in which the channel can be assigned to the mobile unit.

8.4 Definition of Channel Assignment

8.4.1 Channel assignment to the cell sites—fixed channel assignment

In a fixed channel assignment, the channels are usually assigned to the cell site for relatively long periods. Two types of channels are assigned: set-up channels and voice channels.

Set-up channels.   There are 21 set-up channels assigned each cell in a $K = 4$, $K = 7$, or $K = 12$ frequency-reuse pattern. If the set-up channel antennas are omnidirectional, then each cell only needs one set-up channel. This leaves many unused set-up channels. However, the set-up channels of blocks A and B are adjacent to each other. In order to avoid interference between two systems, the set-up channels in the neighborhood of Channel 333 (block A) and Channel 334 (block B) are preferably unused.

Voice channels.   One way of dividing the total voice channels into 21 sets is exemplified in Sec. 8.1. The assignment of certain sets of voice channels in each cell site is based on causing minimum cochannel and adjacent-channel interference. Cochannel and adjacent channel interference can be calculated from equations in Chaps. 6 and 7.

Supervisory audio tone (SAT).   This consists of three SATs. Based on the assignment of each SAT in each cell, we can show the method for further reducing cochannel interference, as mentioned in Sec. 7.7.2.

8.4.2   Channel assignment to traveling mobile units

This situation always ocurs in the morning, when cars travel into the city, and at night, when the traffic pattern reverses. If the traffic density is uniform, the unsymmetrical mobile-unit antenna pattern (assuming large backward energy from the motion of the vehicle) does not affect the system operation much. However, when the traffic becomes heavier as more cars approach the city, the traffic pattern becomes nonuniform and the sites closest to the city, or in the city, cannot receive the expected number of calls or handoffs in the morning because of the mobile unit antenna patterns. At night, as the cars move out of the city, the cell sites closest to the city would have a hard time handing off calls to the sites away from the city.

To solve these problems, we have to use less transmitted power for both set-up and voice channels for certain cell sites. We also have to raise the threshold level for reverse set-up channels and voice channels at certain cell sites in order to control the acceptance of incoming calls and handoff calls. Three methods can be used.

Underlay-overlay.[1]   The traffic capacity at an omnidirectional cell or a directional cell (see Fig. 8.3) can be increased by using the underlay-overlay arrangement. The underlay is the inner circle, and the overlay is the outer ring. The transmitted powers of the voice channels at the site are adjusted for these two areas. Then different voice frequencies

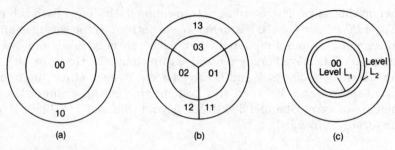

(a)   (b)   (c)

Figure 8.3 Underlaid-overlaid cell arrangements. (*a*) Undelay-overlay in omni-cell; (*b*) underlay-overlay in sectorized cells; (*c*) two-level handoff scheme.

are assigned to each area. In an omnidirectional cell, the frequency-reuse distance of a seven-cell reuse pattern is $D = 4.6R$, where R is the radius of the cell. One overlay and one underlay are shown in Fig. 8.3*a*. Because of the sectorization in a directional cell, the channel assignment has a different algorithm in six regions (Fig. 8.3*b*), i.e., three overlay regions and three underlay regions. A detailed description is given in Sec. 8.5.4.

Frequency assignment.   We assign the frequencies by a set of channels or any part of a set or more than one set of the total 21 sets. Borrowed-frequency sets are used when needed. On the basis of coverage prediction, we can assign frequencies intelligently at one site or at one sector without interfering with adjacent cochannel sectors or cochannel cells.

Tilted antenna.   The tilted directional antenna arrangement can eliminate interference. Sometimes antenna tilting is more effective than decreasing antenna height, especially in areas of tall trees or at high sites. When the tilting angles become 22° or greater, the horizontal pattern creates a notch in the front of the antenna, which can further reduce the interference (see Fig. 6.10).

8.5   Fixed Channel Assignment

8.5.1   Adjacent-channel assignment

Adjacent-channel assignment includes neighboring-channel assignment and next-channel assignment. The near-end–far-end (ratio) interference, as mentioned in Sec. 7.3.1, can occur among the neighboring channels (four channels on each side of the desired channel). Therefore, within a cell we have to be sure to assign neighboring channels in an omnidirectional-cell system and in a directional-antenna-

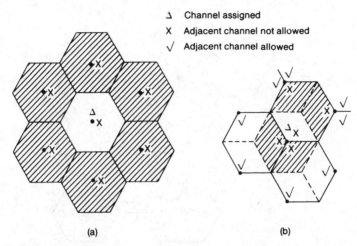

Figure 8.4 Adjacent channel assignment. (*a*) Omnidirectional-antenna cells; (*b*) directional-antenna cells.

cell system properly. In an omnidirectional-cell system, if one channel is assigned to the middle cell of seven cells, next channels cannot be assigned in the same cell. Also, no next channel (preferably including neighboring channels) should be assigned in the six neighboring sites in the same cell system area (Fig. 8.4*a*). In a directional-antenna-cell system, if one channel is assigned to a face, next channels cannot be assigned to the same face or to the other two faces in the same cell. Also, next channels cannot be assigned to the other two faces at the same cell site (Fig. 8.4*b*). Sometimes the next channels are assigned in the next sector of the same cell in order to increase capacity. Then performance can still be in the tolerance range if the design is proper.

8.5.2 Channel sharing and borrowing[2,3]

Channel sharing. Channel sharing is a short-term traffic-relief scheme. A scheme used for a seven-cell three-face system is shown in Fig. 8.5. There are 21 channel sets, with each set consisting of about 16 channels. Figure 8.5 shows the channel set numbers. When a cell needs more channels, the channels of another face at the same cell site can be shared to handle the short-term overload. To obey the adjacent-channel assignment algorithm, the sharing is always cyclic. Sharing always increases the trunking efficiency of channels. Since we cannot allow adjacent channels to share with the nominal channels in the same cell, channel sets 4 and 5 cannot both be shared with channel sets 12 and 18, as indicated by the grid mark. Many grid marks are indicated in Fig. 8.5 for the same reason. However, the

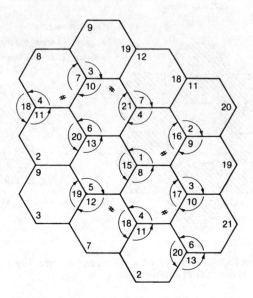

\# possible interference area

Figure 8.5   Channel-sharing algorithm.

upper subset of set 4 can be shared with the lower subset of set 5 with no interference.

In channel-sharing systems, the channel combiner should be flexible in order to combine up to 32 channels in one face in real time. An alternative method is to install a standby antenna.

Channel borrowing.   Channel borrowing is usually handled on a long-term basis. The extent of borrowing more available channels from other cells depends on the traffic density in the area. Channel borrowing can be implemented from one cell-site face to another face at the same cell site.

In addition, the central cell site can borrow channels from neighboring cells. The channel-borrowing scheme is used primarily for slowly-growing systems. It is often helpful in delaying cell splitting in peak traffic areas. Since cell splitting is costly, it should be implemented only as a last resort.

8.5.3   Sectorization

The total number of available channels can be divided into sets (subgroups) depending on the sectorization of the cell configuration: the 120°-sector system, the 60°-sector system, and the 45°-sector system.

A seven-cell system usually uses three 120° sectors per cell, with the total number of channel sets being 21. In certain locations and special situations, the sector angle can be reduced (narrowed) in order to assign more channels in one sector without increasing neighboring-channel interference. This point is discussed in Sec. 10.6. Sectorization serves the same purpose as the channel-borrowing scheme in delaying cell splitting. In addition, channel coordination to avoid cochannel interference is much easier in sectorization than in cell splitting. Given the same number of channels, trunking efficiency decreases in sectorization.

Comparison of omnicells (nonsectorized cells) and sectorized cells

Omnicells. If a $K = 7$ frequency-reuse pattern is used, the frequency sets assigned in each cell can be followed by the frequency-management chart shown in Fig. 8.1. However, terrain is seldom flat; therefore, $K = 12$ is sometimes needed for reducing cochannel interference. For $K = 12$, the channel-reuse distance is $D = 6R$, or the cochannel reduction factor $q = 6$.

Sectorized cells. There are three basic types.

1. The 120°-sector cell is used for both transmitting and receiving sectorization. Each sector has an assigned a number of frequencies. Changing sectors during a call requires handoffs.

2. The 60°-sector cell is used for both transmitting and receiving sectorization. Changing sectors during a call requires handoffs. More handoffs are expected for a 60° sector than a 120° sector in areas close to cell sites (close-in areas).

3. The 120°- or 60°-sector cell is used for receiving sectorization only. In this case, the transmitting antenna is omnidirectional. The number of channels in this cell is not subdivided for each sector. Therefore, no handoffs are required when changing sectors. This receiving-sectorization-only configuration does not decrease interference or increase the D/R ratio; it only allows for a more accurate decision regarding handing off the calls to neighboring cells.

8.5.4 Underlay-overlay arrangement[1]

In actual cellular systems cell grids are seldom uniform because of varying traffic conditions in different areas and cell-site locations.

Overlaid cells. To permit the two groups to reuse the channels in two different cell-reuse patterns of the same size, an "underlaid" small cell is sometimes established at the same cell site as the large cell (see Fig. 8.3). The "doughnut" (large) and "hole" (small) cells are treated as two different cells. They are usually considered as "neighboring cells."

The use of either an omnidirectional antenna at one site to create two subring areas or three directional antennas to create six subareas is illustrated in Fig. 8.3b. As seen in Fig. 8.3, a set of frequencies used in an overlay area will differ from a set of frequencies used in an underlay area in order to avoid adjacent-channel and cochannel interference. The channels assigned to one combiner—say, 16 channels—can be used for overlay, and another combiner can be used for underlay.

Implementation. The antenna of a set-up channel is usually omnidirectional. When an incoming call is received by the set-up channel and its signal strength is higher than a level L, the underlaid cell is assigned; otherwise, the overlaid cell is assigned. The handoffs are implemented between the underlaid and overlaid cells. In order to avoid the unnecessary handoffs, we may choose two levels L_1 and L_2 and $L_1 > L_2$ as shown in Fig. 8.3c.

When a mobile signal is higher than a level L_1 the call is handed off to the underlaid cell. When a signal is lower than a level L_2 the call is handed off to the overlaid cell. The channels assigned in the underlaid cell have more protection against cochannel interference.

Reuse partition. Through implementation of the overlaid-cell concept, one possible operation is to apply a multiple-K system operation, where K is the number of frequency-reuse cells. The conventional system uses $K = 7$. But if one K is used for the underlaid cells, then this multiple-K system can have an additional 20 percent more spectrum efficiency than the single K system with an equivalent voice quality. In Fig. 8.6a, the $K = 9$ pattern is assigned to overlaid cells and the $K = 3$ pattern is assigned to underlaid cells. Based on this arrangement the number of cell sites can be reduced, while maintaining the same traffic capacity. The decrease in the number of cell sites which results from implementation of the multiple K systems is shown in Fig. 8.6b. The advantages of using this partition based on the range of K are

1. The K range is 3 to 9; the operational call quality can be adjusted and more reuse patterns are available if needed.

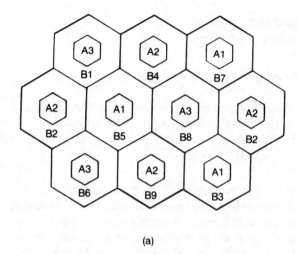

(a)

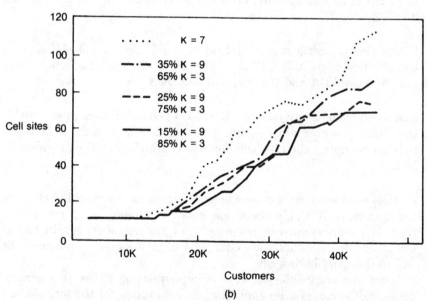

Customers

(b)

Figure 8.6 Reuse-partition scheme. (*After Whitehead, Ref. 1.*) (*a*) Reuse partition $K_A = 3$; $K_B = 9$. (*b*) Reuse-partitioning performance.

2. Each channel set of old $K = 9$ systems is the subset of new $K = 3$ systems. Therefore, the amount of radio retuning in each cell in this arrangement is minimal.

3. When cell splitting is implemented, all present channel assignments can be retained.

8.6   Nonfixed Channel Assignment Algorithms[4-9]

8.6.1   Description of different algorithms

Fixed channel algorithm.   The fixed channel assignment (FCA) algorithm is the most common algorithm adopted in many cellular systems. In this algorithm, each cell assigns its own radio channels to the vehicles within its cell.

Dynamic channel assignment.   In dynamic channel assignment (DCA), no fixed channels are assigned to each cell. Therefore, any channel in a composite of 312 radio channels can be assigned to the mobile unit. This means that a channel is assigned directly to a mobile unit. On the basis of overall system performance, DCA can also be used during a call.

Hybrid channel assignment.   Hybrid channel assignment (HCA) is a combination of FCA and DCA. A portion of the total frequency channels will use FCA and the rest will use DCA.

Borrowing channel assignment.   Borrowing channel assignment (BCA) uses FCA as a normal assignment condition. When all the fixed channels are occupied, then the cell borrows channels from the neighboring cells.

Forcible-borrowing channel assignment.[9]   In forcible-borrowing channel assignment (FBCA), if a channel is in operation and the situation warrants it, channels must be borrowed from the neighboring cells and at the same time, another voice channel will be assigned to continue the call in the neighboring cell.

There are many different ways of implementing FBCA. In a general sense, FBCA can also be applied while accounting for the forcible borrowing of the channels within a fixed channel set to reduce the chance of cochannel assignment in a reuse cell pattern.

The FBCA algorithm is based on assigning a channel dynamically but obeying the rule of reuse distance. The distance between the two cells is *reuse distance*, which is the minimum distance at which no cochannel interference would occur.

Very infrequently, no channel can be borrowed in the neighboring cells. Even those channels currently in operation can be forcibly borrowed and will be replaced by a new channel in the neighboring cell

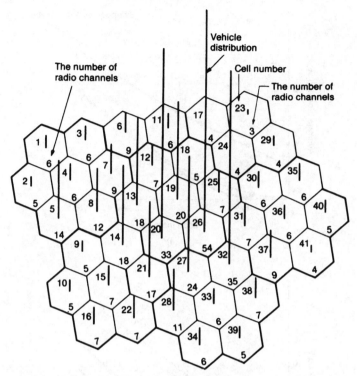

Figure 8.7 Cellular system. Vehicle and radio-channel distribution in the busy rush hour. (*After Sekiguchi et al., Ref. 9.*)

or the neighboring cell of the neighboring cell. If all the channels in the neighboring cells cannot be borrowed because of interference problems, the FBCA stops.

8.6.2 Simulation process and results

On the basis of the FBCA, FCA, and BCA algorithms, a seven-cell reuse pattern with an average blocking of 3 percent is assumed and the total traffic service in an area is 250 erlangs. The traffic distributions are (1) uniform traffic distribution—11 channels per cell; (2) a nonuniform traffic distribution—the number of channels in each cell is dependent on the vehicle distribution (Fig. 8.7). The simulation model is described as follows:

1. Randomly select the cell (among 41 cells).
2. Determine the state of the vehicle in the cell (idle, off-hook, on-hook, handoff).

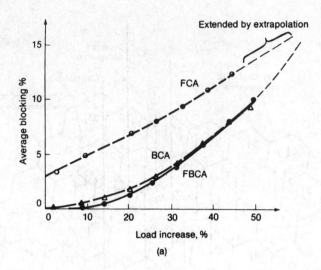

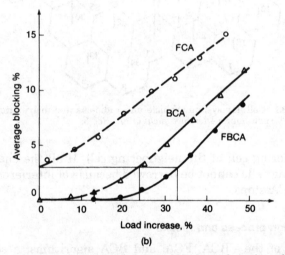

Figure 8.8 Comparison of average blockings from three different schemes. (*After Sekiguchi et al., Ref. 9*) (*a*) Average blocking in spatially uniform traffic distribution; (*b*) average blocking in spatially nonuniform traffic distribution.

3. In off-hook or handoff state, search for an idle channel. The average number of handoffs is assumed to be 0.2 times per call. However, FBCA will increase the number of handoffs.

Average blocking. Two average blocking cases illustrating this simulation are shown in Fig. 8.8. In a uniform traffic condition (Fig. 8.8*a*), the 3 percent blocking of both BCA and FBCA will result in a load increase of 28 percent, compared to 3 percent blocking of FCA. There is no difference between BCA and FBCA when a uniform traffic condition exists.

In a nonuniform traffic distribution (Fig. 8.8b), the load increase in BCA drops to 23 percent and that of FBCA increases to 33 percent, as at an average blocking of 3 percent. The load increase can be utilized in another way by reducing the number of channels. The percent increase in load is the same as the percent reduction in the number of channels.

Handoff blocking.   Blocking calls from all handoff calls occurring in all cells is shown in Fig. 8.9. Handoff blocking is not considered as the regular cell blocking which can only occur at the call setup stage. In both BCA and FBCA, load is increased almost equally to 30 percent, as compared to FCA at 3 percent handoff blocking in uniform traffic (Fig. 8.9a). For a nonuniform traffic distribution, the load increase of both BCA and FBCA at 4 percent blocking is about 50 percent (Fig. 8.9b), which is a big improvement, considering the reduction in interference and blocking. Otherwise, there would be multiple effects from interference in several neighboring cells.

8.7 How to Operate with Additional Spectrum

On July 24, 1986 the FCC announced that a totally new additional spectrum of 10 MHz would be allocated to the cellular mobile industry. This spectrum provides 166 voice channels, with 83 channels for each carrier. The new spectrum allocation is shown in Fig. 8.2.

In the future, cellular systems must serve both the old mobile units, which operate 666 channels, and the new mobile units, which operate 832 channels. The new mobile units will have less blocked calls then the old mobile units when they are used in areas of heavy traffic. However, because the additional spectra for bands A and B are discretely and alternately allocated, the neighboring channels between bands A and B occur at two points (one point between channels 666 and 667, and the other between channels 716 and 717) in the frequency spectrum. At these two points, the tendency for neighboring-channel interference is high.

According to the analysis given in Sec. 7.3.1, the "neighboring channels" can consist of four channels on each side of two systems. Therefore, these eight channels must be used with extreme caution. Unless we know the frequency channel assignments of the other system, or coordinate with the other system, it is not wise to use these channels.

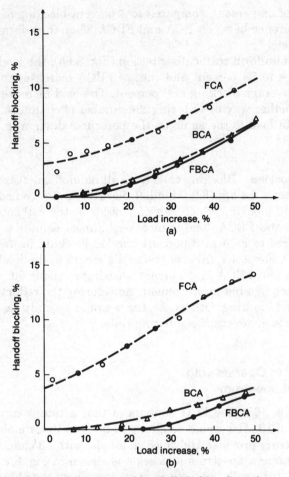

Figure 8.9 Comparison of handoff blocking from three different schemes. (*After Sekiguchi et al., Ref. 9*) (*a*) Handoff blocking in spatially uniform traffic distribution (*b*) handoff blocking in spatially nonuniform traffic distribution.

The ratio of the new additional spectrum to the present spectrum is 5/20 MHz = 25 percent, which means that the effective increase in the spectrum is 25 percent if we can fully use it.

The new additional spectrum utilization factor η at any given period of time can be calculated from

$$\eta = \frac{B}{A + B}$$

where A is the number of customers who are using old mobile units and B is the number of customers using new mobile units. If B is increasing very slowly, then η can be very small. This would defeat the purpose of implementing the new additional spectrum. Therefore, the new mobile units should outdate the old mobile units such that A remains the same and B is increasing. Assume that the number of new subscribers per year is

$$\frac{B}{A} = \frac{1}{10}$$

Then the spectrum-utilization factor for the first year that the new system is implemented would be

$$\eta = \frac{0.10A}{A + 0.10A} = 9\%$$

Then for the second year, the B/A ratio would be

$$\frac{B}{A} = \frac{1}{5}$$

and the spectrum-utilization factor η would be

$$\eta = \frac{0.2A}{A + 0.2A} = 17\%$$

These calculations are based on the assumption that new mobile units are assigned only to new additional channels so that the traffic capacity using the old spectrum will not worsen. After η exceeds 20 percent, the new mobile units have to be assigned to all the 395 voice channels. Implementation of the new additional spectrum is discussed further in Chap. 10.

8.8 Traffic and Channel Assignment

The vehicular traffic density of a coverage area is a critical element and must be determined before a system is designed. This traffic pattern in busy hours can be confined to different zones within the service area. This traffic-density information should be converted to the number of cars per 1000- × 1000-ft grid (or 2000- × 2000-ft grid) and stored in the grids of the contour map provided in Sec. 4.7.

If the traffic pattern predominates over the simple signal coverage pattern, cell-site selection will be based on the traffic pattern.

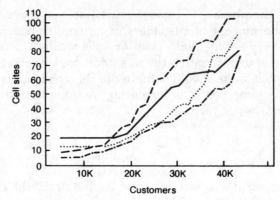

Figure 8.10 City-to-city variation. (*After Whitehead, Ref. 1.*)

Choice of the initial cell sites should be based on the signal covered in zones of heavy vehicular traffic. This means that the cell site would most likely be located at the center of those zones.

After call traffic data are collected while the system is operating, we can update the call traffic data at each cell site to correlate with the vehicular traffic data. This information will be useful for determining whether new cell splitting is needed. If it is, then we must determine how many radios should be installed at the new site and where it is to be located. These decisions are all related to frequency channel assignment. A typical chart illustrating the variation from city to city is shown in Fig. 8.10. A city may have twice as many cell sites to handle the same number of customers in the busy hours. This means that the number of cars per unit area is much higher in one city than that in the other city. Many techniques for implementing the high-capacity cellular systems are discussed in Chap. 10.

8.9 Perception of Call Blocking from the Subscribers

The regular blocked calls are counted when those calls are requested through the setup channel but no voice channels are available. If the setup channel is very busy or has poor coverage, the calls then can not be got through the setup channel. In the cases, the system operator does not know those unrecorded dropped calls. However as the subscribers are concerned, those calls are also blocked calls and named setup channel blockage in Sec. 13.1.1.

References

1. J. F. Whitehead, "Cellular Spectrum Efficiency Via Reuse Planning," *35th IEEE Vehicular Technology Conference Record,* May 21–23, 1985, Boulder, Colorado, pp. 16–20.
2. S. W. Halpern, "Reuse Partitioning in Cellular Systems," *Conference Record of the 33rd IEEE VTS Conference,* May 1983.
3. D. L. Huff, "AT&T Cellular Technology Review," *Conference Record of 1985 IEEE Military Communications Conference,* October 1985, Boston, pp. 490–494.
4. V. H. MacDonald, "The Cellular Concept," *Bell System Technical Journal,* Vol. 58, No. 1, January 1979, pp. 15–42.
5. D. C. Cox and D. O. Reudink, "Increasing Channel Occupancy in Large-Scale Mobile Radio Systems: Dynamic Channel Reassignment," IEEE *Transactions on Vehicular Technology,* Vol. VT-22, November 2, 1973, pp. 218–222.
6. L. G. Anderson, "A Simulation Study of Some Dynamic Channel Assignment Algorithms in a High Capacity Mobile Telecommunications System," *IEEE Transactions on Vehicular Technology,* Vol. VT-22, November 1973, pp. 210–217.
7. T. J. Kahwa and N. D. Georganas, "A Hybrid Channel Assignment Scheme in Large-Scale, Cellular-Structured Mobile Communication Systems," *IEEE Transactions on Communication,* Vol. COM-26, April, 1978, pp. 432–438.
8. G. Nehme and N. D. Georganas, "A Simulation Study of High-Capacity Cellular Land-Mobile Radio-Communication Systems." *Canadian Electrical Engineering Journal,* Vol. 7, No. 1, 1982, pp. 36–39.
9. H. Sekiguchi, H. Ishikawa, M. Koyama, and H. Sawada, "Techniques for Increasing Frequency Spectrum Utilization in Mobile Radio Communication Systems," *35th IEEE Vehicular Technology Conference Record,* Boulder, Colorado, May 21–23, 1985, pp. 26–31.

9

Handoffs and Dropped Calls

9.1 Value of Implementing Handoffs

9.1.1. Why handoffs[1-5]

Once a call is established, the set-up channel is not used again during the call period. Therefore, handoff is always implemented on the voice channel. The value of implementing handoffs is dependent on the size of the cell. For example, if the radius of the cell is 32 km (20 mi), the area is 3217 km^2 (1256 mi^2). After a call is initiated in this area, there is little chance that it will be dropped before the call is terminated as a result of a weak signal at the coverage boundary. Then why bother to implement the handoff feature? Even for a 16-km radius cell handoff may not be needed. If a call is dropped in a fringe area, the customer simply redials and reconnects the call.

Handoff is needed in two situations where the cell site receives weak signals from the mobile unit: (1) at the cell boundary, say, -100 dBm, which is the level for requesting a handoff in a noise-limited environment; and (2) when the mobile unit is reaching the signal-strength holes (gaps) within the cell site as shown in Fig. 9.1.

9.1.2 Two types of handoff

There are two types of handoff: (1) that based on signal strength and (2) that based on carrier-to-interference ratio. The handoff criteria are different for these two types. In type 1, the signal-strength threshold level for handoff is -100 dBm in noise-limited systems and -95 dBm in interference-limited systems. In type 2, the value of C/I at the cell boundary for handoff should be 18 dB in order to have toll quality voice. Sometimes, a low value of C/I may be used for capacity reasons.

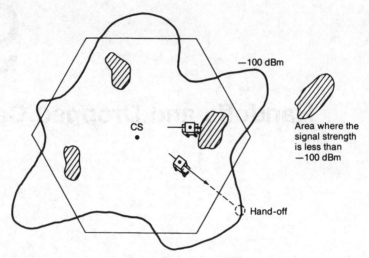

Figure 9.1 Occurrence of handoff.

Type 1 is easy to implement. The location receiver at each cell site measures all the signal strengths of all receivers at the cell site. However, the received signal strength (RSS) itself includes interference.

$$\text{RSS} = C + I \tag{9.1-1}$$

where C is the carrier signal power and I is the interference. Suppose that we set up a threshold level for RSS; then, because of the I, which is sometimes very strong, the RSS level is higher and far above the handoff threshold level. In this situation handoff should theoretically take place but does not. Another situation is when I is very low but RSS is also low. In this situation, the voice quality usually is good even though the RSS level is low, but since RSS is low, unnecessary handoff takes place. Therefore it is an easy but not very accurate method of determining handoffs. Some systems use SAT information together with the received signal level to determine handoffs (Sec. 13.1.2).

Handoffs can be controlled by using the carrier-to-interference ratio C/I, which can be obtained as described in Sec. 6.3.

$$\frac{C + I}{I} \approx \frac{C}{I} \tag{9.1-2}$$

In Eq. (9.1-2), we can set a level based on C/I, so C drops as a function of distance but I is dependent on the location. If the handoff is dependent on C/I, and if the C/I drops, it does so in response to increase in (1) propagation distance or (2) interference. In both cases, handoff

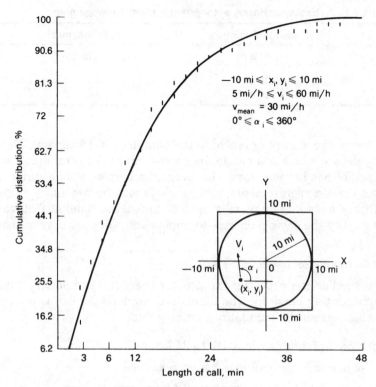

Figure 9.2 The probability of requiring handoff.

should take place. In today's cellular systems, it is hard to measure C/I during a call because of analog modulation. Sometimes we measure the level I before the call is connected, and the level $C + I$ during the call. Thus $(C + I)/I$ can be obtained. Another method of measuring C/I is described in Sec. 6.3.

9.1.3 Determining the probability of requirement for handoffs[6]

To find the probability of requiring a handoff, we can carry out the following simulation. Suppose that a mobile unit randomly initiates a call in a 16-km (10-mi) cell. The vehicle speed is also randomly chosen between 8 and 96 km/h (5 to 60 mi/h). The direction is randomly chosen to be between 0 and 360°; then the chance of reaching the boundary is dependent on the call holding time.

Figure 9.2 depicts the probability curve for requiring handoff. Table 9.1 summarizes the results. If the call holding time is 1.76 min, the only chance of reaching the boundary is 11 percent, or the chance that a handoff will occur for the call is 11 percent. If the call holding time

TABLE 9.1 Probability of Having a Handoff in a 10-mi Coverage Area

Handoff probability, %	Call length, min
11.3	1.76
18	3
42.6	6
59.3	9

is 3 min, the chance of reaching the boundary is 18 percent. Now we may debate whether a handoff is needed or not. In rural areas, handoffs may not be necessary. However, commercial mobile units must meet certain requirements, and handoffs may be necessary at times. Military mobile systems may opt not to use the handoff feature and may apply the savings in cost to implement other security measures.

9.1.4 Number of handoffs per call

The smaller the cell size, the greater the number and the value of implementing handoffs. The number of handoffs per call is relative to cell size. From the simulation, we may find

0.2 handoff per call in a 16- to 24-km cell

1–2 handoffs per call in a 3.2- to 8-km cell

3–4 handoffs per call in a 1.6-to 3.2-km cell

9.2 Initiation of a Handoff

At the cell site, signal strength is always monitored from a reverse voice channel. When the signal strength reaches the level of a handoff (higher than the threshold level for the minimum required voice quality), then the cell site sends a request to the mobile telephone switching level (MTSO) for a handoff on the call. An intelligent decision can also be made at the cell site as to whether the handoff should have taken place earlier or later. If an unnecessary handoff is requested, then the decision was made too early. If a failure handoff occurs, then a decision was made too late.

The following approaches are used to make handoffs successful and to eliminate all unnecessary handoffs. Suppose that -100 dBm is a threshold level at the cell boundary at which a handoff would be taken. Given this scenario, we must set up a level higher than -100 dBm—say, -100 dBm $+ \Delta$ dB—and when the received signal reaches this level, a handoff request is initiated. If the value of Δ is fixed and large, then the time it takes to lower -100 dBm $+ \Delta$ to -100 dBm is longer. During this time, many situations, such as the mobile unit turning back toward the cell site or stopping, can occur as a result of the direction and the speed of the moving vehicles. Then the signals will never drop below -100 dBm. Thus, many unnecessary handoffs may occur simply because we have taken the action too early. If Δ is small, then there is not enough time for the call to hand off at the cell site and many calls can be lost while they are handed off. Therefore, Δ should be varied according to the path-loss slope of the received signal strength (Sec. 4.2) and the level-crossing rate (LCR) of the signal strength (Sec. 1.63) as shown in Fig. 9.3.

Let the value of Δ be 10 dB in the example given in the preceding paragraph. This would mean a level of -90 dBm as the threshold level for requesting a handoff. Then we can calculate the velocity V of the mobile unit based on the predicted LCR[7] at a -10-dB level with respect to the root-mean-square (rms) level, which is at -90 dBm; thus

$$
V = \begin{cases} \dfrac{n\lambda}{\sqrt{2\pi}\ (0.27)} & \text{ft/s} \\ n\lambda & \text{mi/h} \end{cases} \quad \text{at } -10\text{-dB level} \qquad (9.2\text{-}1)
$$

where n is the LCR (crossings per second) counting positive slopes and λ is the wavelength in feet. Equation (9.2-1) can be simplified as

$$
V(\text{mi/h}) \approx n(\text{crossings/s}) \text{ at 850 MHz and a } -10\text{-dB level} \qquad (9.2\text{-}2)
$$

Here, two pieces of information, the velocity of vehicle V and the path-loss slope γ, can be used to determine the value of Δ dynamically so that the number of unnecessary handoffs can be reduced and the required handoffs can be completed successfully.

There are two circumstances where handoffs are necessary but cannot be made: (1) when the mobile unit is located at a signal-strength hole within a cell but not at the boundary (see Fig. 9.3) and (2) when the mobile unit approaches a cell boundary but no channels in the new cell are available.

In case 1, the call must be kept in the old frequency channel until it is dropped as the result of an unacceptable signal level. In case 2, the new cell must reassign one of its frequency channels within a reasonably short period or the call will be dropped.

The MTSO usually controls the frequency assignment in each cell and can rearrange channel assignments or split cells when they are necessary. Cell splitting is described in Sec. 10.4.

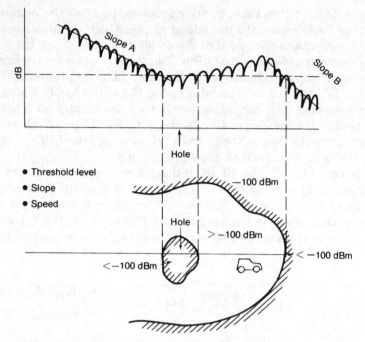

Figure 9.3 Parameters for handling a handoff.

9.3 Delaying a Handoff

9.3.1 Two-handoff-level algorithm

In many cases, a two-handoff-level algorithm is used. The purpose of creating two request handoff levels is to provide more opportunity for a successful handoff. A handoff could be delayed if no available cell could take the call.

A plot of signal strength with two request handoff levels and a threshold level is shown in Fig. 9.4. The plot of average signal strength is recorded on the channel received signal-strength indicator (RSSI) which is installed at each channel receiver at the cell site. When the signal strength drops below the first handoff level, a handoff request is initiated. If for some reason the mobile unit is in a hole (a weak spot in a cell) or a neighboring cell is busy, the handoff will be requested periodically every 5 s. At the first handoff level, the handoff takes place if the new signal is stronger (see case I in Fig. 9.4). However, when the second handoff level is reached, the call will be handed off with no condition (see case II in Fig. 9.4).

The MTSO always handles the handoff call first and the originating calls second. If no neighboring calls are available after the second

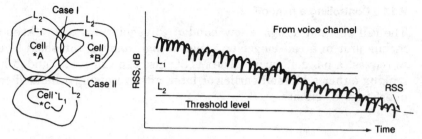

Figure 9.4 A two-level handoff scheme.

handoff level is reached, the call continues until the signal strength drops below the threshold level; then the call is dropped. If the supervisory audio tone (SAT) is not sent back to the cell site by the mobile unit within 5 s, the cell site turns off the transmitter.

9.3.2 Advantage of delayed handoffs

Consider the following example. The mobile units are moving randomly and the terrain contour is uneven. The received signal strength at the mobile unit fluctuates up and down. If the mobile unit is in a hole for less than 5 s (a driven distance of 140 m for 5 s, assuming a vehicle speed of 100 km/h), the delay (in handoff) can even circumvent the need for a handoff.

If the neighboring cells are busy, delayed handoff may take place. In principle, when call traffic is heavy, the switching processor is loaded, and thus a lower number of handoffs would help the processor handle call processing more adequately. Of course, it is very likely that after the second handoff level is reached, the call may be dropped with great probability.

The other advantage of having a two-handoff-level algorithm is that it makes the handoff occur at the proper location and eliminates possible interference in the system. Figure 9.4, case I, shows the area where the first-level handoff occurs between cell A and cell B. If we only use the second-level handoff boundary of cell A, the area of handoff is too close to cell B. Figure 9.4, case II, also shows where the second-level handoff occurs between cell B and cell C. This is because the first-level handoff cannot be implemented.

9.4 Forced Handoffs

A *forced handoff* is defined as a handoff which would normally occur but is prevented from happening, or a handoff that should not occur but is forced to happen.

9.4.1 Controlling a handoff

The cell site can assign a low handoff threshold in a cell to keep a mobile unit in a cell longer or assign a high handoff threshold level to request a handoff earlier. The MTSO also can control a handoff by making either a handoff earlier or later, after receiving a handoff request from a cell site.

9.4.2 Creating a handoff

In this case, the cell site does not request a handoff but the MTSO finds that some cells are too congested while others are not. Then the MTSO can request cell sites to create early handoffs for those congested cells. In other words, a cell site has to follow the MTSO's order and increase the handoff threshold to push the mobile units at the new boundary and to hand off earlier.

9.5   Queuing of Handoffs

Queuing of handoffs is more effective than two-threshold-level handoffs. The MTSO will queue the requests of handoff calls instead of rejecting them if the new cell sites are busy. A queuing scheme becomes effective only when the requests for handoffs arrive at the MTSO in batches or bundles. If handoff requests arrive at the MTSO uniformly, then the queuing scheme is not needed. Before showing the equations, let us define the parameters as follows.

$1/\mu$ average calling time in seconds, including *new calls* and *handoff calls* in each cell

λ_1 arrival rate (λ_1 calls per second) for originating calls

λ_2 arrival rate (λ_2 handoff calls per second) for handoff calls

M_1 size of queue for originating calls

N number of voice channels

a $(\lambda_1 + \lambda_2)/\mu$

b_1 λ_1/μ

b_2 λ_2/μ

The following analysis can be used to see the improvement. We are analyzing three cases.[8]

1. *No queuing on either the originating calls or the handoff calls.* The blocking for either an originating call or a handoff call is

$$B_o = \frac{a^N}{N!} P(0) \qquad (9.5\text{-}1)$$

where

$$P(0) = \left(\sum_{n=0}^{N} \frac{a^N}{n!} \right)^{-1} \qquad (9.5\text{-}2)$$

2. *Queuing the originating calls but not the handoff calls.* The blocking probability for originating calls is

$$B_{oq} = \left(\frac{b_1}{N} \right)^{M_1} P_q(0) \qquad (9.5\text{-}3)$$

where

$$P_q(0) = \left[N! \sum_{n=0}^{N-1} \frac{a^{n-N}}{n!} + \frac{1 - (b_1/N)^{M_1+1}}{1 - (b_1/N)} \right]^{-1} \qquad (9.5\text{-}4)$$

The blocking probability for handoff calls is

$$B_{oh} = \frac{1 - (b_1/N)^{M_1+1}}{1 - (b_1/N)} P_q(0) \qquad (9.5\text{-}5)$$

3. *Queuing the handoff calls but not the originating calls.* The blocking probability for handoff calls is

$$B_{hq} = \left(\frac{b_2}{N} \right)^{M_2} P_q(0) \qquad (9.5\text{-}6)$$

where $P_q(0)$ is as shown in Eq. (9.5-4). The blocking probability for originating calls is

$$B_{ho} = \frac{1 - (b_2/N)^{M_2+1}}{1 - (b_2/N)} P_q(0) \qquad (9.5\text{-}7)$$

Example 9.1 The following parameters are given. The number of channels at the cell site $N = 70$. The call holding time is 101 s = 0.028 h. The number of originating calls attempted per hour is expressed as $\lambda_1 = 2270$. The number of handoff calls attempted per hour is expressed as $\lambda_2 = 80$. Then

$$A = \frac{\lambda + \lambda_2}{\mu} = (2270 + 80)\, 0.028 = 65.80$$

$$b_1 = \frac{\lambda_1}{\mu} = 2270 \times 0.028 = 63.60$$

$$b_2 = \frac{\lambda_2}{\mu} = 2.24$$

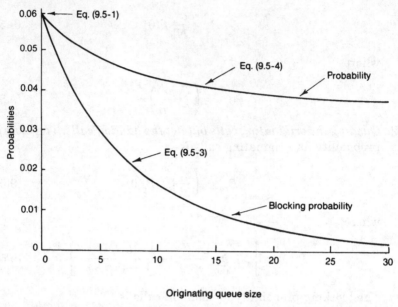

Figure 9.5 Probability and blocking probability graph showing blocking probability for originating calls queuing for originating calls (N = 70).

Given these parameters, Eqs. (9.5-1), (9.5-3), (9.5-5), (9.5-6), and (9.5-7) have been plotted in Figs. 9.5, 9.6, 9.7, and 9.8 respectively.

We have seen (Figs. 9.5 and 9.6) with queuing of originating calls only, the probability of blocking is reduced. However, queuing of originating calls results in increased blocking probability on handoff calls, and this is a drawback. With queuing of handoff calls only, blocking probability is reduced from 5.9 to 0.1 percent by using one queue space (see Fig. 9.7). Therefore it is very worthwhile to implement a simple queue (one space) for handoff calls. Adding queues in handoff calls does not affect the blocking probability of originating calls in this particular example (see Fig. 9.8). However, we should always be aware that queuing for the handoff is more important than queuing for those initiating calls on assigned voice channels because call drops upset customers more than call blockings.

9.6   Power-Difference Handoffs

A better algorithm is based on the power difference (Δ) of a mobile signal received by two cell sites, home and handoff. Δ can be positive or negative. The handoff occurs depending on a preset value of Δ.

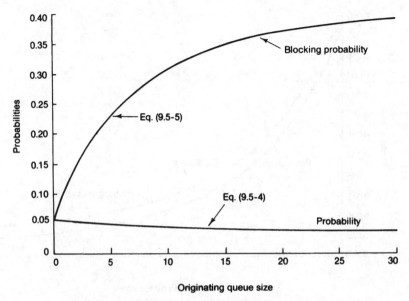

Figure 9.6 Probability and blocking probability graph showing blocking probability for handoff calls (queuing for originating calls) ($N = 70$).

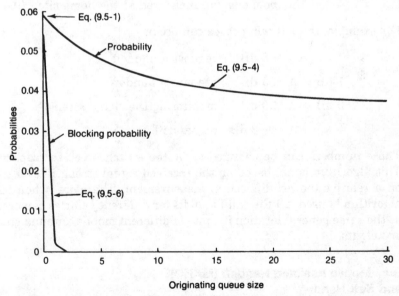

Figure 9.7 Probability and blocking probability graph showing blocking probability for handoff calls (queuing for handoff calls)($N = 70$).

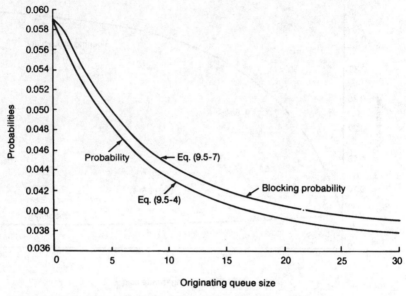

Figure 9.8 Probability and blocking probability graph showing blocking probability for originating calls (queuing for handoff calls) ($N = 70$).

Δ = the mobile signal measured at the candidate handoff site

  – the mobile signal measured at the home site   (9.6-1)

For example, the following cases can occur.

$$\Delta > 3 \text{ dB} \qquad \text{request a handoff}$$
$$1 \text{ dB} < \Delta < 3 \text{ dB} \qquad \text{prepare a handoff}$$
$$-3 \text{ dB} < \Delta < 0 \text{ dB} \qquad \text{monitoring the signal strength}$$
$$\Delta < -3 \text{ dB} \quad \text{no handoff}$$

Those numbers can be changed to fit the switch processor capacity. This algorithm is not based on the received signal strength level, but on a relative (power difference) measurement. Therefore, when this algorithm is used, all the call handoffs for different vehicles can occur at the same general location in spite of different mobile antenna gains or heights.

9.7 Mobile Assisted Handoff (MAHO) and Soft Handoff

In a normal handoff procedure, the request for a handoff is based on the signal strength or the SAT range of a mobile signal received at the cell site from the reverse link. In the digital cellular system, the mobile receiver is capable of monitoring the signal strength of the setup channels of the neighboring cells while serving a call. For instance, in a TDMA system, one time slot is used for serving a call, the rest of the time slots can be used to monitor the signal strengths of setup channels. When the signal strength of its voice channel is weak, the mobile unit can request a handoff and indicate to the switching office which neighboring cell can be a candidate for handoff. Now the switching office has two peices of information, the signal strengths of both forward and reverse setup channels, of a neighboring cell or two different neighboring cells. The switching office, therefore, has more intelligent information to choose the proper neighboring cell to handoff to.

The soft handoff is applied to one kind of digital cellular system named CDMA. In CDMA systems, all cells can use the same radio carrier. Therefore, the frequency reuse factor K approaches one. Since the operating radio carriers of all cells are the same, no need to change from one frequency to another frequency but change from one code to another code. Thus there is no hard handoff. We call this kind of handoff a soft handoff. If sometimes there are more than one CDMA radio carrier operating in a cell, and if the soft handoff from one cell to another is not possible for some reason, the intra-cell hard handoff may take place first, then go to the inter-cell soft handoff.

9.8   Cell-Site Handoff Only

This scheme can be used in a noncellular system. The mobile unit has been assigned a frequency and talks to its home cell site while it travels. When the mobile unit leaves its home cell and enters a new cell, its frequency does not change; rather, the new cell must tune into the frequency of the mobile unit (see Fig. 9.9). In this case only the cell sites need the frequency information of the mobile unit. Then the aspects of mobile unit control can be greatly simplified, and there will

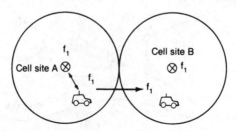

Figure 9.9 Cell-site handoff-only scheme.

be no need to provide handoff capability at the mobile unit. The cost will also be lower.

This scheme can be recommended only in areas of very low traffic. When the traffic is dense, frequency coordination is necessary for the cellular system. Then if a mobile unit does not change frequency on travel from cell to cell, other mobile units then must change frequency to avoid interference.

Therefore, if a system handles only low volumes of traffic, that is, if the channels assigned to one cell will not reuse frequency in other cells, then it is possible to implement the cell-site handoff feature as it is applied in military systems.

9.9   Intersystem Handoff

Occasionally a call may be initiated in one cellular system (controlled by one MTSO) and enter another system (controlled by another MTSO) before terminating. In some instances, *intersystem handoff* can take place; this means that a call handoff can be transferred from one system to a second system so that the call be continued while the mobile unit enters the second system.

The software in the MTSO must be modified to apply this situation. Consider the simple diagram shown in Fig. 9.10. The car travels on a highway and the driver originates a call in system A. Then the car

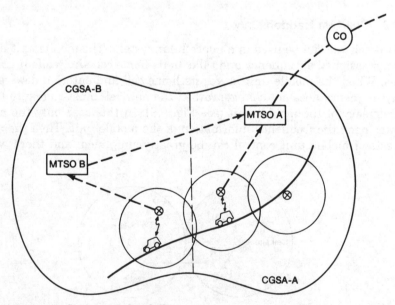

Figure 9.10   Intersystem handoffs.

leaves cell site A of system A and enters cell site B of system B. Cell sites A and B are controlled by two different MTSOs. When the mobile unit signal becomes weak in cell site A, MTSO A searches for a candidate cell site in its system and cannot find one. Then MTSO A sends the handoff request to MTSO B through a dedicated line between MTSO A and MTSO B, and MTSO B makes a complete handoff during the call conversation. This is just a one-point connection case. There are many ways of implementing intersystem handoffs, depending on the actual circumstances. For instance, if two MTSOs are manufactured by different companies, then compatibility must be determined before implementation of intersystem handoff can be considered. A detailed discussion of this topic appears in Sec. 11.4.

9.10   Introduction to Dropped Call Rate

The definition of dropped call rate.   The definition of a dropped call is after the call is established but before it is properly terminated. The definition of "the call is established" means that the call is setup completely by the setup channel. If there is a possibility of a call drop due to no available voice channels, this is counted as a blocked call not a dropped call.

If there is a possibility that a call will drop due to the poor signal of the assigned voice channel, this is considered a dropped call. This case can happen when the mobile or portable units are at a standstill and the radio carrier is changed from a strong setup channel to a weak voice channel due to the selective frequency fading phenomenon.

The perception of dropped call rate by the subscribers can be higher due to:

1. The subscriber unit not functioning properly (needs repair).

2. The user operating the portable unit in a vehicle (misused).

3. The user not knowing how to get the best reception from a portable unit (needs education).

Consideration of dropped calls.   In principle, dropped call rate can be set very low if we do not need to maintain the voice quality. The dropped call rate and the specified voice quality level are inversely proportional. In designing a commercial system, the specified voice quality level is given relating to how much C/I (or C/N) the speech coder can tolerate. By maintaining a certain voice quality level, the dropped call rate can be calculated by taking the following factors into consideration:

1. Provide signal coverage based on the percentage (say 90%) that all the received signal will be above a given signal level.

2. Maintain the specified co-channel and adjacent channel interference levels in each cell during a busy hour, i.e., the worst interference case.

3. Since the performance of the call dropped rate is calculated as possible call dropping in every stage from the radio link to the PSTN connection, the response time of the handoff in the network will be a factor when the cell becomes small, the response time for a handoff request has to be shorter in order to reduce the call dropped rate.

4. The signaling of the handoff and the MAHO algorithm will also impact the call dropped rate.

5. The relationship among the voice quality, system capacity and call dropped rate can be expressed through a common parameter C/I.

Relationship among capacity, voice quality, dropped call rate. Radio Capacity m is expressed as follows:

$$m = \frac{B_T/B_c}{\sqrt{\frac{2}{3}\,(C/I)_S}} \tag{9.10-1}$$

where B_T/B_c is the total number of voice channels. B_T/B_c is a given number, and $(C/I)_S$ is a required C/I for designing a system. The above equation is obtained based on six co-channel interferers which occur in busy traffic, i.e., a worst case. In an interference limited system, the adjacent channel interference has only a secondary effect. The derivation of Eq. (9.10-1) will be expressed in Chap. 13. Eq. (9.10-1) can be changed to the following form:

$$(C/I)_S = \frac{3}{2}\left(\frac{B_T/B_c}{m}\right)^2 = \frac{3}{2}\left(\frac{B_T}{B_c}\right)^2 \cdot \frac{1}{m^2} \tag{9.10-2}$$

Since the $(C/I)_S$ is a required C/I for designing a system, the voice quality is based on the $(C/I)_S$. When the specified $(C/I)_S$ is reduced, the radio capacity is increased. When the measured (C/I) is less than the specified $(C/I)_S$, both poor voice quality and dropped calls can occur.

Coverage of 90% equal-strength contour. The coverage in cellular cells always uses the coverage of 90% equal-strength contour. The prediction tool (Lee Model) described in Chap. 4 is used to predict the equal-strength contour at level C with 50% time and 50% area in a cell. For example, let $C = -102$ dBm, which is 18 dB above the ambient noise -120 dBm. If $C = -102$ dBm is 50% equal-strength contour, then increase the level to $C + 10$ dB contour which can be calculated from the following equation:

$$P(x' < A) = \int_{-\infty}^{A} \frac{1}{\sqrt{2\pi}\sigma} \exp\left[\frac{(y' - \overline{m})^2}{2\sigma^2}\right] dy'$$

$$= P\left(x < \frac{A - \overline{m}}{\sigma}\right) \tag{9.10-3}$$

Eq. (9.10-3) is the cumulative distribution function where A is the desired signal level and $\overline{m}$ is the mean level. σ is long-term fading due to terrain contour. If $A = C + 10 = -92$ dB and $\sigma = 8$ dB:

$$S = P\left(x < \frac{-92 - (-102)}{8}\right) = P\left(x < \frac{10}{8}\right) = 0.9082 \tag{9.10-4}$$

Eq. (9.10-4) can also be interpreted as being at a -92 dBm contour, the signal above the level of -102 dBm is 90.8%. Of course, the level of -102 dBm is determined to be 18 dB above -120 dBm which is the ambient noise level. The $(C/N)_S$ of 18 dB is the required level for getting a voice quality.

9.11   Formula of Dropped Call Rate

The dropped call rate can be calculated either using general formula or by a commonly used formula.

General formula of dropped call rate.   The general formula of dropped call rate P in a whole system can be expressed as:

$$P = 1 - \left[\sum_{n=0}^{N} \alpha_n X^n\right] = \sum_{n=0}^{N} \alpha_n \cdot P_n \tag{9.11-1}$$

where

$$P_n = 1 - X^n \tag{9.11-2}$$

P_n is the probability of a dropped call when the call has gone through n handoffs and

$$X = (1 - \delta)(1 - \mu)(1 - \theta\nu)(1 - \beta)^2 \tag{9.11-3}$$

δ = Probability that the signal is below the specified receive threshold (in a noise-limited system).

μ = Probability that the signal is below the specified cochannel interference level (in an interference-limited system).

τ = Probability that no traffic channel is available upon handoff attempt when moving into a new cell.

θ = Probability that the call will return to the original cell.

β = Probability of blocking circuits between BSC and MSC during handoff.

α_n = The weighted value for those calls having n handoffs, and $\Sigma_{n=0}^{N} \alpha_n = 1$

$N = N$ is the highest number of handoffs for those calls.

Eq. (9.11-3) needs to be explained clearly as follows:

(1) z_1 and z_2 are two events, z_1 is the case of no traffic channel in the cell, z_2 is the case of no-safe return to original cell. Assuming that z_1 and z_2 are independent events, then

$$P(z_2|z_1) \cdot P(z_1) = P(z_1) \cdot P(z_1) = \theta \cdot \tau$$

(2) $(1 - \beta)$ is the probability of a call successfully connecting from the old BSC to the MSC. Also, $(1 - \beta)$ is the probability of a call successfully connecting from the MSC to the new BSC. Then the total probability of having a successful call connection is:

$$\left. \begin{array}{ll} \text{BSC (old)} \to \text{MSC} & (1 - \beta) \\ \text{MSC} \to \text{BSC (new)} & (1 - \beta) \end{array} \right\} (1 - \beta)^2$$

(3) The call dropped rate P expressed in Eq. (9.11-1) can be specified in two cases:

1. In a noise limited system (startup system): there is no frequency reuse, the call dropped rate P_A is based on the signal coverage. It can also be calculated under busy hour conditions. In a noise-limited environment (for worst case)

$$\delta = \delta_1$$

$$\mu = \mu_1$$

$$\left. \begin{array}{l} \tau = \tau_1 \\ \theta = \theta_1 \\ \beta = \beta_1 \end{array} \right\} \text{ the conditions for the noise limited case}$$

2. In an interference-limited system (mature system): frequency reuse is applied, and the dropped rate P_B is based on

the interference level. It can be calculated under busy hour conditions.

In an interference-limited environment (for worst case)

$$\delta = \delta_2$$

$$\mu = \mu_1$$

$$\left.\begin{array}{l} \tau = \tau_2 \\ \theta = \theta_2 \\ \beta = \beta_2 \end{array}\right\} \text{ the conditions for the interference limited case}$$

Eq. (9.11-1) has to make a distinguished difference between P_A and P_B. The cases of P_A and P_B do not occur at the same time. When capacity is based on frequency reuse, the interference level is high, the size of the cells is small, and coverage is not an issue. The call dropped rate totally depends on interference.

Commonly used formula of dropped call rate. In a commonly used formula of dropped call rate, the values of τ, θ, and β are assumed to be very small and can be neglected. Then Eq. (9.11-3) becomes:

$$X = (1 - \delta)(1 - \mu) \tag{9.11-4}$$

Furthermore, in a noise-limited case, $\mu \to 0$, Eq. (9.11-1) becomes:

$$P_A = \sum_{n=0}^{N} \alpha_n P_n = \sum \alpha_n [1 - (1 - \delta)^n] \tag{9.11-5}$$

and in an interference-limited system, $\delta \to 0$, Eq. (9.11-1) becomes:

$$P_B = \sum_{n=0}^{N} \alpha_n P_n = \sum \alpha_n [1 - (1 - \mu)^n] \tag{9.11-6}$$

Handoff distribution of calls, α_n. The α_n is the weight value for those calls having n handoffs. Then the handoff distribution of all α_n's is needed for calculating Eq. (9.11-1), or Eq. (9.11-5), or Eq. (9.11-6). The relationship of all α_n's is:

$$\sum_{n=0}^{N} \alpha_n = 1$$

The handoff distribution of calls α_n can be assumed as follows:
The α_n in macrocells is used for calculating the dropped call rate P_A:

Kinds of Units	n Handoffs Per Call	Percent of Units	α_n
Handset Units	$n = 0$	100%	$\alpha_0 = 1$
Mobile Units	$n = 0$	20%	$\alpha_0 = 0.2$
	$n = 1$	60%	$\alpha_1 = 0.6$
	$n = 2$	20%	$\alpha_2 = 0.2$

The α_n in microcells is used for calculating the dropped call rate P_B:

Kinds of Units	n Handoffs Per Call	Percent of Units	α_n
Handset Units	$n = 0$	80%	$\alpha_0 = 0.8$
	$n = 1$	20%	$\alpha_1 = 0.2$
Mobile Units	$n = 0$	20%	$\alpha_0 = 0.2$
	$n = 1$	60%	$\alpha_1 = 0.6$
	$n = 2$	20%	$\alpha_2 = 0.2$

The values of α_n are used for calculating the dropped call rate. For instance, calculating the general formular of dropped call rate (Eq. (9.11-1)) in macrocells (noise-limited system) for mobile units.

$$P_A = 1 - [0.2X^0 + 0.6X^1 + 0.2X^2]$$

$$= 0.2P_0 + 0.6P_1 + 0.2P_2 \qquad (9.11\text{-}7)$$

where X is expressed in Eq. (9.11-3). In Eq. (9.11-3), the values of τ, θ, and β are usually small. Therefore, the value of X is heavily dependent on δ and μ.

9.12   Finding the Values of δ and μ

The values of δ and μ can be derived for a single cell case and in the case of a handoff. The single cell case solution is used for estimating the blocked calls. The reason behind this is that the probability of δ and μ in a single case is used for the blocked call rate of setting up calls. Assuming that after a call is set up, the call will not be dropped in a cell until the mobile unit travels into the handoff region.

Formula for δ and μ. We first find the value of δ in a single cell by integrating Eq. (9.10-3) over a whole cell to find the area Q in which the measured x will be greater than $A(r) - \overline{m}/\sigma$. The mean value $\overline{m}$ is a specified receive level. A is the signal level which is a function of $A(r)$ that exceeds $\overline{m}$ at the distance r which is less or equal to the cell radius R.

$$Q = \int_0^R P\left(x > \frac{A(r) - \overline{m}}{\sigma}\right) \cdot 2\pi r dr \qquad (9.12\text{-}1)$$

The probability δ that the signal is below a specified receive threshold $\overline{m}$ in a noise-limited environment system is

$$S = \frac{\pi R^2 - Q}{\pi R^2}$$

$$= 1 - \frac{1}{\pi R^2} \int_0^R \left(1 - P\left(x < \frac{A(r) - \overline{m}}{\sigma}\right)\right) \cdot 2\pi r dr \quad (9.12\text{-}2)$$

The probability μ that the signal is below the specified signal level C over the interference level I in an interference-limited system can also be expressed as:

$$\mu = \frac{\pi R^2 - Q}{\pi R^2}$$

$$= 1 - \frac{1}{\pi R^2} \int_0^R \left(1 - P\left(x < \frac{A(r) - C}{\sigma}\right)\right) \cdot 2\pi r \cdot dr \quad (9.12\text{-}3)$$

we may use the numerical calculation to solve Eq. (9.12-2) and Eq. (9.12-3) for dropped calls due to handoffs.

Calculation of δ and μ in a single cell. δ is calculated numerically in a noise-limited case. The cell can be divided into five rings as shown in Fig. 9.11. Eq. (9.12-2), then can be expressed as:

$$\delta = 1 - \frac{\sum_{i=1}^{5} p_i\left(x > \frac{A_i(r_i) - \overline{m}}{\sigma}\right) \cdot a_i}{\pi R^2} \quad (9.12\text{-}4)$$

where

$$1 - P_i\left(x < \frac{A_i(r_i) - \overline{m}}{\sigma}\right) = p_i\left(x > \frac{A_i(r_i) - \overline{m}}{\sigma}\right)$$

$$a_i = \pi[2i - 1]r_1^2$$

$$\sum_{i=1}^{5} a_i = \pi R^2 \quad (9.12\text{-}5)$$

in a single cell. $A_5(r_5 = R)$ is the desired signal level at the cell radius $R = 5r_1$. Let

$$p_i \left(x > \frac{A_i(r_i) - \overline{m}}{\sigma} \right) = p_i$$

for simplicity. Eq. (9.12-4) can also be expressed as:

$$\delta = \frac{\sum\limits_{i=1}^{5} (1 - p_i) \cdot a_i}{\pi R^2} \tag{9.12-6}$$

Eq. (9.12-6) is also the equation for obtaining the value of μ in the interference case.

δ_h and μ_h are improved due to the natural two-site diversity in the handoff region. Due to natural situations providing equivalent two-site diversity in the handoff region, in region a_5, the probability of dropping a call is reduced by $1 - (1 - p_5)^2$ as compared with p_5. In region a_4, the probability of dropping a call is $1 - (1 - p_4)(1 - p_6)$ as compared with p_4. p_6 is the probability of a dropped call due to the fact that the handoff takes place in a_4 by the new cell coverage. Therefore, δ_h and μ_h are expressed as:

$$\left. \begin{array}{l} \delta_h \\ \mu_h \end{array} \right\} =$$

$$\frac{(1 - p_5)^2 a_5 + (1 - p_4)(1 - p_6)a_4 + (1 - p_3)a_3 + (1 - p_2)a_2 + (1 - p_1)a_1}{\pi R^2}$$

$$\tag{9.12-7}$$

Be aware that p_i is the probability of having a successful call and P_i is the probability of a dropped call.

Example 9.1 Given $\sigma = 6$, $\overline{m} = -104$ dBm, $A_5 = -96$ dBm, find the value of δ_h during a handoff? (See Fig. 9.11.)
Based on the 40 dB/dec rule, we can obtain $A_4 = -92$ dBm, $A_3 = -87$ dBm, $A_2 = -80$ dBm, $A_1 = -68$ dBm, $A_6 = -99$ dBm and also

$$p_5 \left(x < \frac{-96 - (-104)}{6} \right) = 0.9082, \, p_4 = 0.948, \, p_3 = 0.9977,$$

$$p_2 = 1, \, p_1 = 1, \, p_6 = 0.7967$$

Then applied to Eq. (9.12-7), we obtain

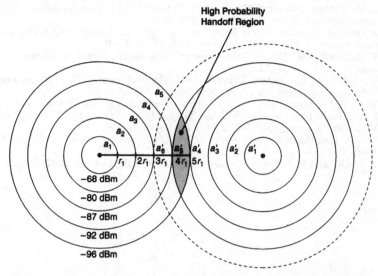

Figure 9.11 The diagram for calculating the dropped calls due to handoffs.

$$\delta_h =$$

$$\frac{(1 - p_5)^2 a_5 + (1 - p_4)(1 - p_6)a_4 + (1 - p_3)a_3 + (1 - p_2)a_2 + (1 - p_1)a_1}{\pi R^2}$$

$$= 0.64\%$$

Example 9.2 Given $\sigma = 6$, $I = -104$ dBm, $C/I = 12$ dB, and the signal received is requested to be 8 dB above the average C/I, find the value of μ_h during a handoff? Based on the 40 dB/dec rule, $C = -92$ dBm and $A_5 = -84$ dBm, we obtain $A_1 = -50$ dBm, $A_2 = -62$ dBm, $A_3 = -75$ dBm, and $A_4 = -80$ dBm, $A_6 = -87$ dBm. Then applying Eq. (9.12-7), we find:

$$\mu_h =$$

$$\frac{(1 - p_5)^2 a_5 + (1 - p_4)(1 - p_6)a_4 + (1 - p_3)a_3 + (1 - p_2)a_2 + (1 - p_1)a_1}{\pi R^2}$$

$$= 1.45\%$$

References

1. "Advanced Mobile Phone Services," Special Issue, *Bell System Technical Journal*, Vol. 58, January 1979.

2. F. H. Blecher, "Advanced Mobile Phone Service," *IEEE Transactions on Vehicular Technology,* Vol. ST-29, May 1980, pp. 238–244.
3. J. Oetting, "Cellular Mobile Radio—an Emerging Technology," *IEEE Communications Magazine,* Vol. 21, No. 8, November 1983, pp. 10–15.
4. V. H. MacDonald, "The Cellular Concept," *Bell System Technical Journal,* Vol. 58, January 1979, pp. 15–43.
5. W. C. Y. Lee, "Elements of Cellular Mobile Radio Systems," *IEEE Transactions on Vehicular Technology,* Vol. VT-35, May 1986, pp. 48–56.
6. W. C. Y. Lee and H. Smith, "A Computer Simulation Model for the Evaluation of Mobile Radio Systems in the Military Tactical Environment," *IEEE Transactions on Vehicular Technology,* Vol. VT-32, May 1983, pp. 177–190.
7. W. C. Y. Lee, *Mobile Communications Design Fundamentals,* Howard W. Sams & Co., 1986, p. 101.
8. D. R. Cox and W. L. Smith, *Queues,* Chapman & Hall Book Co., 1961, Chap. 2.

Chapter 1

Operational Techniques and Technologies

10.1 Adjusting the Parameters of a System

10.1.1 Increasing the coverage for a noise-limited system

In a noise-limited system, there is no cochannel interference or adjacent-channel interference. This means that either (1) no cochannels and adjacent channels are used in the system or (2) channel reuse distance is so large that the interference would be negligible. The following approaches are used at the cell site to increase the coverage.

Increasing the transmitted power. Usually, increasing the transmitted power of each channel results in coverage of a larger area. When the power level is doubled, the gain increases by 3 dB. Increase in covered area can be found as follows. The received power P_r can be obtained from the transmitted power P_t (see Chap. 4), where P_r is a function of the cell radius. Let the received power P_r be the power received in an original cell of a radius of r_1

$$P_{r_1} = \alpha\, P_{t_1} r_1^{-4} \qquad (10.1\text{-}1)$$

Area covered then is

$$A_1 = \pi r_1^2$$

where α is a constant and P_{r_1} can be obtained from P_{t_1}.

Case 1. The transmitted power remains unchanged but the received power changes. If the received power is to be strong, the cell radius should be smaller. The relation is

$$\frac{P_{r_1}}{P_{r_2}} = \frac{r_1^{-4}}{r_2^{-4}} = \frac{r_2^4}{r_1^4} \tag{10.1-2}$$

or
$$r_2 = \left(\frac{P_{r_1}}{P_{r_2}}\right)^{1/4} r_1 \tag{10.1-3}$$

If $P_{r_2} = 2P_{r_1}$, and the transmitted power remains the same, the radius reduces to

$$r_2 = (0.5)^{1/4} \quad r_1 = 0.84 r_1$$

and the area reduces to

$$\frac{A_2}{A_1} = \frac{\pi r_2^2}{\pi r_1^2} = \frac{r_2^2}{r_1^2} = \frac{(0.84 r_1)^2}{r_1^2} = 0.71 \tag{10.1-4}$$

Case 2. The transmitted power changes but the received power doesn't; then the 1-mi reception level changes if the transmitted power changes. From Eq. (10.1-1) we obtain

$$P_{r_1} = \alpha P_{t_1} r_1^{-4} \quad P_{r_2} = \alpha P_{t_2} r_2^{-4}$$

In this case, since $P_{r_1} = P_{r_2}$, it follows that

$$r_2 = \left(\frac{P_{t_2}}{P_{t_1}}\right)^{1/4} r_1 \tag{10.1-5}$$

If the transmitted power P_{t_2} is 3 dB higher than P_{t_1}, then

$$r_2 = (2)^{1/4} \quad r_1 = 1.19 r_1$$

and the area increase is

$$\frac{A_2}{A_1} = \frac{r_2^2}{r_1^2} = (1.19)^2 = 1.42 \tag{10.1-6}$$

A general equation should be expressed as

$$r_2 = \left(\frac{P_{r_1} P_{t_2}}{P_{r_2} P_{t_1}}\right)^{1/4} r_1 \tag{10.1-7}$$

or
$$A_2 = \left(\frac{P_{r_1} P_{t_2}}{P_{r_2} P_{t_1}}\right)^{1/2} A_1 \tag{10.1-8}$$

Increasing cell-site antenna height. In general, the 6 dB/oct rule applies to the cell-site antenna height in a flat terrain, that is, doubling the antenna height causes a gain increase of 6 dB. If the terrain contour is hilly, then an effective antenna height should be used, depending on the location of the mobile unit. Sometimes doubling the actual antenna height results in a gain increase of less than 6 dB and sometimes more. This phenomenon was described in Chap. 4.

Using a high-gain or a directional antenna at the cell site. The gain and directivity of an antenna increase with the received level—the same effect seen with an increase of transmitted power.

Lowering the threshold level of a received signal. When the threshold level is lowered, the acceptable received power is lower and the radius of the cell increases [Eq. (10.1-3) applies]. The increase in service area due to a lower received level can be obtained from Eq. (10.1-8). Let $P_{t_2} = P_{t_1}$, and $P_{r_2} = 0.25P_{r_1}$ (i.e., -6 dB). Then $A_2 = 2A_1$. The received level is reduced by 6 dB, and the service area is doubled.

A low-noise receiver. The thermal noise kTB level (see Sec. 7.6.6) is -129 dBm. In a noise-limited environment, if the front-end noise of the receiver is low and the received power level remains the same, the carrier-to-noise ratio becomes large in comparison to a receiver with a high front-end noise. This low-noise receiver can receive a signal from a farther distance than can a high-noise receiver.

Diversity receiver. A diversity receiver is very useful in reducing the multipath fading. When the fading reduces, the reception level can be increased. Diversity receiver performance is discussed in further detail in Sec. 10.2.3.

Selecting cell-site locations. With a given actual antenna height and a given transmitted power, coverage area can be increased if we can select a proper site. Of course, in principle, for coverage purposes, we always select a high site if there is no risk of interference. However, sometimes we need to cover an important area within the coverage area; in such cases it is necessary to move around the site location.

Using repeaters and enhancers to enlarge the coverage area or to fill in holes. This is discussed in Sec. 10.2.

Engineering the antenna patterns. The technique of engineering the antenna patterns mentioned in Sec. 7.6.4 can be used to cover a desired service area.

10.1.2   Reducing the interference

In most situations, the methods mentioned in Sec. 10.1.1 for increasing the coverage area would cause interference if cochannels or adjacent channels were used in the system. Methods for reducing the interference are as follows.

1. *A good frequency-management chart.*   As shown in Fig. 8.1, there are 21 sets of channels in the chart. In each channel set, the neighboring frequency is 21 channels away. No interference can be caused within a set of 16 channels.

2. *An intelligent frequency assignment.*   In order to assign the 21 sets in a $K = 7$ frequency reuse pattern and to avoid the interference problems from adjacent-channel or cochannel interference, an intelligent frequency assignment in real time is needed.

3. *A proper frequency among a set assigned to a particular mobile unit.*   Depending on the current situation, some idle channels may be noisy, some may be quiet, and some may be vulnerable to channel interference. These factors should be considered in assignment of frequency channels.

4. *Design of an antenna pattern on the basis of direction.*   In some directions a strong signal may be needed; in other directions no signal may be needed. The design tool should include the findings of signal requirements on the basis of antenna direction.

5. *Tilting-antenna patterns.*   To confine the energy within a small area, we may use an umbrella-pattern omnidirectional antenna or downward tilting directional antenna.

6. *Reducing the antenna height.*   We can use this method because reducing interference is more important than radio coverage.

7. *Reducing the transmitted power.*   In certain circumstances, reducing transmitted power can be more effective in eliminating interference than reducing the height of the antenna.

8. *Choosing the cell-site location.*   The propagation prediction model described in Chap. 4 can be used to select cell-site locations for eliminating interference.

10.1.3   Increasing the traffic capacity

Small cell size.   If we can control the radiation pattern, we can reduce the size of the cell and increase the traffic capacity. This approach is based on the assumption that all the mobile units are identical, including the mobile antennas and their mounting.

Increasing the number of radio channels in each cell. Either omnidirectional or directional antennas can be used in each cell. Sometimes the channel combiner can process only 16 channels. Thus, if we need 96 channels, we need six transmitted antennas. Also, if 6 frequency sets are used, then the total of 21 sets is divided by 6. The closest neighboring channels would be only four channels away. A good channel assignment method is needed (see Chap. 8).

Enhanced frequency spectrum. Cellular mobile industries have been allocated an additional 166 voice channels. With an enhanced frequency spectrum, traffic capacity is increased.

Queuing. Queuing of handoff calls can increase traffic capacity, as discussed in Chap. 9.

Dynamic channel assignemnt. Dynamic, rather than fixed, channel assignment is another means of increasing traffic capacity. As mentioned in Chap. 8, external environmental factors, such as traffic volume, are considered in dynamic channel assignment.

10.2   Coverage-Hole Filler

Because the ground is not flat, many water puddles form during a rainstorm; for the same reason, many holes (weak spots) are created in a general area during antenna radiation. There are several methods for filling these holes.

10.2.1   Enhancers (repeaters)[1]

An enhancer is used in an area which is a hole (weak spot) in the serving cell site. There are two types of enhancer: wideband and channelized enhancers.

The wideband enhancer is a repeater. It is designed for either block A or block B channel implementation. All the signals received will be amplified. Sometimes it can create intermodulation products; therefore, implementation of an enhancer in an appropriate place to fill the hole without creating interference is a challenging job. One application is shown in Fig. 10.1. The amplifier requires only low amplification. The signal is transmitted from the cell site and received at the enhancer site by a higher directional antenna which is mounted at a high altitude. The signal received in the forward channel will be radiated by the lower antenna, which is either an omnidirectional or a directional antenna at the enhancer. The mobile units in the vicinity of the enhancer site will receive the signal. The mobile unit uses the

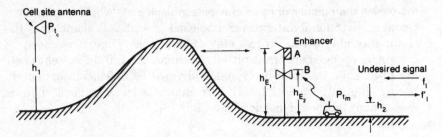

Figure 10.1 Enhancer.

reverse channel to respond to calls (or originate calls) through the enhancer to the cell site.

However, the amplifier amplifies both the signal and the noise, as discussed in Sec. 7.4.2. Therefore, the enhancer cannot improve the signal-to-noise (S/N) ratio. The function of enhancers is actually a relay, receiving at a lower height h_2 and transmitting to a higher height h_1 or vice versa. The gain of the enhancer can be adjusted from 10 to 70 dB, and the range is from 0.5 to 3 km.[1] The received signal at the mobile units and at the cell site with an enhancer placed in the middle can be expressed as

$$P_{Rm} = P_{t_r} + g_c - L_a + (G + g_{E_1} + g_{E_2}) - L_b + g_m \qquad (10.1\text{-}9)$$

and
$$P_{Rc} = P_{t_m} + g_m - L_b + (G + g_{E_1} + g_{E_2}) - L_a + g_c \qquad (10.1\text{-}10)$$

where
P_{t_c} = transmitted power at cell site
P_{t_m} = transmitted power at mobile unit
g_c = antenna gain at cell site
g_m = antenna gain at mobile unit
g_{E_1}, g_{E_2} = antenna gain at enhancer
G = amplification gain at enhancer
P_{Rc}, P_{R_m} = received power at cell site and at mobile unit, respectively
h_1 = antenna height at cell site
h_2 = antenna height at mobile unit
h_{E_1}, h_{E_2} = antenna heights at enhancer
L_a = path loss between cell site and enhancer
L_b = path loss between enhancer and mobile unit

The general formula of path loss in a mobile radio environment [see Eq. (4.2-18)] can be used to calculate both L_a and L_b. Equation (4.2-18) contains an expression of a function of antenna height which would vary in different situations.

If the undesired signal received by the antenna at height h_{E_1} is transmitted back to the cell site, cochannel or adjacent-channel interference may result. This could also occur when an undesired signal is received by the antenna at height h_{E_1} because of poor design and is repeatedly transmitted by the antenna at height h_{E_2}, causing interference in a region in which undesired signal enhancement should not occur.

The channelized enhancer should amplify only the channels that it selected previously with a good design. Therefore, it is a useful apparatus for filling the holes.

Caution: Three points should be noted in the installation of an enhancer.

1. Ring oscillation might easily occur. The separation between two (upper and lower) antennas at the enhancer is very critical. If this separation is inadequate, the signal from the lower antenna can be received by the upper antenna or vice versa and create a ring oscillation, thus jamming the system instead of filling the hole.

2. The distance between the enhancer and the serving cell site should be as small as possible to avoid spread of power into a large area in the vicinity of the serving site and beyond.

3. Geographic (terrain) contour should be considered in enhancer installation.

10.2.2 Passive reflector

In order to redirect the incident energy, the reflector system should be installed in a field far from both the transmitting antenna and the receiving antenna.[2] The approximate separation between the antenna and the reflector is

$$d_1 > \frac{2A_T}{\lambda} + \frac{2A_1}{\lambda} \quad \text{and} \quad d_2 > \frac{2A_1}{\lambda} + \frac{2A_R}{\lambda} \qquad (10.2\text{-}1)$$

where A_T, A_R = effective aperture of transmitting antenna and receiving antennas, respectively
d_1, d_2 = distance from reflector to transmitting antenna and receiving antenna, respectively
λ = wavelength

If the transmitting and receiving antennas are linear elements, then

$$d_1 > \frac{2L_T^2}{\lambda} + \frac{2A_1}{\lambda} \quad \text{and} \quad d_2 > \frac{2A_1}{\lambda} + \frac{2L_R^2}{\lambda} \qquad (10.2\text{-}2)$$

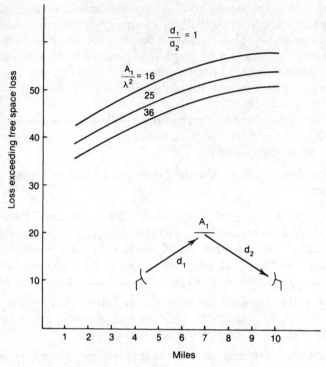

Figure 10.2 Effective use of reflectors $d_1/d_2 = 1$.

where L_T and L_R are, respectively, the transmitted and received lengths of the elements. The incident angle in this case would be less than 70° in order to deflect the energy in another direction (see Fig. 10.2).

The dimension of the reflector should be many wavelengths. Assume that 100 percent of the incident power is reflected; then

$$P_R = P_T \frac{A_T A_R A_1^2}{\lambda^4 d_1^2 d_2^2} \tag{10.2-3}$$

where P_T, P_R = transmitted and received power, respectively, and

$$A_T = \frac{G_T \lambda^2}{4\pi} \tag{10.2-4}$$

$$A_R = \frac{G_R \lambda^2}{4\pi} \tag{10.2-5}$$

Then Eq. (10.2-3) becomes

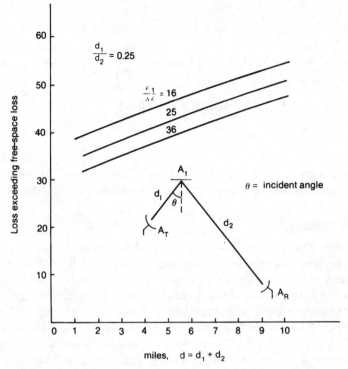

Figure 10.3 Effective use of reflectors $d_1/d_2 = 0.25$.

$$P_R = P_T G_T G_R \frac{A_1^2}{(4\pi)^2 d_1^2 d_2^2}$$

$$= P_T G_T G_R \left[\left(4\pi \frac{d}{\lambda}\right)^2\right]^{-1} \cdot \left(\frac{d^2 (A_1/\lambda^2)^2}{d_1^2 d_2^2}\right)$$

free-space loss (FSL)   excessive loss   (10.2-6)

where $d = d_1 + d_2$ and

$$P_R = 10 \log(\text{FSL}) + 10 \log \left[\left(\frac{d^2 \lambda^2}{d_1^2 d_2^2}\right)\left(\frac{A_1}{\lambda^2}\right)^2\right] \qquad (10.2\text{-}7)$$

The excessive loss in Eq. (10.2-7) is plotted in Fig. 10.2 for the case of $d_1/d_2 = 1.0$ and in Fig. 10.3 for $d_1/d_2 = 0.25$ at 850 MHz. In a mobile radio environment, d_1 can be considered to be in a free space and d_2 to be a mobile radio path from the reflector to the mobile unit.

Then Eq. (10.2-3) can be modified as

$$P_R = P_T \frac{A_T A_R A_1^2}{\lambda^2 d_1^2 d_2^4}$$

$$= P_T G_T G_R \left[\left(4\pi \frac{d_1}{\lambda} \right)^2 \frac{d_2^4}{\lambda^4} \right]^{-1} \cdot \frac{(A_1/\lambda^2)^2}{1} \qquad (10.2\text{-}8)$$

or
$$P_R = 10 \log(\text{FSL}) + 10 \log \left[\frac{d^2 A_1^2 \lambda^2}{d_1^2 d_2^4} \right] \qquad (10.2\text{-}9)$$

excessive loss

Comparing Eq. (10.2-9) with Eq. (10.2-7), we realize that the excessive loss from a reflector is greater in a mobile radio environment.

10.2.3 Diversity

The diversity receiver can be used to fill the holes. Because the diversity receiver can receive a lower signal level, the hole that existed in a normal receiver reception case now becomes a no-hole (or lesser hole) situation with the use of the diversity receiver. An improvement in the signal-to-noise ratio of a two-branch diversity receiver[3] is shown in Fig. 10.4. The diversity schemes can be classified as[4] (1) polarization diversity, (2) field component-energy density,[5] (3) space diversity, (4) frequency diversity, (5) time diversity, and (6) angle diversity.

For any two independent branches the performance obtained from any of the diversity schemes listed above is the same; that is, the correlation coefficient of the two received signals becomes zero. The performance can be degraded if the two signals obtained from the two branches are dependent on a correlation coefficient, as shown in Fig. 10.4. The performance can also vary with different diversity-combiner techniques.[6] The maximal-ratio combiner is the best performance combiner. The equal-gain combiner has a 0.5-dB degradation as compared with the maximal-ratio combiner. The selective combiner has a 2-dB degradation as compared with the maximal-ratio combiner.

The performance increase based on a diversity scheme for a two-branch equal-gain diversity combiner is shown in Fig. 10.5a for the cumulative probability distribution (CPD) and in Fig. 10.5b for the level-crossing rate (LCR). The average duration of fades $\bar{t}$ can be obtained by calculating $\bar{t} = \text{CPD/LCR}$ as shown in Eq. (1.6-9). Also, we can plot the performance of the diversity combined signal with different correlation coefficients between two branches. For example, at the cell site, the correlation coefficient ρ between branches is set to be 0.7 for the reality of physical antenna separation. At the mobile unit, however, the signal correlation of two branches is almost zero with a sep-

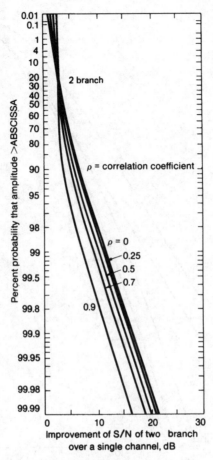

Figure 10.4 Improvement of signal-to-noise ratio of a two-branch signal over a single channel signal. (*After Lee, Ref. 3.*)

aration of $d = 0.5\lambda$. Reductions in fading and in level-crossing rate are shown in Fig. 10.5*a* and *b*, respectively. The improvement in the signal-to-noise ratio of a two-branch signal over a single branch with different values of correlation coefficients between channel signals is shown in Fig. 10.4. The maximum improvement occurs when $\rho = 0$.

10.2.4 Cophase technique

The cophase technique is used to bring all signal phases from different branches to a common phase point. Here, the common phase point is the point at which the random phase in each branch is reduced. There

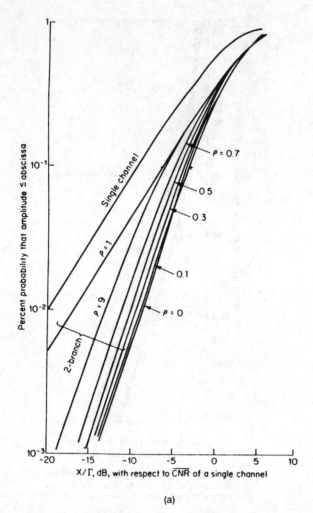

Percent probability that amplitude ≤ abscissa

Single channel

$\rho = 0.7$

0 5

0 3

0.1

$\rho = 1$

$\rho = 9$

$\rho = 0$

2-branch

10^{-1}

10^{-2}

10^{-3}

-20   -15   -10   -5   0   5   10

X/Γ, dB, with respect to $\overline{CNR}$ of a single channel

(a)

Figure 10.5 (*a*) Cumulative probability distribution of a two-branch correlated equal-gain-combining signal. (*After Lee, Ref. 7.*) (*b*) level-crossing rate of a two-branch equal-gain-combining signal. (*After Lee, Ref. 7.*)

are two kinds of cophase techniques: feedforward and feedback[7] (these circuits are shown in Fig. 10.6*a* and *b*, respectively).

The feedforward cophase technique has been used for satellite communication applications. It is simpler than the phase-locked loop. The latter is also called the *Granlund combiner*. The outcome of the feedback technique is always better than that of the feedforward technique provided the two filters in the circuit have been properly designed to avoid any significant time delay.

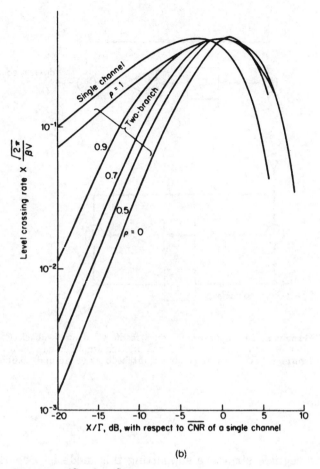

(b)

Figure 10.5 (*Continued*)

10.3 Leaky Feeder

10.3.1 Leaky waveguides

Typically, the velocity of propagation of an electromagnetic wave V_g in the waveguide is greater than the speed of light V_c. However, the carrier frequency in hertz should be the same as in the waveguide and in free space. Thus, if two waves have the same frequency, their wavelengths will be longer in the waveguide than in free space, as seen from the following equation.

$$\lambda_g = \frac{V_g}{f} \qquad (10.3\text{-}1)$$

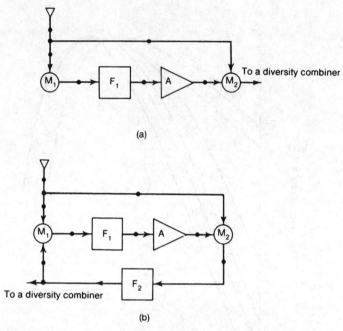

(a)

(b)

Figure 10.6 Two cophase techniques, feedforward and feedback (F = the filter, M = the mixer, and A = the limiting amplifier, as shown in the figure). (*a*) Feedforward combiner; (*b*) feedback (Granlund) combiner.

Therefore,

$$\lambda_g > \lambda_c \tag{10.3-2}$$

If the waveguide structure supporting this mode is properly opened up, then the energy will leak into the exterior region.[8] The opening slots (apertures) will usually be placed along the waveguide periodically. This leaky waveguide is different from a slot antenna. The slot antenna is designed to radiate all the energy into the space at the slot, whereas in the leaky waveguide, fractional energy will be leaking constantly. Because V_g is greater than V_c, the leaky waveguide may sometimes be categorized as a fast-wave antenna.

The general field expression can be written for the interior and exterior regions of the waveguide and matched across the slot boundary. For a circular-shaped waveguide,[9] the internal field is TE_{11}. The attenuation, or the leakage energy, is shown in Fig. 10.7*a*. Figure 10.7*b* shows the dimensions of the circular waveguide. The leakage rate is a function of position in the waveguide, where $\alpha = 0.1$ is the fraction of the input power absorbed in the load, that is, the amount of energy that leaks out.

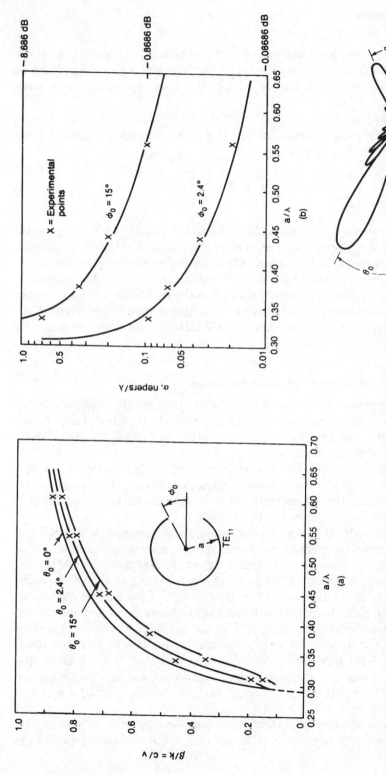

Figure 10.7 Complex propagation constant for a TE_{11} slotted cylinder. (*After Harrington, Ref. 9.*). (a) Leaky energy due to slot size; (b) wave velocity in a slotted cylinder, (c) principal plane pattern of a leaky wave slot; showing forward and backward main beams.

321

The leaky waveguide pattern.   Pattern is also a very important factor in the application of leaky waveguides. We would like the pattern to be similar to that in Fig. 10.7c, which can serve a larger area along the waveguide.

The coaxial cable or leaky coaxial.   The phase velocity V of a wave traveling on the line is given by

$$V = \frac{\omega}{\beta} = \frac{1}{\sqrt{LC}}$$ (10.3-3)

At frequencies below 1000 MHz, the use of coaxial cable is universal because the attenuation per unit length is reasonable and the dimensions are practical for passing the principal modes of leaky waves. At higher frequencies, since the dimensions of coaxial cable cannot be physically reduced in order to suppress the high modes of leaky waves in the coaxial cable, excessively high attenuation might occur. Consequently, for frequencies above 3000 MHz, waveguides are generally used.

10.3.2   Leaky-feeder radio communication

In some areas, such as in tunnels or in other confined spaces such as underground garages or a cell of less than 1-mi radius, leaky-feeder techniques become increasingly important to provide adequate coverage and reduce interference.

In 1956 a "guided radio" was introduced.[10] This "radio" is actually a low-frequency inductive communication device. The proposal included utilization of existing conductors such as power cables and telephone lines to transmit the signal.

Also in 1956, the leaky-feeder principle for propagation of VHF and UHF signals through a tunnel or a confined area was presented.[11] The open-braided type (i.e., containing zigzag slots) of coaxial cable is used in most applications for suppressing any resulting surface-wave interference (Fig. 10.8). However, in this design, if the cable slots are all the same size, then there is nonuniform energy leakage along the cable. A great deal of energy may leak out at the slots which are arrived at first. For instance, a leaky cable can have a loss of 2 dB per 100 ft at 1000 MHz. The "daisy chain" system patented in 1971 avoids the complications and shortcomings of two-way signal boosters along the cable.[12] Therefore the radiation signal level can be within a specified range.

Because of "intrinsic safety" considerations, in order to prevent any incendiary sparks (e.g., as in a coal mine), the RF powers cannot ex-

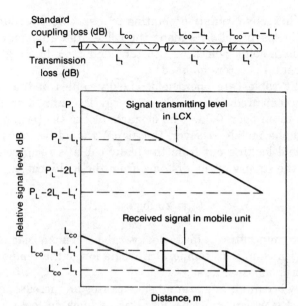

Figure 10.8 Grading technique.

ceed a maximum of 500 mW, and any line-fed power passed over leaky feeders used for boosters (power amplification along the cable) should be limited to a few watts. In urban applications, a 0.25-mi-long leaky cable will be used without the power amplifying stage.

The leaky feeder is characterized by transmission and coupling losses. Transmission loss is expressed in decibels per unit length. Coupling loss is defined by the ratio of power received by a dipole antenna at a distance s equal to 1.5 m away from the cable to the transmitted power in the cable at a given point. The smaller the ratio, the greater the loss. If the distance is other than 1.5 m, the coupling loss (or free-space loss from a leaky cable) L increases as d increases.

$$L \ (s \text{ at } d) = (\text{coupling loss at 1.5 m}) + 10 \log \left(\frac{d}{1.5} \right) \quad (10.3\text{-}4)$$

The free-space loss from a leaky cable is described later. The coupling loss can be controlled by size and slot angle, whereas the transmission loss varies with the coupling loss and cannot be chosen independently of the coupling loss. The principal of leaky-cable operation is

1. Use high-coupling-loss (little energy will leak out) cables near the transmitter end. Usually high-coupling-loss cables have a low-transmission loss and are of greater length in use. We can arrange the lengths of cables due to different coupling losses as shown in Fig. 10.8.

2. The intensive radiation pointing to a specific direction is caused by periodic spacing of slots along the cable. Radiation can be distributed through joint points or boosters and by adjusting the signal phases around boosters as needed.

3. Leaky cables are open fields. Leaky cables in the tunnels are easily implemented because their energy is confined to the tunnel. However, in an open field, if no obstacle blocks the path between the cable and the mobile receiver, the signal should be less varied. The electric field leaking out from the leaky cable is reciprocally proportional to the square of the distance from the leaky cable.

$$L_r = 20 \log s \quad \text{dB} \tag{10.3-5}$$

4. Low temperature affects leaky cable. Transmission-loss levels change with change in temperature. The lower the temperature, the less the transmission loss.

5. Snow accumulation around slots causes an increase in transmission loss. Reflection and path loss due to snow on leaky cable cause an increase of coupling loss.

6. The boosters are power amplifiers. Therefore, many narrowband-modulated carriers passing through common broadband amplifiers generate intermodulation (IM) product power. In order to reduce the IM product to a specified level, the linear amplifiers should be operated at a reduced output level by backing them off from the 1-dB-gain compression point.

The amplification of a fundamental signal and its most dominant IM (i.e., third-order IM) is illustrated in Fig. 10.9. Because the slopes of curves for fundamental signal and third-order IM are always fixed, the higher the intercept point, the lower the IM product interference. Also, we can find the output backoff level from a given IM product suppression, Y_1, as

$$2 (Y_1 + Y_0) = \Delta$$

$$Y_1 = \frac{\Delta}{2} - Y_0 \tag{10.3-6}$$

where Y_0 is the power difference between the intercept point and the 1-dB-gain compression point. The IM product levels and numbers are given in Ref. 13. Other literature references can be found in Refs. 22 and 23.

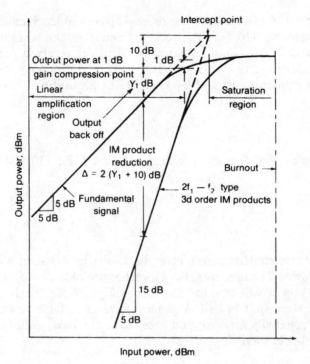

Figure 10.9 Input-output characteristics of a linear amplifier (*After Suzuki et al., Ref. 12*).

10.4 Cell Splitting

When the call traffic in an area increases, we must split the cell so
that we can reuse frequency more often, as we have mentioned in
Chap. 2. This involves reducing the radius of a cell by half and splitting an old cell into four new small cells. The traffic is then increased
fourfold.[14]

10.4.1 Transmitted power after splitting

The transmitted power P_{t_2} for a new cell, because of its reduced size,
can be determined from the transmitted power P_{t_1} of the old cell.

If we assume that the received power at the cell boundary is P_r,
then the following equations (where α is a constant) can be deduced
from Eq. (10.1-1).

$$P_r = \alpha P_{t_0} R_0^{-\gamma} \tag{10.4-1}$$

$$P_r = \alpha P_{t_2} \left(\frac{R_1}{2}\right)^{-\gamma} \tag{10.4-2}$$

Equation (10.4-1) expresses the received power at the boundary of the old cell and Eq. (10.4-2), the received power at the boundary of the new cell $R_1 = (R_0/2)$. To set up an identical received power P_r at the boundaries of two different-sized cells, and dropping the parameter P_r by combining Eqs. (10.4-1) and (10.4-2), we find

$$P_{t_1} = P_{t_0} \left(\frac{1}{2}\right)^{-\gamma} \tag{10.4-3}$$

For a typical mobile radio environment, $\gamma = 4$, Eq. (10.4-3) becomes

$$P_{t_1} = \frac{P_{t_0}}{16} \tag{10.4-4}$$

or
$$P_{t_1} = P_{t_0} - 12 \quad \text{dB} \tag{10.4-5}$$

The new transmitted power must be 12 dB less than the old transmitted power. The new cochannel interference reduction factor q_2 after cell splitting is still equal to the value of q (see Eq. 2.3-1) since both D and R were split in half. A general formula is for a new cell which is split repeatedly n times, and every time the new radius is one-half of the old one; then $R_n = R_0/2^n$.

$$P_{t_n} = P_{t_0} - n(12) \quad \text{dB} \tag{10.4-6}$$

When cell splitting occurs, the value of the frequency-reuse distance q is always held constant. The traffic load can increase four times in the same area after the original cell is split into four subcells. Each subcell can again be split into four subcells, which would allow traffic to increase 16 times. As the cell splitting continues, the general formula can be expressed as

New traffic load = $(4)^n$ × (the traffic load of start-up cell)   (10.4-7)

where n is the number of splittings. For $n = 4$, this means that an original start-up cell has split four times. The traffic load is 256 times larger than the traffic load of the start-up cell.

10.4.2 Cell-splitting technique

The two techniques of cell splitting are described below.

Permanent splitting. Selecting small cell sites is a tough job. The antenna can be mounted on a monopole or erected by a mastless arrangement which will be described later. However, these splittings can be easy to handle as long as the cutover from large cells to small cells

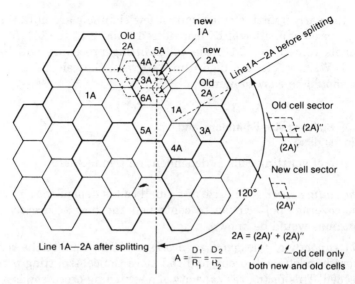

Figure 10.10 Cell-splitting techniques.

takes place during a low traffic period. The frequency assignment should follow the rule (see Sec. 2.3) based on the frequency-reuse distance ratio q with the power adjusted.

Real-time splitting (dynamic splitting). In many situations, such as traffic jams at football stadiums after a game, in traffic jams resulting from automobile accidents, and so on, the idle small cell sites (inactive ones) may be rendered operative in order to increase the cell's traffic capacity.

Cell splitting should proceed gradually over a cellular operating system to prevent dropped calls. Suppose that the area exactly midway between two old 2A sectors requires increased traffic capacity as indicated in Fig. 10.10. We can take the midpoint between two old 2A sectors and name it "new 2A." The new 1A sector can be found by rotating the old 1A–2A line (shown in Fig. 10.10) clockwise[15] 120°. Then the orientation of the new set of seven split cells is determined. To maintain service for ongoing calls while doing the cell splitting, we let the channels assigned in the old 2A sector separate into two groups.

$$2A = (2A)' + (2A)'' \tag{10.4-8}$$

where $(2A)'$ represents the frequency channels used in both new and old cells, but in the small sectors, and $(2A)''$ represents the frequency channels used only in the old cells.

At the early splitting stage, only a few channels are in $(2A)'$. Gradually, more channels will be transferred from $(2A)''$ to $(2A)'$. When no channels remain in $(2A)''$, the cell-splitting procedure will be completed. With a software algorithm program, the cell-splitting procedure should be easy to handle.

10.4.3 Splitting size limitations and traffic handling

The size of splitting cells is dependent on two factors.

The radio aspect.   The size of a small cell is dependent on how well the coverage pattern can be controlled and how accurately vehicle locations would be known.

The capacity of the switching processor.   The smaller the cells, the more handoffs will occur, and the more the cell-splitting process is needed. This factor, the capacity of a switching processor, is a larger factor than the handling of coverage areas of small cells.

10.4.4 Effect on splitting

When the cell splitting is occurring, in order to maintain the frequency-reuse distance ratio q in a system, there are two considerations.

1. Cells splitting affects the neighboring cells. Splitting cells causes an unbalanced situation in power and frequency-reuse distance and makes it necessary to split small cells in the neighboring cells. This phenomenon is the same as a ripple effect.

2. Certain channels should be used as barriers. To the same extent, large and small cells can be isolated by selecting a group of frequencies which will be used only in the cells located between the large cells on one side and the small cells on the other side, in order to eliminate the interference being transmitted from the large cells to the small cells.

10.5 Small Cells (Microcells)

As we mentioned in Sec. 10.4.3, the limitation of a small cell is based on the accuracy of vehicle locations and control of the radiation patterns of the antennas. In this section, we try to find the means of control, the radiation power. The intelligent cell concept and application to microcells are described in Chap. 16.

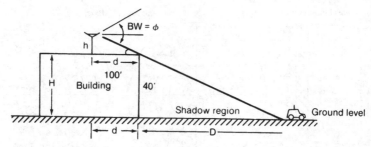

Figure 10.11 Rooftop-mounted antennas.

10.5.1 Installation of a mastless antenna

Use of existing building structures. Building structures can be used to mount cell-site antennas. In such cases the rooftop usually is flat. There should be enough clearance around the antenna post mounted in the middle of the building to avoid blockage of the beam pattern from the edge of the roof (see Fig. 10.11). A formula may be applied for this situation. Given the vertical beamwidth of antenna ϕ and the distance from the antenna post to the edge of the roof d, the height of the post can be determined by

$$h = d \tan \left(\frac{\phi}{2} \right) \tag{10.5-1}$$

If a 6-dB-gain antenna has a vertical beamwidth of about 28° and the distance from the antenna post to the roof edge is 31 m (100 ft), then the required antenna height is 7.5 m (25 ft). The shaded region around the building depends on the height of the building.

$$D = \frac{H + h}{\tan(\phi/2)} - d \tag{10.5-2}$$

If $H = 12$ m (40 ft), $h = 7.5$ m (25 ft), $\phi/2 = 14°$, then Eq. (10.5-2) becomes $D = 49$ m (160 ft).

The shadow region is calculated for a single building only. If there are adjacent buildings, multipath scattered waves are generated, and the shadow region is reduced.

Use of the antenna structures. The panel-type antennas[16] are ideal antenna structures for hanging on each side of the wall. For an omnidirectional configuration, the four-panel antennas mounted on the four sides of the building can be combined as in an omnidirectional antenna.

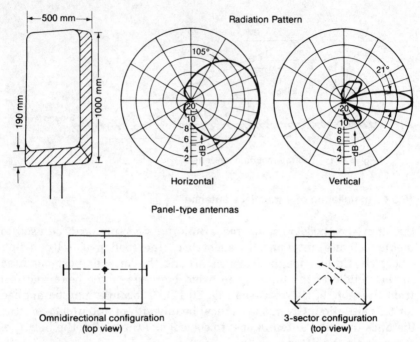

Figure 10.12 Panel-type antennas and their applications (*After Kathrein, Ref. 16*).

For a sectorized configuration, each antenna occupies one sector. If a three-sector configuration is used, two panel antennas should be mounted close to the two corners of the building and one panel mounted on a flat wall of the building as shown in Fig. 10.12.

10.5.2 Tailoring a uniform-coverage cell

We will develop a lightweight, fold-up, portable directional antenna with adjustable beamwidth capability. It can be in the form of a corner reflector or an n-element array. This kind of antenna can be attached to the outer walls of the building in different directions. Perhaps we can attach such antennas to different buildings and form a desired coverage. The transmitted power of the antenna in each direction is also adjusted so that the coverage becomes uniformly distributed around the cell boundary. These antennas may be called *coverage sectored antennas*. The "coverage sectors" and the "frequency sectors" are not necessarily the same. Usually, several coverage sectors represent a frequency sector when the cell size becomes small. Since the power coverage is based on the coverage sectors and the frequency assignment is based on the frequency sectors, the existing software should be modified to incorporate this feature.

10.5.3 Vehicle-locating methods

By locating the vehicle and calculating the distance to it, we can obtain information useful for assigning proper frequency channels.

There are many vehicle-locating methods. In general, we can divide these into two categories: installation of equipment (1) in the vehicles and (2) at the cell site.

Installing equipment in the vehicles

Triangulation. Three or more transmitting antennas are used at different cell sites. Since the locations of the sites are known, the vehicle's location can be based on identification of three or more sites. However, the accuracy is limited by the multipath phenomenon.

In certain areas of the United States, especially near the coasts, Loran-C transmitters are used by the U.S. Coast Guard. These transmitters operate at 500 kHz. A Loran-C receiver installed in a vehicle can facilitate locating the vehicle.

Fifth-wheel and gyroscope equipment. A gyroscope and a fifth wheel are used for determining the direction and distance a vehicle has traveled from a predetermined point at any given time.

The globe-position satellite (GPS). There are seven active GPSs, each of them circling the earth roughly twice a day at an altitude of 1840 km (11,500 mi) and transmitting at a frequency of approximately 1.7 GHz. There are two codes, the C code and the P code. The C code is the coarse code which can be used by the civilian services. The P code is the precision code, which is used only by the military forces. At least three or more GPS satellites should be seen in space at any time, so that a GPS receiver can locate its position according to the known positions of the GPS satellites. Under this condition, we need at least 18 GPSs but only 7 GPSs are in space today. The GPS location is very accurate, generally within 6 m (20 ft). When four GPSs are in space, we can measure three dimensions, i.e., latitude, longitude, and altitude. The cost of a GPS receiver is very high and is not affordable at present for commercial cellular mobile systems.

Installing equipment at the cell site. In general, either of the following three methods alone cannot provide sufficient accuracy for locating vehicles; a combination of two or all three methods is recommended.

Triangulation based on signal strength. Record the signal strength received from the mobile unit at each cell site and then apply the triangulation method to find the location of the mobile unit. The degree of accuracy is very poor because of the multipath phenomenon.

Trangulation based on angular arrival.   Record the direction of signal arrival at each cell site and then apply the triangulation method to find the location of the mobile unit.

Triangulation based on response-time arrival.   Send a signal to the mobile unit. It will return with a time delay or a phase change. Measurement of the time delay or the phase change at each site can indicate the distance from that site.

Two or more distances from different sites can help us determine the location of the vehicle. However, the delay spread in the mobile environment can be 0.5 μs in suburban areas to 3 μs in urban areas. We may need ingenuity to solve this problem if the locating method is based on the response-time arrival.

Present cellular locating receiver.   Each cell site is equipped with a locating receiver which can both scan and measure the signal strengths of all channels. This receiver can be used to continuously scan the frequencies, or to scan on request.

Continuous-scanning scheme.   In continuous scanning of all 333 channels, assume that scanning each channel takes 20 ms. Thus, the time interval between two consecutive measurements of any single channel is 333×20 ms = 6.6 s.

If a car is driven at 30 mi/h (=44 ft/s), the interval between two different measurements on one frequency will be 6.6 s or 290 ft, so we would not expect a drastic change in this interval. However, the time interval of 20 ms for measuring the signal strength on a frequency is too short—only about 1 ft in distance (about 1 wavelength). As discussed in Chap. 1, we need 40 wavelengths to obtain good measurement data.[17] Therefore, if we are using the continuous-scanning scheme, the running mean $M(N)$ at the Nth sample based on the average of N samples should be tracked.

$$M(N) = \frac{\sum_{i=1}^{N} x_i}{N}$$

$$M(N + 1) = \frac{\sum_{i=2}^{N+1} x_i}{N}$$

The advantage of this scheme is that

1. Each cell site "knows" the signal-plus-interference levels $(S + I)$ of all active channels. The cell site can respond without delay to the mobile telephone switching office (MTSO) regarding the $S + I$ level of any one channel for measuring.

2. Each cell site knows the interference (or noise) levels of all idle channels. The cell site can choose a prospective (candidate) channel on the basis of its low interference level.

The selected channel mentioned in item 2 must not only generate a better signal-to-interference ratio but also less interference that affects the other cells. The argument for this is that if no interference is received by a channel, this channel will not cause interference in others. Use of a high-interference channel will not only cause deterioration in voice quality but also generate more interference in other cells.

Scan-under-request scheme. When measurement of a channel's signal strength is requested, the cell site must have enough time to measure it with a locating receiver; there is usually one locating receiver per cell site. Each locating receiver is capable of tuning and measuring all channels. Therefore, actual amount of time spent measuring the signal strength of each individual channel upon request can afford to be relative long for high accuracy. The disadvantages of the continuous-scanning scheme are compensated for by its advantages and vice versa.

10.5.4 Portable cell sites

For rapid addition of new cell sites to an existing system, portable cell sites are used to serve the traffic temporarily while the permanent site is under construction. In other situations, when it has not been determined whether the prospective site will be appropriate, the portable cell site can be used for a short period of time so that real operational data can be collected to determine whether this site will be suitable for a permanent site installation. Construction of a cell site normally requires three primary activities: (1) site acquisition, (2) building and tower construction, and (3) equipment installation and testing. A fair amount of time is needed for each activity. The portable cell site consists of buildings, equipment, and antennas, and all three of these items should be transportable.[18]

10.5.5 Different antenna mountings on the mobile unit

The different antenna mountings used on the mobile units affect system performance. The rooftop-mounted antenna, because of the great antenna height above the ground and the roughly uniform coverage, provides maximum coverage.

On the other hand, antennas mounted on windows, car bumpers, or trunks provide less coverage than do those mounted on roofs. However, more than 70 percent of the cars in cellular systems use glass-mounted antennas. Now the system operator must decide whether the available cellular system has to tune for glass-mounted mobile units or rooftop-mounted mobile units.

If the system is tuned for rooftop-mounted mobile antennas, that is, if the $q = D/R$ ratio is based on the reception of $C/I \geq 18$ dB of the rooftop-mounted antennas, then the cell coverages for rooftop-mounted units provides for no gaps. However, the cell coverage for glass-mounted units does allow gaps to form because of the weaker signals around all cell boundaries, which in turn results in excessive call drops.

If the system is tuned for all mobile units with glass-mounted antennas, then coverage of all cells will be suitable for these units. But the mobile units with rooftop-mounted antennas will travel deeply into the neighboring cells because of a still adequate signal at the cell boundaries and the delay of handoffs taking place. Then the units with rooftop-mounted antennas will experience channel interference such as cross talk and dropped calls due to distorted SAT tones.

Since the system tuning mentioned above cannot satisfy both kinds of mobile antenna usage, the author recommends that if the handoff is based on signal strength level a system should be tuned for those mobile units which have glass-mounted antennas. Then for the rooftop-mounted antenna units, attenuation can be added or the antenna can be tilted at some appropriate angle (loss due to off-vertical position).

If the handoff is based on power-difference schemes (Sec. 9.6), the effect of this scheme on different mobile antenna mountings becomes much less. The handoff occurrence areas for the mobile units with different antennas and different mountings are very much the same.

10.6 Narrowbeam Concept

The narrowbeam-sector concept is another method for increasing the traffic capacity (see Fig. 10.13). For a $K = 7$ frequency-reuse pattern with 120° sectors as a conventional configuration, each sector will con-

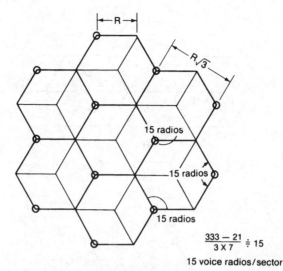

15 voice radios/sector

Figure 10.13 Ideally located cell sites over a flat terrain ($K = 7$).

tain approximately 15 voice channels, a number which is derived from the total 312 voice channels

$$\frac{333 - 21}{3 \times 7} = 15 \text{ channels per } 120° \text{ sector}$$

For a $K = 4$ frequency-reuse pattern [19] with 60° sectors (Fig. 10.14*a*), the number of channels in each 60° sector is

$$\frac{333 - 21}{4 \times 6} = 13 \text{ channels per } 60° \text{ sector}$$

In the $K = 7$ pattern there is a total of 21 sectors with 15 channels in each sector; in the $K = 4$ pattern there is a total of 24 sectors with 13 channels in each sector. The spectrum efficiency of using these two patterns can be calculated using the Erlang B table in Appendix 1.1. With a blocking probability of 2 percent, the results are: an offer load of 189 erlangs for $K = 7$ and 177 erlangs for $K = 4$. This means that the $K = 7$ pattern offers a 7 percent higher spectrum efficiency than the $K = 4$ pattern does. As seen in Fig. 10.14*a* a number of cell sites have been eliminated for $K = 4$ as compared with $K = 7$, assuming the same coverage area. However, the $K = 4$ arrangement results in increased handoff processing. Also the antennas erected in each site with a $K = 4$ pattern should be relatively higher than those with a

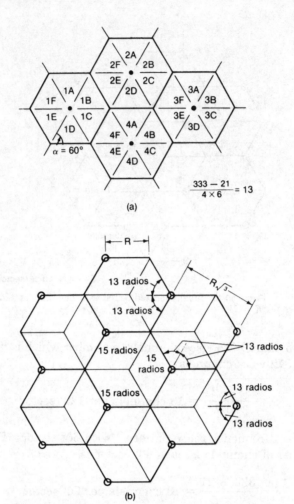

$$\frac{333 - 21}{4 \times 6} = 13$$

(a)

(b)

Figure 10.14 60°-sector cell sites. (*a*) Motorola's plan (*K* = 4): thirteen voice radios per sector in every 60° sector; (*b*) ideally located cell sites (*K* = 7) (mixed 120° and 60° sectors as needed).

$K = 7$ pattern. Otherwise, channel interference among channels will be increased because the wrong frequency channels will be assigned to the mobile units due to the low antenna height in the system. As a result, the actual location of the mobile units in smaller sectors may be incorrect.

Here we could use the scheme in Fig. 10.14*b* for customizing channel distribution; that is, usage of the 120° and 60° sectors can be mixed. Some 120° sectors can be replaced by two 60° sectors in a $K = 7$ pattern. The number of channels can then be increased from 15 to 26 as shown in Fig. 10.14*b*. This scheme would be suitable for small-cell systems. The antenna-height requirement for 60° sectors in small cells is relatively higher than that for 120° sectors. Besides, the 24 subgroups (each containing 13 channels) are used as needed in certain areas. These sector-mixed systems follow a $K = 7$ frequency-reuse pattern, and the traffic capacity is dramatically increased as a result of customizing the channel distribution according to the real traffic condition.

Comparison of narrowbeam sectors with underlay-overlay arrangement.
In certain situations the narrowbeam sector scheme is better in a small cell than the underlay-overlay scheme. In a small cell, it is very difficult to control power in order to make underlay-overlay schemes work effectively. For 60° sectors, the 60° narrowbeam antennas would easily delineate the area for operation of the assigned radio channels. However, choosing the correct narrowbeam sector where the mobile unit is located is hard. As a result, many unnecessary handoffs may take place.

In a 1-mi cell, if the traffic density is not uniformly distributed throughout the cell, the choice of using narrowbeam sectors or an underlay-overlay scheme is as follows; use the former for the angularly nonuniform cells, and use the latter for the radially nonuniform cells.

10.7 Separation between Highway Cell Sites

In generally light traffic areas, signal-strength coverage is a major concern, especially the coverage along the highways. There are two potential conditions to be considered in highway coverage: (1) relatively heavy traffic and (2) light traffic. In condition 1 there would be a high to average human-made noise level and in condition 2, a relatively to very low human-made noise level. Under these conditions, we recommend that the new highway cell-site separation should be much greater than that used for a normal cell site.

Omnidirectional antenna. As it is necessary to cover not only the highway but also areas in the vicinity of the highway, the omnidirectional antenna is used. When the cell sites are chosen and put up along the highway (see Fig. 10.15), the line-of-sight situation is usually assumed.

Although the general area around the highway could be suburban, because of the line-of-sight situation the path loss should be calculated using an open-area curve instead of the suburban-area curve shown in Fig. 4.3.

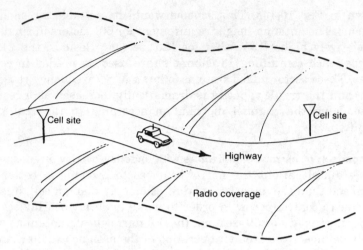

Figure 10.15   Highway cell sites.

The differences between highway cell-site separation and normal cell-site separation using path-loss values from the suburban and the open-area curves (see Fig. 4.3), respectively, are plotted in Fig. 10.16. The curve is labeled "Noise condition of human origin."

Traffic along highways away from densely populated areas is usually light and the level of automotive noise is low, perhaps 2 dB lower than the average human-made noise-level condition. Based on the 2-dB noise quieting assumption, another curve labeled "Low noise condition of human origin" is also shown in Fig. 10.16.

From Fig. 10.16 we can obtain the following data.

Average cell-site separation, mi; normal human-made noise condition	Highway cell-site separation, mi	
	Normal human-made noise condition	Low human-made noise condition
6	9.5	11
10	15	17

The purpose of using the omnidirectional antenna at the cell sites along the highway is to cover the area in the vicinity of the highway where residential areas are usually located.

Two-directional antennas.   In certain areas where only highway coverage is needed, two-directional antennas, such as horns or a pair of yagi antennas placed back to back at the cell site along the highway, could be installed. The directivity can result in a further increase in

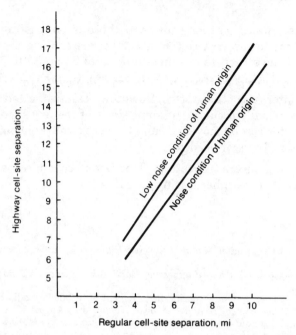

Figure 10.16 Highway cell-site separation.

separation between the sites. Equation (10.7-1) shows the relation between an increase in directivity ΔG in decibels and the increase in the additional separation Δd.

$$\Delta d = d[10\Delta G/20 - 1] \qquad (10.7\text{-}1)$$

This equation can easily be derived from a free-space path-loss condition.

10.8 Low-Density Small-Market Design

In a small market (city) one of the primary concerns is cost, since the low-density subcriber environment is basically suburban. Here, antennas can be lower but cover the same area as would a higher antenna in an urban area. In a noise-limited environment such as a suburban one, there are no problems in frequency assignment. The antenna tower can be constructed according to the following four considerations.

1. Use an existing high tower if available to obtain the maximum coverage in a given area. Because no frequency-reuse scheme will be applied, cochannel interference is of no concern.

2. Use a low-cost antenna tower. If there is no existing tower, then construct a low-cost tower in a farm area, using chicken wire to fasten the structure at a height of 15 to 24 m (50 to 80 ft).

3. Use portable sites. Portable sites can be moved around to serve the best interests of the system. Downtown call traffic and highway call traffic usually occur at different times of the day. One or two portable sites can be moved around to cover two traffic patterns at two different times if needed.

4. Apply enhancers (repeaters). This is an economical way to extend coverage to peripheral fringe areas.

References

1. Astronet Corporation, *"Cellular Coverage Enhancers,"* Issue 3, Astronet Co., April 1985.
2. H. Jasik (ed.), *Antenna Engineering Handbook,* McGraw Hill Book Co., 1961, p. 13-2.
3. W. C. Y. Lee, "Antenna Spacing Requirement for a Mobile Radio Base-Station Diversity," *Bell System Technical Journal,* Vol. 50, July-August 1971, pp. 1850–1876.
4. W. C. Y. Lee, *Mobile Communication Design Fundamentals,* Howard W. Sams & Co., 1986, p. 115.
5. W. C. Y. Lee, "Statistical Analysis of the Level Crossings and Duration of Fades of the Signal from an Energy Density Mobile Radio Antenna," *Bell System Technical Journal,* Vol 46, February 1967, pp. 417–448.
6. M. Schwartz, W. R. Bennett, and S. Stein, *Communication Systems and Techniques,* McGraw Hill Book Co., 1966, Chap. 10.
7. W. C. Y. Lee, *Mobile Communications Engineering,* McGraw-Hill Book Co., 1982, pp. 324–329.
8. E. A. Wolff, *Antenna Analysis,* John Wiley & Sons, 1967, p. 423.
9. R. F. Harrington, "Complex Propagation Constant for a TE_{11} Slotted Cylinder," *Journal of Applied Physics,* Vol.24, 1953, p. 1368.
10. P. N. Wyke and R. Gill, "Applications of Radio Type Mining Equipment at Collieries," *Proceedings of the Institution of Electrical and Mechanical Engineers,* Vol. 36, November, 1955, pp. 128–137.
11. Q. V. Davis, D. J. R. Martin, and R. W. Haining, "Microwave Radio in Mines and Tunnels," *34th IEEE Vehicular Technology Conference Record,* Pittsburgh, May 21–23, 1984, pp. 31–36.
12. T. Suzuki, T. Hanazawa, and S. Kozono, "Design of a Tunnel Relay System with a Leaky Cable Coaxial Cable in an 800-MHz Band Land Mobile Telephone System," *IEEE Transactions on Vehicular Technology,* Vol. 29, August 1980, pp. 305–306.
13. W. C. Y. Lee, *Mobile Communications Design Fundamentals,* Howard W. Sams & Co., 1986, p. 147.
14. V. H. MacDonald, "The Cellular Concept," *Bell System Technical Journal,* Vol. 58, January 1979, pp. 15–42.
15. W. C. Y. Lee, "Elements of Cellular Mobile Radio Systems," *IEEE Transactions on Vehicular Technology,* Vol. VT-35, May 1986, pp. 48–56.
16. Kathrein, "Directional Antennas (a family of Model 740), Kathrein, Inc., Cleveland, Ohio.
17. W. C. Y. Lee, *Mobile Communications Design Fundamentals,* Howard W. Sams & Co., 1986, p. 53.
18. J. Proffitt, "Portable Cell Site," *36th IEEE Vehicular Technology Conference,* Dallas, Texas, May 1986, *Conference Record,* p. 291.
19. Motorola proposal to FCC, 1977.

20. W. C. Y. Lee and Y. S. Yeh, "Polarization Diversity System for Mobile Radio," *IEEE Transactions on Communications,* Vol. COM 20, No. 5, October 1972, pp. 912–923.

21. W. C. Y. Lee, "An Energy Diversity Antenna for Independent Measurement of the Electric and Magnetic Field," *Bell System Technical Journal,* Vol. 46, September 1967, pp. 1587–1599.

22. R. A. Isberg and D. Turrell, "Applying CATV Technology and Equipment in Guided Radio Systems," *34th IEEE Vehicular Technology Conference Record,* Pittsburgh, May 21–23, 1984, pp. 37–42.

23. T. Yuge and S. Sakaki, "Train Radio System Using Leaky Coaxial Cable," *34th IEEE Vehicular Technology Conference Reord,* Pittsburgh, May 21–23, 1984, pp. 43–48.

Switching and Traffic

11.1 General Description

11.1.1 General Introduction

Switching equipment is the brain of the cellular system. It consists of two parts, the switch and the processor. The switch is no different from that used in the telephone central office. The processor used in cellular systems is a special-purpose computer. It controls all the functions which are specific for cellular systems, such as frequency assignment, decisions regarding handoff (including decisions regarding new cells for handoff), and monitoring of traffic. The smaller the cell, the more handoffs involved, and the greater the traffic load required. The processor can be programmed to correct its own errors and to optimize system performance. General (noncellular) telephone switching equipment is described first, and then cellular switching equipment is discussed in detail.

11.1.2 Basic switching[1]

In circuit switching (analog switching), a dedicated connection is made between input and output lines or trunks at the switching office and physical switching begins. Space-division switches have been generally used in circuit switching, but they can also be used for digital switching. However, time-division switches can be used only for digital switching. In large digital switches, both time- and space-division switching are used.

k outputs

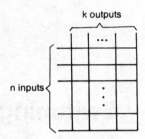

Figure 11.1 Number of crosspoints in space-division switches.

Space-division switching. A rectangular coordinate switch interconnects n inputs from n lines to k outputs of other lines. When interconnection is possible at every cross-point, the switch is nonblocking.

$$\text{Number of cross-points} = nk \qquad (11.1\text{-}1)$$

These switches are laid out in space as shown in Fig. 11.1.

For simplicity in analyzing the following nonblocking system, let the total number of input lines N equal the total number of output lines. An N-input group is fed into N/n switch arrays, and each switching network has an nk switch as shown in Fig. 11.2a. The three-stage switching network (S-S-S) is shown in Fig. 11.2b with N outputs. The total number of cross-points is[2]

$$C = 2\left(\frac{N}{n}\right)(nk) + k\left(\frac{N}{n}\right)^2 = 2\,Nk + k\left(\frac{N}{n}\right)^2 \qquad (11.1\text{-}2)$$

If $k = 2n - 1$, the nonblocking rule, then the number of cross-points C can be obtained from Eq. (11.1-2).

Time-division switching. For time switching to be carried out, all call messages must be first slotted into time samples, such as in pulse-code modulation (PCM) for the digital form of voice transmission. Voice quality is sampled at 8000 samples per second. Each sample (125-μs frame) must have eight levels (2^3); then each voice channel requires 64 kbps (kilobits per second) of transmission.

The time-slot switchings limit the number of channels per frame that may be multiplexed. Multiplexing of more channels requires more memory storage. Take a t_f-microsecond frame and let t_c be the memory cycle time in microseconds required for both write-in and readout of a memory sample. Then the maximum number of channels that can be supported is

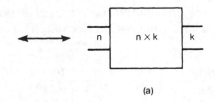

(a)

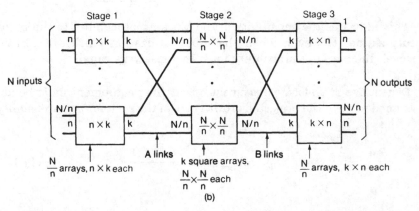

(b)

Figure 11.2 A switching network. (a) An nK switch; (b) a three-stage switching network.

$$N = \frac{t_f}{2\, t_c} \qquad (11.1\text{-}3)$$

For $t_c = 50\text{-nsec}$ logic (write-in–readout time) and $t_f = 125\ \mu s$, the number N is 1250.

Blocking probability analysis of multistage switching.[3]   In three-stage switching, the assumption of an S-S-S case (traversing two links) is the same as that of a T-S-T case. In the latter case, time division and space division are combined. The blocking probability that no free path from input channel to output channel is available is

$$P_B = [1 - (1 - p)^2]^k \qquad (11.1\text{-}4)$$

where
$$p = \frac{An}{k} \qquad (11.1\text{-}5)$$

where k = number of slot frames or number of outputs of each stage
   n = number of input time slots or input lines of each stage
   $A = \lambda/(\lambda + \mu)$ is the inlet channel utilization

We may also treat A as the probability that a typical channel is busy. The parameters λ and μ are explained in Sec. 9.5. If $n = 120$, $k = 128$, and

$$A = \begin{cases} 0.7, \\ 0.9, \\ 0.95, \end{cases} \quad \text{then} \quad P_b = \begin{cases} 10^{-7} \\ 0.042 \\ 0.214 \end{cases}$$

This indicates the sensitivity of the blocking probability to the input utilization. However, when the input utilization reaches a certain level, the blocking probability must increase very rapidly.

Dimensions of switching equipment. Switching equipment must be designed to meet the projected growth rate of the system. The probability of loss P_b is

$$P_b \text{ (calls lost)} = \frac{(a^N/N!)}{\sum\limits_{n=1}^{N} (a^n/n!)} \tag{11.1-6}$$

$$P_b \text{ (lost calls held)} = \frac{(a^N/N!)[N/(N-a)]}{\sum\limits_{n=1}^{N} (a^n/n!)} \tag{11.1-7}$$

where N is number of trunks and a represents the traffic load, as in Sec. 9.6. Using either Eq. (11.1-6) or Eq. (11.1-7), we can find the number of trunks needed in a given demanding situation.

11.1.3 System congestion[4,5]

Time congestion. Consider a generic model in a circuit-switched exchange with a simple output trunk group. Each M input either is idle for an exponential length of time $1/\lambda$ or generates a call with a holding time of $1/\mu$. Each arriving call will be assigned to one of the outgoing trunks. For the probability of the number of calls in progress p_n, we obtain

$$p_n = \frac{(\lambda/\mu)^n \binom{M}{n}}{\sum\limits_{n=0}^{N} (\lambda/\mu)^n \binom{M}{n}} \quad 0 \leq n \leq N \tag{11.1-8}$$

If the system is fully occupied, then

$$P_B = P_N \qquad (11.1\text{-}9)$$

This is called "time congestion."

Call congestion. The other way to measure congestion is to count the total number of calls arriving during a long time interval and record those calls that are lost because of a lack of resources, such as busy trunks.

The probability of call loss is P_L. Let $p(a)$ be the unconditional probability of arrival of a call, P_B be the probability that the system is blocked, and $p_N(a)$ be the probability that a call arrives when the system is blocked. Then

$$P_L = \frac{p_N(a)}{p(a)} P_B \qquad (11.1\text{-}10)$$

If the conditional probability $p_N(a)$ is independent of the state of system blocking, then $p_N(a) = p(a)$, Eq. (11.1-10) becomes $P_L = P_B$, and the two measurements—time congestion and call congestion—are the same.

11.1.4 Ultimate system capacity

There are two limits on system capacity: (1) the amount of traffic that the switches can carry and (2) the amount of control that the processor can exercise without the occurrence of unacceptable losses. Limit 2, the amount of control that the processor can exercise without excessive losses, can be broken down as follows.

1. *Ultimate capacity due to traffic load.* Traffic capacity consists of two parameters, the number of calls per hour λ and their duration $1/\mu$. The average call duration is about 100 s (i.e., $1/\mu = 100$ s). The physical limits of switching capacity are reflected in the number of trunk interfaces N and the traffic load a.

2. *Ultimate capacity due to control.* Processor control operates on a delayed basis when requests are queued or scanned at regular intervals. There are two levels of control. At level 1, processor control is involved in scanning and interfacing with customers. At level 2, central processors are involved when all the data for a call request are received. The delay on the processor can be calculated as

$$\text{Average call delay (s)} = \frac{1/\mu}{2(1 - a')} \qquad (11.1\text{-}11)$$

where a' is the traffic load on the processor and $1/\mu$ is the holding time that the processor takes to handle a call.

Ultimate system capacity limitations are reflected in limits on control, such as the number of calls that the system can handle, including handoffs, scanning and locating, paging, and assigning a voice channel. Therefore, the processing capacity for cellular mobile systems is much greater than that for noncellular telephone systems. In noncellular telephone switching the duration of the call is irrelevant, but in a cellular system it is a function of frequency management and the number of handoffs.

Assigning a value to the processor traffic

Level 2 control only (centralized system). It is extremely difficult to estimate accurately how a system will perform under real traffic conditions. A traffic simulation can be used. For total capacity, let

	P, %	$1 - P$, %
For eventualities	5	95
For false traffic		
(i.e., call abandoned before completion)	30	70
For peak traffic	30	70

Thus, $(1 - 0.05)(1 - 0.3)(1 - 0.3) = 0.465$ or a total capacity of 46.5 percent (or in this case 0.465 erlangs) for call processing.

It is assumed that no level 1 control is operating in the processor. Assume that the processor holding time is 100 ms; then applying this to Eq. (11.1-11), we find that the average call delay is

$$t_d = \frac{100 \text{ ms}}{2(1 - 0.465)} = 93.46 \text{ ms}$$

The average delay on call processing during the busy call equals

$$93.46 \times 0.465 = 43.46 \text{ ms}$$

Level 1 and level 2 control (decentralized system). Assume that level 1 control in a system reduces the load on level 2 control by absorbing most of the false traffic at level 1 control. Then at level 2 control the call-processing occupancy rate is $(1 - 0.05)(1 - 0.3) = 0.665$ (or 66.5 percent).

Assume that the total processor holding time is less than 100 ms for the average call (the total time taken to handle a call is the sum of the total number of instructions required during the call); then

$$t_d = \text{average call delays} = \frac{100 \text{ ms}}{2\,(1 - 0.665)} = 150 \text{ ms}$$

The average delay on call processing during the busy call equals 150 ms $\times$ 0.665 = 100 ms.

When comparing the centralized system with the decentralized system, we find that in the centralized system the average call delay time is much shorter but its processing capacity is much less. However, the decentralized system is more flexible, is easier to install, and has a greater potential for expansion.

11.1.5 Call drops

Call drops are caused by factors such as (1) unsuccessful completion from set-up channel to voice channel, (2) blocking of handoffs (switching capacity), (3) unsuccessful handoffs (processor delay), (4) interference (foreign source), and (5) improper setting of system parameters.

The percentage of call drops is expressed as

$$P_{cd} = \frac{\text{number of call drops before completion}}{\text{total number of accepted calls handled by set-up channels}}$$

$$= \frac{\sum\limits_{i=1}^{5} C_{d_i}}{C_t}$$

Because this percentage is based on many parameters, there is no analytic equation. But when the number of call drops increases, we have to find out why and take corrective action. The general rule is that unless an abnormal situation prevails, call drops usually should be less than 5 percent.

11.2   Cellular Analog Switching Equipment

11.2.1   Description of analog switching equipment

Most analog switching equipment consists of processors, memory, switching network, trunk circuitry and miscellaneous service circuitry. The control is usually centralized, and there is always some degree of redundancy. A common control system is shown in Fig. 11.3. The programs are stored in the memory that provides the logic for controlling telephone calls. The processor and the memory for programming and calls are duplicated. The switching network provides a means for in-

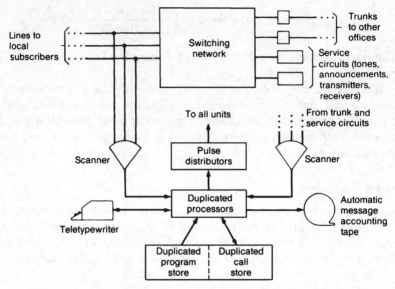

Figure 11.3 A typical analog switching system. (*After Chadha et al., Ref. 6.*)

terconnecting the local lines and trunks. The scanners are read under the control of the central processor. The changes in every connection at the line side and at the trunk side are also controlled by the processor. The central processor sends the order to all the units (switching network, trunk, service circuits) through pulse distributions. The automatic message accounting (AMA) tapes are used for recording the call usage. Three programs are stored in most switching equipment: (1) call processing (set up, hand off, or disconnect a call), (2) hardware maintenance (diagnose failed or suspected failed units), and (3) administration (collect customer records, trunk records, billing data, and traffic count).

11.2.2   Modification of analog switching equipment

The local line side has to change to the trunk side as shown in Fig. 11.4 because the mobile unit does not have a fixed frequency channel. Therefore, the mobile unit itself acts as a trunk line. In addition, the processors have to be modified to handle cellular call processing, the locating algorithm, the handoff algorithm, the special disconnect algorithm, billing (air time and wire line), and diagnosis (radio, switching, and other hardware failure).

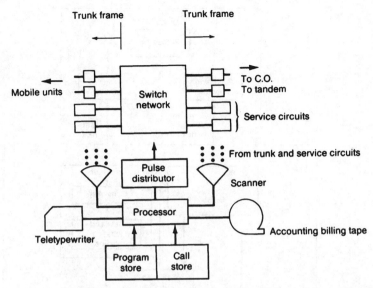

Figure 11.4 A modified analog switching system for cellular mobile systems.

11.2.3 Cell-site controllers and hardware

Mobile telephone switching office (MTSO) system manufacturers designed their own cell-site controllers and transceivers (radios). Cell-site equipment is shown in Fig. 11.5. The cell site can be rendered "smarter," that is, programmed to handle many semiautonomous functions under the direction of the MTSO. Cell-site equipment consists of two basic frames.

1. Data frame—consists of controller and both data and locating radios
    a. Providing RF radiation, reception, and distribution
    b. Providing data communication with MTSO and with the mobile units
    c. Locating mobile units
    d. Data communication over voice channels
2. Maintenance and test frame
    a. Testing each transmitting channel for
        i. Incident and reflected power to and from the antenna
        ii. Transmitter frequency and its deviation
        iii. Modulation quality
    b. Testing each receiving channel for
        i. Sensitivity
        ii. Audio quality

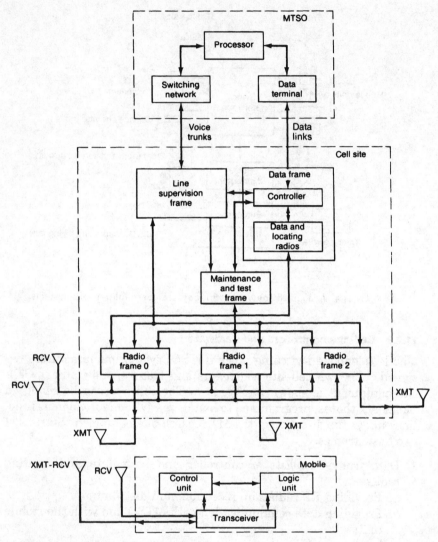

Figure 11.5 Cellular mobile major systems.

11.3   Cellular Digital Switching Equipment

11.3.1   General concept

The digital switch, which is usually the message switch, handles the digitized message. The analog switch, which is the circuit switch, must hold a call throughout the entire duration of the call. The digital (message) switch can send the message or transmit the voice in digital form; therefore, it can break a message into small pieces and send it at a fast rate. Thus, the digital switch can alternate between ON and OFF modes periodically during a call. During the OFF mode, the switch can handle other calls. Hence, the call-processing efficiency of digital switching is higher than that of analog switching.

The future digital switch could be a switching packet which would send digital information in a nonperiodic fashion on request. There are other advantages to digital switching besides its greater efficiency. Digital switches are always small, consume less power, require less human effort to operate, and are easier to maintain. Digital switching is flexible and can grow modularly. Digital switching equipment can be either centralized[7,8] (Fig. 11.6a) or decentralized[9,10] (Fig. 11.6b). A centralized system of a digital system has an architecture similar to that of an analog system. Motorola's EMX2500, Ericsson's AXE-10, and Northern Telecom International's DMS-MTXM are large centralized digital systems. A decentralized system is described here.

A decentralized system is slightly different from a remote-control switching system. In the remote-control switching system, a main switch is used to control a remote secondary switch as shown in Fig. 11.6c. In a decentralized system, all the switches are treated equally; there is no main switch.

11.3.2 Elements of switching

One decentralized switching system is introduced here for illustration. It is the American Telephone & Telegraph (AT&T) Autoplex 1000,[9] which consists of an executive cellular processor (ECP), digital cellular switches (DCS), an interprocess message switch (IMS), RPC (ring peripheral control), and nodes

1. ECP transports messages from one processor to another.

2. IMS attached to a token ring (IMS uses a token ring technology) provides interfaces between ECP, DCS, cell sites, and other networks. The RPC attached to the ring permits direct communication among all the elements through the ring.

3. DCS, which are digital cellular switches, function as modules to allow the system to grow.

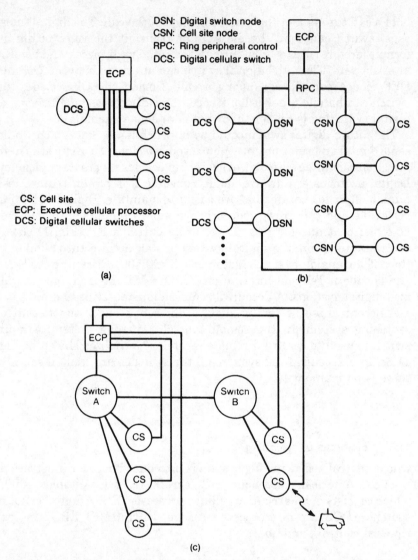

Figure 11.6 Cellular switching equipment. (*a*) A centralized system; (*b*) a decentralized system; (*c*) remote-controlled switching.

4. RPC forms a ring that connects two types of nodes: CSN (cell-site node) and DSN (digital switch nodes).

5. Nodes are: RPC nodes for connecting the ECP to the ring, CSN nodes for connecting cell-site data to the ring, and DSN nodes for connecting data links to the ring.

11.3.3   Comparison between centralized and decentralized systems

The analysis of overall system capacity given in Sec. 11.1.3 can be used here for comparison. In general, a centralized system has only one control level, whereas the decentralized system has more than one. In a one-level control system, the utilization of call processing is less than that in a multilevel control system. Moreover, the delay time for a central control system is always shorter than that for a decentralized system; thus, the more levels of control, the greater the call-processing utilization and the longer the delay time. However, in principle, decentralized systems always have room to grow and are flexible in dealing with increasing capacity. Centralized systems deal with large traffic loads.

11.4   Special Features for Handling Traffic

The switching equipment of each cellular system has different features associated with the radios (transceivers) installed at the cell sites.

11.4.1   Underlay-overlay arrangement

The switching equipment treats two areas as two cells, but at a cocell site (i.e., two cells sharing the same cell site). Therefore, the algorithms have to be worked out for this configuration. In Sec. 8.5.4 we discussed the underlay-overlay arrangement in terms of channel assignment. Here we discuss this arrangement in terms of MTSO control.

To initiate a call, the MTSO must know whether the mobile unit is in an overlay or an underlay area. The MTSO obtains this information from the received signal strength transmitted by the mobile unit. To hand off a call, the MTSO must know whether this is a case of handoff from (1) an overlay area to an underlay area or (2) vice versa. In case 1 the signal strengthens and exceeds a specific level, and then the unconventional handoff takes place. In case 2 the signal strength weakens and falls below a specific level, and then the conventional handoff takes place.

11.4.2   Direct call retry

Direct call retry is applied only at the set-up channel. When all the voice channels of a cell are occupied, the set-up channel at that cell can redirect the mobile unit to a neighboring-cell set-up channel. This is the order used by the original set-up channel to override the "pick the strongest signal" algorithm. In this scheme, the MTSO received all the call traffic information from all the cells and thus can distribute the call capacity evenly to all the cell sites.

11.4.3 Hybrid systems utilizing high sites and low sites

The high site is always used for coverage, and it can also be used to fill many holes which may be created by the low site. Therefore, if in some areas the mobile unit cannot communicate through the low site, the high site will take over, and as soon as the signal reception gets better at the low site, the call will hand off to the low site. The algorithms include computation of the following configuration.

1. When the signal strength received at the cell site weakens, the handoff is requested and the high site picks it up.

2. The MTSO continues to check the signal of this particular mobile unit from all the neighboring low sites. If one site receives an acceptable signal from the mobile unit, the handoff will be forwarded to that site.

11.4.4 Intersystem handoffs

Intersystem handoffs were described in Chap. 9. The processor requires particular software to utilize this feature. There are four conditions of intersystem handoff, as shown in Fig. 11.7.

1. A long-distance call becomes a local call while a home mobile unit becomes a roamer (Fig. 11.7a).

2. A long-distance call becomes a local call while a roamer becomes a home mobile unit (Fig. 11.7b).

3. A local call becomes a long-distance call while a home mobile unit becomes a roamer (Fig. 11.7c).

4. A local call becomes a long-distance call while a roamer becomes a home mobile unit (Fig. 11.7d).

All four cases have to be implemented.

11.4.5 Queuing feature

When a nonuniform traffic pattern prevails and the call volume is moderate, the queuing feature can help to reduce the blocking prob-

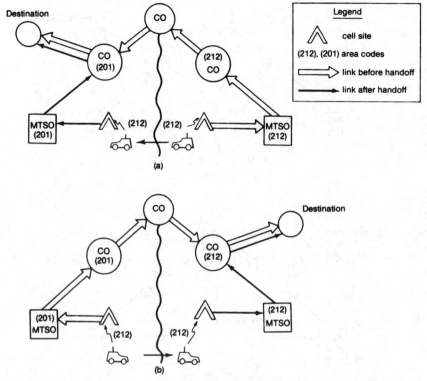

Figure 11.7 Four conditions of intersystem handoffs. (*a*) A toll call becomes a local call and the home mobile unit becomes a roamer; (*b*) a toll call becomes a local call and a roamer becomes a home mobile unit; (*c*) a local call becomes a toll call and the home mobile unit becomes a roamer; (*d*) a local call becomes a toll call and a roamer becomes a home mobile unit.

ability. The improvement in call origination and handoffs as a result of queuing is described in Chap. 9. The switching system has to provide memory or buffers to queue the incoming calls if the channels are busy. The number of queue spaces does not need to be large. There is a finite number beyond which the improvement due to queuing is diminished, as described in Chap. 9.

11.4.6 Roamers

Initiating the call. If two adjacent cellular systems are compatible, a home mobile unit in system A can travel into system B and become a "roamer." The switching MTSO can identify a valid roamer and offer the required service. The validation can be the mobile unit's MIN or ESN (see Chap. 3).

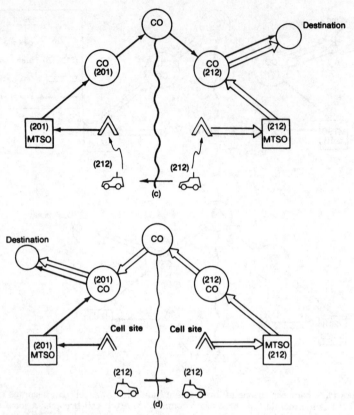

Figure 11.7 (*Continued*)

Handing off the call.   The feature of intersystem handoffs can be applied in order to continue the call. Intersystem handoffs are described in Sec. 11.4.4.

Clearinghouse concept.   Because of the increase of roamers in each system, checking the validation of each roamer in the roamer's own system becomes a complex problem for an automatic roaming system. The cellular system "clearinghouses" (several nationwide companies) provide a central file of the validation of all users' MIN and ESN in every system. There are two files, positive and negative validation. Positive validation is done by checking whether the user's number is on the active customer list. The negative validation file lists the numbers of users whose calls should be rejected from the automatic roaming system. The payment for transmitting validation data to and from the clearinghouse plus the service charge has to be justified against the revenue lost through delinquent users (those who do not pay on time).

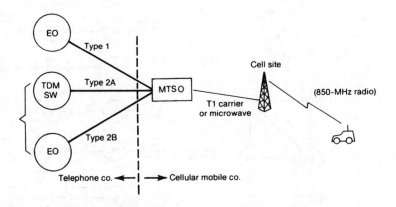

Type 1 interconnects an MTSO to an LEC end office

Type 2A interconnects an MTSO to an LEC tandem office

Type 2B interconnects to an LEC end office in conjunction with type 2A on a high-usage alternate-routing basis.

Figure 11.8 Three types of interconnection linkage.

11.5 MTSO Interconnection

11.5.1 Connection to wire-line network

The MTSO operates on a trunk-to-trunk basis. The MTSO intercon-nection arrangement is similar to a private-branch exchange (PBX) or a class 5 central office (a tandem connection) (see Fig. 11.8). The MTSO has three types of interconnection links.

Type 1—interconnects a MTSO to a local-exchange carrier (LEC) end office

Type 2A—interconnects an LEC tandem office

Type 2B—interconnects to an LEC end office in conjunction with type 2A on a high-usage alternate-routing basis

The three-level hierarchy of a public telephone network is shown in Fig. 11.9. With this diagram, we can illustrate the three types of calls: (1) a local call, (2) an intra-LATA (local access and transport area) call, and (3) an inter-LATA call.

11.5.2 Connection to a cell site

Two types of facility are used.

1. Cell-site *trunks* provide a voice communication path. Each trunk is physically connected to a cell-site voice radio. The number of trunks

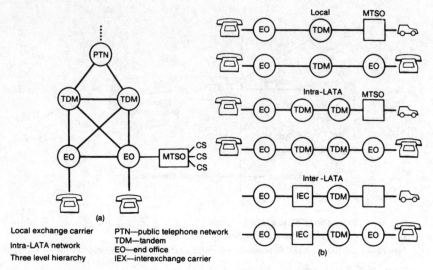

Figure 11.9 Three-level hierarchy. (*a*) Interconnection of MTSO; (*b*) three types of call.

is decided on the basis of the traffic and the desired blocking probability (grade of service).

2. The cell site acts as a *traffic concentrator* for the MTSO. For instance, we may design an average busy-hour radio channel occupancy of at least 60 to 70 percent for high-traffic cells.

Both T1-carrier cables and microwave links are used. The duplication is needed for reliability.

11.6 Small Switching Systems[11-13]

Small switching equipment can be used in a small market (city). This switching equipment can usually be developed modularly. It consists of (1) a transmitter and a receiver, (2) a cell-site controller, (3) a local switch (a modified PABX should be used), (4) a channel combiner, and (5) a demultiplexer.

Small switches should be low-cost items. A high existing tower can be used for a cell-site antenna to cover a large area.

11.7 System Enhancement

Consider the following scenarios.

1. Each trunk is now physically connected to each voice radio as mentioned previously. But if the trunk can be switched to different

radios, then the dynamic frequency assignment scheme can be accomplished.

2. Let a cell site pick up a switch in a decentralized multiple switching equipment system. This is a different concept than usual. Normally the switching equipment controls a cell site. For the land-initiated calls, the switching equipment picks up a cell site through paging. For mobile-originated calls, the cell site handling the call can select the appropriate switching equipment (DCS) from among the two or three units of switching equipment, assuming the cell site can detect the traffic conditions.

We can construct an analogy here. In a supermarket, everyone is waiting in line to pay the cashier for their merchandise. In order to reduce the waiting time, the store manager (central switching office) can direct the customer to the cashier (switching equipment) with the shortest line or the customer (cell site) can select the cashier with the shortest line. Both methods can work equally well assuming both the manager and the customer have the ability to choose well.

This system enhancement may be made in the future to all systems when artificial intelligence techniques become fully developed and can be used to implement the enhancements.

References

1. N. Schwartz, *Telecommunication Network,* Addison-Wesley, 1987, Chap.10.
2. C. Y. Lee, "Analysis of Switching Network," *Bell System Technical Journal,* Vol. 34, November 1955, pp. 1287–1315.
3. J. G. Pearce, *Telecommunications Switching,* Plenum Press, 1981, Chap. 5.
4. M. J. Hills, *Telecommunications Switching Principles,* MIT Press, 1979.
5. C. Clos, "A Study of Non-Blocking Switching Network," *Bell System Technical Journal,* Vol. 32, No. 2, March 1953, pp. 406–424.
6. K. J. S. Chadha, C. F. Hunnicutt, S. R. Peck, and J. Tebes, Jr., "Mobile Telephone Switching Office," *Bell System Technical Journal,* Vol. 58, January 1979, pp. 71–96.
7. Northern Telecom Cellular Switches MTX/MTXCX.
8. Ericsson Cellular Switches AXE 10 and CMS 8800.
9. AT&T Cellular Switches Autoplex 10 and Autoplex 1000.
10. Motorola Cellular Switches DMX 500 and DMX 2500.
11. Astronet Small Cellular Switches, Astronet Corp., Lake Mary , Florida.
12. Quintron Small Cellular Switches, "Vision Series Cellular System," Quintron Corp., Quincy, Illinois.
13. CRC Freedom-2000 (small cellular switches), Cellular Radio Corp., Vienna, Virginia.

12

Data Links and Microwaves

12.1 Data Links

Implementation of data links is an integral part of cellular mobile system design, and the performance of data links significantly affects overall cellular system performance.

The cell site receives the data from the MTSO to control the call process of mobile units. It also collects data from the reverse set-up channel from active mobile units and attempts to send it to the MTSO. There are three types of data links available: (1) wire line, (2) 800-MHz radios, and (3) microwaves. The following discussion describes each alternative and its advantages and disadvantages. The wireline connection[1] uses the telephone company's T1 carrier. Regular telephone wire can transmit only at a low rate (2.4 kbps); therefore, a high-data-rate cable must be leased. The T1 carrier has a wideband transmission (1.5 Mbps) that consists of 24 channels, and each channel can transmit at a rate of 64 kbps. For handling the data, a digital terminal converts the incoming analog signals to a digital form suitable for application in a digital transmission facility. Many digital terminals are multiplexed to form a single digital line called a *digital channel bank*.

Digital channel banks multiplex many voice-frequency signals and code them into digital form. The sampling rate is 8 kHz. Each channel is coded into 7-bit words. A signaling bit indicates the end of each 7-bit sample. After 24 samples, one sample for each channel, a frame bit is sent again. The total number of bits per frame is

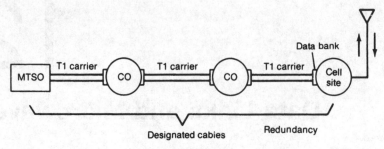

Figure 12.1 Data-link connection through cable.

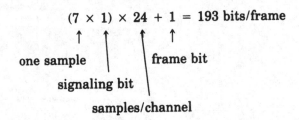

Since 8000 frames per second and 193 bits per frame are specified, a digital capacity of 1.54 Mbps is required.

The data link has to have a data bank at each end of the T1 carrier cable. The data bank has to convert all the information into the 1.54 Mbps before sending it out through the cable. The number of T1 carriers required is determined by the number of radios installed at the cell site, e.g., if 60 radios are installed, then three T1-carrier cables are needed. The T1 carrier cables are installed in duplicate to provide redundancy (see Fig. 12.1). A major disadvantage to using wire-line data links is that the T1-carrier route may be rearranged by the telephone company at any time without notice. Therefore, it is not totally under the user's control. In addition, the leasing cost should be compared to the long-run cost of using the microwave link if owned by the cellular operator.

The data could also be sent by 800-MHz radios. However, this would cause interference among all the channels, and since every radio channel can handle a signaling rate of only 10 kbps, we would need an additional 666 channels just to handle this data link from the cell sites to the MTSO. This can be a good idea for low-capacity systems.

Microwave links seem to be most economical and least problematical. Details of their installation are given below. However in a rural area, capacity is not a problem. We can use half of the cellular channels for data-link use.

12.2   Available Frequencies for Microwave Links

The microwave system is used to cover a large area; it should also be used as the "backbone." Before designing it, we must consider (1) system reliability, (2) economical design, (3) present and future frequency selection, (4) minimization of the number of new microwave sites, and (5) flexible and multilevel systems. The microwave frequencies can be grouped as follows.

Frequency, GHz	Allowed bandwidth, MHz	5-year channel loading	Minimum path length, km
2	3.5	None	5
4	20	900	17
6	30	900	17
11	40	900	5
18	220	None	None
23	100	None	None

As can be seen from this tabular analysis, for the higher frequencies there are fewer restrictions, thus allowing greater flexibility in system design.

The 2-GHz band.   The minimum path length of 5 km (3.1 mi) and the limited 3.5-MHz radio-frequency (RF) bandwidth place several restrictions on the use of the 2-GHz band. Capacity is probably limited to eight T1 span lines. Installation of a 6- or 8-fit dish is required. Because of the limited path length, the limited traffic capacity, larger antennas at cell sites, and the difficulty in obtaining frequency coordination, 2 GHz is not desirable.

The 4- and 6-GHz bands.   The minimum path length of 17 km (10.5 mi) and the minimum channel load of 900 channels for 6 GHz along with the 4-GHz frequency present a restriction.

The 11-GHz band.   The minimum path length of 5 km (3.1 mi) and the minimum channel loadings of 900 voice channels would make this band a poor choice for the final path to the cell sites. However, the greater bandwidth availability and lower frequency congestion would make this an ideal band for high-density routes between the collection points and the MTSO.

The 18- and 23-GHz bands.   The lack of FCC restrictions on minimum path length and the minimum channel loadings would appear to make these two frequency bands ideal for paths to the cell site. These frequency bands are not characterized by the presence of the

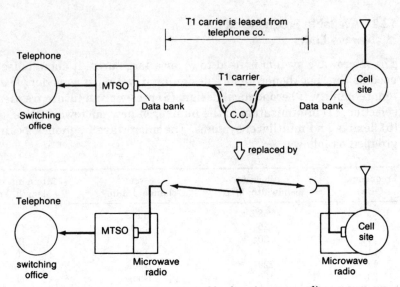

Figure 12.2 Replacement of T1 carrier cables by microwave radios.

RF congestion at lower frequencies. Cell sites can be implemented with 2- or 4-ft dishes, compared to the larger 6- or 8-ft dishes needed at lower frequencies.

12.3 Microwave Link Design and Diversity Requirement

There are three basic considerations here. First, the microwave propagation path length is always longer than the cellular propagation path, say, 25 mi or longer. Second, the path is always 100 or 200 ft above the ground. Third, the microwave transmission is a line-of-sight radio-relay link. Figure 12.2 shows the replacement of T1 carrier cable by microwave radios.

However, microwave links will render the system susceptible to one kind of multipath fading, in which the microwave transmission is affected by changes in the lower atmosphere, where atmospheric conditions permit multipath propagation.

Although deep fades are rare, they are sufficient to cause outage problems in high-performance communications systems.[2,3] A signal is said to be in a fade of depth 20 log L dB; that is, the envelope (20 log l) of the signal is below the level L.

$$20 \log l \leq 20 \log L$$

Usually, we are interested only in fades deeper than -20 dB.

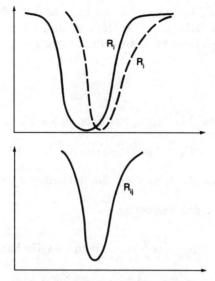

Figure 12.3 Formation of a simultaneous fade from two over-lapped fades.

$$20 \log L < -20 \text{ dB} \quad \text{or} \quad L < 0.1$$

From the experimental data, the number of fades can be formulated as

$$N = \begin{cases} 6410L/60.88 \text{ days} = 105.29L/\text{day} & \text{(in the 6-GHz band)} \\ 3670L/60.88 \text{ days} = 60.28L/\text{day} & \text{(in the 4-GHz band)} \end{cases}$$

$$(12.3\text{-}1)$$

The average deviation of fades is

$$\bar{t} = \begin{cases} 490L \text{ s} & \text{(in the 6-GHz band)} \\ 408L \text{ s} & \text{(in the 4-GHz band)} \end{cases} \qquad (12.3\text{-}2)$$

In order to reduce the fades, two methods can be used: a spaced diversity and a frequency diversity. In order to use diversity schemes, we have to gather some additional data. The two signals obtained individually from two channels can be used to measure the number of simultaneous fades, as shown in Fig. 12.3.

If the frequency separation or the spaced separation is very large, the number of simultaneous fades can be drastically reduced. Therefore, it follows that the number of simultaneous fades of two signals must be small to obtain good diversity reception.

A parameter F_N is defined as the ratio of N_i to N_{ij}, where N is the number of fades from a single channel and N_{ij} is the number of simultaneous fades from two individual channels.

$$F_N = \frac{N_i}{N_{ij}} \tag{12.3-3}$$

The ratio F_N should be large. For deep fades, F_N is

$$F_N = \tfrac{1}{2}qL^{-2} \quad \text{for} \quad L < 0.1 \tag{12.3-4}$$

The q is a parameter defined by the following equations.

1. For separations in frequency,

$$q = \frac{1}{4}\left(\frac{\Delta f}{f}\right) \quad \text{(in the 6-GHz band)} \tag{12.3-5}$$

$$q = \frac{1}{2}\left(\frac{\Delta f}{f}\right) \quad \text{(in the 4-GHz band)} \tag{12.3-6}$$

where Δf is the frequency separation and f is the operational frequency.

2. For separation in space,

$$q = (2.75)^{-1} \left(\frac{s^2}{\lambda d}\right) \tag{12.3-7}$$

where s is the vertical antenna separation, λ is the wavelength, and d is the path length. All values are measured in the same units.

The term improvement F has been used to describe the ratio of the total time T_i spent in fades to the total time T_{ij} spent in simultaneous fades. For deep fades

$$F = \frac{T_i}{T_{ij}} \approx 2F_N \quad L < 0.1 \tag{12.3-8}$$

The total fading time T_i in a year in a 6-GHz propagation can be obtained from Eqs. (12.3-1) and (12.3-2) as

$$T_i = N\bar{t} = \begin{cases} (37904.4L/\text{year}) \, (490L \text{ s}) \\ 1857 \text{ s/year} \quad \text{at} \quad L = 0.01 \text{ (or } -40 \text{ dB)} \\ 31 \text{ min/year} \quad \text{at} \quad L = 0.01 \text{ (or } -40 \text{ dB)} \end{cases}$$

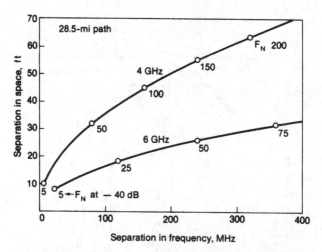

Figure 12.4 Separations in space and in frequency that provide equal values of F_N, the ratio of the number of fades to the number of simultaneous fades. (*After Vigants, Ref. 2.*)

We can use the same step to obtain the total fading time T_i in 4-GHz propagation.

$$T_i = N\bar{t} = \begin{cases} (22002L/\text{year})\,(408L \text{ s}) \\ 897.68 \text{ s/year} & \text{at} & L = 0.01 \text{ (or } -40 \text{ dB)} \\ 15 \text{ min/year} & \text{at} & L = 0.01 \text{ (or } -40 \text{ dB)} \end{cases}$$

The values of F_N at -40 dB are shown along the curves in Fig. 12.4 for 4 and 6 GHz, respectively. To achieve $F_N = 5$ in 4 GHz, we must use a separation in vertical spacing of *10 ft*, or the frequency separation should be *8 MHz*.

To achieve $F_N = 5$ in 6 GHz, we must use a vertical antenna separation of 9 ft, or the frequency separation should be 25 MHz. The improvement F can be obtained from Eq. (12.3-8) as $F = 2 \times 5 = 10$, or the total fading time after a diversity scheme at $L = 0.01$ is reduced to

$$T_{ij} = \begin{cases} 3.1 \text{ min/year} & \text{(in the 6-GHz band)} \\ 1.5 \text{ min/year} & \text{(in the 4-GHz band)} \end{cases}$$

12.4   Ray-Bending Phenomenon[4]

This phenomenon occurs because air is denser at lower levels than at higher levels. Starting with Snell's law for two layers with different refractive indices n_1 and n_2, we obtain

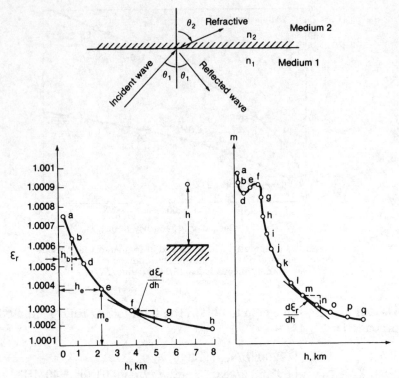

Figure 12.5 Illustration of Snell's law and finding dn/dh from a measured piece of data. (*After Hund, Ref. 4.*)

$$\frac{\sin \theta_1}{\sin \theta_2} = \frac{n_2}{n_1} = \frac{\sqrt{\mu_2 \varepsilon_2}}{\sqrt{\mu_1 \varepsilon_1}} = \frac{C_1}{C_2} \qquad (12.4\text{-}1)$$

The other parameters are explained as follows.

1. If θ_2 is the reflection angle (Fig. 12.5) and $n_1 = n_2$, then

$$\sin \theta_1 = \sin \theta_2 \qquad (12.4\text{-}2)$$

Snell's law indicates that the incident angle θ_1 is equal to the reflected angle θ_2. Also assume that there are no conductivity effects in the atmosphere and that the troposphere is not magnetic: $\mu_1 = \mu_2 = \mu_0$. For layer 1, the dielectric constant is ε_1 and the corresponding velocity C_1 of wave propagation is

$$C_1 = \frac{1}{\sqrt{\mu_0 \varepsilon_1}}$$

and for layer 2, the dielectric constant is ε_2 and the velocity of wave propagation is C_2.

2. If the θ_2 is the refraction angle (see Fig. 12.5) for the wave transmission into medium 2, then

$$n_1 \sin \theta_1 = n_2 \sin \theta_2 \qquad (12.4\text{-}3)$$

Assume that the refraction indexes n_1, n_2, n_3 decrease as the altitude h_1, h_2, h_3 increases. Then

$$\frac{\sin \theta_1}{\sin \theta_2} = \frac{n_2}{n_1} \qquad (12.4\text{-}4)$$

$$\frac{\sin \theta_2}{\sin \theta_3} = \frac{n_2}{n_3} \qquad (12.4\text{-}5)$$

for a gradient $dn\,/\,dh$ of n, and $n \sin \theta = $ constant. Then at a certain altitude h, the index of refraction is

$$n = \sqrt{\varepsilon_r} \qquad (12.4\text{-}6)$$

At the altitude $h + dh$

$$n + \left(\frac{dn}{dh}\right) dh = \sqrt{\varepsilon_r + \left(\frac{d\varepsilon_r}{dh}\right) dh} \qquad (12.4\text{-}7)$$

The $d\varepsilon_r/dh$ can be found from a measured (or statistically predicted) curve at a given location (see Fig. 12.5). Then $dn\,/\,dh$ can be found. The equation for ray bending is expressed as

$$\frac{1}{\rho} = -\frac{1}{n}\frac{dn}{dh} \qquad (12.4\text{-}8)$$

Now ρ is the radius of curvature which can be calculated as dn/dh changes as a result of $d\varepsilon_r/dh$. The radius ρ can be plotted by computer as shown in Fig. 12.6.

12.5 System Reliability

The microwave radio link is a stand-alone system. A typical system layout is shown in Fig. 12.7 for a transmitter and in Fig. 12.8, for a receiver.

12.5.1 Equipment reliability

All radio equipment should be redundant (i.e., equipped with duplicates) with a standby and automatic switchover in case of failure. An

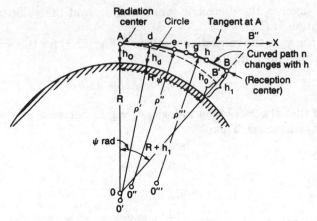

Figure 12.6 The radius of curvature of the wave path. (*After Hund, Ref. 4.*) Curved path *AB* requires the least time to move electromagnetic wave energy from a through angle ψ to line *OB″*.

alarm system is available for reporting an emergency condition at any microwave site to the central alarm station at the MTSO. Redundant power converters have been included at each cell site. A space diversity can be implemented for further increasing system reliability.

12.5.2 Path reliability

The microwave path should be a clear line-of-sight path between two points. Each path should be calculated and studied by field survey. Sometimes larger antenna size, higher tower, shorter distance, more

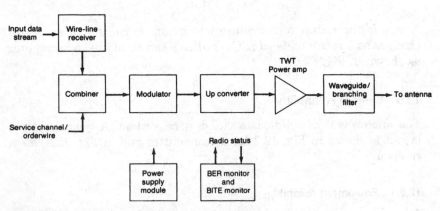

Figure 12.7 Microwave radio transmit block diagram.

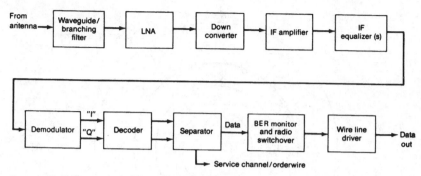

Figure 12.8 Microwave radio receive block diagram.

diversity, or greater capacity are required to increase the path reliability. An important consideration is elimination or reduction of the
multipath reception at the receiver as a result of the reflection along
the path. The reflected energy would be negligible if the reflector were
out of the first Fresnel zone, which is H

$$H \geqslant \sqrt{\frac{\lambda d_1 (d - d_1)}{d}}$$ (12.5-1)

where d_1 and d_2 are as shown in Fig. 12.9 and λ is the operating
wavelength. Five special cases are as follows.

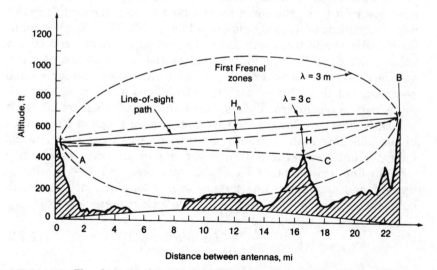

Figure 12.9 The clearance distance from the closest objects. (*From Ref. 8, p. 439.*)
Typical profile plot showing first Fresnel zones for 100 MHz and 10 GHz.

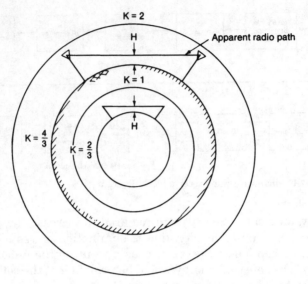

Figure 12.10 Illustration of earth bulge conditions.

Hyperreflectivity. Hyperreflectivity may occur, such as in wave propagation over water, metal objects, and large flat surfaces. In these cases, additional path clearance is recommended.

Bending. Since the earth is curved and because dielective permittivity varies with height, bending occurs. On the average, the radio wave is bent downward, i.e., the earth radio wave acts as if the earth's radius were four-thirds of its real value (see Fig. 12.10). The effective radius for K (ratio of effective earth radius to true earth radius) can be any value other than $K = \frac{4}{3}$ and can be treated as a function of atmospheric conditions. Sometimes it can be as low as one-half for a small percentage of time. If we base our calculations on $K = \frac{4}{3}$, then the ray is curved downward. This is called an "earth bulge" condition. It will cause the path loss to increase over a wide range of frequencies unless adequate path clearance is provided. A value of $K = 1$ would indicate that the earth is completely flat or the ray travels in a straight line. On the other hand, the wave based on $K = \frac{2}{3}$ will tilt upward and may cause interference over a long distance. An earth bulge factor of $\frac{2}{3}$ or $\frac{4}{3}$ is used to provide a clear area.

$$H \geqslant \frac{d_1 d_2}{2} \quad \text{(for a factor of } \frac{4}{3}\text{)} \qquad (12.5\text{-}2)$$

$$H \geqslant d_1 d_2 \quad \text{(for a factor of } \frac{2}{3}\text{)} \qquad (12.5\text{-}3)$$

where d_1 and d_2 are in miles and H is in feet.

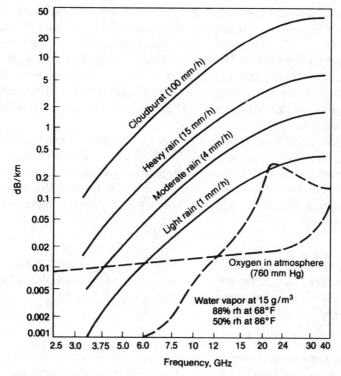

Figure 12.11 Estimated atmospheric absorption. (*From Ref. 9, p. 443.*)

High microwave frequencies. For a microwave frequency above 10 GHz, the oxygen, water vapor, and rain attenuate (or scatter) the microwave beam. To determine the total path attenuation, we must add the free-space loss (FSL) to the rain loss while considering the anticipated rain rates. The rain rate is measured by millimeters per hour. Usually, a rain rate of 15 mm/h or greater will be considered as heavy rain. Some areas, such as Florida, may have a great deal of precipitation. Some areas, such as southern California, are arid. The history of rain-rate data can be obtained from the U.S. Weather Bureau's annually accumulated rain statistics collected since 1953 in 263 cities. The author was the first to suggest this method at Bell Laboratories in 1972. (Several rain-rate models are given in Refs. 5, 6, and 7.) Once the rain statistics of each city are known, the decibels per kilometer for different rain rates can be found at each operating frequency as shown in Fig. 12.11. The rain rate is governed by the size and shape of the raindrops. The path loss varies with both the raindrop size and rain rate.

The effects of haze, fog, snow, and dust are insignificant. The size of the rain cell (a rain-occupied area) will be considered along the microwave link. The heavier the rain, the smaller the rain cell. Also, the rain-rate profile will be nonuniform. To calculate a microwave link, we need to know the (1) link gain—power, antenna size, antenna height, and receiver sensitivity; (2) free-space loss; (3) attenuation due to a predicted rain rate; and (4) given availability—allowable downtime, such as 1 h in a year or 10^{-4}.

The transmission rate of a signal over a microwave link is limited to the time-delay spread, and the time-delay spread is based on the distance. Usually we design the link primarily on the basis of the rain effect; therefore, the link is usually short because of the rain attenuation, and the time-delay spread at the shorter distance is not considered.

Power system reliability. Battery systems or power generators are needed for the microwave systems in case of power failure. Usually a 24-V dc battery system will be installed with 8 to 10 h of reserve capacity.

Microwave antenna location. Sometimes the reception is poor after the microwave antenna has been mounted on the antenna tower. A quick way to check the installation before making any other changes is to move the microwave antenna around within a 2 to 4 ft radius of the previous position and check the reception level. Surprisingly favorable results can be obtained immediately because multipath cancellation is avoided as a result of changing reflected paths at the receiving antenna.

Also, at any fixed microwave antenna location, the received signal level over a 24-h time period varies.

12.6 Microwave Antennas[8]

12.6.1 Characteristics of microwave antennas

Microwave antennas can afford to concentrate their radiated power in a narrowbeam because of the size of the antenna in comparison to the wavelength of the operating frequency; thus, high antenna gain is obviously desirable. Some of the more significant characteristics are discussed in the following paragraphs.

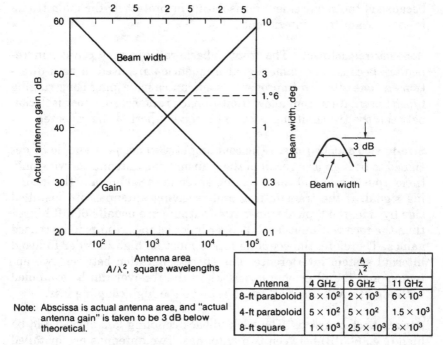

Figure 12.12 Approximate antenna gain and beamwidth. (*From Ref. 8, p. 445.*)

Beamwidth.   The greater the size of the antenna, the narrower the beamwidth. Usually the beamwidth is specified by a half-power (3-dB) beamwidth and is less than 10° at higher microwave frequencies. The beamwidth sometimes can be less than 1°. The narrowbeam can reduce the chances of interference from adjacent sources or objects such as adjacent antennas. However, a narrowbeam antenna requires a fair amount of mechanical stability for the beam to be aimed at a particular direction. Also, the problem of antenna alignment due to the ray-bending problems discussed earlier restricts the narrowbeam antenna to a certain degree. The relationship between gain and beamwidth is depicted in Fig. 12.12.

Sidelobes.   The sidelobes of an antenna pattern would be the potential source of interference to other microwave paths or would render the antenna vulnerable to receiving interference from other microwave paths.

Front-to-back ratio.   This is defined as the ratio of the maximum gain in the forward direction to the maximum gain in the backward direction. The front-to-back ratio is usually in the range of 20 to 30 dB because of the requirement for isolating or protecting the main transmission beam from interference.

Repeater requirement. The front-to-back ratio is very critical in repeaters because the same signal frequencies are used in both directions at one site. An improper design can cause a ping-pong ringing type of oscillation from a low front-to-back ratio or from poor isolation between the transmitting port and receiving port of the repeater.

Side-to-side coupling loss. The coupling loss, in decibels, should be designed to be high as a result of the transmitting antenna carrying only the output signal and the receiving antenna receiving only the incoming signal. If the transmitting and receiving antennas are installed side by side, the typical transmitter outputs are usually 60 dB higher than the receiver input level. Longer link distance results in increased values. Therefore, the coupling losses must be high in order to avoid internal system interference. The space separation between two antennas and the filter characteristics in the receiver can be combined with a given antenna pattern to achieve the high coupling loss.

Back-to-back coupling. The back-to-back coupling loss also should be high (e.g., 60 dB) between two antennas. Two antennas are installed back to back, one transmitting and one receiving. However, it is much easier to reach a high back-to-back coupling loss than a side-to-side coupling loss.

12.6.2 Polarization and space diversity in microwave antennas

Polarization. To reduce adjacent-channel interference, microwave relay systems can interleave alternate radio-channel frequencies from a horizontal polarized wave to a vertical polarized wave.

The same approach can be applied to the left- and right-handed circularly polarized waves, but the beamwidths of antennas for this orthogonal system are relatively large and therefore are not attractive.

In the polarization system, the *cross-coupling loss* is specified. This loss is defined as the ratio of the power received in the desired polarization to the power coupling into other polarization. The cross-coupling loss (isolation) should be as high as possible. Usually 25 to 30 dB is required for one hop.

Space diversity.[9] The two antennas separated vertically or horizontally as described in Sec. 12.3 can be used for a two-branch space-diversity arrangement. In a space-diversity receiver, the required re-

TABLE 12.1 Horn-Reflector Antenna Characteristics

Frequency polarization	4 GHz Vertical	4 GHz Horizontal	6 GHz Vertical	6 GHz Horizontal	11 GHz Vertical	11 GHz Horizontal
Midband gain, dB	39.6	39.4	43.2	43.0	48.0	47.4
Front-to-back ratio, dB	71	77	71	71	78	71
Beamwidth (azimuth), degrees	2.5	1.6	1.5	1.25	1.0	0.8
Beamwidth (elevation), degrees	2.0	2.13	1.25	1.38	0.75	0.88
Sidelobes, dB below main beam	49	54	49	57	54	61
Side-to-side coupling, dB	81	89	120	122	94	112
Back-to-back coupling, dB	140	122	140	127	139	140

ception level is relatively low so that the transmitted power on the other end of the link can be reduced. This is also an effective method for increasing the coupling loss between the transmitting antenna and receiving antenna.

12.6.3 Types of microwave-link antenna

Two kinds of antenna are used for microwave links.

1. A parabolic dish, used for short-haul systems. Antennas sizes range from 1.5 m (5 ft) to 3 m (10 ft) in diameter.
2. A horn-reflector antenna, to trap the energy outward from the focal point. The advantages of using this antenna are
   a. Good match—return loss 40–50 dB.
   b. Broadband—a horn antenna can work at 4, 6, and 11 GHz.
   c. One horn can be used for two polarizations with high cross-coupling loss.
   d. Small sidelobes—high back-to-back coupling loss.

The gains, coupling losses, and beamwidths are listed in Table 12.1 for different frequencies and different polarizations.

12.6.4 Installation of microwave antennas

A microwave antenna cannot be installed at any arbitrary location. Selection of an optimum position is very important. In many situations if we cannot move horizontally, we can move vertically. In a microwave-link setup, there are two fixed effective antenna heights, one at each end based on each reflection plane where the reflection point is incident on it. The gain of the received signal also relates to the two effective antenna heights if they are low. The antenna location can be moved around to find the best reception level. Sometimes it is worthwhile to take time to search for the location that gives the best reception.

References

1. Bell Telephone Laboratories, *Engineering and Operations in the Bell System,* 1977, Bell Telephone Labs, Inc., Chap. 10.
2. A. Vigants, "Number and Duration of Fades at 6 and 4 GHz," *Bell System Technical Journal,* Vol. 50, March, 1971, pp. 815–842.
3. A. Vigants, "The Number of Fades in Space-Diversity Reception," *Bell System Technical Journal,* Vol. 49, September 1970, pp. 1513–1554.
4. A. Hund, *Short-Wave Radiation Phenomenon,* Vol. 2, McGraw-Hill Book Co., 1952, pp. 980–985.
5. W. C. Y. Lee, "No-Cost and Fast Time in Obtaining the Signal Attenuation Statistics due to Rainfall in Major U.S. Cities," Bell Labs Internal Report, May 10, 1974.
6. S. H. Lin, "More on Rain Rate Distributions and Extreme Value Statistics," *Bell System Technical Journal,* Vol. 57, May–June, 1978, pp. 1545–1568.
7. W. C. Y. Lee, "An Approximate Method for Obtaining Rain Rate Statistics for use in Signal Attenuation Estimating," *IEEE Transactions on Antenna and Propagation,* Vol. AP-27, May 1979, pp. 407–413.
8. Bell Telephone Laboratories, *Transmission Systems for Communications,* 4th ed., Western Electric Company, 1970, Chap. 18.
9. H. Yamamoto, "Future Trends in Microwave Digital Radio," *IEEE Communications,* Vol. 25, February 1987, pp. 40–52.

13

System Evaluations

13.1 Performance Evaluation

13.1.1 Blockage

There are two kinds of blockage: set-up channel blockage and voice-channel blockage.

Set-up channel blockage B_1. Information regarding set-up channel blockage cannot be obtained at the cell site because the mobile unit will be searching for the busy/idle bit of a forward set-up channel in order to set up its call. If the busy bit does not change after 10 call attempts in 1 s, a busy tone is generated, and no mobile transmit takes place. In another case the mobile transmit takes place as soon as the idle bit is shown. Several initiating cells can intercollide at the same time. When it occurs, the mobile unit counts it as one seizure attempt. If the number of seizure attempts exceeds 10, then the call is blocked. This kind of blockage can be detected only by mobile phone users. If the occurrence of blockage of the system is in doubt, each of the three specified set-up channels can be assigned in each of the three sectors of a cell, and the total number of incoming calls among the three sectors can be compared with that from a single set-up channel (omni). It should be determined whether there is a difference between two call-completion numbers, one from a single set-up channel and the other from three set-up channels. This is one way to check the blockage if the single set-up channel seems too busy. The set-up channel blockage should be at least less than half of the specified blockage (usually 0.02) in the mobile cellular system.

If all the call-attempt repeats are independent events, then the resultant blocking probability B_1 after n attempts is related to the blocking probability of the single call attempt B, as

$$B_1 = 1 - (1 - B) \sum_{i=0}^{n} B^i = B^n \qquad (13.1\text{-}1)$$

Example 13.1   Assume that the blocking probability of a set-up channel is .005, and the holding time at the set-up channel is 175 ms per call. There is only one channel; then the offered load (from Appendix 1.1) a is .005. Thus the number of set-up calls being handled is

$$C = \frac{.005 \times 3600 \times 1000}{175} = 120 \text{ calls} \qquad \text{(one call attempt)}$$

Example 13.2   All parameters are the same as in Example 13.1, except that the offered load a changes to $a = 0.02$. Then the number of set-up calls is

$$C = \frac{.02 \times 3600 \times 1000}{175} = 480 \text{ calls} \qquad \text{(one call attempt)}$$

Example 13.3   Given the number of set-up calls per hour, find the blocking probability B_t after 10 call attempts in 1 s.

Consider the following cases.

Case 1.   Assume that there are two set-up calls per second or 7200 calls per hour. Since each set-up call takes 175 ms, the offered load a is

$$a = \frac{175 \times 2}{1000} = .35$$

The blocking probability B (see Appendix 1.1) is $B = .25$ (assuming one call attempt).

Since the average interval for each attempt is 100 ms, 10 attempts have to be completed in 1 s. It is a kind of conditional probability problem. In the worst case, a mobile unit has to fail the tenth call attempt before giving up. During this period, because of the failure of all call attempts, the two set-up calls from other mobile units should have been successful with a probability of 1. The length of two set-up calls is 350 ms, which is roughly the time interval required for four attempts; i.e., these four attempts are definitely blocked with a blocking probability of 1 and should not be counted as attempts. Therefore, only six attempts count. Using Eq. (13.1-1), we obtain

$$B_1 = (.25)^6 = .00012$$

which is quite low and, of course, acceptable.

Case 2. Assume that there are three set-up calls per second or 10,800 calls per hour. Then

$$a = \frac{175 \times 3}{1000} = .525 \quad \text{(offered load)}$$

$$B = .342 \qquad \text{(blocking probability; see Appendix 1.1)}$$

Since three set-up calls take 525 ms, roughly six out of ten attempts are definitely blocked following the same argument stated in case 1. Only four attempts count, then the resultant blocking probability is

$$B_t = (.342)^4 = .013$$

which is too high for the set-up channel.

Voice-channel blockage B_2. Voice-channel blockage can be evaluated at the cell site. When all calls come in, some are refused for service because there are no available voice channels. Suppose that we are designing a voice channel blockage to be .02. On this basis, $B_2 = .02$, and after determining the holding time per call[1] and roughly estimating the total number of calls per hour at the site,[2] we can find the number of radios required.

Example 13.4 Assume that 2000 calls per hour are anticipated. The average holding time is 100 s per call, and the blocking probability is .02 (2 percent). Then the offered load is

$$a = \frac{2000 \times 100 \text{ s}}{60 \times 60 \text{ s}} = 55.5 \text{ erlangs}$$

Use $a = 55.5$ and $B_2 = 0.02$ to find $N = 66$ channels required (refer to Appendix 1.1).

The actual blocking probability data must be used to check the outcome from the Erlang B model (Appendix 1.1). Although the difference can be up to 15 percent, the Erlang B model is still considered as a good model for obtaining useful estimates.

End-office trunk blockage B_3. The trunks connecting from the MTSO to the end office can be blocked. This usually occurs when the call traffic starts to build up and the number of trunks connected to the end office becomes inadequate. Unless this corrective action is taken, the blockage during busy periods increases. An additional number of trunks could be provided at the end office when needed.

The total blockage B_t. As the total call blockage is the result of all three kinds of blockage, the total blockage is

$$B_t = B_1 + B_2 (1 - B_1) + B_3 (1 - B_1) (1 - B_2)$$

$$= 1 - (1 - B_1) (1 - B_2) (1 - B_3) \qquad (13.1\text{-}2)$$

Example 13.5 Assume that $B_1 = .01$ and $B_2 = B_3 = .02$. Then the total blockage is

$$B_t = .01 + .0198 + .0194 = .0492 \approx 5\%$$

The result in Example 13.5 indicates that even when each individual blockage (i.e., B_1, B_2, and B_3) is small, the total blockage becomes very large. Therefore, the resultant blockage is what we are determining.

13.1.2 Call drops (dropped-call rate)

Call drops are defined as calls dropped for any reason after the voice channel has been assigned. Sometimes call drops due to weak signals are called *lost calls*. The dropped-call rate is partially based on the handoff-traffic model and partially based on signal coverage. The calculation of dropped call rate is shown in Sec. 9.10 through Sec. 9.12. The evaluation of call drops is stated in this section.

The handoff traffic model. A new handoff cell site treats handoffs the same way as it would an incoming call. Therefore, the blockage for handoff calls is also $B = .02$. Some MTSO systems may give priority to handoff calls rather than to incoming calls. In this case the blocking probability will be less than .02.

A warning feature can be implemented when the call cannot be handed off and may be dropped with high probability, enabling the customer to finish the call before it is dropped. Then the dropped-call rate can be reduced.

The loss of SAT calls. If the mobile unit does not receive a correct SAT in 5 s, the mobile-unit transmitter is shut down. If the mobile unit does not send back a SAT in 5 s, the transmitter at the cell site is shut down. In both cases the call is dropped. If the correct SAT cannot be detected at the cell site, as in cases of strong interference, then (1) the SAT can be offset by more than 15 Hz (see section entitled "The total dropped-call rate," below) or (2) the SAT tone generator in the mobile unit may not produce the desired tone.

Calculation of SAT interference conditions. The desired SAT is $\cos w_1 t$, and the undesired SAT is $\rho \cos w_2 t$. When $\rho \ll 1$, the SAT detector at the cell site can easily detect w_1. When ρ is greater and starts to approach 1, SAT interference occurs. The following analysis shows the degree of the interference due to the value of ρ.

$$\cos w_1 t + \rho \cos w_2 t = A(t) \cos \theta(t) \qquad (13.1\text{-}3)$$

where

$$A(t) = \sqrt{1 - \rho^2 + 2\rho \cos (w_1 - w_2)t} \qquad (13.1\text{-}4)$$

$$\psi(t) = w_2 t - w_1 t$$

$$\theta = w_1 t + \tan^{-1} \frac{\rho \sin \psi(t)}{1 + \rho \cos \psi(t)} \qquad (13.1\text{-}5)$$

$$w = \frac{d\theta}{dt} = w_1 + \frac{w_2 - w_1}{([1 + \rho \cos \psi(t)]/\{\rho[\rho + \cos \psi(t)]\}) + 1} \qquad (13.1\text{-}6)$$

Let $\cos \psi(t) = 1$ in Eq. (13.1-6), the extreme condition of w which is the offset frequency from a desired SAT.

$$w = w_1 + \left(1 + \frac{1}{\rho}\right)^{-1} (w_2 - w_1) \qquad (13.1\text{-}7)$$

For

$$\rho = \begin{cases} .3 & w = w_1 + .22(w_2 - w_1) \\ .5 & w = w_1 + .333(w_2 - w_1) \\ .75 & w = w_1 + .45(w_2 - w_1) \end{cases}$$

$$(13.1\text{-}8)$$

If $w_2 - w_1 = 30$ Hz, for two adjacent SATs

$$\rho = \begin{cases} .5 & \omega = \omega_1 \pm 9.95 & \text{(acceptable)} \\ .75 & \omega = \omega_1 \pm 12.9 & \text{(marginal)} \end{cases}$$

If $\omega_2 - \omega_1 = 60$ Hz, for two ends of SATs

$$\rho = \begin{cases} .3 & \omega = \omega_1 \pm 13.8 & \text{(marginal)} \\ .5 & \omega = \omega_1 \pm 19.9 & \text{(unacceptable)} \\ .75 & \omega = \omega_1 \pm 25.8 & \text{(unacceptable)} \end{cases}$$

Adjacent SATs cannot interfere with the desired SAT for an undesired SAT level below $\rho = .75$ (meaning a level of -2.5 dB). However, when two SATs are not adjacent to each other, the undesired SAT level should at least be lower than $\rho = 0.3$ (-10.5 dB) in order for no interference to occur.

Unsuccessful complete handoffs. Because of the limitations in processor capacity, the duration of the handoff process may occasionally be too long, and the mobile unit may not be informed of a new channel to be handed off.

The total dropped-call rate. Assume that the handoff blocking is B_4, the probability of lost SAT calls is B_5, and the probability of an unsuccessful complete handoff is B_6. Then the total drop call rate is

$$B_d = B_4 + B_5 (1 - B_4) + B_6 (1 - B_4)(1 - B_5) \qquad (13.1\text{-}9)$$

Usually, the dropped-call rate should be less than 5 percent.

13.1.3 Voice quality

It is very important that the voice quality of a channel be tested by subjective means. Some engineers try to use the signal-to-noise-plus-distortion ratio (SINAD) to evaluate voice quality. Although SINAD is an objective test, using it to test voice quality may result in misleading conclusions. Worst of all, engineers are always proud of the apparatus they have designed, and they tend to ignore others' opinions. To serve the public interest, we must survey consumers for their opinions.

Evaluation of system performance based on a subjective test can be used to set performance criteria. As was discussed in Chap. 1, 75 percent of cellular phone users report that voice quality is good (CM4; i.e., circuit merit 4) or excellent (CM5) over 90 percent of the service area. These numbers can vary depending on how well the service is performed—in other words, the cost. First we must know what kind of service we are providing to the public. We should then let the customers judge the voice quality.

A typical curve of subjective tests was shown in Fig. 7.1b for different conditions. For this particular run, the mobile speed is 56 km/h (35 mi/h).

13.1.4 Performance Evaluation

Often we encounter a situation where two systems are being installed in two different areas; one system is deployed in a flat area where the average measured bit error rate (BER) is low at the cell site on the basis of bit-stream data received from mobile-units, and the other system is deployed in a hilly area where BER is relatively high. Of course, if the same degree of skill was used to install both systems, the system in the hilly area would otherwise be inferior and would provide the poorer performance. But if each system claims to be better than the other, how can we judge these two systems fairly? One way is to compensate for variations in performance due to geographic location.[2] Assume that system A is deployed in a flat terrain and system B, in a hilly area. Both systems use the same kind of antennas and run the measured data with the same kind of mobile units. Then the handicap of building a system in a hilly area can be compensated for before comparing this system's performance with others.

Another way is to set the BER to, say, 10^{-3} and find the percentage of areas where the measured BER is greater than 10^{-3} in relation to the area of the whole system, then compare the percentages from the two different systems in two different areas.

Here the two pieces of raw field-strength recorded data in the area should be received from two mobile units moving around in two systems, respectively. From these data, the two local means (the envelope of the raw field-strength data) can be obtained. Then the two cumulative probability distributions (CPD), $P(x < X)$ of two local means, can be plotted. Let us first normalize the average power at a 50 percent level because the local means have a log-normal distribution, and the differences in transmitted power in the two systems are factored out. There will be two straight lines on log-normal scale paper (see Fig. 13.1). The standard deviations of two log-normal curves can be found from Fig. 13.1 by determining a level X where $P(x \le X) = 90$ percent.

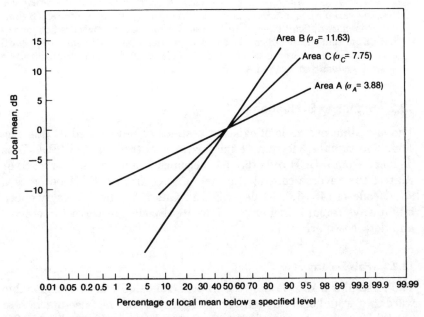

Figure 13.1 Different local mean statistics in different areas.

$$X = \begin{cases} 5 \text{ dB} & \text{for system A} \\ 15 \text{ dB} & \text{for system B} \end{cases}$$

A normal distribution can be found from published mathematical tables indicating that $P(x \leq 1.29) = 90$ percent, where

$$x \leq \frac{X - m}{\sigma} = \frac{X}{\sigma} = 1.29 \qquad (13.1\text{-}10)$$

where $m = 0$ dB and σ is the standard deviation. For

$$X = \begin{cases} 5 \text{ dB} & \sigma_A = \dfrac{5}{1.29} = 3.88 \text{ dB} \\ 15 \text{ dB} & \sigma_B = \dfrac{15}{1.29} = 11.6 \text{ dB} \end{cases}$$

The standard deviations of A and B indicate that system B covers a relatively hillier area than system A.

Example 13.6 From these measurements we find that the BER in 10 percent of area A in system A is greater than 10^{-3}. The BER in 25 percent of area B in system B is greater than 10^{-3}. The local means of two systems are shown in Fig. 13.1. We would like to judge which system has a better performance. From Fig. 13.1, 10 percent of the area in system A corresponds to 10 percent of the total signal below -5 dB. If system A is deployed in area C ($\sigma_c = 7.75$ dB), 25 percent of area C will have a BER greater than 10^{-3} below the -5-dB level. Now the variance σ_B of area B is greater than the variance σ_C of area C; thus, area B is more difficult to serve than area C. However, the measurement shows that 25 percent of area B in system B is greater than 10^{-3} below the -9-dB level. Therefore, according to the criterion BER $\geq 10^{-3}$, system B proves to be a better system than system A.

13.2 Signaling Evaluation

The signaling protocols of existing systems are evaluated in this section. The signaling format of the forward control channel (FOCC) as deduced from a BCH code (63, 51) becomes a short code of (40, 28) or that of the reverse control channel (RECC) from a BCH becomes a short code of (48, 36), as described in Chap. 3. The 12 parity-check bits always remain unchanged. This BCH code can correct one error and detect two errors.

13.2.1 False-alarm rate

The false-alarm rate is the rate of occurrence of a false recognizable word that would cause a malfunction in a system. The false-alarm rate should be less than 10^{-7}. Now we would like to verify that the BCH code (40, 28) can meet this requirement. The Hamming distance d of BCH (40, 28) is 5. This means that in every different code word at least 5 out of 40 bits are different. Then the false-alarm rate *FAR* can be calculated as

$$FAR = p_e^d (1 - p_e)^{L-d} \qquad (13.2\text{-}1)$$

where p_e is the BER and d is the length of a word in bits.

Assume that in a noncoherent frequency-shift-keying (FSK) modulation system, the average BER of a data stream in a Rayleigh fading environment is

$$\langle p_e \rangle = \frac{1}{2 + \Gamma} \qquad \text{(noncoherent FSK)} \qquad (13.2\text{-}2a)$$

and the average BER of a data stream received by a differential phase-shift-keying (DPSK) modulation system in the same environment is

$$\langle p_e \rangle = \tfrac{1}{2} \left(\frac{1}{\Gamma + 1} \right) \qquad \text{(DPSK)} \qquad (13.2\text{-}2b)$$

where Γ is the carrier-to-noise ratio. Let* $\Gamma = 15$ dB; then we obtain BER from Eq. (13.2-2) as

$$\langle p_e \rangle = .03 \qquad \text{(noncoherent FSK)}$$

$$\langle p_e \rangle = .015 \qquad \text{(DPSK)}$$

Substituting $p_e = .03$, which is the higher BER, into Eq. (13.2-1), we obtain

$$FAR \approx (.03)^5 = 2.43 \times 10^{-8} \leqslant 10^{-7}$$

This meets the requirement that $FAR < 10^{-7}$.

13.2.2 Word error rate consideration

The word error rate (WER) plays an important role in a Rayleigh fading environment. The length of a word of an FOCC is $L = 40$ and the transmission rate is 10 kbps. Then the transmission time for a 40-bit word is

$$T = \frac{40}{10,000} = 4 \text{ ms}$$

---

*The C/N ratio of a data channel can be lower than 18 dB of a voice channel.

From Sec. 1.6.6, the average duration of fades can be obtained from the following assumptions: frequency = 850 MHz, vehicle speed = 15 mi/h, and threshold level = −10 dB (10 dB below the average power level); then the average duration of fades is

$$\bar{t} = 0.33 \times \left(\sqrt{2\pi} \, \frac{V}{\lambda} \right)^{-1} = 7 \text{ ms} \qquad (13.2\text{-}3)$$

Equation (13.2-3) shows that the transmission time of one word is shorter than the average duration of fades while the vehicle speed is 15 mi/h; that is, the whole word can disappear under the fade. Therefore, redundancy schemes are introduced. From Chap. 3, the FOCC format is

200 bits	word A (40 bits × 5 times), 28 information bits	
200 bits	word B (40 bits × 5 times), 28 information bits	
10 bits	bit synchronization	
11 bits	word synchronization	
42 bits	Busy/Idle-status bits	
463 bits		56 information bits

The throughput can be obtained from

$$\frac{56}{463} = \frac{1200 \text{ bps}}{10{,}000 \text{ bps}}$$

Therefore, the throughput is 1200 bps (baseband rate).

13.2.3 Word error rate calculation

The WER can be calculated as follows. We may use a DPSK system because it has a general but simple analytic formula, more general than Eq. (13.2-2*b*)

$$\langle p_e \rangle = \tfrac{1}{2} \left(\frac{1}{\Gamma + 1} \right)^M \qquad (13.2\text{-}4)$$

where M is the number of diversity branches. It is difficult to obtain the WER from the correlation coefficient in the bit stream at a specific vehicle speed because the correlation coefficients of any two bits among all the bits in a word at that particular speed form a correlation coefficient matrix which is difficult to handle. Fortunately, we can find two extreme values, one at the speed $V \to \infty$ and the other at the speed $V - \to 0$. The calculations are described in detail in Ref. 3. Here we are simply illustrating the results.[4]

The performance of word error rates is shown in Fig. 13.2 for two cases: (1) no error correction and (2) one error correction. We have noticed that without redundancy (no repeat), the WER of a fast-fading case is worse than that of a slow-fading case. The WER obtained from a finite speed will lie between these two curves.

When a redundancy scheme is applied (Fig.13.2a), i.e., repeating K times and making a majority voting on each bit, the WER of a slow-fading case becomes worse than that of a fast-fading case. This change in WER provides a great improvement in performance. Therefore, a redundancy scheme is of value in a mobile radio environment with a variable vehicle speed. This phenomenon is also illustrated in Fig. 13.2b for a 1-bit error correction code. Figure 13.3a and b shows the

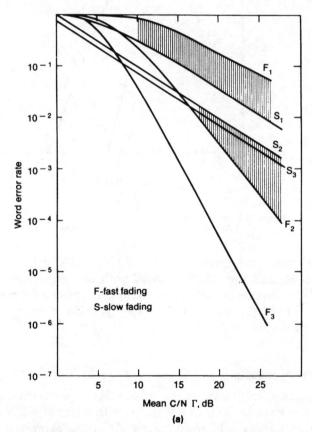

Figure 13.2 Word error rate for $N = 40$ bits. Number of branches $M = 1$; S_1, F_1, no repeat ($K = 1$); S_2, F_2 two-thirds voting ($K = 3$); S_3, F_3 three-fifths voting ($K = 5$). (a) Case 1: $M = 1$, $t = 0$. No error correction. (b) Case 2: $M = 1$, $t = 1$. One-bit error correction.

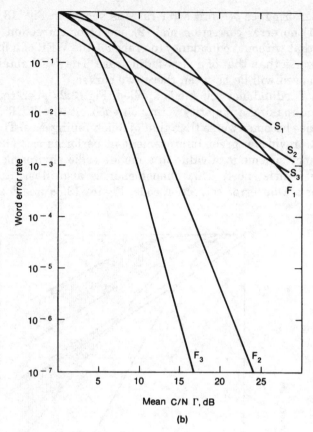

Figure 13.2   (*Continued*)

WER for two-branch diversity. Figure 13.3a is WER with no error correction code, and Fig. 13.3b is WER with one error correcton code.
Further improvements are seen in the figure. The error correction code
and the diversity plus the redundancy provide a desired signaling
performance.

13.2.4   Parity check bits

In this section, we illustrate the generation of parity check bits in a
word. Let a word of (7, 4) be generated with 4 bits of information and
a 3-bit parity check. The word matrix $[C]$ can be expressed as

$$[C] = [x_m] [G] \tag{13.2-5}$$

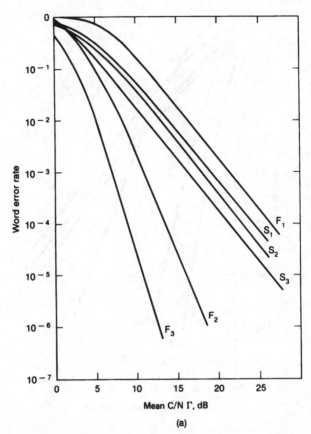

Figure 13.3 Word error rate for N = 40 bits. Number of branches M = 2; S_1, F_1, no repeat (K = 1); S_2, F_2 two-thirds voting (K = 3); S_3, F_3 three-fifths voting (K = 5). (a) Case 3: M = 2, t = 0. No error correction. (b) Case 4: M = 2, t = 1. One-bit error correction.

where $[x_m]$ is the information matrix and $[G]$ is the generation matrix. Let $[G]$ have the following form.

$$[G] = \begin{bmatrix} 1000 & 101 \\ 0100 & 111 \\ 0010 & 110 \\ 0001 & 001 \end{bmatrix} = [IP] \qquad (13.2\text{-}6)$$

$$\text{Identity} \qquad \text{Parity}$$

The parity matrix of three bits can be arranged in any order and have

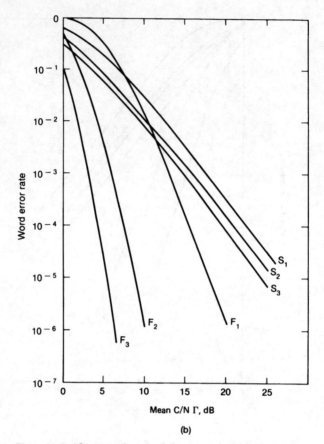

Figure 13.3 (*Continued*)

any combination of 0s and 1s. For example, let $[x_m]$ = 1001, then substituting $[x_m]$ and Eq. (13.2-6) into Eq. (13.2-5) yields

$$[C] = [1001] \begin{bmatrix} 1000 & 1 & 0 & 1 \\ 0100 & 1 & 1 & 1 \\ 0010 & 1 & 1 & 0 \\ 0001 & 0 & 1 & 1 \end{bmatrix} = [1001 \ C_5 C_6 C_7]$$

5th 6th 7th

Then C_5 can be obtained by multiplying the fifth column by $[x_m]$ and applying modulo 2 addition* as

---

*Modulo 2 additions;
$$1 + 1 = 0 \qquad 1 + 0 = 1 \qquad 0 + 0 = 0 \qquad 0 + 1 = 1$$

$$C_5 = 1 \cdot 1 + 0 \cdot 1 + 0 \cdot 1 + 1 \cdot 0 = 1 + 0 + 0 + 0 = 1$$

The same process is applied to C_6 and C_7 as

$$C_6 = 0 + 0 + 0 + 1 = 1 \quad C_7 = 1 + 0 + 0 + 1 = 0$$

Therefore the information fits (1001), along with the three parity check bits, become a word (1001110).

13.3 Measurement of Average Received Level and Level Crossings

13.3.1 Calculating average signal strength[5]

The signal strength can be averaged properly to represent a true local mean $m(x)$ to eliminate the Rayleigh fluctuation and retain the long-term fading information due to the terrain configuration. Let $\hat{m}(x)$ be the estimated local mean. If a length of data L is chosen properly, $\hat{m}(x)$ will approach $m(x)$ as

$$\hat{m}(x) = \frac{1}{2L} \int_{x-L}^{x+L} r(y)\,dy = \frac{1}{2L} \int_{x-L}^{x+L} m(y)r_0(y)\,dy$$

$$= m(x) \left[\frac{1}{2L} \int_{x-L}^{x+L} r_0(y)\,dy \right] = m(x) \tag{13.3-1}$$

or
$$\frac{1}{2L} \int_{x-L}^{x+L} r_0(y)\,dy \rightarrow 1 \tag{13.3-2}$$

where $r_0(y)$ is a Rayleigh distributed variable. If the value of Eq. (13.3-2) is close to 1, then $\hat{m}(x)$ is close to $m(x)$. The spread of $\hat{m}(x)$, denoted as $\sigma_{\hat{m}}$, can be expressed as

$$1\sigma_{\hat{m}} \text{ spread} = 20 \log \frac{m(x) + \sigma_{\hat{m}}}{m(x) - \sigma_{\hat{m}}} \quad \text{in dB} \tag{13.3-3}$$

Equation (13.3-3) is plotted in Fig. 13.4. The $1\sigma_{\hat{m}}$ spread is used to indicate the uncertainty range of a measured mean value from a true mean value if the length of the data record is inadequate.

The proper length 2L. If we are willing to tolerate $1\sigma_{\hat{m}}$ spread in a range of 1.56 dB, then $2L = 20\lambda$. If the tolerated spread is in a range of 1 dB, then $2L = 40\lambda$.

For length $2L$ less than 20 wavelengths, the $1\sigma_{\hat{m}}$ spread begins to

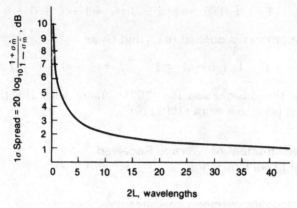

Figure 13.4 The value of $1\sigma_{\hat{m}}$ spread.

increase quickly. When length $2L$ is greater than 40λ, the $1\sigma_{\hat{m}}$ spread decreases very slowly.

In addition, the mobile radio signal contains two kinds of statistical distributions: $m(y)$ and $r_0(y)$. If a piece of signal data $r(y)$ is averaged, we find that if the length is shorter than 40λ, the unwanted $r_0(y)$ may be retained whereas at lengths above 40λ smoothing out of long-term fading $m(y)$ information may result. Therefore, 20 to 40λ is the proper length for averaging the Rayleigh fading signal $r(y)$.

Sampling average.* As mentioned previously, when using the averaging process with a filter, it is difficult to control bandwidth even when the length of the data to be integrated is appropriate. Therefore, the sample values of $r(t)$ are used for sampling averaging instead of analog (continuous waveform) averaging. Then we must determine how many samples need to be digitized across a signal length of $2L$ (see Fig. 13.5). The number of samples taken for averaging should be as small as possible. However, we have to calculate how many sample points are needed for adequate results. We set a confidence level of 90 percent and determine the number of samples required for the sampling average. The general formula is

$$P\left(-1.65 \leq \frac{\bar{r}_j - \hat{m}_j}{\hat{\sigma}_j} \leq 1.65\right) = 90\% \qquad (13.3\text{-}4)$$

---

*The detailed derivation is shown in W. C. Y. Lee, *Mobile Communications Design Fundamentals,* John Wiley & Sons, 1993, Sec. 2.2.2.

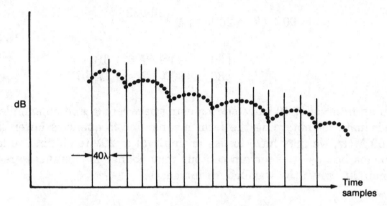

Figure 13.5 Sample average over a $2L = 40 \lambda$ of data.

Let $\hat{m}_j$ and $\hat{\sigma}_j$ be the mean and the standard deviation of ensemble average* $\bar{r}_j$ of jth interval $(2L)$ and r_j be a gaussian variable

$$\hat{\sigma} = \frac{\sigma_r}{N} \qquad \hat{m} = m$$

where m and σ_r are the mean and the standard deviation of a Rayleigh sample r. N is the number of samples. Therefore,[5]

$$m = \frac{\sqrt{\pi}}{2} \sqrt{\overline{r^2}} \tag{13.3-5}$$

$$\sigma_r = \frac{\sqrt{4 - \pi}}{2} \sqrt{\overline{r^2}} \tag{13.3-6}$$

$$\frac{\sigma_r}{m} = \sqrt{\frac{4 - \pi}{\pi}} \tag{13.3-7}$$

Substitution of Eqs. (13.3-5) to (13.3-7) into Eq. (13.3-4) yields

$$P\left(\left(1 - \frac{0.8625}{\sqrt{N}}\right) m \le \bar{r}_j \le \left(1 + \frac{0.8625}{\sqrt{N}}\right) m\right) = 90\% \tag{13.3-8}$$

Then the 90 percent confidence interval CI expressed in decibels is

---

*Time average in a mobile radio environment is an ergodic process in statistics. Therefore the values from a time average with a proper interval and an ensemble average are the same.

$$90\% \; CI = 20 \log \left(1 + \frac{0.8625}{\sqrt{N}}\right)$$

$$\text{(13.3-9)}$$

$$N = \begin{cases} 50 & 90\% \; CI = 1 \text{ dB} \\ 36 & 90\% \; CI = 1.17 \text{ dB} \end{cases}$$

In an interval of 40 wavelengths using between 36 and 50 samples is adequate for obtaining the local means. For frequencies lower than 850 MHz, we may have to use an interval of 20λ to obtain the local means because the terrain contour may change at distances greater than 20λ when the wavelength increases.

13.3.2 Estimating unbiased average noise levels[6]

Usually the sampled noise in a mobile environment contains high-level impulses that are generated by the ignition noise of the gasoline engine. Although the level of these impulses is high, the pulse width of each impulse generally is very narrow (see Fig. 13.6). As a result, the energy contained in each impulse is very small and should not have any noticeable effect on changing the average power in a 0.5 s interval.

However, in a normal situation averaging a sampled noise is done by adding up the power values of all samples, including the impulse samples, and dividing the sum by the number of samples. This is called the *conventionally averaged noise power* and is denoted n_c. In

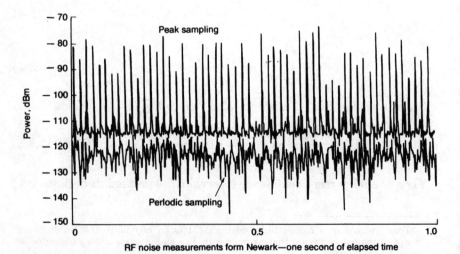

Figure 13.6 Environmental noise traces.

this case, these impulse samples dominate the average noise calculation and result in a mean value which is not representative of the actual noise power. Use of this value affects the design requirements of the system (signaling and voice). Hence, before the following new technique was introduced, it was not known why there was no correlation between BER and signal-to-noise ratio measured in certain geographic areas. In a new statistical method the average noise is estimated by excluding the noise impulses while retaining other forms of interference. This technique is compatible with real-time processing constraints.

Description of the method. A counter in the mobile unit counts the instantaneous noise measurements which fall below a preset threshold level X_t and sends a message containing the number of counts n to the database for recording. From the database data, we can calculate the percentage of noise samples x_i below the present level X_t

$$P(x_i \leq X_t) = \frac{n}{N} \qquad (13.3\text{-}10)$$

where in our case N is the total number of samples. Once we know the percentage of noise samples below level X_t, we can obtain the average "noise" X_0 exclusive of the noise spikes from the Rayleigh model. Furthermore the level X_t can be appropriately selected for both noise and signal measurements, because both band-limited noise and mobile radio fading follow the same Rayleigh statistics.

Estimating the average noise X_0. For a Rayleigh distribution (band-limited noise), the average noise power exclusive of the noise spikes X_0 can be obtained from

$$X_0 = 10 \log \left\{ -\frac{1}{\ln\,[1 - P(x_i \leq X_t)]} \right\} + X_t \quad \text{dBm} \quad (13.3\text{-}11)$$

This technique can be illustrated graphically using Rayleigh paper. Since $P(x_i \leq X_t)$ is known for a given X_t, we can find a point P_t on the paper as illustrated in Fig. 13.7. Through that point, we draw a line parallel to the slope of the Rayleigh curve and meet the line of $P = 63$ percent. This crossing point corresponds to the X_0 level (unbiased average power in decibels over 1 mW, dBm).

Example 13.7 If a total number of samples is 256 and 38 samples are below a level -119 dBm, then the percentage is

$$\frac{38}{256} = 15\%$$

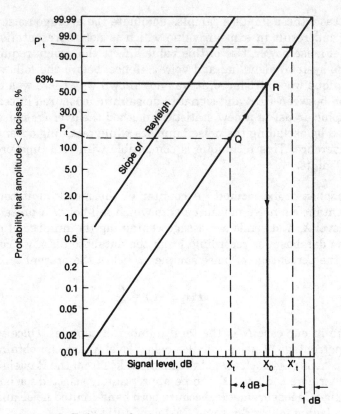

Figure 13.7 Technique of estimating average noise.

Draw a line at 15 percent and meet at Q on the Rayleigh curve (see Fig. 13.7). Assume that the X_t is -119 dBm. The X_0 is the average power because 63 percent of the sample is below that level. Then

$$X_0 = X_t + 4 \text{ dB} = -119 + 4 = -115 \text{ dBm}$$

Example 13.8 The total number of samples is 256. Three noise spikes are 20 dB above the normal average. Find the errors, using the following two methods. Compare the results.

Use geometric average method. Let the power value of each sample (of 253 samples) after normalization be 1, i.e., the average is 1. Then the measured average of 256 samples, including three spikes, is

$$\text{Measured average} = \frac{\sum\limits_{1}^{253} x_i + 100 \sum\limits_{1}^{3} x_i}{256} = 2.16 \qquad (\text{assume } x_i = 1)$$

$$= 3.3 \text{ dB} \qquad \text{above the true average}$$

Statistical average method

$$63\% \text{ of samples } = 256 \times 0.63 = 161 \text{ samples}$$

This means that 161 samples should be under the average power level. Now three noise spikes added to the 161 samples increases the number of samples to 164.

$$\frac{164}{256} = 64\%$$

The power levels at 63 and 64 percent show almost no change (see Fig. 13.7). Typical data averaging using the geometric and statistical average methods is illustrated in Fig. 13.8. The corrected value is approximately -118 to -119 dBm. The geometric average method biases the average value and causes an unacceptable error as shown in the figure.

13.3.3 Signal-strength conversion

Confusion arises because the field strength (in decibels above 1 μV, dBμ) is measured in free space, and the power level in decibels above 1 mW (dBm) is measured at the terminal impedance of a given receiving antenna. Furthermore, the dimensions of the two units are different. The signal field strength measured on a linear scale is in microvolts per meter (μV/m), and the power level measured on a linear scale is in milliwatts (or watts).

Further confusion arises because of the notation "dBμ." Sometimes dBμ means the number of decibels above 1 μV measured at a given

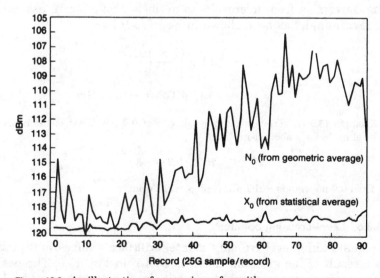

Figure 13.8 An illustration of comparison of n_c with x_e.

voltage. Sometimes, it represents the number of decibels referred to microvolts per meter when field strength is being measured.

The conversion from decibels (microvolts per meter, dBμ) to decibels above 1 mW (dBm) at 850 MHz is shown in Eq. (5.1-13) using the relationship between induced voltage and effective antenna length.[7] The conversion at a frequency other than 850 MHz can be obtained as follows.[8]

$$P_{max}(\text{in dBm at } f_1 \text{ MHz}) =$$

$$P_{max}(\text{in dBm at 850 MHz}) + 20 \log\left(\frac{850}{f_1}\right) \quad (13.3\text{-}12)$$

where f_1 is in megahertz. The details of this conversion are given in Sec. 5.1.3.

13.3.4 Receiver sensitivity

The sensitivity of a radio receiver is a measure of its ability to receive weak signals. The sensitivity can be expressed in microvolts or in decibels above 1 μV.

$$Y \quad dB\mu V = 20 \log (x \quad \mu V) \quad (13.3\text{-}13)$$

Also, the sensitivity can be expressed in milliwatts or dBm.

$$y \quad dBm = 10 \log (x \quad mW) \quad (13.3\text{-}14)$$

The conversion from microvolts to decibels above 1 mW, assuming a 50-Ω terminal, has been shown in Eq. (5.1-15) as

$$0 \text{ dB}\mu V = 10 \log \frac{(1 \times 10^{-6})^2}{50}$$

$$= -137 \text{ dBW} = -107 \text{ dBm} \quad (13.3\text{-}15)$$

Example 13.9 A receiver has a sensitivity of 0.7 μV. What is the equivalent level in decibels above 1 mW?

$$20 \log 0.7 = -3$$

Then 0.7 μV equals -107 dBm $-3 = -110$ dBm.

13.3.5 Level-crossing counter[9]

A signal fading level crossing counter will face a false-count problem as a result of the granular noise as shown in Fig. 13.9. The positive slope crossing count should be 3, but the false counts may be 12. A

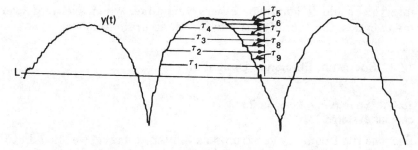

Figure 13.9 An algorithm for a level-crossing counter.

proposed level-crossing counter can eliminate the false counts. First, by sampling the fading signal at an interval of T seconds, we can choose the interval T such that $1/T$ is small in comparison to the fading rate. The duration of stay τ_i is measured for every sample time which is above level L and the time span until the signal drops below level L. We can also use a device for measuring the percentage of time that $y(t)$ is above L. This device may be called a level crossing counter.

Let us define

$$p = P_r[y(t) \geq L] \tag{13.3-16}$$

and

$$q = 1 - p = P[y(t) < L] \tag{13.3-17}$$

We count the number of times M that τ_i is above L and then sum the duration of total stays $T_s(T_s = \Sigma_{i=1}^{M} \tau_i)$ above L. The average duration of upward fading is

$$\tau_p = \frac{2T_s}{M} \tag{13.3-18}$$

The average duration of fades τ_q where $y(t)$ is below L is

$$\tau_q = \tau_p \frac{1 - p}{p} \tag{13.3-19}$$

and the level crossing rate n at level L is

$$n = \frac{1}{\tau_p + \tau_q} = \frac{p}{\tau_p} \tag{13.3-20}$$

The advantage of this method is that we stop our time count whenever $y(t)$ crosses L, thus avoiding false counts due to noise when $y(t)$ is close to level L. Obviously noise can give an incorrect measure of the "duration of stay" for a single interval; however, noise shortens many intervals while it lengthens others, and thus, when averaged over many cycles, the "duration of stay" is an accurate number.

13.4   Spectrum Efficiency Evaluation[10]

13.4.1   Spectrum efficiency for cellular systems

Because the frequency spectrum is a limited resource, we should utilize it very effectively. In order to approach this goal, spectrum efficiency should be clearly defined from either a total system point of view or a fixed point-to-point link perspective. For most radio systems, spectrum efficiency is the same as channel efficiency, the maximum number of channels that can be provided in a given frequency band. This is true for a point-to-point system that does not reuse frequency channels such as a cellular mobile radio. An appropriate definition of spectrum efficiency for cellular mobile radio is the number of channels per cell. Therefore, in cellular mobile radio systems:

$$\text{Spectrum efficiency} \neq \text{channel efficiency}$$

The system capacity is directly related to spectrum efficiency but not to channel efficiency.

13.4.2   Advantages and impact of FM

In 1936 E. H. Armstrong published a paper entitled *"A Method of Reducing Disturbance in Radio Signaling by a System of Frequency Modulation."*[11] This paper explored the tradeoffs between noise and bandwidth in FM radio. Since then, engineers have understood the concept of reducing noise by increasing bandwidth in system design.

The parameters for system comparison are *voice quality, transmitted power,* and *cell size.* Satisfactory voice quality is generally accepted as governed by the carrier-to-noise ratio $C/N_{\text{FM}} = 18$ dB.* This is the level at which 75 percent of the users state that voice quality is either good or excellent in 90 percent of the service area on a 30-kHz FM channel in a multipath fading environment.[12] For point-to-point radio links, the wider the channel bandwidth, the lower the required level of transmitted power.

To maintain voice quality when channel bandwidth is reduced, it is necessary to increase the signal-to-noise ratio in order to improve the reception. Transmission power is then also increased.

---

*If we use values other than 18 dB, the analysis used in this section remains the same.

Every time power is increased, interference problems are created. For point-to-point radio links, these problems are manageable because no frequency reuse is involved. This is not the case, however, for a frequency-reuse system such as a cellular radio. A cellular radio telephone system includes many mobile-unit customers and, depending on demand at any given time in a given system, identical channels will be operating simultaneously in different geographic locations. As the number of cells increases in a given area, interference may appear in one of several forms: cochannel, adjacent-channel, or multichannel at colocations; thus the probability of its occurrence increases. Interference may also result from received power-level differences. In a frequency-reuse system, however, cochannel separation is more critical to the system than adjacent-channel interference because adjacent-channel interference may be eliminated by the use of sharp filters.

13.4.3 Number of frequency-reuse cells K

The formula for determining the number of frequency-reuse cells in a standard cellular configuration is derived by combining Eqs. (2.4-3) and (2.4-5) with $\gamma = 4$ based on the 40 dB/dec path-loss rule.[13]

$$\frac{C}{I} = \frac{1}{6}\left(\frac{D}{R}\right)^4 = \frac{(3K)^2}{6} = \frac{3K^2}{2} \tag{13.4-1}$$

or
$$K = \sqrt{\frac{2}{3}\frac{C}{I}} \tag{13.4-2}$$

The number of frequency-reuse cells is a function of the required carrier-to-interference ratio.

A higher required carrier-to-interference ratio at the boundary of a cell results in the need for more frequency-reuse cells. The pattern of reuse cells can then be determined.

13.4.4 Number of channels per cell m

The next factor to be determined is the number of channels per cell, which is a function of the total number of channels available (amount of available spectrum divided by channel bandwidth) and the required carrier-to-interference ratio. The formula for this factor is[10]

$$m = \frac{B_t}{B_c K} = \frac{B_t}{B_c \sqrt{(2/3)(C/I)}} \tag{13.4-3}$$

for $M = mK$ total number of channels,

where m = number of channels per cell, also called radio capacity by
Lee[10]

K = number of frequency-reuse cells (see Eq. 13.4-2)

B_t = total bandwidth (transmitted or received)

B_c = channel bandwidth

13.4.5 Rayleigh fading environment

The Rayleigh fading environment is the mobile radio environment
caused by multipath fading, which is the cellular system environment.
Therefore, it is more realistic to determine the spectrum efficiency of
a cellular mobile radio in a Rayleigh fading environment.

In a multipath fading environment, a simple FM system which may
not have either preemphasis-deemphasis or diversity schemes would
receive its baseband signal-to-noise ratio (S/N), which is converted
from the carrier-to-interference ratio (C/I) but S/N is 3 dB lower than
C/I.[14]

System advantages. The FCC has released specifications which result
in advantages vis à vis the signal-to-noise ratio for transceivers in the
existing FM cellular system. The first advantage is preemphasis-
deemphasis, which equalizes the baseband signal-to-noise ratio over
the entire voice band (f_1 to f_2). We make the assumption of gaussian
noise because the interference obtained from all six cochannel interferers behaves in a noiselike manner. The improvement factor ρ_{FM},
that is, the improvement of FM with preemphasis or deemphasis over
FM without them, can be calculated as follows.[15,16]

$$\rho_{FM} = \frac{(f_2/f_1)^2}{3} = \frac{(3000\ \text{hZ}/300\ \text{Hz})^2}{3} = 33.3\ (=)\ 15.2\ \text{dB} \quad (13.4\text{-}4)$$

Another advantage is the two-branch diversity combining receiver,
which is very suitable for FM and reduces multipath fading. The advantage of the two-branch diversity receiver is that the baseband signal-to-noise ratio S/N of a two-branch FM receiver shows a 8-dB improvement over the signal-to-noise ratio of a single FM channel.

$$\left(\frac{S}{N}\right)_{2brFM} = 8\ \text{dB} + \left(\frac{S}{N}\right)_{FM} \quad (13.4\text{-}5)$$

The existence of compandors is assumed for compressing the signal
bandwidth and taking advantage of the quieting factor during pauses.
However, the voice quality improvement due to the quieting factor

cannot be expressed mathematically. It is understood that all the analog modulation systems use compandors.

Present FM system. The subjective required carrier-to-interference ratio for FM is

$$\left(\frac{C}{I}\right)_{FM} = 18 \text{ dB } (=) \ 63.1 \qquad (13.4\text{-}6)$$

The baseband signal-to-noise ratio can be obtained from the previous analysis as follows.

$$\left(\frac{S}{N}\right)_{2brFM} = -3 \text{ dB } + \text{ deemphasis gain } + \text{ diversity gain } + \left(\frac{C}{I}\right)_{FM}$$

$$= 38.23 \text{ dB } (=) \ 6652.73 \qquad (13.4\text{-}7)$$

The signal-to-noise value $S/N = 38$ dB is a reasonable figure for obtaining good quality at the baseband.[17] The notation $(=)$ means a conversion between decibels and a linear ratio.

SSB systems. Single-sideband receivers, best case, have a carrier-to-interference ratio equal to the signal-to-noise ratio at baseband since SSB is a linear modulation.[17] The term "best case" means that the signal fades are completely removed, that is, the environment approaches a gaussian.* There is no advantage in using diversity schemes in a gaussian environment. If the environment is Rayleigh, the carrier-to-interference ratio must always be higher than that in a gaussian environment in order to obtain the same voice quality at the baseband. An explanation is given in Sec. 13.4.7. To obtain a similar voice quality, the signal-to-noise ratio of both FM and SSB systems at the baseband should be the same.[18] The formula used to determine the required C/I of SSB is

$$\left(\frac{C}{I}\right)_{SSB} = \left(\frac{S}{N}\right)_{SSB} = \left(\frac{S}{N}\right)_{2brFM} = 38.23 \text{ dB } (=) \ 6652.73$$

This means that the required 38.23-dB carrier-to-interference ratio of 38.23 dB for an SSB system (see Sec. 13.4.7) is equivalent to the required carrier-to-interference ratio of 18 dB for an FM system for

---

*SSB systems at 800 MHz have not been commercially available because of technical difficulties. It is assumed here that an ideal SSB at 800 MHz can be built for mobile radios.

TABLE 13.1   Channels per Cell

(Rayleigh Fading Environment)

System	Bandwidth B_c, kHz	Cells per set K	Total number of channels B_t/B_c	Channels per cell m
FM	30.0	7	333	47.57
SSB	7.5	66	1333	20.00
SSB	5.0	66	2000	30.0
SSB	3.0	66	3333	50.05

equivalent voice quality. The number of channels per cell and the number of channels per square mile for the system can then be calculated.

Number of channels per cell m.   The preceding analysis gives us the information needed to determine the number of channels per cell. Assuming $B_t = 10$ MHz, the formula is [from Eq. (13.4-3)]

$$m = \frac{B_t}{B_c K} = \frac{10 \text{ MHz}}{B_c \sqrt{(2/3)(C/I)}} \tag{13.4-8}$$

Given a total bandwidth (B_t) of 10 MHz, $C/I = 18$ dB for FM, and $C/I = 38.23$ dB for SSB, it is possible to determine the number of channels per cell m by the substitution of the above values in Eq. (13.4-8) and shown in Table 13.1. As this table shows, FM cellular systems need fewer cells than do SSB systems to provide quality voice service.

There is a dispute as to whether at 800 MHz, an SSB system needs a C/N of 38 dB or less to provide an S/N of 38 dB at the baseband.[19–22] Since there is no commercial 800-MHz SSB system, no subjective test can be used for SSB voice quality. Comparing the performance of an existing FM system to that of a nonexisting SSB system is difficult. Also, it is not proper to use the results from a 150-MHz SSB system without making a thorough subjective test in a Rayleigh fading environment* and applying it to a 800-MHz SSB system.

13.4.6   Determination of cell size

It is possible to determine the size of comparable cells for 30-kHz FM, 3-kHz SSB, 5-kHz SSB, and 7.5-kHz SSB once the number of frequency-reuse cells and the number of channels per cell have been calculated. These values are related to the level of carrier power required

---

*Air-to-ground communications media do not exhibit Rayleigh fading behavior. Also the required C/I of SSB at 800 MHz would be different from that at 150 MHz.

at reception to maintain similar voice quality. Since SSB has a relatively narrow bandwidth, the noise level is also lower. The SSB noise level must be adjusted to the FM noise level in order to determine the power required for SSB.

Required power in each SSB system. The SSB required carrier-to-interference ratio must be 38.23 dB. Therefore, the power required by the SSB system after the noise level (including interference) has been adjusted can then be determined. The voice quality of an SSB system having a carrier-to-interference ratio of 38.23 dB is equivalent to an FM system having a carrier-to-interference ratio of 18 dB. This assumes that in-band pilot tones can smooth out the fading signal and causes no distortion in SSB reception. The noise levels of different SSB bandwidths can be shown as follows.

$$\left(\frac{C}{I}\right)_{SSB} = 38.23 \text{ dB}$$

$$\left(\frac{C}{I}\right)_{SSB3kHz} = 10.23 \text{ dB} + 18 \text{ dB} + 10 \text{ dB}$$

$$\left(\frac{C}{I}\right)_{SSB5kHz} = 12.45 \text{ dB} + 18 \text{ dB} + 7.74 \text{ dB}$$

$$\left(\frac{C}{I}\right)_{SSB7.5kHz} = 14.21 \text{ dB} + 18 \text{ dB} + 6.02 \text{ dB}$$

These figures are illustrated clearly in Fig. 13.10.

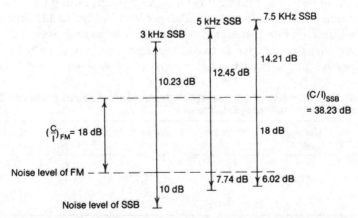

Figure 13.10 The interference-plus-noise levels for 30-kHz FM and different bandwidths of SSB.

SSB cell size determined for the required additional power. For a given transmitted power, the cell size can be determined. Assuming an FM cell radius of 10 mi and applying the 40 dB/dec path-loss rule, the cell sizes for an SSB system may be determined as follows.

For a 3-kHz SSB system

$$10 \log \left(\frac{10^{-4}}{R^{-4}} \right) = -10.23 \text{ dB} \qquad R = 5.55 \text{ mi}$$

For a 5-kHz SSB system

$$10 \log \left(\frac{10^{-4}}{R^{-4}} \right) = -12.45 \text{ dB} \qquad R = 4.88 \text{ mi}$$

For a 7.5-kHz SSB system

$$10 \log \left(\frac{10^{-4}}{R^{-4}} \right) = -14.21 \text{ dB} \qquad R = 4.41 \text{ mi}$$

The higher the carrier-to-interference ratio for a given power, the closer the mobile unit is to the cell. Assuming the same voice quality at the boundary of the cell, a 10-mi-radius FM 30-kHz cell is equivalent to a 5.5-mi 3-kHz SSB cell, a 4.88-mi 5-kHz SSB cell, or a 4.41-mi 7.5-kHz SSB cell. Therefore, an FM system permits larger cells and an SSB system requires smaller cells to provide for the same voice quality in the same area.

Comparison of cochannel cell separation and radius of FM and SSB systems in a Rayleigh fading environment. The cochannel cell separations and the radii of both FM and SSB cells in a Rayleigh fading environment are summarized in Table 13.2 and expressed in Fig. 13.11. Therefore, in a cellular mobile radio environment, a FM system *permits* larger cells with less separation between cochannel cells and a SSB system requires smaller cells with greater separation between cochannel cells.

TABLE 13.2 Comparison of Cochannel Cell Separation and Radius of FM and SSB Systems in a Rayleigh Fading Environment

System	Cells per set K	Radius R, mi	Diameter D, mi	D/R	Bandwidth, kHz
FM	7	10	46	4.6	30
SSB	66.6	5.5	77.77	14.14	3
		4.88	69	14.14	5
		4.41	62.36	14.14	7.5

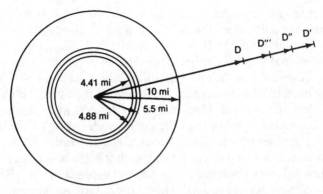

Figure 13.11 Cochannel cell separations and radii of FM and SSB systems in a Rayleigh fading environment.

Channels per square mile.   Table 13.3 shows a comparison of the channels per square mile and the spectrum efficiency of each system in the Rayleigh fading environment based on equivalent voice quality and a given transmitted power. This comparison shows that the existing 30-kHz FM cellular system is about as spectrally efficient as the hypothetical 3-kHz SSB system and much more efficient than either of the two SSB (5- or 7.5-kHz) systems offering a voice quality similar to that proposed for commercial service.

13.4.7   Considerations of SSB systems in a Rayleigh fading mobile radio environment

The voice signal requires about a S/N ratio of 40-dB output at the baseband for high quality.[23] Our baseband S/N ratio is calculated to be 38 dB, which is very close to 40 dB.

TABLE 13.3 Comparison of System Efficiency and Spectrum Efficiency in an Area of 10-mi Radius

		System efficiency		Spectrum efficiency	
System	Bandwidth, mHz	Cell radius, mi	No. of cells required	Channels per cell, m	No. of channels per square mile
FM	30	10	1.0	47.5	0.15
SSB	7.5	4.4	$\left(\dfrac{10}{4.4}\right)^2 = 5.17$	20.0	0.06
SSB	5	4.8	$\left(\dfrac{10}{4.8}\right)^2 = 4.34$	30.0	0.10
SSB	3	5.5	$\left(\dfrac{10}{5.5}\right)^2 = 3.31$	50.0	0.16

In a gaussian environment, if the output S/N of a SSB signal at the baseband is 38 dB, then the S/N at the RF is also 38 dB because SSB is linearly modulated.[17] Now, in a Rayleigh fading environment, the received signal can be further degraded as a result of the multipath fading. Therefore, in order to maintain the baseband S/N at 38 dB, we may need to receive a signal much higher than 38 dB if no diversity scheme is implemented. How high the C/I level at RF should be cannot be determined unless the SSB mobile radio at 800 MHz is realized and a subjective test is done. Since a single-branch SSB cannot be used in a mobile radio environment at 800 MHz because of rapid fading,[24] the value of applying a two-branch diversity to the SSB at 800 MHz also must be questioned. Our reasoning is as follows.

Let S_0 be a voice signal, r_1 and r_2 the envelopes of the fading, θ_1 and θ_2 the random phases received by two spaced antennas, and ω_0 the carrier angular frequency. Where an equal-gain combined receiver is considered, in an SSB system the combined signal envelope is

$$r = \left| r_1 S_0 e^{j(\omega_0 t + \theta_1)} \right| + \left| r_2 S_0 e^{j(\omega_0 t + \theta_2)} \right|$$

$$= r_1 S_0 + r_2 S_0 = (r_1 + r_2) S_0 \qquad (13.4\text{-}9)$$

which is the same as the baseband signal representation. The term $(r_1 + r_2)$ does reduce the fading as compared to either individual r_1 or r_2 to a certain degree, but it also acts as a distortion term to S_0. The distortion of voice S_0 on $(r_1 + r_2)$ received by a two-branch diversity receiver for SSB in a ground mobile radio (Rayleigh) environment is still quite high at 800 MHz. This is because the effect of fading is multiplicative and produces intermodulation products with the signal modulation that cannot be eliminated by filtering.[25] In air-to-ground transmission, the direct-wave path dominates, the fading phenomenon (rician) is not severe, and a two-branch diversity does help in improving the voice quality at the reception.

An amplitude companding single sideband (ACSB) with an in-band pilot tone is considered[26] under the assumption that this kind of SSB can in principle completely remove the Rayleigh fading; therefore, no diversity scheme* is needed. Then we do not require an increase in the received power level at the RF, but rather simply retain the same level as the baseband S/N ratio of 38.23 dB.

Preemphasis and deemphasis are not widely used in SSB systems. The disadvantage of the use for SSB systems in a nonfading environment is discussed by Schwartz[27] and Gregg.[28]

---

*The diversity scheme is used to eliminate fading.

In a mobile radio environment, in order to transmit a predistorted SSB signal using preemphasis to suppress the noise level, the signal cánnot be completely restored because the effect of fading is multiplicative and it produces intermodulation products in the voice band that cannot be eliminated by filtering as mentioned before. Therefore, the use of preemphasis and deemphasis in an 800-MHz SSB mobile radio system is questionable.

Reference 18 shows that an RF signal with required C/I = 38 dB received by a SSB system through a gaussian environment results in a baseband signal where S/N = 38 dB also. In a Rayleigh fading environment, the C/I must definitely be higher than 38 dB for S/N to be 38 dB. How high is not known since 800-MHz SSB equipment has not been manufactured. We may use the information obtained from a FM system for maintaining the same voice quality in different environments.

Required for FM in a gaussian environment
Required for FM in a Rayleigh fading environment
$$\frac{C}{I} \geq \begin{cases} 10 \text{ dB} \\ 18 \text{ dB} \end{cases}$$

The difference in C/I for the FM system between the two kinds of environment is 8 dB. We may use the 8-dB difference and add it to C/I = 38 dB for an SSB system. Then

For SSB in a Rayleigh fading environment $$\frac{C}{I} = 46 \text{ dB}$$

Suppose that a diversity scheme is used in SSB as Shivley[21] suggested. Then C/I for SSB in a less-fading (or no-fading) environment is

For SSB in a less-fading (or no-fading) environment

$$\frac{C}{I} - 8 \text{ dB} = 38 \text{ dB}$$

The same result is obtained if we assume that the fading is completely removed, $C/I = S/I$, and that the environment becomes gaussian. In a gaussian environment, we cannot reduce C/I below 38 dB, because S/N is 38 dB. Also, in a gaussian environment, the diversity scheme does not apply and adds no value.

Of course, the analysis shown here remains to be proved if and when an 800-MHz SSB mobile unit is developed in the future. After all, the methodology of solving this problem remains unchanged.

13.4.8 Narrowbanding in FM

Relationship between C/I **at IF and** S/N **at baseband.** In Ref. 12 we defined acceptable voice quality as existing when 75 percent of customers say that the voice quality is good or excellent in a 90 percent coverage area. When these numbers change, voice quality changes accordingly. The changes reflect the cost of deploying a cellular system. As the percentages specified above increase, the cost of designing the system to meet these requirements also increases. For now, we will use the numbers specified above as our criteria.

We let the customer listen to the voice quality of a 30-kHz FM two-branch diversity receiver with preemphasis-deemphasis and companding features at the cell site while the mobile transmitter is traveling at speeds ranging from 0 to 60 mi/h in a Rayleigh environment. Judging by the preceding subjective criterion, the C/I level at the input of the receiver is 18 dB.

The baseband signal-to-noise ratio (S/N) has been calculated in Eq. (13.4-7).

$$S/N = \underbrace{18 - 3}_{\text{Rayleigh environment}} + \underbrace{15.23}_{\text{deemphasis advantage}} + \overset{\text{diversity advantage}}{8} = 38.24 \text{ dB}$$

For a 15-kHz FM channel, the bandwidth is half as broad and affects the S/N; the other features, such as diversity and preemphasis-deemphasis, remain the same. We can find from Fig. 13.12 (Ref. 29) that in order to maintain the same voice quality of $(C/I)_{30 \text{ kHz}} = 18$ db, then

$$\left(\frac{C}{I}\right)_{15 \text{ kHz}} = 24 \text{ dB}$$

We may follow the same steps used in Sec. 13.4.6 along with the diagram shown in Fig. 2.5.

$$\frac{C}{N+1} = \frac{C}{(kTB + NF) + \sum_{i=1}^{6} I_i}$$

$$= 24 \text{ dB (15-kHz FM) or 18 dB (30-kHz FM)} \quad (13.4\text{-}10)$$

Using the techniques from previous sections, we can perform further calculations.

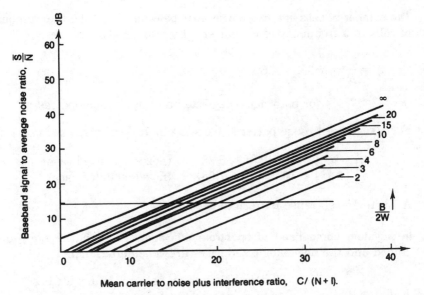

Figure 13.12 Baseband signal-to-noise ratio versus average carrier-to-interference ratio (Rayleigh fading). (*From Ref. 29.*)

The interference-plus-noise levels. From the preceding calculations Eq. (13.4-10), we can obtain the interference-plus-noise levels $(N + I)$ at the boundary of a 10-mi-radius cell as follows.

$$\text{For 30-kHz FM system} \quad N + I = -117 \text{ dBm}$$
$$\text{For 15-kHz FM system} \quad N + I = -123 \text{ dBm}$$

Since the received signal strength is -99 dBm for both 30- and 15-kHz, the $C/(N + I)$ ratio must be 18 dB and 24 dB for 30- and 15-kHz FM, respectively, in order to maintain the same voice quality. This relationship is shown in Fig. 13.13.

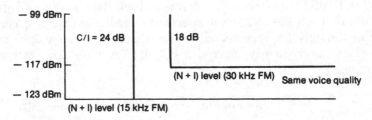

Figure 13.13 The interference-plus-noise levels for 30- and 15-kHz FM channels.

The number of cells in a frequency-reuse pattern. Let K be the number of cells in a frequency-reuse pattern; then in 30-kHz systems,

$$q = \frac{D}{R} \overset{\Delta}{=} \sqrt{3K} = 4.6$$

$K = 7$ (for both noise-neglected and noise-included cases)

For a seven-cell reuse pattern operating in 15-kHz systems,

$$K = \frac{q^2}{3} \begin{cases} = 13 & (q = 6.23) & \text{(noise-neglected case)} \\ = 16 & (q = 7.08) & \text{(noise-included case)} \end{cases}$$

A 13- to 16-cell reuse pattern is needed.

Intersystem comparison of spectrum efficiency. If both the noise-included and the noise-neglected cases are considered, then

$$K = \begin{cases} 7 & \text{(30-kHz FM system, both noise-included} \\ & \text{and noise-neglected cases)} \\ 16 & \text{(15-kHz FM system, noise-included case)} \\ 13 & \text{(15-kHz FM system, noise-neglected case)} \end{cases}$$

The spectrum efficiency for both systems can be shown to be about the same as follows:

$$\frac{333}{7} = 47.5 \text{ channels per cell} \qquad \text{(30-kHz FM)}$$

$$\frac{666}{16} = 41.63 \text{ channels per cell} \qquad \text{(15-kHz FM, noise-included case)}$$

$$\frac{666}{13} = 51.2 \text{ channels per cell} \qquad \text{(15 kHz FM, noise-neglected)}$$

Increasing spectrum efficiency by degrading voice quality. If we accept $C/I = 18$ dB for 15-kHz FM, this means that voice quality is degraded by 6 dB. Then the cochannel interference reduction factor q becomes approximately 4.6, corresponding to a frequency-reuse pattern of $K = 7$. Since the frequency channel is doubled, in a 10-MHz system (666 channels)

$$\frac{666}{7} = 2 \times \text{(number of 30-kHz FM channels per cell)}$$

Therefore, voice quality is sacrificed to gain spectrum efficiency.

Another approach is to increase the transmitted power of the 15-kHz FM system by 6 dB. Then the *C/I* remains the same because the interference is also increased by 6 dB. Therefore, no advantage to spectrum efficiency can be obtained by either increasing or reducing power.

13.5 Portable Units

All of today's system design tools are designed to improve the performance of cellular mobile units. Therefore, portable units become a secondary byproduct of the cellular system. Since very few systems in the United States are designed mainly for portable units, it will be necessary to study each existing system individually to determine whether portable units are suitable for it. Portable unit usage can be adopted easily by some existing systems but not by others for many reasons. If we find that portable units become very popular in an existing mobile cellular system, then we have to reconsider some parameters used for the mobile cellular system to adapt to portable unit usage. There are two parts to the calculation.

13.5.1 Loss due to building penetration

The loss (attenuation) when propagating an 800-MHz wave due to building penetration is very high.[31-36] Also, the structure-related attenuation varies, depending on the geographic area. In Tokyo, the path-loss difference inside and outside buildings at first-floor level is about 26 dB. But in Chicago, the path-loss difference under the same conditions is 15 dB as shown in Fig. 13.14.

These variations are attributable to differences in building construction. In Tokyo, many supporting metal frames (mesh configuration) are used in the building structures to allow the buildings to withstand earthquakes. In Chicago, fewer supporting metal frames are needed as there is less risk of earthquakes. Therefore, the *building penetration* is far less severe in Chicago than in Tokyo.

The same would apply to buildings in California. For instance, the loss due to building penetration in Los Angeles would be higher than that in Chicago but lower than that in Tokyo, because Los Angeles has only a few high-rise buildings. In Los Angeles the penetration at the first-floor level would be around −20 dB, as compared with the received signals at mobile units outside at street level.

Signal attenuation at the building basement level in Chicago is 30 dB below the signal received from the street-level mobile unit. This indicates that the signals do not penetrate basement structures easily. There are two ways to solve this problem.

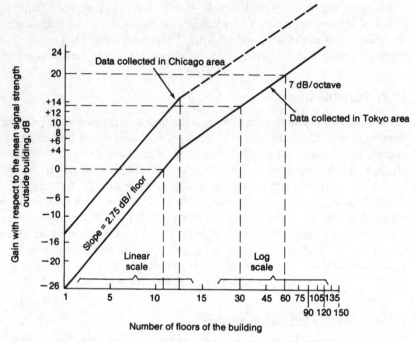

Figure 13.14 The building penetration losses in the Tokyo area and in the Chicago area.

1. Select the cell site closest to those downtown buildings used for conventions. Then calculate the received power for the portable units. Let the cell coverage be at a radial distance R, with the receiving level of $C/I = 18$ dB at the cell boundary as designed. The same site will be used for portable units. The receiving level of portable units is $C/I = 10$ dB because of the slow-motion or no-motion environment. At the basement, the reception level of a portable unit is $-30 + (18 - 10) = -22$ dB, i.e., 22 dB weaker than that received at the cell boundary R. Applying Eq. (4.2-10a), we obtain

$$40 \log \frac{R_1}{R} = -22 \text{ dB} \quad \text{or} \quad R_1 = R(10^{-(22/40)}) = 0.28R \quad (13.5\text{-}1)$$

From this calculation, the region in which the portable unit can be used in the basement is confined to an area of $0.28R$.

2. Install the repeaters (or enhancers) or leaky feeders to enhance the signal strength inside the building.

13.5.2 Building height effect

Usually, the signal reception level increases as the height of the building where the antenna is located increases. Comparing the two measurements from Chicago and Tokyo, we find that the slope (gain increase) of 2.7 dB per floor (i.e., 2.7 dB gain per each floor-level increment) can be a good value on which to base calculations regarding suitability of portable unit usage. However, this value is valid only up to the thirteenth floor. After that, a logarithmic 7 dB/oct scale is used. Now we start at the first floor.

First-floor region. Following the same procedure as when calculating the signal strength in the basement region, we find the signal strength requirements at the first floor inside the building for the portable unit to be

$$-15 + (18 - 10) = -7 \text{ dB} \qquad \text{(Chicago)}$$

$$-26 + (18 - 10) = -18 \text{ dB} \qquad \text{(Tokyo)}$$

$$-20 + (18 - 10) = -12 \text{ dB} \qquad \text{(intermediate value)}$$

Applying these findings to Eq. (13.5-1), we obtain

$$40 \log \frac{R_1}{R} = \begin{cases} -7 \text{ dB} \\ -18 \text{ dB} \\ -12 \text{ dB} \end{cases} \qquad R_1 = \begin{cases} 0.668R & \text{(Chicago)} \\ 0.35R & \text{(Tokyo)} \\ 0.5R & \text{(intermediate value)} \end{cases}$$

Nth-floor region. The area serviced increases as a function of height of 2.7 dB per floor below the thirteenth floor, where we see a gain increase of 2.7 dB per floor, and above the thirteenth floor, 7 dB/oct is used. We can calculate the service area in Chicago as follows.

$$40 \log \frac{R_1}{R} = -7 + N(2.7) \qquad \text{(in Chicago)}$$

where N is the number of floors. The same procedures apply to Los Angeles and Tokyo. In Table 13.4, we see that the service region increases at higher floor levels. However, after the thirteenth floor, the increase in gain is very small. The difference between the two floors from two cities becomes smaller for heights beyond the thirtieth floor (see Fig. 13.14).

TABLE 13.4 Building Penetration Loss

Condition	Building penetration loss	Shadow loss*
Building penetration	+27 dB (Tokyo)	27 dB (Chicago)
	+15 dB (Chicago)	(regardless of floor height)
Window area	+6 dB	
1st–13th floors	2.75 dB/floor (Tokyo)	
	2.67 dB/floor (Chicago)	
13th–30th floors	7 dB/oct (Tokyo and Chicago)	

Shadow loss is defined as the loss due to a building standing in the radio-wave path.

13.5.3 Interference caused by portable units

Interference to the other portable units. The portable unit has a transmitting power of 600 mW (28 dBm). The interference at the cell site from two different portable units can be determined as follows. We now can consider interference at higher floor levels (see Fig. 13.14). We find that reception at the sixth floor is the same as that at street level in Chicago and that reception at the eleventh floor is the same as that at street level in Tokyo. Reception at the thirtieth floor in Tokyo is 13 dB higher than that at street level.

A portable unit transmitter can transmit a signal to a cell site following line-of-sight propagation. A signal from a portable unit on the thirtieth floor that is received by the cell site could interfere with the reception of a signal from a portable unit on the eleventh floor. The interference level for the portable unit on the eleventh floor is 13 dB. The interference range becomes

$$20 \log \frac{R_1}{R} = -13 \text{ dB} \qquad \text{then} \qquad R_1 = 0.22R$$

Since $R_1 = 0.22R$, the portable unit used on the thirtieth floor at the cell boundary R will not interfere with cell-site reception from a portable unit on the eleventh floor at 0.22 R away. If the power of both units can be controlled at the cell site, the near-end to far-end ratio interference of 13 dB can be reduced. We must be aware of this interference and find ways to eliminate it once we know its cause. A method for this was described in Chap. 10. The selection of a method for eliminating interference is based on environmental factors such as building height and density, which vary from area to area.

Interference to the mobile units. Now assume that at the cell boundary (the cell radius is R) the mobile unit received a signal at −100 dBm and the reception level of a portable unit at the thirtieth floor is −87 dBm (-100 + 13 dB). If the cell site has a 10-W transmitter and a 6-dB-gain antenna, then the transmitting site has an effective radiated power (ERP) of 46 dBm. The path loss on a thirtieth floor of the cell boundary becomes 133 dB (46 + 87). Now for calculating a reverse path, the portable unit has a 600-mW (28-dBm) transmitter; and the mobile unit has a 3-W (35-dBm) transmitter. Then the signal received at the 100-ft cell-site antenna is (28 $-$ 133) = -105 dBm for the portable unit and (35 $-$ 133 $-$ 13) = -111 dBm for the mobile unit. The difference in received levels is 6 dB. This near-end to far-end ratio interference can be eliminated by the frequency assignment and power control of the portable units at the cell site.

13.5.4 Difference between mobile cellular and portable cellular systems

It is very interesting to point out the differences in characteristics, coverage charts, and system design aspects for mobile and portable cellular systems.

Different characteristics

Mobile units	Portable units
Two-dimensional system	Three-dimensional system
Needs handoffs	No handoffs
Severe signal fading due to vehicle movement	No fading or mild fading if walking
Gain changes with ground elevation	Gain changes with building height
Loss due to multipath reflection	Loss due to building penetration
Required $C/I \geq 18$ dB	Required $C/I \geq 10$ dB
Power consumption is not an issue	Power consumption is a key issue
Attractive because of various features	Attractive because of their small size and light weight

Different coverage charts. Using the Philadelphia path-loss curve shown in Fig 4.3 with the standard condition parameters listed in Sec. 4.2.1, we can illustrate the differences in the coverage charts of mobile units and portable units in urban areas, as shown in Fig. 13.15. The receiver sensitivity of both units is assumed to be -117 dBm. Also assume that the required C/I of a portable unit is 10 dB, and the required C/I of a mobile unit is 18 dB. The mobile-unit coverage is about 6 mi, while the portable coverages differ with building height, as shown in Fig. 13.15. The higher the building, the greater the coverage range.

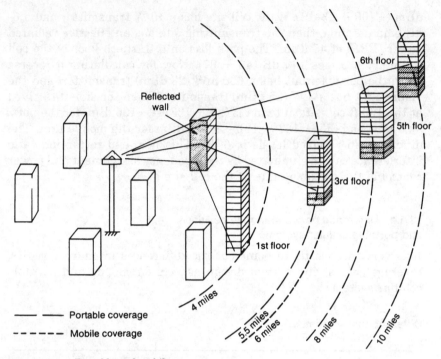

6th floor

Reflected
wall

5th floor

3rd floor

1st floor

———— Portable coverage

— — — — — Mobile coverage

4 miles

5.5 miles
6 miles

8 miles

10 miles

Figure 13.15 Portable and mobile coverage.

Different design concepts. In mobile cellular systems, we try to cover
the area with an adequate signal from a cell site; then transmitted
power, antenna height and gain, and location are the parameters involved. Reduction of both multipath fading and cochannel interference
is described in Chap. 6.

In a portable cellular system, the coverage range increases with the
height of the building. Therefore no fixed coverage range can be given
for the portable units. Also the cochannel interference reduction ratio
q [$q = D/R$ see Eq. (2.3-1)] for the portable cellular systems has no
meaning since the cell radius R changes with building height. If we
try to apply the design techniques for mobile cellular systems to portable cellular systems, the results are not very good. One way to look
at this problem is that each building structure offers less interference
inside the building. The lower the building, the greater the protection
from interference. Therefore we should not raise the transmitted
power and try to penetrate the building; rather we should take advantage of this natural shielding environment. Thus we should link
to each repeater (enhancer) mounted at the top of each building. Since
the buildings are tall, reception will be good because of the building's
height, and only a small amount of transmitted power is needed at

the cell site (see Fig. 13.15). If leaky cables or cables with antennas (shown in Fig. 13.16) from the repeater are connected to each floor, the signal in the whole building will be covered. This is the proper arrangement, but it is also a different concept of designing a portable cellular system.

13.6   Evaluation of Data Modem

13.6.1   Requirement

The data modem used in the current analog system must meet the following requirements.

1. Data transmissions have to use 30 kHz voice channels.
2. The SAT tone must be maintained at around 6000 Hz in voice channels. Then the transmission rate has to be either lower or higher, but it must be clear from the 6000-Hz SAT.
3. The transmission rate cannot be lower than a rate[37] which lies in the dominant random-FM region of $f_{\text{rfm}} < 2(V/\lambda)$. This specification is based on the vehicle speed V and the wavelength of the operating frequency. For instance, if $V = 104$ km/h (65 mi/h), and $\lambda = 1$ ft (at 850 MHz), then $f_{\text{rfm}} = 2(V/\lambda) = 190$ Hz. This means that the data modem transmission rate cannot be below 190 Hz because of

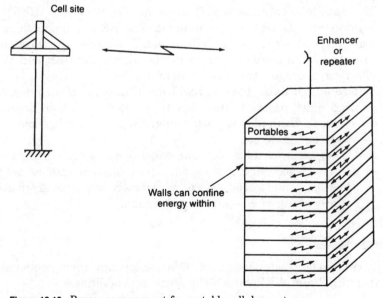

Figure 13.16 Proper arrangement for portable cellular system.

the unique random-FM characteristics in the mobile radio environment. When the mobile unit stops, the random FM disappears.

4. The same severe fading requires us to use redundancy, coding, automatic-repeat request (ARQ) scheme, diversity, and so on when transmitting data. The current mobile cellular signaling transmission rate is 10 kbps. Using the BCH code with five repeats, the data throughput is 1200 bits. However, the power spectrum density of a 10-kbps data stream with a Manchester coding (a biphase waveform) is spread out over the 6000-Hz region.[38] If the SAT cannot be detected because of the data modem, then the data modem cannot be used.

5. Mobile unit (vehicular) speed is a significant factor.

    a. Suppose that a car can be driven slowly while approaching a stop but that it never stops. In this case, the average duration of fades is very long. Most of the time, if the word is in the fade, the whole word is undetected.

    b. Suppose that a car can travel at a rate of 104 km/h (65 mi/h). The number of level crossings at -10 dB (10 dB below average power) is 65 crossings per second, and the average duration is 1.54 ms, as mentioned in Sec. 13.2.

    Thus, a word length must be designed to fit in these two cases.

6. Handoff action is another factor in data modem design. Whenever a handoff occurs, a piece of data information is lost. The average would be 200 ms. The ARQ scheme would be useful for this purpose.

7. The data modem must satisfy a specific bit error rate (BER) and word error rate (WER) requirement. The BER is independent of vehicle speed, whereas the WER is not. The higher the throughput rate (baseband transmission rate), the higher the BER and WER. However, there are two data modem markets.

    a. Use of fast data rate—in real-time situations, a customer may need quick access to data but may not need a high degree of accuracy. Examples of such customers are police agencies, real estate agencies, etc.

    b. Use of accurate data—for transmitting figures requiring a high degree of accuracy, a slow data transmission rate is needed. These customers need data accuracy more than fast acquisition, as in banking or computer applications.

13.6.2   Testing

Any data modem operating in a cellular mobile unit must demonstrate that its BER and WER satisfy the following conditions.

1. The data modem should be tested while the mobile unit is parked (stationary) at the side of the highway. Because the mobile radio environment is very noisy, the spike noise resulting from ignition-induced combustion and the sharp fades caused by the noise of passing trucks would also affect data transmission.

2. The data modem should be tested while the mobile unit is driven at different speeds, say 5, 10, 45, and 60 mi/h at the boundary of the cell.

3. The data modem should be tested during handoff conditions.

References

1. S. W. Halpren, "Techniques for Estimating Subjective Opinion in High-Capacity Mobile Radio," Microwave Mobile Symposium, Boulder, Colorado, 1976.
2. W. C. Y. Lee, *Mobile Communications Engineering*, McGraw-Hill Book Co., 1982, p. 390.
3. W. C. Y. Lee, *Mobile Communications Engineering*, McGraw-Hill Book Co., 1982, pp. 398–400.
4. W. C. Y. Lee, "The Advantages of Using Repetition Code in Mobile Radio Communications," *36th IEEE Vehicular Technology Conference Record* (May 1986, Dallas, Texas), pp. 157–161.
5. W. C. Y. Lee, "Estimate of Local Average Power of a Mobile Radio Signal," *IEEE Transactions on Vehicular Technology*, Vol. VT-34, February 1986, pp. 22–27.
6. W. C. Y. Lee, "Estimating Unbiased Average Power of Digital Signal to Presence of High-level Impulses," *IEEE Transactions on Instrumentation and Measurement*, Vol. IM-32, September 1983, pp. 403–409.
7. E. C. Jordan (Ed.), *Reference Data for Engineers, Radio Electronics Computer, and Communications*, 7th Ed., Howard W. Sams & Co., 1986.
8. W. C. Y. Lee, "Convert Field Strength to Received Power for Use in Systems Design," *Mobile Radio Technology*, April 1987.
9. W. C. Y. Lee, D. O. Reudink, and Y. S. Yeh, "A New Level Crossing Counter," *IEEE Transactions on Instrumentation and Measurement*, March 1975, pp. 79–81.
10. W. C. Y. Lee, "Spectrum Efficiency: A Comparison Between FM and SSB in Cellular Mobile Systems," presented at the Office of Science and Technology, Federal Communications Commission, Washington, D.C., August 2, 1985.
11. E. H. Armstrong, "A Method of Reducing Disturbances in Radio Signaling by a System of Frequency Modulation," *Proceedings of the IRE*, Vol. 24, May 1936, pp. 689–740.
12. V. H. MacDonald, "The Cellular Concept," *Bell System Technical Journal*, Vol. 58, January 1979, pp. 15–42.
13. W. C. Y. Lee, *Mobile Communications Design Fundamentals*, Howard W. Sams & Co., 1986, p. 144.
14. W. C. Y. Lee, *Mobile Communications Engineering*, McGraw-Hill Book Co., p. 248.
15. P. F. Panter, *Modulation Noise, and Spectral Analysis*, McGraw-Hill Book Co., 1965, p. 447.
16. A. B. Carlson, *Communications Systems*, McGraw-Hill Book Co., 1968, p. 279.
17. A. B. Carlson, *Communications Systems*, McGraw-Hill Book Co., 1968, p. 267.
18. A. B. Carlson, *Communications Systems*, McGraw-Hill Book Co., 1968, p. 288.
19. T. L. Dennis, "Mission the Point," *Telephony*, January 20, 1986, p. 13.
20. W. C. Y. Lee, "The Point Is," *Telephony*, January 20, 1986, p. 13.
21. N. R. Shivley, "ACSB vs. FM," *Communications Magazine*, February 1986, p. 10.

22. Joran Hoff, "Digital Mobile," presented at CTLA meeting in Phoenix, Arizona, January 19–21, 1987. (Hoff drew the same conclusion as the author did in Ref. 10.)
23. H. Taub and D. L. Schilling, *Principles of Communication Systems,* McGraw-Hill Book Co., 1971, p. 290.
24. M. J. Gans and Y. S. Yeh, "Modulation, Noise and Interference," in W. C. Jakes (ed.), *Microwave Mobile Communications,* John Wiley and Sons, 1974, Chap. 4, p. 201.
25. M. J. Gans and Y. S. Yeh, "Modulation, Noise and Interference," in W. C. Jakes (ed.), *Microwave Mobile Communications,* John Wiley and Sons, 1974, Chap. 4, p. 207.
26. J. P. McGeeham and A. J. Bateman, "Theoretical and Experimental Investigation of Feed Forward Signal Regeneration as a Means of Combating Multi-Path Propagation Effects in Pilot-base SSB Mobile Radio Systems," *IEEE Transactions on Vehicular Technology,* Vol. 32, February 1983, pp. 106–120.
27. M. Schwartz, *Information Transmission, Modulation, and Noise,* McGraw-Hill Book Co., 1970, p. 486.
28. W. D. Gregg, *Analog and Digital Communications,* John Wiley & Sons, 1977, p. 257.
29. W. C. Y. Lee, *Mobile Communications Engineering,* McGraw-Hill Book Co., 1982, p. 248.
30. W. C. Y. Lee, *Mobile Communications Engineering,* McGraw-Hill Book Co., 1982, p. 108.
31. S. Kozono and K. Watanabe, "Influence of Environmental Buildings on UHF Land Mobile Radio Propagation," *IEEE Transactions on Communications,* Vol. COM-25, October 1977, pp. 1113–1143.
32. E. H. Walker, "Penetration of Radio Signals into Building in the Cellular Radio Environment," *Bell System Technical Journal,* Vol. 62, No. 9, Part I, November 1983, pp. 2719–2734.
33. M. Sakamoto, S. Kozono, and T. Hattori, "Basic Study on Portable Radio Telephone System Design," *IEEE Vehicular Technology Conference Record* (San Diego, Calif., 1982), pp. 279–284.
34. D. C. Cox, R. R. Murray, A. W. Norris, "Measurements of 800 MHz Radio Transmission into Buildings with Metallic Walls," *Bell System Technical Journal,* Vol. 62, November 1983, pp. 2695–2718.
35. D. Parsons, *The Mobile Cellular Propagation Channel,* Halsted Press, a division of John Wiley & Sons, 1992, Chap. 4.
36. K. Fujimoto and J. R. James (eds.), *Mobile Antenna Systems Handbook,* Artech House, 1994, Chap. 2.
37. W. C. Y. Lee, *Mobile Communications Design Fundamentals,* Howard W. Sams & Co., 1986, p. 111.
38. W. C. Y. Lee, *Mobile Communications Engineering,* McGraw-Hill Book Co., 1982, p. 343.

Introduction to
Digital Systems

14.1 Why Digital?

14.1.1 Advantages of digital systems

In an analog system, the signals applied to the transmission media are continuous functions of the message waveform. In the analog system, either the amplitude, the phase, or the frequency of a sinosoidal carrier can be continuously varied in accordance with the voice or the message.

In digital transmission systems, the transmitted signals are discrete in time, amplitude, phase, or frequency, or in a combination of any two of these parameters. To convert from analog form to digital form, the quantizing noise due to discrete levels should be controlled by assigning a sufficient number of digits for each sample,and a sufficient number of samples is needed to apply the Nyquist rate for sampling an analog waveform.

One advantage of converting message signals into digital form is the ruggedness of the digital signal. The impairments introduced in the medium in spite of noise and interference can always be corrected. This process, called *regeneration,* provides the primary advantages for digital transmission. However, a disadvantage of this ruggedness is increased bandwidth relative to that required for the original signal.

The increased bandwidth is used to overcome the impairment introduced into the medium. An analogy to this is given in Sec. 14.1.2. In addition to the cost advantage, power consumption is lower and digital equipment is generally lighter in weight and more compact. In mobile cellular systems, there are more advantages in applying digital technology.

14.1.2. Analogy to modulation schemes

We may apply the analogy of moving books to compare analog and digital transmission. We may use a daily event to describe the different modulation schemes in the communications field. Consider the following scenario. On one side of a large hall, 20 books have been piled up (signal to be sent). We want to move these books to the other side of the hall (receiving end). Assume that the floor of the hall is not slippery and is flat (transmission medium). Then we would hire a strong, muscular person who could carry the 20 books with both hands and safely transport them to the other side without dropping a single book. This is analogous to a double sideband. The space which this person is occupying is equal to the width of the person's shoulders, as shown in Fig. 14.1a. If the person is very strong, one hand can be used to carry the 20 books while the other hand can be used to carry something else. This is analogous to a single sideband. If the floor in the middle of the hall contains many water puddles and (or) bumps, this is analogous to a rough medium. We might not hire a strong adult to carry the books because we might be afraid that the person would fall and drop all the books in the middle of the hall. Therefore, we might hire 10 children, each one carrying two books as seen in Fig. 14.1b. Among 10 children, perhaps only 1 or 2 might fall in the middle of the hall and drop the books. But most of the books will be carried to the other side of the hall. The 10 children, each carrying two books, is analogous to frequency modulation (FM). Spreading the signal into a wide spectrum (children) applies the first principle of using "spread spectrum" in FM.

A similar analogy is shown in Fig. 14.1c. Besides puddles and bumps, there are three guard jammers to stop anyone from bringing books over to the other side of the hall. In such a case we would have to hire as many as 20 children, each carrying one book and running against the guards. Some of the children may fall because of the puddles and bumps and some may be blocked by the guards, but most of the books will arrive on the other side of the hall.

If a force of 20 children is not sufficient to carry the books to the other side, then we must hire at least 40 children, each carrying half a book. This is the concept of spread spectrum. The spread-spectrum technique is used for combatting rough media and enemy jamming in military communications.

In each modulation scheme, energy must be confined within a specified bandwidth. In this chapter we will demonstrate the regular mod-

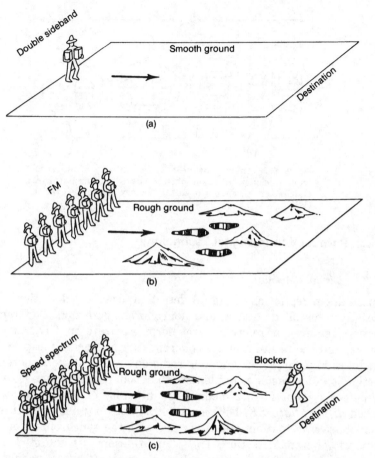

Figure 14.1 Analogy between radio transmission and moving blocks. (a) Walking over smooth ground; (b) ground with rough conditions; (c) ground with rough conditions plus blocks.

ulation schemes, phase shift keying (PSK) and frequency shift keying (FSK). We will also demonstrate how the schemes, MSK, GMSK, and GTFM can be applied to confine energy when transmitting a signal in the medium. The analogy can be extended to the adult or child who saves space while carrying books (Fig. 14.2a). The adult carrying the books (Fig. 14.2b) is taking too much space, and this method is not recommended. Common sense indicates that if the person wants to carry books through a crowd, the approach illustrated in Fig. 14.2a is more suitable than that illustrated in Fig. 14.2b. This analogy has been presented in the hopes of stimulating the readers to think about modulation schemes.

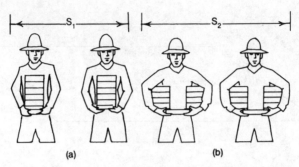

Figure 14.2 Analogy between space utilization and spectrum efficiency. (*a*) A satisfactory space utilization. (MSK, GMSK, or GTFM), (*b*) unsatisfactory space utilization (PSK).

14.2 Introduction to Digital Technology

14.2.1 Digital detection

There are three forms of digital detection: coherent detection, differentially coherent detection, and noncoherent detection. Coherent detection requires a reference waveform accurate in frequency and phase, and the use of a phase-coherent carrier tracking loop for each modulation techique. In a mobile radio environment, noncoherent detection is much easier to implement than coherent detection is. A form of detection that is intermediate in difficulty of implementation is called *differential PSK*. Differential PSK does not need absolute carrier phase information, and it circumvents the synchronization problems of coherent detection. The phase reference is obtained by the signal itself, which is delayed in time by an exact bit of spacing. This system maintains a phase reference between successive symbols and is insensitive to phase fluctuation in the transmission channel as long as these fluctuations are small during each duration of a symbol interval T. In differntial binary-phase shift-keying (DBPSK) a symbol is a bit. The weak point of this scheme is that whenever there is an error in phase generated by the medium, two message error bits will result.

There are several aspects of digital detection.[1]

Carrier recovery. Carrier recovery for the suppressed carrier signal $A(t) \sin (\omega_0 t)$ plus noise $n(t)$ can be obtained by two methods. A squaring or frequency-doubling loop can be used (see Fig. 14.3). The loop contains a phase-locked loop as shown in Fig. 14.4. The phase-locked loop maintains a constant phase ϕ_n of $\cos(2\pi f_0 t + \phi_n)$, which is the recovered carrier. Another carrier-recovery technique uses the Costas loop, which generates a coherent phase reference independent of the

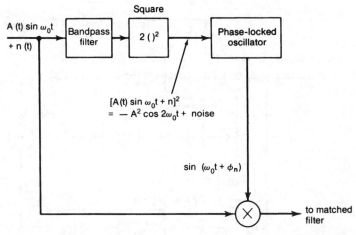

Figure 14.3 Block diagram of the square-law carrier-recovery technique.

binary modulation by using both in-phase and quadrature channels. The Costas loop (Fig. 14.5) is often preferred over the squaring loop because its circuits are less sensitive to center-frequency shifts and are generally capable of wider bandwidth operation. In addition, the Costas loop results in circuit simplicity.

Carrier-phase tracking (phase-locked loop). Carrier-tracking accuracy depends on several system parameters, including the phase noise in the carrier introduced by various oscillator short-term stabilities, carrier-frequency drifts, carrier-tracking-loop dynamics, transient response, the acquisition-performance requirement, and the signal-to-

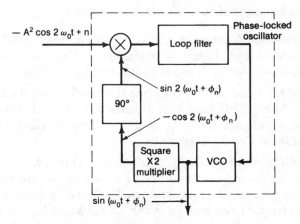

Figure 14.4 Phase-locked oscillator.

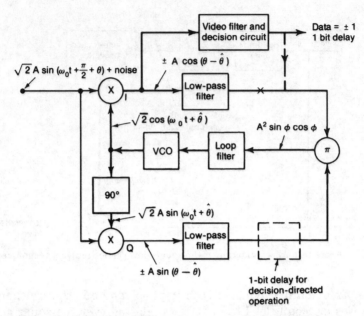

Figure 14.5 Costas loop OSK carrier-recovery circuit and bit detector; the phase error is defined as $\phi \triangleq \theta - \hat{\theta}$. The decision-directed configuration is shown in dashed lines.

noise ratio S/N in the carrier-tracking loop. The phase-locked loop in the carrier-recovery tracking loop must have sufficient noise bandwidth to track the phse noise of the carrier. For a given carrier-phase–noise spectrum, one can compute the phase-locked-loop noise bandwidth B required to track the carrier. Clearly, too large a noise bandwidth can permit the occurrence of the thermal noise effect.

Phase-equalization circuits—for cophase combining

1. *Feedforward.* A circuit using two mixers to cancel random FM can be used (see Fig. 14.6a) as a phase-equalization circuit in each branch of an N-branch equal-gain diversity combiner.

2. *Feedback.* A modified circuit from the feedforward circuit is shown in Fig. 14.6b. This circuit is also used for each branch of an N-branch equal-gain diversity combiner. The feedback combiner is also called a Granlund combiner.

3. *The total combining circuit.* As shown in Fig. 14.7, either a feedback or a feedforward circuit can be used in the combiner to form a two-branch equal-gain combiner. The circuit connects to a coherent match-filter receiver for BPSK as shown in Fig. 14.8.

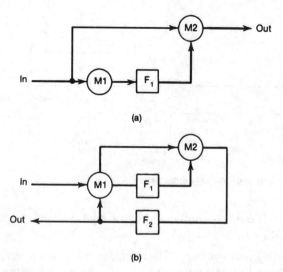

Figure 14.6 Cophase combining techniques. (*a*) Feed-forward cophase combining; (*b*) feedback cophase combining.

Bit synchronization. Power-efficient digital receivers require the installation of a bit synchronizer. Bit synchronization commonly applies self-synchronization techniques, that is, it extracts clock time directly from a noisy bit stream. There are four classes of bit synchronizer.

1. *Nonlinear-filter synchronizer.* This open-loop synchronizer is commonly used in high-bit rate links which normally operate at high signal-to-noise ratios.

2. *The data-transition tracking synchronizer.* This closed-loop synchronizer combines the operations of bit detection and bit synchro-

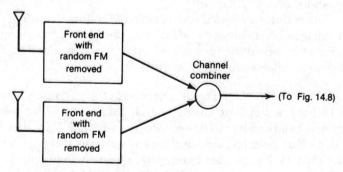

Figure 14.7 A two-branch diversity receiver.

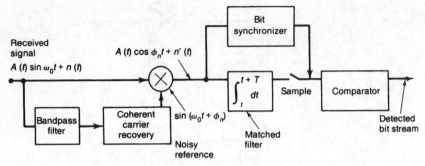

Figure 14.8 Coherent matched-filter receiver for BPSK.

nization. It can be employed at low signal-to-noise ratios and medium data rates.

3. *Early-late synchronizer.* This synchronizer uses early- and late-gate integral and dump channels, which have absolute values. It is simpler to implement than the data-transition tracking synchronizer and less sensitive to dc offsets.

4. *Optimum synchronizer.* This synchronizer provides an optimal means of searching for the correct synchronization time slot during acquisition. However, this approach generally is not practical.

14.2.2   Modulation for digital systems

There are several aspects of digital modulation, and they are described below.

Requirements. Basic digital modulation techniques are amplitude-shift keying (ASK), frequency-shift keying (FSK), phase-shift keying (PSK), and hybrid modulation techniques involving amplitude, frequency, and phase-shift keying.

In mobile cellular systems, the selection of a digital modulation for radio transmission involves satisfaction of the following requirements: (a) narrower bandwidths, (b) more efficient power utilization, and (c) elimination of intermodulation products.

Narrower bandwidths.   For all the forms of modulation, it is desirable to have a constant envelope and, therefore, utilize relatively narrower bandwidths. In these cases, FSK and PSK are recommended. For example, multiphase-shift-keying (MPSK) for large values ($M > 4$) has greater bandwidth efficiency than does BPSK or QPSK but power use is less efficient.

More efficient power utilization.   It is preferable to provide more channels for a given power level. Therefore, enhanced power utilization is essential. Besides, the FCC has limited the total power (100 W) to be radiated from each base-station antenna. This limitation governs the number of channels which can be served given the power allowed for each channel.

Elimination of intermodulation products.   QPSK is commonly used with a transmission efficiency of about 1 to 2 bps/Hz. This value has been found to offer satisfactory trade-off between efficient frequency utilization and transmitter power economy. However, in mobile radio links, when nonlinear class C power amplifiers are used, any spurious radiation should be suppressed. For reducing the spurious signals, we are selecting a constant or low-fluctuation envelope property. There are two types of broadly classified modulations.

Modulation schemes.   There are several modulation schemes.

1. *Modified QPSK.* There are two kinds of QPSK besides a regular QPSK with restricted phase-transition rules.
   a. *QPSK.* The conventional QPSK shown in Fig. 14.9a has phase ambiguity. The ideal QPSK signal waveform

$$A \sin[\omega_0 t + \theta_m(t)] = \pm \frac{A}{\sqrt{2}} \sin\left(\omega_0 t + \frac{\pi}{4}\right) \pm \frac{A}{\sqrt{2}} \cos\left(\omega_0 t + \frac{\pi}{4}\right)$$

$$(14.2\text{-}1)$$

   where $\theta_m = (0, \pi/2, \pi, 3\pi/2)$ and the value of θ_m should match the sign of Eq. (14.2-1); that is, $\theta_m = 0$ for $(+, +)$, $\theta_m = \pi/2$ for $(+, -)$, $\theta_m = \pi$ for $(-, +)$, and $\theta_m = 3\pi/2$ for $(-, -)$.

Bit pair	Absolute phase
0 0	0
0 1	$\pi/2$
1 1	π
1 0	$3\pi/2$

   b. *Offset QPSK (OQPSK).* This scheme is a QPSK, but the even-bit stream is delayed by a half-bit interval with respect to the odd 1 bit as shown in Fig. 14.9b.
   c. $\pi/4$-*shift QPSK.* A phase increment of $\pi/4$ is added to each symbol shown in Fig. 14.9c.

   Both OQPSK and $\pi/4$ shift QPSK have no π-phase transition; therefore, no phase ambiguity would occur as in QPSK. However,

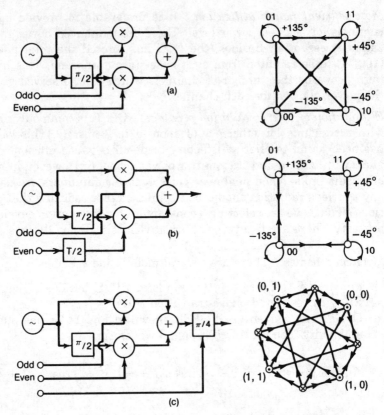

Figure 14.9 Modulator constitution and signal-space diagrams of (*a*) conventional QPSK, (*b*) offset QPSK, and (*c*) π/4 shift. (*After Hirade et al., Ref. 2, p. 14.*)

intrinsically they produce a certain amount of residual envelope fluctuation. Sometimes, a phase-locked loop (PLL) is inserted at the modulation output to remedy this problem.

2. *The differential encoding of QPSK (DQPSK).* This is the same as in DBPSK, but the differential encoding of the bit pairs selects the phase change rather than the absolute phase. However, DQPSK has phase ambiguity just like QPSK.

Bit pair	Phase changes
0 0	0
0 1	π/2
1 1	π
1 0	3π/2

The QPSK carrier recovery would be slightly different than that of BPSK.[2]

3. *Modified FSK—continuous-phase frequency-shift-keying (CP-FSK) with low modulation index*
   a. Minimum-shift-keying (MSK)—also called *fast FSK* (FFSK)
   b. Sinusoidal FSK(SFSK)
   c. Tamed FSK (TFSK) or tamed frequency modulation (TFM)
   d. Gaussian MSK (GMSK)
   e. Gaussian TFM (GTFM)

Since all the schemes listed above are CP-FSK (and have a low modulation index, they intrinsically have constant envelope properties, unless severe bandpass filtering is introduced to the modulator output. In MSK the frequency shift precisely increases or decreases the phase by 90° in each T second. Thus, the signal waveform is

$$s(t) = \sin\left(\omega_0 t + 2\pi \int_0^t s_i \, d\tau + \frac{n\pi}{2}\right) \qquad 0 < t < T$$

where

$$s_i = \begin{cases} s_1 = \dfrac{1}{4T} & \text{for a data bit 1} \\[2mm] s_2 = -\dfrac{1}{4T} & \text{for a data bit 0} \end{cases} \qquad (14.2\text{-}2)$$

or

$$s(t) = \sin\left(\omega_0 t + \frac{n\pi}{2} \pm \frac{\pi t}{2T}\right) \qquad (14.2\text{-}3)$$

or

$$s(t) = \cos\left(\pm\frac{\pi t}{2T}\right)\sin\left(\omega_0 t + \frac{n\pi}{2}\right) + \sin\left(\pm\frac{\pi t}{2T}\right)\cos\left(\omega_0 t + \frac{n\pi}{2}\right)$$

Comparing Eq. (14.2-1) with Eq. (14.2-3), we find that the two equations are very similar. In fact, the phase-modulation waveforms of the *I*- and *Q*-channel modulations of OQPSK are modulated by sine and cosine waveforms, and thus the output will be identical to that of MSK. Note that it is necessary to modulate both the *I* and *Q* channels during each bit interval to retain the constant envelope of $s(t)$.

Because the phase is continuous from bit to bit, the spectral sidebands of MSK or OQPSK fall off more rapidly than in BPSK or QPSK (see Fig. 14.10). Although MSK demonstrates a superior property in

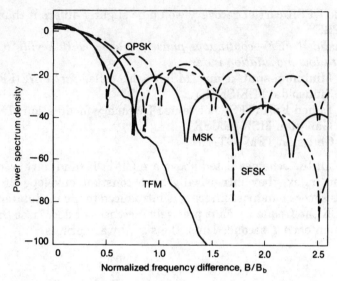

Figure 14.10 Power spectrum density functions of QPSK, MSK, SFSK, and TFM. *(After Hirade et al., Ref. 2, p. 15.)*

terms of its out-of-band spurious spectrum suppression without any filtering, its out-of-band spurious spectrum suppression will not satisfy the severe requirements in the single carrier per channel (SCPC) communications. The sharp edges in MSK phase-transition trajectories (Fig.14.11) can be smoothed by some premodulation baseband filtering. The SFSK shows a smoother phase transition than does the MSK but little improvement in the suppression of out-of-band spurious spectrum. The TFM is a modified MSK using the partial response encoding rule as the phase-transition rule. The smoothed phase trajectory of TFM is shown in Fig.14.11. The outstanding suppressions for the out-of-band spectrum of TFM is shown in Fig. 14.10. However, this outstanding suppression of the out-of-band spurious spectrum can be achieved by using a suitable premodulation baseband filtering on MSK, such as a baseband gaussian filtering, as shown in Fig. 14.12. The GMSK with $B_b T = 0.2$ has the same power spectrum density curves as does TFM, where B_b is the baseband bandwidth. Furthermore, GMSK is easier to implement than TFM.

The following parameters are defined and are in the figures that follow.

B_i ideal bandpass

B_s channel separation (bandwidth)

f_b bit rate of voice coding = $1/T$

$1/T$ transmission bit rate (16 kbps)

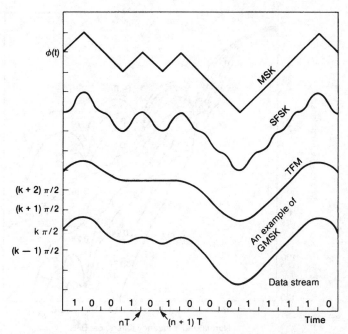

Figure 14.11 Phase-transition trajectories of MSK, SFSK, TFM, and GMSK *(After Hirade et al., Ref. 2, p. 15.)*

$B_i T$ normalized bandwidth of the ideal bandpass filter
$b_b T$ normalized bandwidth of a gaussian filter

Figure 14.13 shows the fractional power in percentage of GMSK signal exceeding the normalized bandwidth $B_i T$ with different values of $B_b T$ ($B_b T = \infty$ means no filter). This becomes conventional MSK. Figure 14.14 shows that the relative power radiated in the adjacent channel for $B_s T$ is equal to 1.5. Let $B_s = 30$ kHz, then the bit rate $f_b = (1/T) = 20$ kbps. For a normalized filter bandwidth $B_b T = 0.24$ or $B_b = 4.8$ kHz, and the relative power in the adjacent channel is -60 dB.

4. 16-QAM and code modulation—these two modulations are not constant envelop modulations, but both can achieve better spectrum efficiency. The readers can read Proakis' book[19] if interested.

Demodulation. When the signal is received, we would like to know the performance of the various demodulation schemes. Some demodulation schemes are better than others regardless of what the modulated signal is like. The orthogonal coherent detector proposed by de Buda[3] can be used for both MSK and TFM.

BER performance is always a good criterion for the comparison of different modulation schemes. Measured BER performance is shown

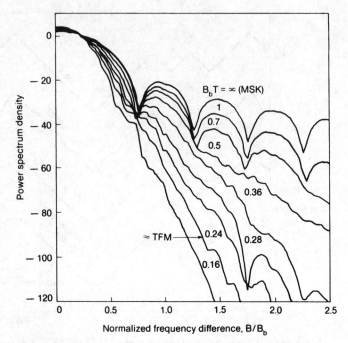

Figure 14.12   Power spectrum density functions of GMSK. *(After Hirade et al., Ref. 2, p. 15.)*

in Fig. 14.15 with a normalized channel bandwidth $B_iT = 0.75$. The family of curves show the BERs for the different filter bandwidths. Also, TFSK (TFM) is plotted for comparison. GMSK with the filter bandwidth of $B_bT = 0.19$ is superior to TFSK. However, one disadvantage of narrowing the channel spectrum is increasing the BER, that is, degrading performance. Sometimes we have to consider whether it is worthwhile to use a gaussian filter.

14.3   ARQ Techniques

14.3.1   Different techniques[4,5]

Automatic-repeat-request (ARQ) techniques include the coding and retransmission request strategy for delivering a message. Preceding the message is a header which contains the source and destination address and useful routing information. Every ARQ message must have a header. There are two principal ARQ techniques.

1. *Stop-and-wait ARQ.* The message originator stops at the end of each transmission to wait for a reply from the receiver (see Fig. 14.16a). Then the following steps can be taken.

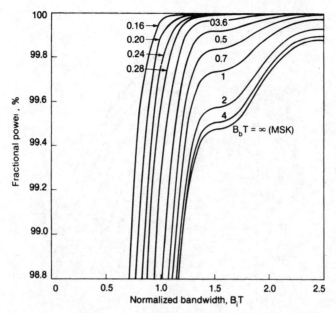

Figure 14.13 Fractional power of GMSK signal. *(After Hirade et al., Ref. 2, p. 15.)*

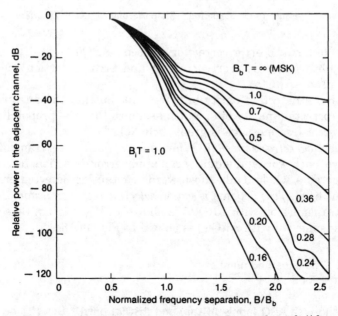

Figure 14.14 Relative power radiated in the adjacent channel. *(After Hirade et al., Ref. 2, p. 16.)*

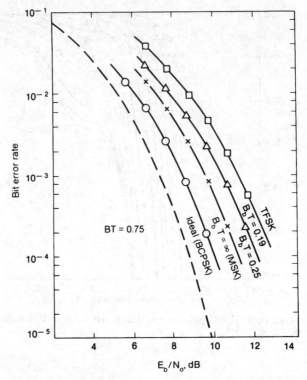

Figure 14.15 Measured BER performance. *(After Hirade et al., Ref. 2, p. 17.)*

a. No forward error correction is used—ARQ(*a*).

b. Both forward error correction and error detection coding are used—ARQ(*b*).

c. Error-detection parity bits are sent, but not the forward error-correction parity bits, which assumes that the probability of an error-free message is great—ARQ(*c*).

2. *Selective retransmission.* When many words are transmitted at once, each word individually can apply error detection, not the message as a whole. Only those words containing detected errors are sent back. This scheme is called *selective retransmission.* Selective retransmission with ARQ(*b*) is shown in Fig. 14.16*b*. Selective retransmission with ARQ(*c*) is shown in Fig. 14.16*c*.

14.3.2 The expected number of transmissions

Stop-and-wait ARQ [apply ARQ(*a*) and ARQ(*b*) only]. Let P_{ew} be a word error rate (WER), and let a message consist of N words. Now the re-

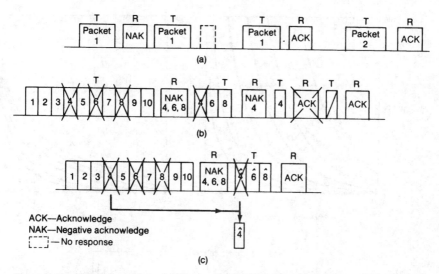

Figure 14.16 ARQ transactions. (*a*) Stop-and-wait ARQ; (*b*) selective retransmission with ARQ(*b*); (*c*) selective retransmission with ARQ(*c*).

quired number of transmissions depends on all N words being successfully transmitted. The expected number of transmissions is

$$E_N = \frac{1}{(1 - P_{ew})^N} \qquad (14.3\text{-}1)$$

assuming independence errors between words and that all words have the same P_{ew}. This assumption can be considered valid for cases where vehicle speed is high. Equation (14.3-1) indicates that the number of transmissions E_N increases more quickly with increasing the message length, that is, as N increases. Equation (14.3-1) is plotted in Fig. 14-17*a*.

Selective retransmission (ST) with ARQ(*b*). Assume that the number of transmissions of one word is independent of the number of transmissions of any other word. The expected number of transmissions required for sending an N-word message with fewer than i transmissions is

$$E_N = \sum_{i=1}^{\infty} [1 - (1 - P_{ew}^{i-1})^N] \qquad (14.3\text{-}2)$$

Equation (14.3-2) is plotted in Fig. 14.17*b*. By comparing Fig. 14.17*a* with Fig. 14.17*b*, we see that stop-and-wait ARQ would require a greater number of transmissions to deliver a message than would selective retransmission with the same block error probability.

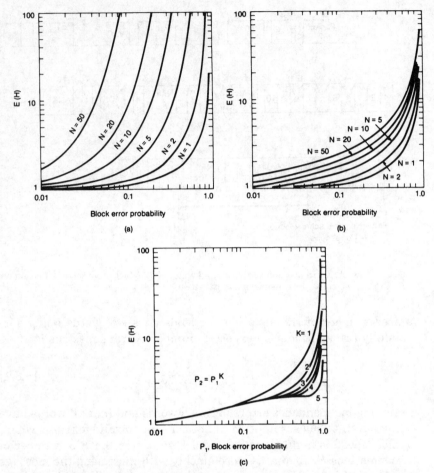

Figure 14.17 Expected number of transmissions for different kinds of ARQ. (*a*) *E*(*H*), expected number of transmissions to deliver an *N*-block message for stop-and-wait ARQ. (*b*) *E*(*H*), expected number of transmissions to deliver an N-block message for SRT ARQ(*b*). (*c*) *E*(*H*), expected number of transmissions to deliver a 10-block message for SRT ARQ(*c*) for various retransmission-failure probabilities.

Selective retransmission (SRT) with ARQ(c). In this scheme, ARQ(*c*) defined in the stop-and-wait ARQ is applied to the selective retransmission scheme. Let the first-transmission probability be P_1 and a retransmission probability be P_2. The expected number of transmissions required for sending an *N*-word message with fewer than *i* transmissions is

$$E_N = 1 + \sum_{i=2}^{\infty} [1 - (1 - P_1 P_2^{i-2})^N] \qquad (14.3\text{-}3)$$

Because the forward error correction is added to retransmission, P_2 is smaller than P_1. Let P_2 be a positive integer power of P_1; that is, $P_2 = P_1^k$, where k represents the various powers of the feedforward error correction code. Equation (14.3-3) is plotted in Fig. 14.17c for $N = 10$. When $k = 1$, the curve shown in Fig. 14.17c is the same as $N = 10$ in Fig. 14.17b.

14.3.3 Transmission Efficiency R

The transmission efficiency is the ratio of the number of information bits to the total number of transmission bits. Let a message consist of N words, where B is number of bits per word, and

$$B = H + L \qquad (14.3\text{-}4)$$

where H is the header bits and L is the information bits. Then

Stop-and-wait ARQ technique

$$R = \frac{NL}{(H + NB)E_N} \qquad (14.3\text{-}5)$$

where E_N is as shown in Eq. (14.3-1).

Selective retransmission

$$R = \frac{NL}{HE_N + NBE_1} \qquad (14.3\text{-}6)$$

where E_N is as shown in Eq. (14.3-2) and

$$E_1 = \frac{1}{(1 - P_{\mathrm{ew}})} \qquad (14.3\text{-}7)$$

Selective retransmission with ARQ(c)

$$R = \frac{NL}{HE_N + NBE_1} \qquad (14.3\text{-}8)$$

where E_N is as shown in Eq. (14.3-3), and E_1 is

$$E_1 = 1 + \sum_{i=2}^{\infty} [1 - (1 - P_1 P_2^{i-2})] \qquad (14.3\text{-}9)$$

14.3.4 Undetected error rates

In the previous sections, we saw that each transmission results in one of two outcomes, success or failure. This is called "hard detection." There is also a "soft detection." In soft detection three probabilities per transmission are denoted: P_c (success), P_d (detected error), and P_u (undetected error) per word. When each of these probabilities is identified for each transmission, then we can perform further calculations.

Stop-and-wait ARQ [apply ARQ(a) and ARQ(b)]. We can find the undetected error probability per single-word message.

$$P_{um} = \frac{P_u}{P_c + P_u} \tag{14.3-10}$$

where

$$P_c = 1 - P_{ew} \tag{14.3-11}$$

and

$$P_u \leq P_{ew}\, 2^{-m} \tag{14.3-12}$$

where m is the number of error-detection parity bits. The undetected error' probability per N-word message is

$$P_{um} = \frac{1 - (1 - P_{ew})^N}{1 + (1 - P_{ew})^N (2^m - 1)} \tag{14.3-13}$$

Selective retransmission. In this technique, we consider b parity bits per each word instead of m parity bits for the whole message. The probability of undetected error for a single-word message is

$$P_{um} = \frac{P_{ew}}{P_{ew}(1 - 2^b) + 2^b} \tag{14.3-14}$$

The probability of undetected error for a N-word message is

$$P_{um} = 1 - \left[1 - \frac{P_{ew}}{P_{ew}(1 - 2^b) + 2^b}\right]^N \tag{14.3-15}$$

The word error rate P_{ew} is a function of vehicle speed. Since the word error rate P_{ew} is derived from the bit error rate, each bit in error can be dependent on or independent of the adjacent bit error related to the vehicle speed. Therefore, the word error rate is not easy to obtain. Any oversimplified model may give incorrect answers. We can obtain the word error rate from two extreme cases, one assuming that the speed approaches infinity and the other assuming that the speed approaches zero as described in Chap. 13.

14.4 Digital Speech[6]

Since digital technologies have evolved, an important study focuses on efficient methods for digitally encoding speech. Speech quality implies a measure of fidelity, which is difficult to specify qualitatively because human perception is involved. The two criteria used are

What is being said (low fidelity accepted by military systems)

Who says it (high fidelity important to commercial systems)

For instance, a military system examiner who comments that someone's speech quality is excellent may be referring to intelligibility and low system noise. It is irrelevant who speaks on the other side (of the examiner) or that the examiner has never spoken to this person before.

14.4.1 Transmission rates in speech coding

These rates are totally dependent on quality characterizations such as toll quality, commentary quality, communications quality, synthetic quality. We may use the mean opinion score (MOS) referred in Sec. 1.5.1 to grade the voice quality.

1. *Toll quality* ($4 < MOS < 4.5$). An analog speech signal is of toll quality when its frequency range is 200 to 3200 Hz; its signal-to-noise ratio is greater than or equal to 30 dB; and its harmonic distortion is less than or equal to 2.3 percent. Digital speech has to have a quality comparable to that of the toll quality of an analog speech signal.

Toll-Quality Transmission

Coder	kbps
Log PCM	56
ADM	40
ADPCM	32
Sub-band	24
Pitch Predictive ADPCM	24
APC, ATC, ΦV, VEV	16

2. *Commentary quality* ($MOS > 4.5$). In general, the signal at bit rates exceeding 64 kbps generates a commentary-quality speech

signal which is better than toll quality, but the input bandwidths are significantly wider than in a noncellular telephone system (up to 7 kHz).

3. *Communications quality ($3 < MMOS < 4$).* At rates below 16 kbps, the signal in the range of 7.2 to 9.6 kbps is a communications-quality speech signal. The signal is highly intelligible but has noticeable reductions in quality and speaker recognition.

Communincations-Quality Transmission

Coder	kbps
Log PCM	36
ADM	24
ADPCM	16
Sub-band	9.6
APC, ATC, ΦV, VEV	7.2

4. *Synthetic quality ($2.5 < MOS < 3$).* At 4.8 kbps and below, the signal provides synthetic quality and speaker recognition is substantially degraded.

Synthetic-Quality Transmission

Coder	kbps
CV, LPC	2.4
Orthogonal	1.2
Formant	0.5

14.4.2 Classes of coder

There are two classes of coder: waveform coders and source coders.

Waveform coders. The speech waveform can be characterized by

1. Amplitude distribution (in time domain)
2. Autocorrelation function (in time domain)
3. Power spectral density (in frequency domain)
4. Spectral flatness measure (removing redundancy in speech waveform)
·5. Fidelity criteria for waveforms

$$\text{Coding noise} = \frac{1}{T} \int_0^T (\text{coding error})^2 \, dt$$

where the coding error is equal to the amplitude difference (samples of a coded waveform minus the original input waveform).

The signal-to-noise ratio is expressed as

$$\frac{S}{N} = \overline{\left[\frac{(\text{input waveform})^2}{\text{coding noise}} \right]}$$

There are two types of speech waveform coders.

1. *Time-domain coders.* Pulse code modulation (PCM), differential pulse code modulation (DPCM), and delta modulation (DM) are commonly used. Adaptive predictive coding (APC) in time-domain coding systems is limited to linear predictors with changing coefficients based on one of the following three types:

   a. Spectral fine structure—in more periods

   b. Short-time spectral envelope—determined by the frequency response of the vocal tract and by the spectrum of the vocal-cord sound pulses

   c. Combination of types *a* and *b*

   In time-domain coders, speech is treated as a single full-band signal: in time-domain predictive coders, speech redundancy is removed prior to encoding by prediction and inverse filtering so that the information rate can be lower.

2. *Frequency-domain coders.* The speech signal can be divided into a number of separate frequency components, and each of these components can be encoded separately. The bands with little or no energy may not be encoded at all. There are two types of coding:

   a. Subband coding (SBC). Each subband can be encoded according to perceptual criteria that are specific to that band.

   b. Adaptive transform coding (ATC). An input signal is segmented and each segment is represented by a set of transform coefficients which are separately quantized and transmitted.

Source coders—vocoders. The synthetic quality of source vocoder speech is not appropriate for commercial telephone application. It is designed for very low bit-rate channels. Vocoders use a linear, quasi-stationary model of speech production.

Sound source characteristics. The sound can be generated by voiced sounds, fricatives, or stops. The source for voiced sounds is represented by a periodic pulse generator. The source for unvoiced sounds is represented by a random noise generator. They are mutually exclusive.

System characterization. The acoustic resonances of the vocal tract modulate the spectra of the sources. Different speech sounds correspond uniquely to different spectral shapes. Vocoders depend on a parametric description of the vocal-tract transfer functions.

1. Channel vocoder—speech signal evaluated at specific frequencies

2. LPC (linear prediction code) vocoder—linear prediction coefficients that describe the spectral envelope

3. Formant vocoder—specified frequency values of major spectral resonances

4. Autocorrelation vocoder—specified short-time autocorrelation function of the speech signal

5. Orthogonal function vocoder—specifies a set of orthonormal functions

Frequency-domain vocoders. A single coder is called a *channel vocoder.* Instead of transmitting the telephone signal directly, only the spectrum of each speech signal is transmitted; 16 values along the frequency axis are needed. Each takes 20 ms and requires a bandwidth of $1/(2 \times 20 \text{ ms}) = 25$ Hz and the total frequency requirement is (16×25) or 400 Hz, which is one-tenth of the bandwidth of the speech signal itself.

Time-domain vocoders. Speech samples would have to be spaced $1/(2 \times 4000) = 0.125$ ms apart, which would require 30 samples to ensure a good quality. Then the frequency requirement is $30/(2 \times 0.125 \text{ ms})$ or 120 kHz. For digital transmission, the number of bits per correlation sample used by a time-domain vocoder should be about twice as high for spectral samples in frequency-domain vocoders. Therefore, time-domain vocoders are not desirable. Yet, one of the most successful innovations in speech analysis and synthesis is linear predictive coding (LPC), which is based on autocorrelation analysis.

LPC vocoders. LPC vocoders constitute an APC system in which the prediction residual has been replaced by pulse and noise sources. For the telephone band, the number of predictor coefficients is 8. For low-quality voice, the number can be as small as 4. The RELP (regular-pulse excited LPC) used by GSM, the VSELP (Vector-sum excited LPC) used by TDMA (IS-54) and the modified VSELP used by CDMA all have ten coefficients.

Hybrid waveform coders-vocoders. A hybrid arrangement of SBC, APC, and LPC is coming into vogue where a portion (lower-frequency band) of the transmission is accomplished by waveform techniques and a portion (upper-frequency band) by voice-excited vocoder techniques.

14.4.3 Complexity of coders

A relative count of logic gates is used to judge the complexity of the coders as follows:

Relative complexity		Coder
1	ADM:	adaptive delta modulator
1	ADPCM:	adaptive differential PCM
5	Sub-band:	subband coder (with CCD filters)
5	P-P ADPCM:	pitch-predictive ADPCM
50	APC:	adaptive predictive coder
50	ATC:	adaptive transform coder
50	ΦV:	phase vocoder
50	VEV:	voice-excited vocoder
100	LPC:	linear-predictive coefficient (vocoder)
100	CV:	channel vocoder
200	Orthogonal:	LPC vocoder with orthogonalized coefficients
500	Formant:	formant vocoder
1000	Articulatory:	vocal-tract synthesizer; synthesis from printed English text

Of these coders, LPC is attractive because of its performance and degree of complexity.

14.5 Digital Mobile Telephony

14.5.1 Digital voice in the mobile cellular environment

Since voice communication is the key service in cellular moble systems, when we think of the digital systems, we must think of a digital voice.

In present-day mobile cellular systems, transmission of a digital voice in a multipath fading environment is a challenging job. The major considerations in implementing digital voice in cellular mobile systems are discussed below, along with a tentatively recommended transmission rate for the cellular mobile system.

Digital voice in the mobile radio environment

1. The criterion for judging a good digital voice through a wire line is employed in three existing digital voice schemes.
    a. In a continuously variable step delta (CVSD) modulation scheme, the present transmission rate is 16 kbps. This is not toll-quality voice transmission and is commonly used by the military.

b. In a LPC scheme, the present transmission rate of 2.4 kbps provides a synthetic quality voice, but a rate of 4.8 kbps using vector quantization[8] may provide a communications-quality voice. A rate of 16 kbps can provide a toll-quality voice.[19]

c. In a pulse code modulation (PCM) scheme, the present transmission rates of 32 kbps and 64 kbps are commonly used; 32 kbps is used by the military while 64 kbps is used commercially.

Of the three schemes, LPC seems most attractive because of its low transmission rate. However, LPC is more vulnerable in terms of distortion to the mobile fading environment.

2. Digital voice has to be processed in real time, which imposes constraints on the digital processing time. This adversely affects LPC but not CVSD.

3. When sending a digital stream (voice) through a radio channel in a fading environment, in general, an LPC scheme needs more code protection than CVSD scheme does because LPC is not implemented in a continuous waveform in either the frequency domain or the time domain while CVSD is implemented in a continuous waveform in the time domain.

4. Because the mobile unit is moving, sometimes rapidly, sometimes slowly, insertion of extra synchronization bits is needed in the normal digital stream.

Considerations for a digital voice transmission in cellular mobile system. The following factors are significant.

1. *Digital transmission rate*
   a. *Present cellular signaling rate.* The present signaling format is designed on the assumption that the mobile unit moves at an average of 30 mi/h and that the transmission rate is 10 kbps. The 21 synchronization bits (10 synchronization bits and 11 frame bits) occur in front of every code word of 48 bits to ensure that the bits are not falling out of synchrony before the resynchronization takes place.
   b. *Consideration of LPC scheme.* If a rate of 4.8 kbps using LPC for a communications-quality voice is accepted its rate is almost half of the present transmission rate, and at this transmission rate a 48-bit word would be acceptable in a fading environment. The resynchronization scheme for a mobile receiver should take place in front of every code word of 48 bits [(21 synchronization bits) + (a code word of 48 bits) = 69 bits]. The number of synchronization bits is almost half the number of bits in a code word. Therefore, the transmission rate would be approximately $(4.8 \times 1.5) = 7.2$ kbps.

c. *Redundancy of transmission.*   The protection of synchronization in a mobile radio environment is not sufficient. If the digital stream were to occur in a signal fade, partial or whole code words would be lost. In order to prevent fading, redundancy of transmission is often used. We would take a minimum redundancy scheme; for example, we would transmit the same message bits three times and take a "2-out-of-3 majority vote" on each bit to minimize the fading impairment of the message bits. For LPC of 4.8 kbps, an RF transmission rate of (4.8 kbps $\times$ 1.5) 3 = 21.6 kbps is needed.* It is reasonable for a 30-kHz channel to carry a transmission rate of 21.6 kbps over a severe fading medium. When an RF transmission rate is given, the channel bandwidth can be narrower with a trade-off of transmitted powers. This point has been described in Sec. 13.4.8.

d. *Modulation, diversity, coding, ARQ, and scrambling.*   Diversity and modulation can help in reducing the RF transmission rate for the digital voice. However, ARQ schemes, fancy coding schemes, and complicated scrambling schemes cannot be implemented for voice transmission. This is because the digital voice must be processed in real time, and these three schemes usually require a fair amount of time for processing. These schemes can be used for data transmission.

2. *Word error rate.*   In the multipath fading environment, the bit error rate P_e is not the only parameter for voice-quality measurement; the word error rate P_w is also important and varies with vehicle speed. However, information on the word error rate for transmission of digital voice over a mobile radio environment only appears in two extreme cases (see Sec. 13.2.3). Assume that we know the required P_e and P_w. We can convert P_e and P_w to a required carrier-to-noise ratio C/N. If a two-branch diversity scheme is applied after a 2-out-of-3 majority-vote redundancy scheme has been used, the bit error rate of 10^{-3} in a relatively slow fading case requires a C/N level of approximately 15 dB. With the C/N level, a word error rate of a 40-bit word is about 10^{-3} (see Fig.13.3a). In general, if the word error rate is the same as or lower than the bit-error rate for a given C/N, the C/N level is acceptable. In our case, P_w and P_e are the same at $C/N = 15$; therefore, the $C/N = 15$ dB is justified.

3. *Relationship between C/N and E_b/N_0.*   The relationship between the carrier-to-noise ratio C/N, the energy-per-bit-to-noise-per-hertz ratio E_b/N_0, the transmission rate R, and the bandwidth B can be expressed as

---

*Applying diversity schemes can reduce this rate.

$$\frac{C}{N} = \frac{E_b}{N_0} \frac{R}{B} \qquad (14.5\text{-}1)$$

When the number of levels C/N increases, the bandwidth decreases. Keeping E_b/N_0 constant, we see that when the bandwidth decreases, the required carrier-to-noise ratio C/N increases. Previously we calculated that $C/N = 15$ dB works for a two-level (binary) system. If the number of levels increases, the C/N will be higher than 15 dB.

Example 14.1 Let $E_b/N_0 = 15$ dB for a two-level system and R_0 and B_0 be the transmission rate and transmission bandwidth, respectively, of the two-level system. Now if we reduce the bandwidth $B_1 = 0.5B_0$, then

$$\left(\frac{C}{N}\right)_1 = (31.6)\frac{R_0}{0.5B_0} = 2\left(\frac{C}{N}\right)_0$$

$$= \left(\frac{C}{N}\right)_0 + 3 \text{ dB}$$

This means that the power increases by 3 dB. If the transmitted power was 50 W, now it is 100 W.

14.5.2 Evaluation of digital voice quality

In general, there are two methods for evaluating digital voice quality.

1. *Listener's opinion.* Use one 16-kbps voice coder and one 8-kbps voice coder in a specified digital system. Then find the two required carrier-to-interference ratios C/I based on the listener's opinion in a Rayleigh fading environment. Then compare the same voice quality with that from an analog FM system at $C/I \geq 18$ dB.

2. *Diagnostic rhyme test (DRT).* The voice quality of a digital format is often tested by DRT. Using the DRT score of 90 as a criterion, above 90 means acceptable for synthetic-quality voice and below 90 means unacceptable. Thus, the bit error should be less than 10^{-3} for an LPC of 2400 bps in a gaussian noise environment. Voice evaluation for an LPC of 4.8 kbps does not appear in the literature. The voice quality in CVSD based on the same DRT criterion requires a bit error rate of only 4 percent or less. The DRT is not designed for toll-quality voice test.

14.6 Practical Multiple-Access Schemes

14.6.1 High-capacity FSK in FDMA

The use of frequency-division multiple access (FDMA) with the present channel bandwidth of 30 kHz is a conventional approach. The high-capacity FSK modulation used in FDMA is based on the rate of digital voice. From the quality transmission of vocoders (see Sec. 14.5.1) we can see that a digital voice must have a 16-kbps coding rate to produce full telephone quality, although the voice quality of a 4.8-kbps LPC with vector quantization may be acceptable for a communications-quality voice (see Sec. 14.5). Now we have to determine the C/I which will provide optimal quality within a given digital system.

Swerup and Uddenfeldt compared a narrowband coherent digital modulation with gaussian MSK to an analog FM system.[11] Two 16-kbps voice coders were used. Residual excited linear predicted codes and subband codes were tested. The digital unit performance can be reduced by 5 dB to obtain the same performance as an analog unit. This 5-dB reduction advantage means a large coverage area and a closed frequency-reuse distance for each cell can be served in a cellular system. This is, in turn, an example of high spectral efficiency usage (described in Sec. 13.4). Consider the following calculations.

1. In an omnidirectional-cell system, assume that $C/I = 13$ dB, i.e.,

$$\frac{C}{I} = \frac{q^4}{6} > 10^{1.3} = 20$$

Solving for q and using Eq. (2.4-5), we obtain

$$q = 3.31 = \sqrt{3K}$$

$$K \approx 4 \quad \text{(frequency-reuse pattern)}$$

In this case the total number of channels is 333; then

$$m = \frac{333}{4} = 83 \text{ channels/cell}$$

which is higher than the 47 channels per cell for $C/I \geq 18$ dB.

2. In 120° direction cells, we also compare two sets of C/I levels. As shown in Fig. 6.6b, there are only two interference cells. Then we can estimate the distance between the mobile unit and the two interfering sites to be $D + 0.5R$, as mentioned in Sec. 6.5.1.

a. $C/I \geq 18$ dB.

$$\frac{(q + 0.5)^4}{2} \geq 10^{1.8} = 63$$

or $\qquad\qquad\qquad\qquad q = 2.85$

The number of frequency-reuse cells is

$$K = \frac{q^2}{3} = 2.71 \sim 3$$

The number of sectors is $(3 \times 3) = 9$; then

$$m = \frac{333}{9} = 37 \text{ channels/sector}$$

b. $C/I \geq 13$ dB

$$\frac{(q + .05)^4}{2} = 10^{1.3} = 20$$

or $q = 2$. The number of frequency-reuse cells is

$$K = \frac{q^2}{3} = \frac{4}{3} = 1.3 \sim 2$$

The number of sectors = $(2 \times 3) = 6$; then

$$m = \frac{333}{6} = 55.5 \text{ channels/sector}$$

3. In 60° directional cells (see Fig. 6.6c and Sec. 6.5.1)
   a. $C/I \geq 18$ dB

$$\frac{(q + 0.7)^4}{1} \geq 10^{1.8} = 63$$

or $\qquad q = 2.12$

$$K = \frac{q^2}{3} = 1.5 \sim 2$$

$$m = \frac{333}{2 \times 6} = 27.75 \text{ channels/sector}$$

b. $C/I \geq 13$ dB

$$\frac{(q + 0.7)^4}{1} \geq 10^{1.3} = 20$$

or $\qquad q = 2.11 - 0.7 = 1.41$

$$K = \frac{q^2}{3} = 0.67 \sim 1$$

$$m = \frac{333}{6} = 55.5 \text{ channels/sector}$$

Apparently, spectrum efficiency is increased by using digital technology.

Discussion.   The preceding analysis is based on an ideal situation. It needs to be verified by measurement in a real cellular system. In the future we may achieve a digital system which can narrow the channel bandwidth and increase the transmission rate. The success of such a system would be proved if the reception at the receiving end were the same as that which would be achieved if the medium were nonfading. Modulation schemes, diversity, coding, redundancy, and ARQ can help to achieve this goal.

The signal can be designed using 4-, 8-, or 16-level MFSK or MPSK. Of course, the higher the number of levels, the narrower the channel bandwidth. However, the increase in transmitted power (or the reduction in range) is the key factor.

In an FDMA system, the transmit data rate needs to serve only one voice channel and the rate can be 10 kbps or less. In this circumstance, the equalizer used to reduce the intersymbol interference (ISI) due to high-data-rate transmission is not required in FDMA.

14.6.2   TDMA system[11]

The use of time-division multiple access (TDMA) could be another approach for increasing spectrum channel efficiency. It also has the potential to reduce the cost of both cell-site and mobile terminal equipment. The bulky analog radio equipment can be replaced by very large scale integrated-circuit (VLSI) digital signal processing. This applies to the use of digital equipment in any system; TDMA is no exception. In addition, TDMA will replace the analog duplex filter with a time switch. A zero IF receiver can be used, that is, there can be direct conversion without superheterodyne. The number of radio transceivers can be reduced, and the size of cell site can be much smaller. Assume that a cell site in a TDMA system has a single transceiver and a simple network interface. This is a very attractive feature. The main drawback of TDMA is the accurate clock requirement.

An inaccurate clock results in time jittering due to the instability of the frequency synthesizer at the transmitter and the clock at the receiver. The variable time delay resulting from the change of vehicle positions and from random FM could affect the synchronization and tracking of the bit streams. The most adverse effect of the synchronization problem is the delay spread mentioned in Chap. 1. In TDMA transmission, severe time dispersion occurs. In urban areas of the United States, the mean delay spread can be as much as 3 µs. Then the transmission rate must be limited to

$$R_b \leq \frac{1}{2\pi \times 3 \times 10^{-6}} = 50 \text{ kbps}$$

By using diversity schemes,[12] the transmission rate can be increased if the same BER is assumed. Another effective scheme is an adaptive decision feedback equalizer[13] which adaptively sums up the multipaths so that the time-delay spread at the receiver is reduced and the sum of multipaths is in the form of a diversity.

The adaptive equalizer can work for narrowband TDMA also, so TDMA is not based on wideband spread spectrum technology. Since TDMA is very immune to interference in the cellular system (one user is designated in one time slot) and allows a simple handoff procedure, it may be suitable for microcell systems. *Microcells* are loosely defined as cells whose radii are less than 1 mi. However in a heavy traffic area, TDMA performance may degrade faster than FDMA performance. Simulation can be used to decide which transmission method is best.

For economic and spatial reasons, a 300-kHz-bandwidth TDMA carrying 300 kbps can provide 10 voice channels. Since a low-cost base station with one radio transceiver can make the TDMA system seem very attractive, this possibility is worth pursuing. The critical issue is building the adaptive decision feedback equalizer properly.

Since the same digital system working in different areas will result in different levels of performance, we should be aware that the same system might test well in a light traffic area but poorly in a heavy traffic area.

14.6.3 Spread spectrum systems in the '80s

Many suggested systems appeared in the literature; we may mention two of them. The uses of a frequency hopping scheme in FDMA and a direct sequence scheme in both TDMA and CDMA are described as follows.

Frequency-division multiple access (FDMA). There are many spread-spectrum systems proposed for use in mobile radios. Cooper and Nettleton proposed[14] a frequency hopping system. In each hop a binary DPSK-modulated system with error-correction coding is implemented. Comparing the relative spectral efficiencies of this proposed system with today's cellular 30-kHz system, we see that they are almost the same. Goodman et al.[15] proposed using multilevel frequency shift keying (MFSK) plus frequency hopping for land mobile radios. Their theoretical results, based on certain assumptions, show a better spectral efficiency.

Time-division multiple access (TDMA).   A digital TDMA system has been proposed by SEL in Germany,[16] which uses a bandwidth of 4 MHz and only one fixed radio frequency for all subscribers in each cell. Therefore it is a spread-spectrum system. In a wideband system, each channel does not need to tune to a particular frequency. All the channels share the same bandwidth and thus have the same type of radio set.

Method of operation.   The signal of each channel in TDMA is coded differently to provide identification and proper channel separation. The difference between a digital TDMA system and an analog FDMA system is shown in Fig. 14.18. In a digital TDMA system, each channel has 32 code patterns (5 bits) and one sign bit; thus, a total 6 bits are transmitted.

In each cell, 60 TDMA channels are available in three channel groups. The merit of this system is shown in Fig. 14.19. Each base station covers three cell sectors in succession by means of an electronic scan-and-stay antenna pattern. Twenty channels are allocated to each sector. Once a cell sector has received its 20 channels (time slots), the

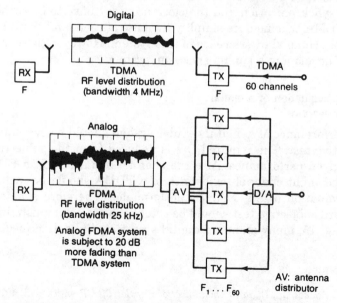

Figure 14.18   Digital TDMA mobile radio and analog FDMA system. *(After Bohm, Ref. 16. © Horizon House—Microwave, Inc.)*

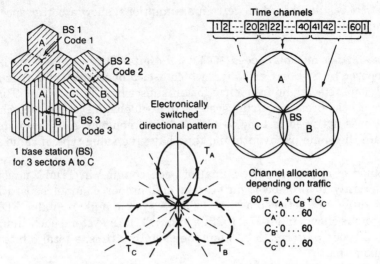

Figure 14.19 Cell structure of wideband TDMA systems. *(After Ref. 16.* © *Horizon House—Microwave, Inc.)*

antenna pattern of the base station is switched over to the next sector. The 20 channels per sector is an arbitrary number. Any number of channels can be used among the total 60 channels.

The synchronization and the clock accuracy have to be considered. This TDMA scheme does simplify the base station, and there is hopefully less channel interference in this system. However, there is a high risk of developing a spread spectrum system.

14.6.4 Evaluation of a digital cellular system[17,18]

The performance of a digital cellular system should be evaluated by a subjective test. The required C/I is deduced from the subjective test for a given performance. Also the channel bandwidth of a FDMA or the equivalent channel bandwidth of a TDMA should be included in the evaluation of the system's performance. Therefore a nationwide standard subjective test should be established to accomplish this task. In Chap. 15, many standard digital systems will be described.

References

1. J. J. Spilker, Jr., *Digital Communications by Satellite,* Prentice-Hall, 1977, Chap. 14.
2. K. Hirade and K. Murota, "A Study of Modulation for Digital Mobile Telephony," *29th IEEE Vehicular Technology Conference Record* (Arlington Heights, Illinois, March 27–30, 1979), pp. 13–19.

3. R. de Buda, "Coherent Demodulation of Frequency Shift Keying with Low Deviation Ratio," *IEEE Transactions on Communications,* Vol. COM-20, June 1972, pp. 466–470.
4. R. A. Comroe and D. J. Costello, Jr., "ARQ Schemes for Data Transmission in Mobile Radio Systems," *IEEE Transactions on Vehicular Technology,* Vol. VT-33, August 1984, pp. 88–97.
5. P. J. Mabey, "Mobile Radio Data Transmission-Coding for Error Control," *IEEE Transactions on Vehicular Technology,* Vol. VT-27, August 1978, pp. 99–109.
6. J. L. Flanagan, M. R. Schroeder, B. S. Atal, R. E. Crochiere, N. S. Jayant, and J. M. Tribolet, "Speech Coding," *IEEE Transactions on Communications,* Vol. COM-27, April 1979, pp. 710–737.
7. J. Makhoul, "Linear Prediction: A Tutorial Review," *Proceedings of the IEEE,* Vol. 63, April 1975, pp. 560–580.
8. J. Makhoul, S. Roucos, and H. Gish, "Vector Quantizations in Speech Coding," *Proceedings of the IEEE,* Vol. 33, November 1985, pp. 1551–1588.
9. T. C. Bartee, *Digital Communications,* Howard H. Sams & Co., 1986, Chap. 8.
10. J. M. Gilmer, "CVSD Intelligibility Testing," 1985 IEEE Milcom Conference, Boston, October 20–23, 1985, *Conference Record,* pp. 181–186.
11. J. Swerup and J. Uddenfeldt, "Digital Cellular," *Personal Communiations Technology,* May 1986, pp. 6–12.
12. P. A. Bello and B. D. Nelin, "Prediction Diversity Combining with Selectivity Fading Channels," *IEEE Transactions on Communications,* Vol. COM-10, March 1962, pp. 32–42 (correction in Vol. COM-10, December 1962, p. 466).
13. P. Monsen, "Theoretical and Measured Performance of a DFE Modem on a Fading Multipath Channel," *IEEE Transactions on Communications,* Vol. COM-25, October 1977, pp. 1144–1152.
14. G. R. Cooper and R. W. Nettleton, "A Spread Spectrum Technique for High Capacity Mobile Communication," *IEEE Transactions on Vehicular Technology,* Vol. 27, November 1978, pp. 264–275.
15. D. J. Goodman, P. S. Henry, and V. K. Prabhu, "Frequency-Hopped Multilevel FSK for Mobile Radio," *Bell System Technical Journal,* Vol. 59, September 1980, pp. 1257–1275.
16. M. Bohm, "Mobile Telephone for Everyone through Digital technology," *Telecommunications,* July 1985, pp. 68–72.
17. W. C. Y. Lee, "How to Evaluate Digital Cellular Systems," presented to the FCC, Washington, D.C., September 3, 1987.
18. W. C. Y. Lee, "How to Evaluate Digital Cellular Systems," *Telecommunications,* December 1987, p. 45.
19. J. Proakis, *Digital Communications,* McGraw-Hill, 1983.

15

Digital Cellular Systems

Many digital cellular and cordless phone systems have been developed. The cellular systems are GSM, NA-TDMA, CDMA, PDC, and 1800-DCS, and the cordless phone systems are DECT and CT-2 schemes. Although analog cellular systems are limited to using frequency-division multiple-access (FDMA) schemes, digital cellular systems can use FDMA, time-division multiple-access (TDMA), and code-division multiple-access (CDMA). When a multiple-access scheme is chosen for a particular system, all the functions, protocols, and network are associated with that scheme. This chapter covers GSM, NA-TDMA, and CDMA in great detail and briefly describes other systems.

15.1 Global System for Mobile (GSM)[1-5]

CEPT, a European group, began to develop the Global System for Mobile TDMA system in June 1982. GSM has two objectives: pan-European roaming, which offers compatibility throughout the European continent, and interaction with the integrated service digital network (ISDN), which offers the capability to extend the single-subscriber-line system to a multiservice system with various services which are currently offered only through diverse telecommunications networks.

System capacity was not an issue in the initial development of GSM, but due to the unexpected, rapid growth of cellular service, 35 revisions have been made to GSM since the first issued specification. The first commercial GSM system, called D2, was implemented in Germany in 1992.

15.1.1 GSM Architecture

GSM consists of many subsystems, such as the mobile station (MS), the base station subsystem (BSS), the network and switching subsystem (NSS), and the operation subsystem (OSS) (see Fig. 15.1).

The mobile station. The MS may be a stand-alone piece of equipment for certain services or support the connection of external terminals, such as the interface for a personal computer or fax. The MS includes mobile equipment (ME) and a subscriber identity module (SIM). ME does not need to be personally assigned to one subscriber. The SIM is a subscriber module which stores all the subscriber-related information. When a subscriber's SIM is inserted into the ME of an MS, that MS belongs to the subscriber, and the call is delivered to that MS. The ME is not associated with a called number—it is linked to the SIM. In this case, any ME can be used by a subscriber when the SIM is inserted in the ME.

Base station subsystem. The BSS connects to the MS through a radio interface and also connects to the NSS. The BSS consists of a base transceiver station (BTS) located at the antenna site and a base station controller (BSC) which may control several BTSs. The BTS consists of radio transmission and reception equipment similar to the ME in an MS. A transcoder/rate adaption unit (TRAU) carries out encoding and speech decoding and rate adaptation for transmitting data. As a subpart of the BTS, the TRAU may be sited away from the BTS, usually at the MSC. In this case, the low transmission rate of speech code channels allows more compressed transmission between the BTS and the TRAU, which is sited at the MSC.

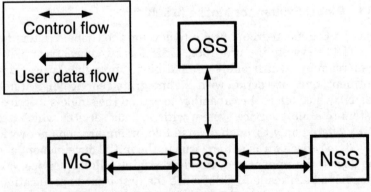

Figure 15.1 The external environment of the BSS.

GSM uses the open system interconnection (OSI). There are three common interfaces based on OSI (Fig. 15.2,): a common radio interface, called *air interface,* between the MS and BTS, an interface A between the MSC and BSC, and an A-bis interface between the BTS and BSC. With these common interfaces, the system operator can purchase the product of manufacturing company A to interface with the product of manufacturing company B. The difference between interface and protocol is that an interface represents the point of contact between two adjacent entities (equipment or systems) and a protocol provides information flows through the interface. For example, the GSM radio interface is the transit point for information flow pertaining to several protocols.

Network and switching subsystem. NSS (see Fig. 15.3) in GSM uses an intelligent network (IN). The IN's attributes will be described later. A signaling NSS includes the main switching functions of GSM. NSS manages the communication between GSM users and other telecommunications users. NSS management consists of:

Mobile service switching center (MSC). Coordinates call set-up to and from GSM users. An MSC controls several BSCs.

Interworking function (IWF). A gateway for MSC to interface with external networks for communication with users outside GSM, such

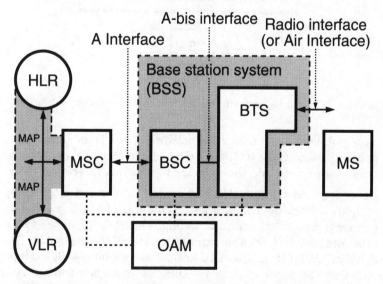

Figure 15.2 Functional architecture and principal interfaces.

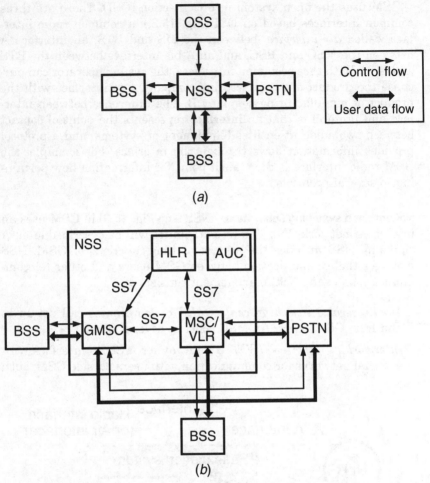

Figure 15.3 NSS and its environment. (*a*) The external environment; (*b*) the internal structure.

as packet-switched public data network (PSPDN) or circuit-switched public data network (CSPDN). The role of the IWF depends on the type of user data and the network to which it interfaces.

Home location register (HLR). Consists of a stand-alone computer without switching capabilities, a database which contains subscriber information, and information related to the subscriber's current location, but not the actual location of the subscriber. A subdivision of HLR is the authentication center (AUC). The AUC manages the security data for subscriber authentication. Another sub-division of HLR is the equipment identity register (EIR) which stores the data of mobile equipment (ME) or ME-related data.

Visitor location register (VLR). Links to one or more MSCs, tempo-rarily storing subscription data currently served by its correspond-ing MSC, and holding more detailed data than the HLR. For ex-ample, the VLR holds more current subscriber location information than the location information at the HLR.

Gateway MSC (GMSC). In order to set up a requested call, the call is initially routed to a gateway MSC, which finds the correct HLR by knowing the directory number of the GSM subscriber. The GMSC has an interface with the external network for gatewaying, and the network also operates the full Signaling System 7 (SS7) signaling between NSS machines.

Signaling transfer point (STP). Is an aspect of the NSS function as a stand-alone node or in the same equipment as the MSC. STP op-timizes the cost of the signaling transport among MSC/VLR, GMSC, and HLR.

As mentioned earlier, NSS uses an intelligent network. It separates the central data base (HLR) from the switches (MSC) and uses STP to transport signaling among MSC and HLR.

Operation subsystem. There are three areas of OSS, as shown in Fig. 15.4: (1) network operation and maintenance functions, (2) subscrip-tion management, including charging and billing, and (3) mobile

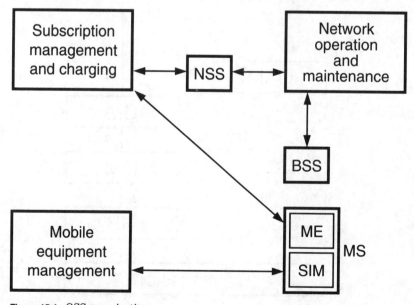

Figure 15.4 OSS organization.

equipment management. These tasks require interaction between some or all of the infrastructure equipment. OSS is implemented in any existing network.

15.1.2 Layer modeling (OSI model)

The Open System Interconnection (OSI) of GSM consists of five layers: transmission (TX), radio resource management (RR), mobility management (MM), communication management (CM), and operation, administration, and maintenance (OAM) (Fig. 15.5). The lower layers correspond to short-time-scale functions, the upper layers are long-time-scale functions.

The TX layer sets up a connection between MS and BTS. The RR layer refers to the protocol for management of the transmission over the radio interface and provides a stable link between the MS and BSC. The BSS performs most of the RR functions. The MM layer (1) manages the subscriber databases, including location data, and (2) manages authentication activities, SIM, HLR, and AUC. The NSS (mainly the MSC) is a significant element in the CM layer. The following functions are parts of the CM layer:

1. *Call control.* The CM layer sets up calls, maintains calls, and releases calls. The CM layer interacts among the MSC/VLR, GMSC,

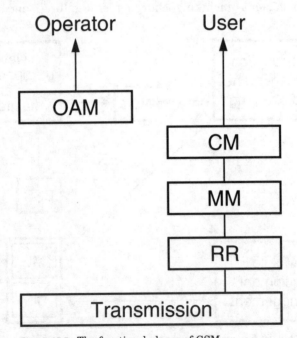

Figure 15.5 The functional planes of GSM.

IWF, and HLR for managing circuit-oriented service, including speech and circuit data.

2. *Supplementary services management.* Allows users to have some control of their calls in the network, and has specific variations from the basic service.

3. *Short message service (SMS).* Related to the point-to-point SMS. A SMS service center (SMS-SC) may connect to several GSM networks. Short message transmission requires setting up a signaling connection between the mobile station and the MSC. The two functions of SMS are

   a. Mobile-originating short message
   b. Mobile-terminating short message

OSS is an integral part of the OAM layer. All the subsystems, such as BSS and NSS, contribute to the OAM operation and maintenance functions.

15.1.3 Transmission

Speech. A 4-kHz analog speech signal converts to a 64-kbps digital signal, then down-converts to 13 kbps before modulation. Using a rate of 13 kbps instead of 64 kbps allows the 13-kbps data rate transmission to occur over a narrowband channel. Since the radio spectrum is a precious and limited resource, using less bandwidth per channel provides more channels within a given radio spectrum.

Digital speech uses:

1. *Regular pulse excitation (RPE).* Generates the impulse noise to simulate the nature of speech.

2. *Linear prediction coding (LPC).* Generates speech waveform by using a filter with eight transmitted coefficients with a speech frame of 20 ms; 260 bits represent a 20-ms speech frame. There are two modes of voice transmission in GSM, continuous (normal mode) and discontinuous.

The discontinuous transmission (DTX) mode decreases effective radio transmission encoding of speech at 13 kbps from a bit rate around 500 bps without speech. In active speech the frame is 260 bits in each 20 ms, and in inactive speech, the frame is 260 bits in 480 ms (24 times longer than normal mode).

A voice activity device (VAD) detects the DTX mode. In the voice protocol, a silence detection (SID) frame precedes the start of DTX. The speech coder provides an additional bit of information indicating whether the speech frame needs to be sent, depending on the VAD algorithm.

An SID starts at every inactivity period and repeats at least twice per second, as long as inactivity lasts. During the inactive speech period determined by VAD, and during every inactive period, artificial noise is generated at the receiver, substituting for background noise.

Data service.   The highest data rate is 9600 bps and has two different modes. A forward error correction mechanism is provided in the transparent (T) mode. In the nontransparent (NT) mode, information is repeated when it is not acknowledged by the other end, and may be called an automatic repeat request (ARQ). Three different users' data rates are employed in the T connection: 2400 bps, 4800 bps, and 9600 bps. After insertion of the auxiliary information bits, the intermediate rates bits become 3.6 kbps, 6 kbps, and 12 kbps, corresponding to the user's 2.4 kbps, 4.8 kbps, and 9.6 kbps, respectively.

The basic GSM data rate is also 12 kbps (6 kbps on the half-rate channel) in an NT connection, but the available throughput varies with the quality of basic transmission and the transmission delay. Generally, the NT mode has less transmission error but also less throughput. The NT mode may be considered for a packet data flow application. The user data stream is sliced into blocks of 200 bits, and, with addition of the redundancy and auxiliary information, the user data stream becomes 240 bits per block. These blocks are used in NT while the ARQ scheme is applied.

An adaptation function called interworking function (IWF) at the network side, and terminal adapting function (TAF) at the terminal, is used to accommodate variable transmission rates (Fig. 15.6). The radio link protocol (RLP) is used for transporting signaling messages between the TAF and IWF.

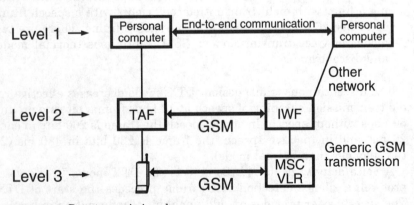

Figure 15.6 Data transmission planes.

Data can transmit over these planes, as shown in Fig. 15.6:

1. End-to-end transmission—direct transmission through hard wire
2. TAF to IWF transmission through subscriber units
3. GSM radio transmission through subscriber units; acts like a voice call in the air.

Although speech interconnection with the ISDN is not a problem, data transmission raises its own problems, as shown in Fig. 15.7. ISDN uses the capacity of a bidirectional 64 kbps/channel, but GSM must use the radio spectrum efficiently, through a bidirectional 13 kbps/channel. Interconnection of data services between GSM and ISDN is not possible without a rate-adapted (RA) box, as shown in Fig. 15.7.

Modulation.   Gaussian minimum-shift keying (GMSK), where $BT = 0.3$ is the normalized bandwidth of a gaussian filter, is the modulation scheme of GSM, where B is the baseband bandwidth, and $1/T$ is the transmission rate. $B = 1/T \times 0.3 = 270$ kbps $\times 0.3 = 81$ kHz. Minimum means the *minimum tone separation*. GMSK utilizes a small spectrum bandwidth to send a GSM carrier channel. The modulation rate of a GSM carrier channel is 270 kbps.

15.1.4   GSM channels and channel modes

Channel structure.   The services offered to users have four radio transmission modes, three data modes, and a speech mode. The radio transmission modes use the physical channels.

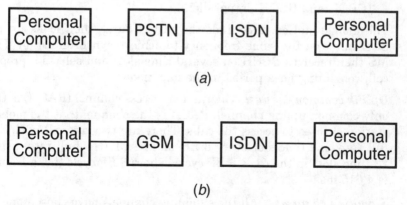

(a)

(b)

Figure 15.7   Interconnection with ISDN. (*a*) PSTN user to ISDN user; (*b*) GSM user to ISDN user.

Physical channels. There are three kinds of physical channels, also called traffic channels (TCHs):

1. *TCH / F (full rate).* Transmits a speech code of 13 kbps or three data-mode rates, 12, 6, and 3.6 kbps.

2. *TCH / H (half rate).* Transmits a speech code of 7 kbps or two data modes, 6 and 3.6 kbps.

3. *TCH / 8 (one-eighth rate).* Used for low-rate signaling channels, common channels, and data channels.

Logical Channels

Common channels. All the common channels are embedded in different traffic channels. They are grouped by the same cycle (51×8 BP), where BP stands for burst period (i.e. time slot), which is 577 μs.

Downlink common channels. There are five downlink unidirectional channels, shared or grouped by a TCH.

- Frequency correction channel (FCCH) repeats once every 51×8 BPs; used to identify a beacon frequency.
- Synchronization channel (SCH) follows each FCCH slot by 8 BPs.
- Broadcast control channel (BCCH) is broadcast regularly in each cell and received by all the mobile stations in the idle mode.
- Paging and access grant channel (PAGCH). Used for the incoming call received at the mobile station. The access grant channel is answered from the base station and allocates a channel during the access procedure of setting up a call.
- Call broadcast channel (CBCH). Each cell broadcasts a short message for 2 s from the network to the mobile station in idle mode. Half a downlink TCH/8 is used, and special CBCH design constraints exist because of the need for sending two channels (CBCH and BCCH) in parallel.

The mobile station (MS) finds the FCCH burst, then looks for an SCH burst on the same frequency to achieve synchronization. The MS then receives BCCH on several time slots and selects a proper cell, remaining for a period in the idle mode.

Uplink common channel. The random-access channel (RACH) is the only common uplink channel. RACH is the channel that the mobile station chooses to access the calls. There are two rates: RACH/F—full rate, one time slot every 8 BP, and RACH/H—half rate, using 23 time slots in the 51×8 BP cycle, where 8 BP cycle (i.e. a frame) is 4.615 ms.

Signaling channels. All the signaling channels have chosen one of the physical channels, and the logical channels names are based on their logical functions:

Slow associated control channel (SACCH). A slow-rate TCH used for signaling transport and used for nonurgent procedures, mainly handover decisions. It uses one-eighth rate. The TCH/F is always allocated with SACCH. This combined TCH and SACCH is denoted TACH/F. SACCH occupies 1 time slot (0.577 ms) in every 26 frames (4.615 ms × 26). The time organization of a TACH/F is shown in Fig. 15.8.

Fast associated control channel (FACCH). Indicates cell establishment, authenticates subscribers, or commands a handover.

Stand-alone dedicated control channel (SDCCH). Occasionally the connection between a mobile station and the network is used solely for passing signaling information and not for calls. This connection may be at the user's demand or for other management operations such as updating the unit's location. It operates at a very low rate and uses a TCH/8 channel.

Radio slots are allocated to users only when call penetration is needed. There are two modes, dedicated and idle. The mode used depends on the uplink and the downlink. In GSM terminology, the downlink is the signal transmitted from the base station to the mobile station, and the uplink is the signal transmitted in the opposite direction. (*Note:* The terrestrial communication terms *uplink* and *downlink* are not to be confused with the same terms used in satellite communications. In many instances the position of the mobile station can be higher than the base station antenna because of the terrain contour. Using the terms in this kind of situation may cause confusion, therefore it is more important for the reader to remember the *definitions* of these terms as used in terrestrial communications. This approach is analogous to using the term *handoff* instead of *handover* in discussions about European cellular telecommunications—as long as the definition itself is clear, the terms will be understood.)

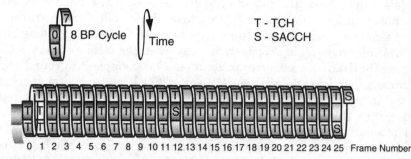

Figure 15.8 Time organization of a TACH/F.

Voice / data channels.   Each time slot of a voice channel contains 260 bits per block. The entire block contains 316 bits. Each time slot of a data channel contains 120 or 240 bits per block.

Channel modes.   Because of the precious value of the radio spectrum, individual users cannot have their own TCH at all times.

Dedicated mode. Uses TCH during call establishment and uses SACCH to perform location updating in the dedicated mode. TCH and SACCH are dedicated channels for both uplink and downlink channels.

Idle mode. During noncall activities, the five downlink channels are in the idle mode: FCCH; SCH; BCCH, which is broadcasting regularly; PAGCH and CBCH, which sends one message every 2 s. During idle mode, the mobile station listens to the common downlink channels, and also uses SDCCH to register a mobile location associated with a particular base station to the network.

15.1.5   Multiple-access scheme

General description.   GSM is a combination of FDMA and TDMA. The total number of channels in FDMA is 124, and each channel is 200 kHz. Both the 935–960 MHz uplink and 890–915 MHz downlink have been allocated 25 MHz, for a total of 50 MHz. Duplex separation is 45 MHz. If TDMA is used within a 200-kHz channel, 8 time slots are required to form a frame, frame duration is 4.615 ms, and the time slot duration burst period is 0.577 ms. There is a DCS-1800 system, which has the same architecture as the GSM, but it is upconverted to 1800 MHz. The downlink is 1805–1880 MHz (base TX) and the uplink is 1700–1785 MHz.

Constant time delay between uplink and downlink.   The numbering of the uplink slots is derived from the downlink slots by a delay of 3 time slots. This allows the slots of one channel to bear the same time slot number in both directions. In this case, the mobile station will not transmit and receive simultaneously because the two time slots are physically separated. Propagation delay when the mobile station is far from the BTS is a major consideration. For example, the round trip propagation delay between an MS and BTS which are 35 km apart is 233 μs. As a result, the assigned time slot numbers of the uplink and downlink channels may not be the same (less than 3 time slots apart). The solution is to let BTS compute a time advance value. The key is to allow significant guard time by taking into account that BCCH is using only even time slots. This avoids the uncertainty of numbering the wrong time slot. Once a dedicated connection is established, the BTS continuously measures the time offset between its own burst schedule and the reception schedule of mobile station bursts on the bidirectional SACCH channel. The time compensation for the propagation delay (sending to the mobile station via SACCH) is 3 time slots minus the time advance.

Frequency hopping. GSM has a slow frequency-hopping radio interface. The slow hopping is defined in bits per hop. Its regular rate is 217 hops/s, therefore, with a transmission rate of 270 kbps, the result is approximately 1200 bits/hop.

If the PAGCH and the RACH were hopping channels, then hopping sequences could be broadcast on the BCCH. The common channel is forbidden from hopping and using the same frequency.

Different types of time slots. Each cell provides a reference clock from which the time slots are defined. Each time slot is given a number (TN) which is known by the base station and the mobile station. The time slot numbering is cyclic. TN0 is a single set broadcast in any given call and repeated every 8 BPs for the confirmation of all common channels. The organization of TN0 (first of eight time slots) in sequence is as follows: FCCH (1), SCH (1), BCCH (4), PAGCH (4), FCCH (1), SCH (1), PAGCH (8), FCCH (1), SCH (1), PAGCH (8), FCCH (1), SCH (1), PAGCH (8), FCCH (1), SCH (1), PAGCH (8).

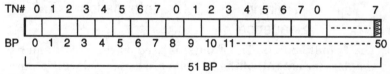

The symbol PAGCH (4) means that the PAGCH channel information appears in consecutive ones of every 8 BP cycle 4 times. Each of the remaining seven TNs (TN1 to TN7) is assigned to one TACH/F channel.

Bursts and training sequences. In TDMA, the signal transmits in bursts. The time interval of the burst brings the amplitude of a transmitted signal up from a starting value of 0 to its normal value. Then a packet of bits is transmitted by a modulated signal. Afterward, the amplitude decreases to zero. These bursts occur only at the mobile station transmission or at the base station if the adjacent burst is not transmitted.

There are tail bits and training sequence bits within a burst. The tail bits are three 0 bits added at the beginning and at the end of each burst which provide the guard time.

The training sequence is a sequence known by the receiver which trains an equalizer, a device which reduces intersymbol interference. The training sequence bits are inserted in the middle of a time slot sometimes called a *midamble,* for the same purpose as a preamble, so that the equalizer can minimize its maximum distance with any useful bit. There are eight different training sequences, with little between any two sequences to distinguish the received signal from the interference signal.

There are several kinds of bursts:

1. The normal burst used in TCH:

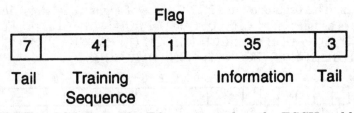

The 1-bit binary information indicating data or signaling is called the *stealing flag.*

2. The access burst used on the RACH in the uplink direction:

	Flag			
7	41	1	35	3
Tail	**Training Sequence**		**Information**	**Tail**

3. The F and S bursts. The F burst is used on the FCCH and has the simplest format. All of the 148 bits are zero, producing a pure sine wave. Five S bursts in each 51×8 BP cycle are used on the SCH. One S burst is shown below:

		Flag			**Flag**		
3	38	1	64		1	38	3
Tail	Information		Training Sequence			Information	Tail

15.1.6 Channel coding and interleaving

Channel coding.   Channel coding improves transmission quality when interference, multipath fading, and Doppler shift are encountered. As a result, the bit error rate and frame error rate or word error rate are reduced, but throughput is also reduced. Four kinds of channel codings are used in GSM:

1. Convolutional codes (L, k) are used to correct random errors: k is the input block bits, and L is the output block bits. Convolutional

codes have three different rates in GSM: (1) the one-half rate $(L / k = 2)$, (2) the one-third rate $(L / k = 3)$, and (3) the one-sixth rate $(L / k = 6)$.

2. Fire codes (L, k) are used as a block code to detect and correct a single burst of errors, where k is the information bits and L is the coded bits.

3. Parity check codes (L, k) are used for error detection. L is the bits of a block, k is the information bits, $L - k$ is the parity check bits.

4. Concatenation codes use convolutional code as an inner code and fire code as an outer code. Both the inner code and the outer code reduce the probability of error and correct most of the channel code. The advantage of using concatenation code is a reduction of the implementation complexity as compared with a single coding operation.

GSM's speech code is sent at a rate of 13 kbps, which represents 260 bits in each 20-ms speech block. After channel coding, each block contains 456 bits and the transmission rate is 22.8 kbps, or 114 bits for time slots. Adding the overhead bits such as tail bits (6), training bits (26), flag bits (2), and guard time bits (8.25), the total bits of a traffic channel is 156 bits in one time slot of 0.577 ms, as shown in Fig. 15.9.

Interleaving. Interleaving scrambles and/or spreads a sequence of bits prior to transmitting them. The sequence of bits is put back in order at the receiving end. Bursts of errors occur during transmission because of signal fading. After being received, these bursts of errors are then converted to random errors and put back in the correct sequence. Interleaving's major drawback is the corresponding delay at the receiving end.

Interleaving schemes are relatively simple in GSM. A code word of 456 bits could be spread into the following format:

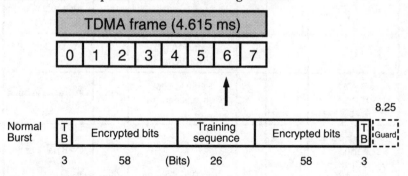

Figure 15.9 TDMA frame and normal burst.

1. Four full bursts—divide 456 bits into 4 parts, each one filling up a whole burst. This interleaving format takes 4.615 ms × 4 = 18.46 ms.

2. Eight half bursts—divide 456 bits into 8 parts, each one filling up half a burst. This interleaving format takes 4.615 ms × 8 = 36.92 ms. Four parts share with the previous and four parts with the new partial code word.

The interleaving and coding for different transmission modes are shown in Table 15.1. Interleaving is a powerful scheme which converts burst errors into random errors, and although it is very effective for data transmission, it is not effective for voice transmission. Voice

TABLE 15.1 Interleaving and Coding for the Different Transmission Modes
(*Reprinted from Ref. 3 p. 246*)

Channel and transmission mode		Input rate (kbit/s)	Input block (in bits)	Coding	Output block (in bits)	Interleaving
TCH/FS	Ia	13	50	Parity (3 bits) convolutional 1/2	456	On 8 half-bursts
	Ib		132	Convolutional 1/2		
	II		78	None		
TCH/F9.6 TCH/H4.8		12 6	240	Convolutional 1/2 punctured 1 bit out of 15	456	Complex, on 22 unequal burst portions
TCH/F4.8		6	120	Addition of 32 null bits Convolutional 1/3	456	Complex, on 22 unequal burst portions
TCH/F2.4		3.6	72	Convolutional 1/6	456	On 8 half bursts
TCH/H2.4		3.6	144	Convolutional 1/3	456	Complex, on 22 unequal burst portions
SCH			25	Parity (10 bits) Convolutional 1/2	78	On 1 S surst
RACH (+ Handover Access)			8	Parity (6 bits) Convolutional 1/2	36	On 1 access burst
Fast associated signalling on TCH/F and /H			184	Fire code 224/184	456	On 8 half bursts
TCH/8, SACCH; BCCH, PAGCH				Convolutional 1/2		On 4 full bursts

transmission operates in real time, and a long delay in response cannot be tolerated.

Without interleaving and overhead bits, the transmit rate for a speech channel is 22.8 kbps, 114 bits per time slot, and 456 bits per four time slots.

15.1.7 Radio resource (RR) management

In a mobile network, radio channels must allocate for call setup, handover and release, on a call bias. This management is additional to the conventional fixed network call handling procedures. There are three management functions; location, handover, and roaming. The implementation of the RR functions require some kind of protocol between the mobile station and the network.

Link protocol. We studied the means of transporting user information in previous sections. But in addition to the user's information, the signaling transfer information exchanges must be sent and understood by every piece of signaling transport equipment. Most information exchange functions are distributed to different kinds of equipment. There are three link protocols to provide information exchanges.

Radio link protocol (RLP), specified in GSM link access protocol over the radio link called LAPDm.

LAPD, the link access protocol (LAP) adapted from ISDN D channel.

Message transfer part (MTP), the protocols used for signaling transport on an SS7 network.

The radio link protocol's signaling message rate is 22.8 kbps. The signaling message rate on the other link protocol is 64 kbps.

Interfaces associated with link protocols

Interface	Link protocol
MS-BTS	LAPDm (GSM spec)
BTS-BSC	LAPD (adopted from ISDN)
BSC-MSC	MTP (SS7 protocol)
MSC/VLR/HLR—SS7 network	MTP (SS7 protocol)
MSC-MSC (call-related signaling)	TUP (telephone user part)
BSC-relay MSC (non-call-related signaling)	ISUP (ISDN user part)
	BSS MAP (MAP/B)
MSC-MSC (non-call-related signaling)	MAP (mobility application part)

Non-call-related signals correspond to protocols in the MSC that are different from those in other MSCs or other HLRs and are grouped together in the MAP. We can distinguish them by MAP/X, where X can be B, C, D, and so forth.

MAP/B   Protocol between BSC and relay MSC

MAP/C   Protocol between GMSC and an HLR

MAP/D   Protocol between another MSC/VLR and HLR

MAP/E   Protocol between MSCs

Figure 15.10 shows the relationships of MAP/X protocols.

15.1.8 Mobility management (MM)

The mobility of cellular system users requires mobility management for location updates, handovers, and roaming. A handover occurs when a voice channel changes as the mobile station enters another cell during a call. Roaming is the ability to initiate a call in one network system and deliver it to another network system by using MM and location update management.

Location update management. The subscription is always associated with its home public land mobile network (PLMN). The roaming customer is associated with visited PLMNs. We may identify whether the call is from PLMN or visited PLMN from the location of the MS.

In the PLMN selection process, the MM normally looks for cells only in the home (serving) PLMN. If no service is available, the user can choose either the automatic mode (the network searches) or the manual mode (the user searches) to search for the desired PLMN. In the limited-service case, the MM continuously monitors only the 30 strongest carriers. Limited service usually takes care of coverage at the border areas of a foreign country.

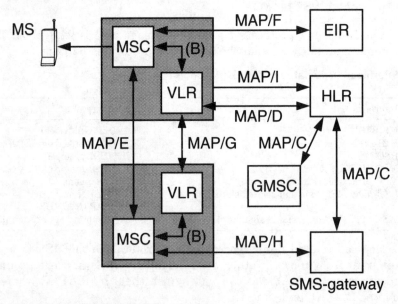

Figure 15.10 MAP/C to MAP/I protocols.

Cell selection. Choosing the best cell from an MS depends on three factors: (1) the level of the signal received by the mobile station, (2) the maximum transmission power of the mobile station, and (3) two parameters p_1 and p_2 specified by the cell. This is called the C_1 criterion.

$$C_1 = A - \max (B, O)$$

$$A = \text{received level average} - p_1$$

$$B = p_2 - \text{maximum RF power of the MS}$$

$$p_1 = \text{a value between } -110 \text{ and } -48 \text{ dBm}$$

$$p_2 = \text{a value between } 13 \text{ and } 43 \text{ dBm}$$

Both values of p_1 and p_2 are broadcast from the cells.

$$\text{MS maximum power} = 29 \text{ to } 43 \text{ dBm}$$

The cell selection algorithm is as follows:

- A SIM must be inserted.
- The strongest C_1 is chosen by obtaining C_1 from candidate cells; the C_1 has to be higher than 0.
- All cells must not be barred from service.

Authentication. Authentication protects the network against unauthorized access.

First Phase. A PIN (personal identification number) code protects the SIM. The PIN is checked by the SIM locally, so the SIM is not sent out over the radio link.

Second phase. The GSM network makes an inquiry by sending a random number (RAND). The 128-bit RAND is sent from the network to the MS, and mixes with the MS's secret parameter, K_i, in an A3 processing algorithm, which produces a 32-bit-long SRES (signed result) number. The SRES is then sent to the network from the MS for verification (see Fig. 15.11).

Encryption. Encryption protects against unauthorized listening. The MS uses the RAND received from the network and mixes K_i through a different algorithm, called A8, and generates K_c (64 bits). The ciphering sequences are generated from the K_c (see Fig. 15.12). The frame number and K_c move to a ciphering algorithm, A5, and generate

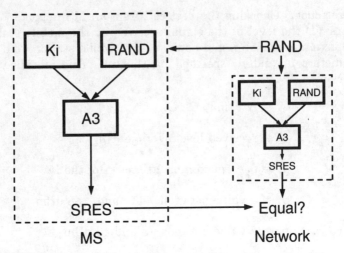

Figure 15.11 The authentication computation.

S_2 (114 bits), which plays an exclusive-or operation between the 114 bits of plain text and ciphering sequence S_2, as shown in Fig. 15.12.

User identity protection—security management. SIM (MS side) and AUC (network side) are the repositories of the subscriber's key K_i. Key K_i never transmits over the air. Both sides perform A3 and A8 computations.

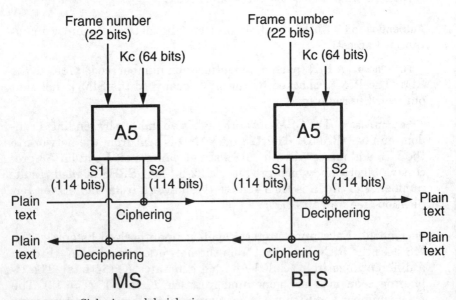

Figure 15.12 Ciphering and deciphering.

15.1.9   Communication management

The CM layer provides telecommunications services such as speech, fax, and data to users via RR and MM layers, as shown in Fig. 15.13. The users include GSM calling party, GSM called party, and the users in both GSM calling and called parties. The management functions of CM are call control, service management, and short message service.

Call control.   CC manages the most circuit-oriented services (speech, circuit data) through the MSC/VLR, GMSC, IWF, and HLR. CC functions set up calls (mobile-originating or base-originating), maintain calls, and release calls. To establish calls, the MS number has to be assigned. MS/ISDN is a mobile station ISDN number, part of the same numbering plan as ISDN numbers. Mobile station roaming number (MSRN) is the routing number, another number which can be a GSM subscriber or third party international mobile subscriber identity (IMSI) and provided by MS to access a foreign network. Figure 15.14 illustrates a domestic call through a GMSC. Figure 15.15 illustrates an international call.

Handover.   The GSM handover algorithm is not specified as a standard. It is a feature of mobile assistance handover (MAHO) and is carried out within the unit. The MS scans for another radio carrier under direction from a base station. It monitors those time slots which are not its own assigned time slots for receiving the signal. In this case, on the request of a base station, the signal strength of a specified

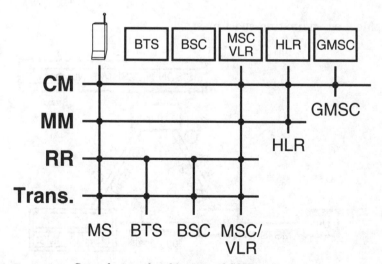

Figure 15.13   General protocol architecture of GSM.

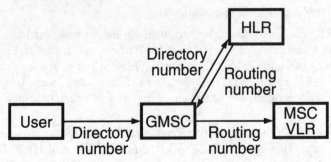

Figure 15.14 The key role of the GMSC for a domestic call.

radio carrier is measured in one time frame, and, on request, the measurements are forwarded to the base station to assist in the handover process. This is called MAHO. The MSC uses two sets of information to decide whether a handover should be initiated and which BTS is the candidate BTS for the handover. The two sets are (1) the signal strengths of the MS as received at the neighboring BTSs and (2) the signal strengths of neighboring BTSs received at the MS. The latter information is from MAHO.

Supplementary services management (SSM). CC provides supplementary services such as call waiting, call forwarding, and automatic answering. SSM is a point-to-point management service. An SSM service center (SSM-SC) may connect to several GSM networks. SSM consists of two functions:

1. Mobile terminating short message

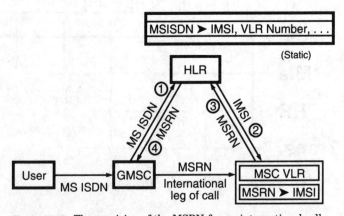

Figure 15.15 The provision of the MSRN for an international call.

2. Mobile originating short message

Short message services. CC provides point-to-point short message services (SMS-PP). GSM is connected to the short message service center. The signaling transmission uses digital audio tones [digital tone multifrequency (DTMF)] to control voice mailbox, answering machine, conferencing, etc.

15.1.10 Network management (NM)

An NM center oversees the following administration tasks:

1. Subscriber management—subscription administration
2. Billing and accounting
3. Maintenance
   a. Minimizing failures
   b. Monitoring operations and indicating by alarm improper operation situations
4. Subscriber administration tasks provide the selected approval code for ME (mobile equipment) within the international mobile equipment identity (IMEI) number. The code totals 15 digits and consists of type approval code (TAC) + final assembly code (FAC) + serial number which stores in EIR.
5. In the GSM telecommunication management network (TMN), all operation and maintenance machines compose a network which is linked to all traffic handling machines. The GSM Q3 is a network management protocol for operation systems functions and traffic handling machines. Two aspects of using GSM Q3 protocol are important. (a) Standardizing data communication protocols on the application level, such as file transfer. (b) Embodying the network modelling in GSM Q3 protocol.

15.1.11 Overview of GSM

Summary of physical layer parameters

TDMA structure	8 time slots per radio carrier
Time slot	0.577 ms
Frame interval	8 time slots = 4.615 ms
Radio carrier number	124 radio carriers (935–960 MHz downlink, 890–915 MHz uplink)
Modulation scheme	Gaussian minimum shift keying with $BT = 0.3$
Frequency Hopping	Slow frequency hopping (217 hops/s)
Equalizer	Equalization up to 16 µs time dispersion

GSM's strength. GSM is the first to apply the TDMA scheme developed for mobile radio systems. It has several distinguishing features:

1. Roaming in European countries

2. Connection to ISDN through RA box

3. Use of SIM cards

4. Control of transmission power

5. Frequency hopping

6. Discontinuous transmission

7. Mobile-assisted handover

15.2   North American TDMA[6-8]

15.2.1   History

North American TDMA (NA-TDMA) is a digital cellular system sometimes called American digital cellular (ADC) or digital AMPS (DAMPS), or North American digital cellular (NADC) or IS-54 system. This TDMA system was approved and design on it was started in 1987 by a group named TR45-3 after the industry debated between frequency-division multiple access and time-division multiple access. The reason those members voted for TDMA was the big influence of European GSM, which is the TDMA system. However, the requirements of designing a digital cellular system in Europe and in North America are different. In Europe, there is a virgin band (935–960 MHz downlink and 890–915 MHz uplink) for the digital cellular system. In North America, there is no new allocated band for the digital cellular system. The digital cellular system has to share the same allocated band with the analog system (AMPS, described in Chap. 3). Also, the digital and the analog systems have to be coexistent. In this circumstance, the low-risk approach is to use the same signal signature as the analog system, i.e., FDMA. Besides, because of the urgent need for large system capacity, the time for designing a new North American system had to be very short. The North American digital system was needed to be available in 1990, in only 3 years. To design a digital FDMA system would be a straightforward task. Since the analog system is a FDMA system, all the physical data gathered for the analog system in the past 20 years could be used for designing the FDMA digital system, and design time would be shortened. On the other hand, to design a TDMA digital system in the same band shared with an FDMA analog system, much more physical data would have to be developed and time would be needed to understand them. Without a good understanding of the limitation of coexistence between two different signal signatures, FDMA and TDMA, it would be very difficult to complete a digital system with good performance in a very short time. If GSM had taken 8 years to develop, NA-TDMA might also need as much time to be revised in order for it to be mature.

Because of the requirement of coexistence, a dual-mode mobile unit was decided on; i.e., the unit can work on both analog and digital systems. In a dual-mode mobile unit, the 21 call set-up channels for the analog system are available in the unit. Why not share the same call set-up channels (analog) for both the analog voice channels and digital voice channels? In this case, no additional spectrum is needed for the digital set-up channels.[9] The spectrum is saved for adding more digital voice. Furthermore, for the sake of speeding up the completion of North American digital systems, the call set-up channels of the digital system could be shared with the analog system to make the call processing the same between the two systems. Thus, the first phase of the NA-TDMA system could be completed earlier.

15.2.2 NA-TDMA architecture

The NA-TDMA architecture is similar to GSM architecture. The only difference is that in NA-TDMA, there is only one common interface, which is the radio interface as shown in Fig. 15.16. The NA-TDMA uses the intelligent network. All the components such as HLR, VLR, AUC, and EIR are the same as used in GSM (see Sec. 15.1.1). In developing the NA-TDMA system, there were two phases:

First phase—to commonly share the 21 set-up channels which are used for the analog system. The first-phase system is only for voice transmission. Both modes, AMPS and digital, are built in the same unit. The handoff procedure has to take care of the following four features:
1. AMPS cell to AMPS cell
2. TDMA cell to TDMA cell
3. AMPS cell to TDMA cell
4. TDMA cell to AMPS cell

Second phase—(1) generate new digital set-up channels (they were in the voice band) to access to TDMA voice channels so that a digital stand-alone unit can be provided and (2) specify a data-service signal protocol for transmitting data.

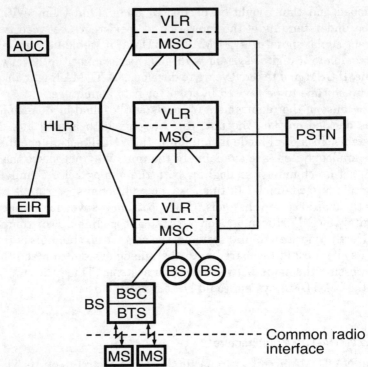

VLR: Visitor location registration
HLR: Home location registration
BS: Base station
AUC: Authentication center
EIR: Equipment identity register
BSC: Base station controller
BTS: Base transceiver station

Figure 15.16   NA-TDMA system architecture.

15.2.3   Transmission and modulation

TDMA structure (digital channels)   In NA-TDMA, the set-up channels are analog channels shared with the AMPS system. One digital channel (a 30 kHz TDMA channel) contains 25 frames per second. Each frame is 40 ms long and has 6 time slots. Each time slot is 6.66 ms long. One frame contains 1944 bits (972 symbols), as shown in Fig. 15.17.

Each slot contains 324 bits (162 symbols) and the duration between bits is 20.57 μs. Therefore, one radio channel is transmitted at 48.6 kbps but only 24,000 symbols per second over the radio path. Each frame consists of 6 time slots. The maximum effect on the signal for

1944 bits (972 symbols)

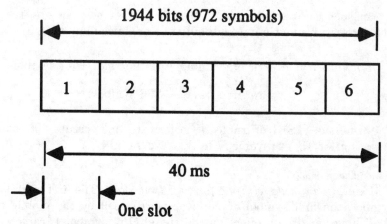

| 1 | 2 | 3 | 4 | 5 | 6 |

40 ms

One slot

Figure 15.17 TDMA frame structure.

a forward time slot is one-half full symbol period and for a reverse time slot is 6 symbol periods (Fig. 15.18).

Frame length. There are two frame lengths, full rate and half rate. Each full-rate traffic channel shall utilize two equally spaced time slots of the frame. The overall length in each slot is shown in Fig. 15.18.

Channel 1 uses time slots 1 and 4

Channel 2 uses time slots 2 and 5

Channel 3 uses time slots 3 and 6

Each half-rate traffic channel shall utilize one time slot of the frame:

Channel 1 uses time slot 1

Channel 2 uses time slot 2

Channel 3 uses time slot 3

Channel 4 uses time slot 4

Channel 5 uses time slot 5

Channel 6 uses time slot 6

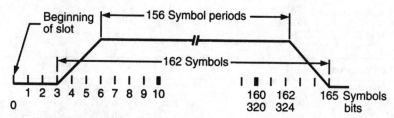

Figure 15.18 Overall length in each slot.

Frame offset. At the mobile station, the offset between the reverse and forward frame timing (without time advanced applied), is

$$\text{Forward frame} = \text{reverse frame} + (1 \text{ time slot} + 44 \text{ symbols})$$

$$= \text{reverse frame} + 206 \text{ symbols}$$

The time slot (TS) 1 of frame N (in forward link) occurs 206 symbol periods after TS 1 of frame N in the reverse link.

Modulation timing

Modulation timing within a forward time slot.   The first modulated symbol (the first symbol of the sync word) used by the mobile unit shall have maximum effect on the signal (156 symbols) transmitted from the base antenna, one-half symbol (1 bit) period after beginning the time slot.

Modulation timing within a reverse time slot.   The first modulated symbol has a maximum effect on the signal transmitted at the mobile unit 6 symbol periods after the beginning of the reverse time slot.

Power level.   In the AMPS system, there are eight power levels. In TDMA there are an additional three levels. Therefore, in total, TDMA has 11 power levels, as shown in Table 15.2. In the carrier-off condition, the output power of the transmitting antenna must fall to -60 dBm within 2 ms. In the carrier-on condition, the output power of the transmitting antenna must come to within 3 dB of the specified level.

TABLE 15.2   Mobile Station Nominal Power Levels

Mobile station power level (PL)	Mobile attenuation code (MAC)	Nominal effective radiated power, dBW, for mobile station power class							
		I	II	III	IV	V	VI	VII	VIII
0	000	6	2	-2	-2	•	•	•	•
1	001	2	2	-2	-2	•	•	•	•
2	010	-2	-2	-2	-2	•	•	•	•
3	011	-6	-6	-6	-6	•	•	•	•
4	100	-10	-10	-10	-10	•	•	•	•
5	101	-14	-14	-14	-14	•	•	•	•
6	110	-18	-18	-18	-18	•	•	•	•
7	111	-22	-22	-22	-22	•	•	•	•
Dual-mode only									
8					$-26 \pm 3 \text{ dB}$	•	•	•	•
9					$-30 \pm 6 \text{ dB}$	•	•	•	•
10					$-34 \pm 9 \text{ dB}$ ·	•	•	•	•

Speech coding (full rate). The NA-TDMA speech coding is a class of speech coding known as code excited linear predictive (CELP) coding. The code is called vector-sum excited linear predictive (VSELP) coding. It uses a codebook to vector-quantize the excitation (residual) signal such that the computation required for the codebook search process at the sender can be significantly reduced. The speech coder sampling rate is 7950 bps. Speech is broken into frames; each frame is 20 ms long and contains 160 symbols. Each frame is further divided into subframes 40 samples (5 ms) long. At the mobile station, the analog speech is converted to uniform pulse-code modulation (PCM) format. The speech coder is preceded by the following voice processing stages: (1) level adjustment, (2) bandpass filter, and (3) analog-to-digital conversion. The VSELP speech decoder is shown in Figure 15.19. The first part is the generation of pulse excitation and the second part is the speech waveform synthesis. Adding the two parts results in quality speech. All the values of parameters H, I, γ, β, L, $\alpha_1 \ldots \alpha_{10}$ for a 20-ms frame of speech are received with a low rate of transmission. Then those parameters are inserted into proper places in the speech decoder, and the speech is recovered at the receiving end.

The delays due to the air interface between the base station and the mobile station may exceed 100 ms, and echo control measures therefore necessary. In a half-rate speech coder, the speech frame of 20 msec may contain 80 symbols.

Modulation. NA-TDMA uses a constant envelope modulation with $\pi/4$-shifted differential quadrature phase shift keying (DQPSK). The modulation scheme uses the phase constellation shown in Fig. 15.20.

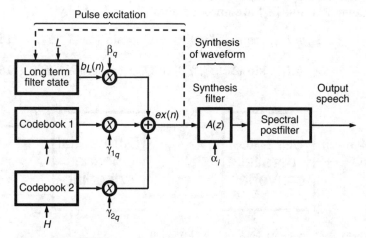

Figure 15.19 VSELP speech decoder.

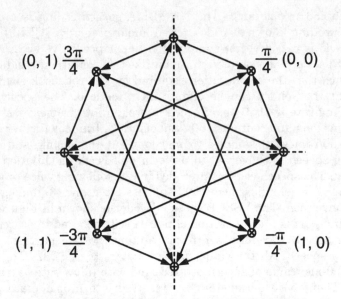

Figure 15.20 Phase constellation of a $\pi/4$-shifted DQPSK.

The rotation of $\pi/4$ occurs alternately at the odd state $\oplus$ and even state $\otimes$. The Gray code is used in the mapping. Every signal phase represents a di-bit symbol as shown in Fig. 15.20. Any two adjacent signal phases differ only in a single bit. The information is encoded differentially; symbols do not correspond to absolute phases, but to the phase difference between two adjacent symbols. A binary data stream b_m is separated into two streams: X_k is the even-numbered bit stream, and Y_k is the odd-numbered bit stream (Fig. 15.21). The streams $\{X_k\}$ and $\{Y_k\}$ are encoded onto $\{I_k\}$ and $\{Q_k\}$ by:

$$I_k = I_{k-1} \cos [\Delta\phi(X_k, Y_k)] - Q_{k-1} \sin [\Delta\phi(X_k, Y_k)] \quad (15.2\text{-}1)$$

$$Q_k = I_{k-1} \sin [\Delta\phi(X_k, Y_k)] + Q_{k-1} \cos [\Delta\phi(X_k, Y_k)] \quad (15.2\text{-}2)$$

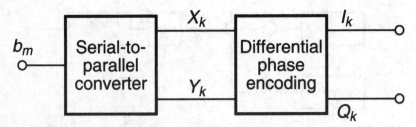

Figure 15.21 A binary data stream conversion.

where I_{k-1}, Q_{k-1} are the amplitudes at the previous pulse time. The phase change $\Delta\phi$ is determined by the following table:

X_k	Y_k	$\Delta\phi$ (even state)	$\Delta\phi$ (odd state)
1	1	$-3\pi/4$	π
0	1	$3\pi/4$	$\pi/2$
0	0	$\pi/4$	0
1	0	$-\pi/4$	$-\pi/2$

The signals I_k and Q_k at the output of the differential phase encoding device can take one of the four values 0, ± 1, $1/\sqrt{2}$.

The baseband filters. The baseband filter shall have (1) linear phase and (2) square-root-raised cosine frequency response as shown in Fig. 15.22 where T is the period that equals 41.1 μs. The QPSK modulation with two components I_k and Q_k is differentially encoded as shown in Fig. 15.21.

The transmitted signal. The resultant transmitted signal $S(t)$ is given by:

$$S(t) = \sum_n g(t - nT) \cos \phi_n \cdot \cos w_c t$$
$$+ \sum_n g(t - nt) \sin \phi_n(t) \sin w_c t \quad (15.2\text{-}3)$$

where $g(t)$ is the pulse shaping with a time response of $H(f)$:

$$g(t) = \frac{1}{2\pi} \int_{-\infty}^{\infty} H(f) e^{jw_c t} \, dt \quad (15.2\text{-}4)$$

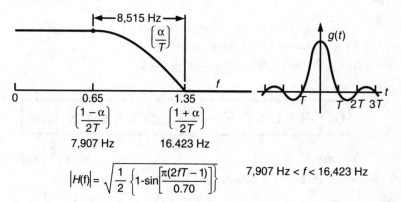

$$|H(f)| = \sqrt{\frac{1}{2} \left\{ 1 - \sin\left[\frac{\pi(2fT - 1)}{0.70} \right] \right\}} \qquad 7{,}907 \text{ Hz} < f < 16{,}423 \text{ Hz}$$

Figure 15.22 Baseband filter characteristics.

and w_c is the radian carrier frequency. ϕ_n is from the differential encoding

$$\phi_n = \phi_{n-1} + \Delta\phi_n \qquad (15.2\text{-}5)$$

15.2.4 Time alignment and limitation of emission

Time alignment. It is necessary to control the TDMA time slot burst (advancing or retarding) transmission from the mobile unit, so that it arrives at the base station receiver in the proper time relationship with respect to other time slot burst transmissions. An error in time alignment causes errors in two signals in the overlap at the head or tail of a time slot.

System access. The mobile station receives an initial traffic channel designation (ITCD) message (order code 01110) which is contained in word 2 (extended address word), then moves to a traffic channel. The mobile station first synchronizes to the forward traffic channel. The time alignment is sent by a physical layer control message over a shortened burst transmission. The mobile station, while operating on a digital traffic channel, is transmitting over a slot interval 324 bits long at certain times. The mobile station continues to transmit a shortened burst at the standard offset reference position until a time alignment message is received from the base station. The mobile station adjusts its transmission time during the next available slot.

Time alignment in handoff message. A mobile handoff message contains estimated time alignment information. Analog-to-digital and digital-to-digital handoff messages contain a shortened burst indicator (SBI) field:

SBI = 00   A handoff to a small-diameter cell
SBI = 01   A handoff from sector to sector
SBI = 10   A handoff to a large-diameter cell.

The shortest burst format

3 symb.   22 symb.

The shortened burst contains:

G1   3 symbol length guard time
R   3 symbol length ramp time

S   14 symbol length sync work

D   6 symbol length coded digital verification color code (CDVCC) (on reverse channel)

G2   22 symbol length guard time

The fields V, W, X, Y consist of

V = 4 zero bits (2 symbols)

W = 8 zero bits (4 symbols)

X = 12 zeros (6 symbols)

Y = 16 zeros (8 symbols)

In the shortened burst format, the symbol interval between any two sync words (total 6 sync words) is a unique interval. After detection of any two or more sync words, the timing alignment is determined at the base station.

Limitations on emissions from digital transmission   The total emission power is shown in Fig. 15.23. This limitation is for the suppression of the energy within the cellular band. In addition, the transmitter emissions in each 30-kHz band anywhere in the mobile station receive band must not exceed −80 dBm at the transmit antenna connector.

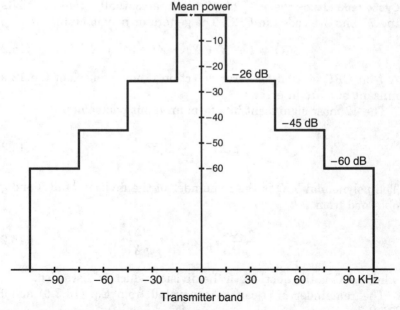

Figure 15.23 Suppression inside cellular band.

15.2.5   Error corrections

Speech data classes.   Channel error correcting for the speech code employs three mechanisms:

1. A rate one-half convolutional code to protect the more vulnerable bits of the speech code.
2. Interleaving the transmitted data for each speech coder frame over two time slots to reduce the burst error due to Rayleigh fading.
3. Use of a cyclic redundancy check (CRC). After the error correction is applied at the receiver, these CRC bits are checked to see if most perceptually significant bits were received properly.

The 159-bit speech coder frame is separated into two classes:

Class 1   77 bits

Class 2   82 bits

Class 1 bits are the important bits to which the convolutional coding is applied. Among the 77 bits, there are 12 most perceptually significant bits in which a 7-bit CRC is used for error detection purposes. Class 2 bits are unimportant bits and are transmitted without any error protection.

Cyclic redundancy check.   The 12 most perceptually significant bits of the 77 bits are coded in CRC. The generator polynomial is

$$g(X) = 1 + X + X^2 + X^4 + X^5 + X^7 \qquad (15.2\text{-}6)$$

A 7-bit CRC is used for error detection if one or more of the 12 significant bits are in error.

The 12 most significant bits form an input polynomial:

$$a(X) = \sum_{k=0}^{11} B_k X^k \qquad (15.2\text{-}7)$$

The polynomial $b(X)$ is the remainder of the division of $a(x)$ and $g(x)$ obtained from

$$\frac{a(X)X^7}{g(X)} = q(X) + \frac{b(X)}{g(X)} \qquad (15.2\text{-}8)$$

where $q(X)$ is the quotient of the division which is discarded.

The remainder $b(X)$ can be generated from Eq. (15.2-6) and Eq. (15.2-7)

$$b(X) = \sum_{k=1}^{7} C_k X^{k-1} \qquad (15.2-9)$$

The input 77 bits $B_1 \cdots B_{77}$ (including 12 significant bits) adding $C_1 \cdots C_7$ and 5 zero bits become 89 important bits.

0-3	4 - 80	81 - 82	84 - 88
4 bits	**77 bits**	**3 bits**	**5 bits (five 0's)**
CRC		**CRC**	**tail bits**

A cyclic redundancy check is performed at the receiving end. After decoding of the class 1 bits, the received CRC $b'(x)$ bits are checked to determine if any errors have been detected. The process of checking the error in CRC uses the received 12 most perceptually significant bits $a'(x)$ in each frame divided by the generator polynomial in Eq. (15.2-6):

$$\frac{a'(X) \cdot X^7}{g(X)} = q'(X) + \frac{b''(X)}{g(X)} \qquad (15.2-10)$$

The received CRC $b'(x)$ are compared with the CRC bits $b''(x)$ generated by Eq. (15.2-10); if $b'(x) \neq b''(x)$, an error has occurred. The causes of error are (1) the data was corrupted by channel errors or (2) an FACCH message was transmitted in place of the speech data. As a result, the speech quality is degraded. A bad frame masking strategy is taking place. There are six states. The state 0 means no error is detected. When each successive speech frame is found to be in error, the state machine moves to the next higher state. Moving to a higher state means more repeats. If two successive frames occur with no detected errors, the state machine is returned to state 0.

15.2.6 Interleaving and coding

Convolutional encoding. The 89 important bits are input to convolutional coder and 176 bits are at the output of the coder. Then adding 82 unimportant bits become 260 bits total for a 20 ms speech frame. Convolutional encoding uses a code rate of 1/2 and memory order 5. The five memory elements generate 32 states in this code. Since the code is 1/2 rate, two outputs alternately come out and are in a sequential order. CC0 is the convolutional code at one output and the other CC1 is the other output.

Interleaving and deinterleaving. The encoded speech after the convolutional code is interleaved over two time slots (Fig. 15.24). Each time slot contains two frames. The encoded speech is placed into a rectangular interleaving array columnwise. The two encoded speech frames are referred to as X and Y.

$$
\begin{array}{lll}
0X & 26X \cdots 234X \\[4pt]
1Y & 27Y & 235Y \\[4pt]
2X \\[4pt]
3Y \\[16pt]
24X & 50X & 258X \\[4pt]
25Y & 51Y & 259Y
\end{array}
$$

The speech code consists of 88 class 1 bits (after CRC coding) and 80 class 2 bits. The class 2 bits are intermixed with the convolutionally coded class 1 bits. The bits in the above array are then transmitted row-wise. The place of the coded class 1 bits and class 2 bits are in a certain mixed order.

Deinterleaving. At the receiving end, each time slot contains the interleaved data from two speech coder frames, X_1 and X_2, which are 20 ms apart. The received data are placed row-wise into a 26×10 deinterleaving array. Once the data from the two time slots are used to fill the deinterleaving array, all the data for frames X are available and can be decoded. After deinterleaving one entire speech coder frame is available.

Delay interval requirement. The mobile station and the base station shall have a delay interval compensation of up to one symbol length.

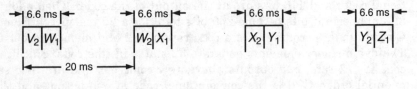

$X = X_1 + X_2$ encoded speech frame

Figure 15.24 Interleaving slots arrangement.

15.2.7 SCM and SID

Station class mark (SCM) must be stored in a mobile station. It formerly used 4 bits for identifying the maximum powers of three different kinds of mobile stations. Now SCM uses 5 bits and can identify eight different power levels.

Power class	Max. power, dBm	Min. power, dBm	Number of power levels	SCM	Transmission	SCM	Bandwidth	SCM
I	6	−22	0–7	0XX00	Continuous	XX0XX	20 MHz	X0XXX
II	2	−22	1–7	0XX01	Discontinuous	XX1XX	25 MHz	X1XXX
III	−2	−22	2–7	0XX10				
IV	−2	−34 ± 9 dB	2–10	0XX11				
V				1XX00				
VI				1XX01				
VII				1XX10				
VIII				1XX11				

Home system identification (SID) is a 15-bit system identification indicator

14 13	12 0
2 bits	System number

00   USA

01   Other countries

10   Canada

11   Mexico

The first two bits indicate the country of origin:

00   United States

01   Other countries

10   Canada

11   Mexico

15.2.8   NA-TDMA channels

In NA-TDMA, there are no common channels such as those used in GSM. The digital call set-up uses the 21 set-up channels which are shared with the analog system.

Supervision of the digital voice channel.   The supervision channels in NADC are similar to those in GSM:

- *Fast associated control channel.* FACCH is a blank and burst channel equivalent to a signaling channel for the transmission of control and supervision messages between the base station and the mobile station. It consists of 260 bits. Mostly FACCH is used for handoff messages.

- *Slot associated control channel.* SACCH is a signaling channel including twelve code bits present in every time slot transmitted over the traffic channel whether these contain voice or FACCH information.

Mobile-assisted handoffs.   The mobile station performs signal quality measurements on two types of channels:

1. Measures the RSSI (received signal strength indicator) and the BER (bit error rate) information of the current forward traffic channel during a call.

2. Measures the RSSI of any RF channel which is identified from the measurement order message from the base station.

MAHO consists of three messages:

1. Start measurement order
   - Measurement order message—sent from the base station to the mobile station.
   - Measurement order acknowledge message—sent from the mobile station to the base station.

2. Stop measurement order
   - Stop measurement order—sent from the base station to the mobile station.
   - Mobile acknowledge—sent from the mobile station to the base station.

3. Channel quality message (mobile to base only)

The mobile transmits the signal quality information over either the SACCH or FACCH. In the case of discontinuous transmission (DTX):

- Whenever the mobile is in the DTX high state, the mobile transmits channel quality information over the SACCH

- When the mobile is in the DTX low state, the mobile transmits the channel quality information over the FACCH

Handoff action. When a handoff order is received, the mobile station is at DTX high state and stays at that state. If the mobile station is at DTX low state it must enter the DTX high state and wait for 200 ms before taking the handoff action. Handoff to a digital traffic channel is described as follows:

1. Turn on signaling tone for 50 ms, turn off signaling tone, turn off transmitter which was operating on the old frequency.

2. Adjust power, tune to new channel, set stored DVCCs to the DVCC field of the received message.

3. Set the transmitter and receiver to digital mode, set the transmit and receive rate based on the message-type field.

4. Set time slot based on the message-type field.

5. Set the time alignment offset to the value based on the TA field.

6. Once the transmitter is synchronized, enter the conversation task of the digital traffic channel.

15.2.9 Discontinuous transmission on a digital traffic channel

In DTX, certain mobile stations can switch autonomously between two transmitter power-level states: DTX high and DTX low. In the DTX high state, the power level of transmitter at the mobile station is indicated by the most recent power-controlling order. In this state, the CDVCC (coded digital verification color code) is sent at all times. CDVCC is used to distinguish the current traffic channel from traffic co-channels. Decoding CDVCC (12, 8) becomes $DVCC_r$. $DVCC_r$ is received and checked with DVCCs for identification. There are 255 codes (2^8, but 0 is not used). In the DTX low state, the transmitter remains off and CDVCC is not sent except for the transmission of FACCH messages. All the SACCH messages will be sent as a FACCH message. After sending all the messages, the transmitter will return to the off state.

15.2.10 Authentication

A secret number PIN (personal identification number) is assigned to each subscriber. The mobile station, on receipt of a random challenge global action message, updates its internal stored RAND variable used as input to the authentication algorithm AUTH1. The mobile uses its PIN, its ESN and the MIN to compute a response to RAND in accordance with AUTH1. The mobile then responds to the page by transmitting its MIN, the output of AUTH1, RANDC (random confirmation), and COUNT (call history parameter). See also Sec. 15.34 for similarity.

15.2.11 Signaling format

A reverse digital traffic channel (RDTC) is used to transport user information and signaling. Two control channels are used: the FACCH is a blank and burst channel, the SACCH is a continuous channel, and interleaving is on the SACCH. The signaling formats of these two channels are shown in Fig. 15.25.

Interleaving on the FACCH bits from 0 to 259 is

Row number	FACCH bits interleaving									
0	215	256	223	258	230	219	257	227	259	189
1	0	25	50	75	231	89	114	139	164	190
2	1	26	51	76	232	90	115	140	165	191
⋮	⋮									
14	13	38	63	88	244	102	127	152	177	203
15	14	39	64	216	245	103	128	153	178	204
⋮	⋮									
25	24	49	74	229	255	113	138	163	118	214

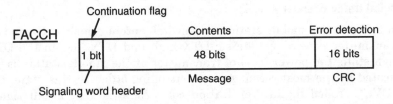

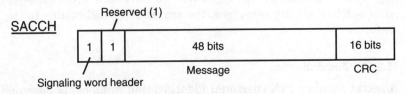

Figure 15.25 Signaling formats of FACCH and SACCH.

The Y row (odd) of this frame combines alternately with the X row (even) of the previous frame to form a FACCH block. In SACCH code, the output of the convolutional coder is diagonally interleaved as the 12 coded bits are transmitted over 12 time slots.

Message structure.   All messages contain:

1. An application message header
2. Mandatory fixed parameters
3. Mandatory variable parameters
4. Remaining length
5. Optional variable parameters

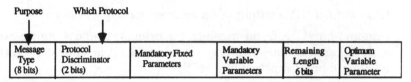

A forward digital traffic channel (FDTC) has same format as the RDTC (reverse digital traffic channel).

15.2.12   Word format

The same word format is used for FACCH and SACCH.

15.3   CDMA[10-12]

CDMA development started in early 1989 after the NA-TDMA standard (IS-54) was established. A CDMA demonstration to test its feasibility for digital cellular systems was held in November 1989. The CDMA "Mobile Station-Base Station Compatibility Standard for Dual Mode Wideband Spread Spectrum Cellular System," was issued as IS-95 (PN-3118, Dec. 9, 1992). CDMA uses the idea of tolerating interference by spread-spectrum modulation. The power control scheme in a CDMA system is a requirement for digital cellular application. However, it was a challenging task and has been solved. Before describing the structure of the system, we list the key terms of CDMA systems.

15.3.1   Terms of CDMA systems

Active set: The set of pilots associated with the CDMA channels containing forward traffic channels assigned to a particular mobile station (MS).

CDMA channel number   An 11-bit number corresponding to the center of the CDMA frequency assignment.

Code channel   A subchannel of a forward CDMA channel. A forward CDMA channel contains 64 code channels. Certain code channels are assigned to different logic channels.

Code channel zero: Pilot channel.

Code channels 1 through 7: Either paging channels or traffic channels.

Code channel 32: A sync channel or a traffic channel.

The remaining code channels are traffic channels.

Code symbol   The output of an error-correcting encoder.

Dim-and-burst   A frame in which the primary traffic is multiplexed with either secondary traffic or signal traffic. It is equivalent to the blank-and-burst function in AMPS.

Forward CDMA channel   Contains one or more code channels.

Frame   A basic timing interval in the system. For the access channel, paging channel, and traffic channel, a frame is 20 ms long. For the sync channel, a frame is 26.666 ms long.

Frame offset   A time skewing of traffic channel frames from system time in integer multiples of 1.25 ms. The maximum frame offset is 18.75 ms.

GPS (Global Position System)   System used for providing location and time information to the CDMA system.

Handoff (HO)   The act of transferring communication with a mobile station from one base station to another.

Hard HO   Occurs when (1) the MS is transferred between disjoint active sets, (2) the CDMA frequency assignment changes, (3) the frame offset changes, and (4) the MS is directed from a CDMA traffic channel to an analog voice channel.

Soft HO   HO from CDMA cell to CDMA cell at the same CDMA frequency.

Idle HO   Occurs when the paging channel is transferred from one base station (BS) to another.

Layering   A method of organization for communication protocols. A layer is defined in terms of its communication protocol to a peer layer.

Layer 1: Physical layer presents a frame by the multiplex sublayer and transforms it into an over-the-air waveform.

Layer 2: Provides for the correct transmission and reception of signaling messages.

Layer 3: Provides the control of the cellular telephone system. The signaling messages orginate and terminate at layer 3.

Long code   A PN (pseudonoise) sequence with period 2^{42}-1 using a tapped n-bit shift register.

Modulation symbol The output of the data modulator before spreading. There are 64 modulation symbols on the reverse traffic channel, 64-ary orthogonal modulation is used, and six code symbols are associated with one modulation symbol. On the forward traffic channel, each code symbol (data rate is 9600 bps) or each repeated code symbol (data rate is less than 9600 bps) is 1 modulation symbol.

Reverse: $\overbrace{110101}^{53}$ $\overbrace{101110}^{48}$ $\longrightarrow$
            (Walsh function 53) (Walsh function 48)
            64 bits   64 bits

            6 code symbols $\longrightarrow$ 1 modulation symbol

Forward   1 code symbol = 1 modulation symbol

Multiplex option   The ability of the multiplex sublayer and lower layers to be tailored to provide special capabilities. A multiplex option defines the frame format and the rate decision rules.

Multiplex sublayer   One of the conceptual layers of the system that multiplexes and demultiplexes primary traffic, secondary traffic, and signaling traffic.

Nonslotted mode   An operating mode of an MS in which the MS continuously monitors the paging channel.

Null traffic data   A frame of sixteen 1's followed by eight 0's sent at the 1200 bps rate. Null traffic channel data serve to maintain the connectivity between MS and BS when no service is active and no signaling message is being sent.

Paging channel   A code channel in a forward CDMA channel used for transmission of (1) control information and (2) pages from BS to MS. The paging channel slot has a 200-ms interval.

Power control bit   A bit sent in every 1.25 ms intervals on the forward traffic channel to the MS that increases or decreases its transmit power.

Primary CDMA channel   A preassigned frequency used by the mobile station for initial acquisition.

Primary paging channel   The default code channel (code channel 1) assigned for paging.

Primary traffic   The main traffic stream between MS and BS on the traffic channel.

Reverse traffic channel   Used to transport user and signaling traffic from a single MS to one or more BSs.

Shared secret data (SSD)   A 128-bit pattern stored in the MS.

> SSD is a concatenation of two 64-bit subsets.
>
> SSD-A is used to support the authentication
>
> SSD-B serves as one of the inputs to generate the encryption mask and private long code.

Secondary CDMA channel   A preassigned frequency (one of two) used by the mobile station for initial acquisition.

Secondary traffic   An additional traffic stream carried between the MS and the BS on the traffic channel.

Slotted mode   An operation mode of MS in which the MS monitors only selected slots on the paging channel.

Sync channel   Code channel 32 in the forward CDMA channel which transports the synchronization message to the MS.

Pilot channel   An unmodulated, direct-sequence (DS) signal transmitted continuously by each CDMA BS. The pilot channel allows a mobile station to acquire the timing of the forward CDMA channel, provides a phase reference for coherent demodulation, and provides a means for signal strength comparisons between base stations for determining when to hand off.

System time   The time reference used by the system. System time is synchronous to universal time coordination (UTC) time and uses the same time origin as GPS time. All BSs use the same system time. MSs use the same system time, offset by the propagation delay from the BS to the MS.

Time reference   A reference established by the MS that is synchronous with the earliest arriving multipath component which is used for demodulation. The time reference establishes transmit time and the location of zero in PN space.

Walsh chip   The shortest identifiable component of a 64-walsh function. On the forward CDMA channel, one chip equals 1/1.2288 MHz or 813.802 ns. On the reverse CDMA channel, one chip equals 4/1.2288 MHz or 3255 ns.

15.3.2 Output power limits and control

Output power. The mean output power of the mobile station shall be less than -50 dBm/1.23 MHz (-111 dBm/Hz) for all frequencies within ± 615 kHz of the center frequency.

Gated output power. The MS shall transmit at nominal controlled levels during gated-on periods. A typical output power in a gated-on period is shown in Figure 15.26. The transmitter noise floor should be less than -60 dBm/1.23 MHz.

Controlled output power. Implementing CDMA power control is a must in the cellular CDMA system for the reverse link transmission in order to eliminate the near-far interference. If all the mobile transmitters

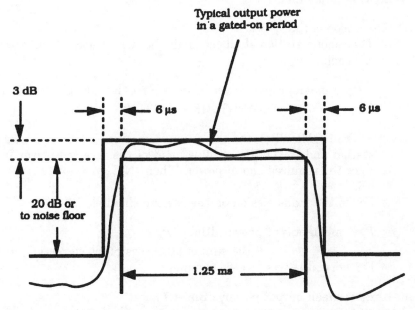

Figure 15.26 Transmission envelope mask (single gated-on power control group).

within a cell site's area of coverage are so controlled, then the total signal power received at the cell site from all mobiles will be equal to the nominal received power times the number of mobiles.

CDMA reverse-link open-loop power control. The mobile station receives a signal suffering both the log-normal and Rayleigh fadings from the forward link, as shown in Fig. 15.27a. The average path loss is obtained as shown in the figure. If the transmitting and receiving ends are sharing the same frequency channel, then reversing the received signal strength as shown in Fig. 15.27b, indicated as the transmit power without smoothing filter, would eliminate the power variation at the cell site. Since CDMA uses duplexing channels, the Rayleigh fading on the forward channel and the reverse channel are not the same. Therefore, the desired average transmit power is sent back on the reverse channel.

At the cell site, the available information on instantaneous value versus the expected value of frame error rate (FER) of the received signal is examined to determine whether to command a particular mobile to increase or decrease its transmit power. This mechanism is called CDMA closed-loop power control. The mobile power received at a cell site after close-loop control is shown in Fig. 15.27c.

In transmission mode, the MS has two independent means for output power adjustment:

1. Open-loop output power
   - The mobile station shall transmit the first probe on the access channel:

   $$\overline{P_A} = \text{mean output power, dBm} = -\text{mean input power, dBm}$$
   $$- 73 + \text{NOMPWR, dB} + \text{INITPWR, dB}$$

   where NOMPWR = the correction of received power at the base station and INITPWP = adjustment of the received power less than the required signal power. When INITPWR = 0, $\overline{P_A}$ = ±6 dB.
   - For initial transmission on the reverse channel,

   $$\overline{P_I} = \text{mean output power, dBm} = \overline{P_A}$$
   $$+ \text{the sum of all access probe correction, dBm}$$
   - For normal reverse traffic channel,

   $$\overline{P_R} = \text{mean output power, dBm} = \overline{P_I}$$
   $$+ \text{the sum of all closed-loop power control correction, dB}$$

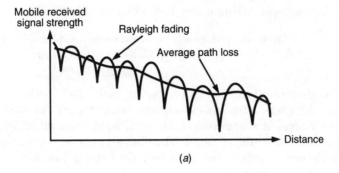

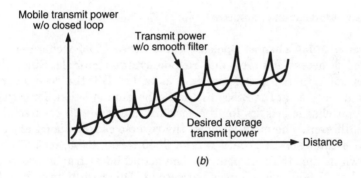

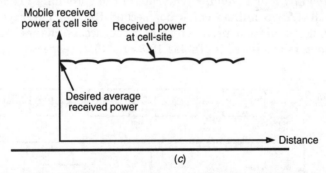

Figure 15.27 Power control mechanism. (*a*) Mobile received signal strength in log-normal shadowing and Rayleigh fading; (*b*) transmit power without closed-loop control and without nonlinear filtering; (*c*) mobile power received at cell site.

- For example, without any correction or adjustment,

$$\text{Mean output power} = - \text{ mean input power} - 73$$
$$= - (-90 \text{ dBm}) - 73$$
$$= + 17 \text{ dBm}$$

2. Closed-loop output power (involving both the mobile station and base station). The mobile station shall adjust its mean output power level in response to each valid power control bit received on the forward traffic channel. The change in mean output power per single power control bit shall be 1 dB nominal, within ±0.5 dB of the nominal change.

15.3.3 Modulation characteristics

Reverse CDMA channel signals. The reverse CDMA channel is composed of access channels and reverse traffic channels. Since the MS does not establish a system time as at the BS, the reverse channel signal received at the BS cannot use coherent detection. Thus the modulation characteristics for the forward channel and reverse channel are different. The modulation of the reverse channel is 64-ary orthogonal modulation at a data rate of 9600, 4800, 2400, or 1200 bps, as shown in Fig. 15.28 at point A. The actual burst transmission rate is fixed at 28,800 code symbols per second. This results in a fixed Walsh chip rate of 307.2 thousand chips per second (kcps). Each Walsh chip is spread by four PN chips. The rate of the spreading PN sequence is fixed at 1.2288 million chips per second (Mcps). The reverse traffic channel modulation parameters and the access channel modulation parameters are listed in Tables 15.3 and 15.4, respectively.

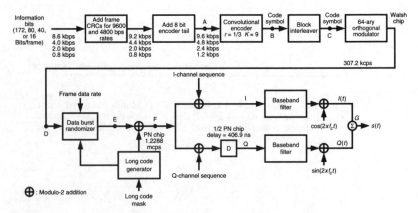

Figure 15.28 Reverse CDMA channel modulation process.

TABLE 15.3   Reverse Traffic Channel Modulation Parameters

Parameter	Data rate, bps				Units
	9600	4800	2400 .	1200	
PN chip rate	1.2288	1.2288	1.2288	1.2288	Mcps
Code rate	⅓	⅓	⅓	⅓	bits/code symbol
Transmit duty cycle	100.0	50.0	25.0	12.5	%
Code symbol rate	28,800	28,800	28,800	28,800	sps
Modulation	6	6	6	6	code symbol/ mod symbol
Modulation symbol rate	4800	4800	4800	4800	sps
Walsh chip rate	307.20	307.20	307.20	307.20	kcps
Mod. symbol duration	208.33	208.33	208.33	208.33	μs
PN chips/code symbol	42.67	42.67	42.67	42.67	PN chip/code symbol
PN chips/mod. symbol	256	256	256	256	PN chip/mod symbol
PN chips/Walsh chip	4	4	4	4	PN chips/Walsh chip

Convolutional encoding. At point B in Fig. 15.28, with a $K = 9$ (9 register) and rate 1/3 convolutional encoder:

1. On the access channel, each code symbol has a fixed data rate of 4800 bps, and each symbol repeats one time consecutively.

2. On the reverse traffic channel, the full data rate is 9600 kbps. For the data rate of 4800 kbps, each symbol repeats one time consecutively. For the data rate of 2400 kbps, each symbol repeats three times consecutively. For the data rate of 1200 kbps, each symbol repeats seven times consecutively.

TABLE 15.4   Access Channel Modulation Parameters

Parameter	Data rate, bps	Units
	4800	
PN chip rate	1.2288	Mcps
Code rate	⅓	bits/code symbol
Code symbol repetition	2	symbols/code symbol
Transmit duty cycle	100.0	%
Code symbol rate	28,800	sps
Modulation	6	code sym/mod. symbol
Modulation symbol rate	4800	sps
Walsh chip rate	307.20	kcps
Mod. symbol duration	208.33	μs
PN chips/code symbol	42.67	PN chip/code symbol
PN chips/mod. symbol	256	PN chip/mod. symbol
PN chips/Walsh chip	4	PN chips/Walsh chip

Interleaving. At point C in Fig. 15.28, the interleaving algorithm will form an array with 32 rows and 18 columns. At 9600 kbps, the interleaver forms a 32×18 matrix as in Table 15.5.

At 9600 bps, the transmission sequence is to send row by row in a sequence order up to row 32. At 4800 bps, the transmission sequence is to send by the unique order of rows as follows:

Row Number →

1 3 2 4 5 7 6 8 9 11 10 12 13 15 14 16 17 19 18 20 21 23 22 24 25 27 26 28 29 31 30 32

Expressed in a formula, the transmission sequence is

$$J, J + 2, J + 1, J + 3$$

for $J = 1 + 4i$ and $i = 0, 1, 2, 3, \ldots, (32/4 - 1)$.

At 2400 bps, the transmission sequence is by a unique order of rows as follows:

$$J, J + 4, J + 1, J + 5, J + 2, J + 6, J + 3, J + 7$$

for $J = 1 + 8i$ and $i = 0, 1, 2 \ldots, (32/8 - 1)$.

At 1200 bps,

$$J, J + 8, J + 1, J + 9, J + 2, J + 10, J + 3, J + 11, J + 4,$$
$$J + 12, J + 5, J + 13, J + 6, J + 14, J + 7, J + 15$$

for $J = 1 + 16i$ and $i = 1, 2$.

For access channel code symbols, the interleaver rows follow this order:

TABLE 15.5 Interleaving Algorithm

Column Row	1	2	3	4	5	6	7	8	9	10	11	12	13	14	15	16	17	18
1	1	33	65	97	129	161	193	225	257	289	321	353	385	417	449	481	513	545
2	2																	
3	3																	
4	4																	
5	5																	
6	6																	
⋮																		
32	32	64	96	128	160	192	224	256	288	320	352	384	416	448	480	512	544	576

$$J, J + 16, J + 8, J + 24, J + 4, J + 20, J + 12, J + 28, J + 2,$$
$$J + 18, J + 10, J + 26, J + 6, J + 22, J + 14, J + 30$$

for $J = 1, 2$.

Orthogonal modulation for reverse channel. As at point D in Fig. 15.28, the 64-ary Walsh codes consist of 64 codes each 64 bits long. They are orthogonal to each other as shown in Table 15.6. Every sixth symbol interpretating each Walsh code of 64 chips is sent out. For example,

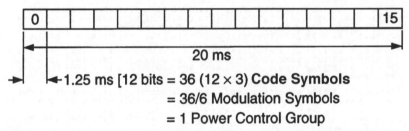

Each 20-ms reverse traffic channel frame shall be divided into 16 equal-length (i.e., 1.25 ms) power control groups numbered from 0 to 15.

```
┌─┬─┬─┬─┬─┬─┬─┬─┬─┬─┬─┬─┬─┬─┬─┬──┐
│0│ │ │ │ │ │ │ │ │ │ │ │ │ │ │15│
└─┴─┴─┴─┴─┴─┴─┴─┴─┴─┴─┴─┴─┴─┴─┴──┘
```

◄────────────── 20 ms ──────────────►

►◄ ◄─ 1.25 ms [12 bits = 36 (12 × 3) Code Symbols
= 36/6 Modulation Symbols
= 1 Power Control Group

The reverse traffic channel and the access channel shall be direct-sequence spread by the long code prior to transmission. The long code shall be periodic with period $2^{42} - 1$ chips and shall satisfy the linear recursion specified by the polynomial

$$p(x) = x^{42} + x^{35} + x^{33} + x^{31} + x^{27} + x^{26} + x^{25} + x^{22} + x^{21}$$
$$+ x^{19} + x^{18} + x^{17} + x^{16} + x^{10} + x^7 + x^6 + x^5 + x^3 + x^2 + x^1 + 1$$

Each PN chip of the long code shall be generated by a 42-shift-register long-code generator.

Data burst randomizing. At point E in Fig. 15.28, the data burst randomizer has generated a masking pattern of 0s and 1s that randomly masks out the redundant data generated by the code repetition. The mask pattern is determined by the data rate of the frame and by a block of 14 bits taken from the long code. These 14 bits shall be the last 14 bits of the long code used for spreading.

Direct sequence spreading. At point F in Fig. 15.28, prior to transmission, the reverse traffic channel and the access channel are direct-sequence spread by the long code. This spreading operation involves modulo-2 addition of the data burst randomizer output stream and

TABLE 15.6   64-ary Orthogonal Symbol Set

```
                11 1111 1111 2222 2222 2233 3333 3333 4444 4444 4455 5555 5555 6666
      0123 4567 8901 2345 6789 0123 4567 8901 2345 6789 0123 4567 8901 2345 6789 0123

 0    0000 0000 0000 0000 0000 0000 0000 0000 0000 0000 0000 0000 0000 0000 0000 0000
 1    0101 0101 0101 0101 0101 0101 0101 0101 0101 0101 0101 0101 0101 0101 0101 0101
 2    0011 0011 0011 0011 0011 0011 0011 0011 0011 0011 0011 0011 0011 0011 0011 0011
 3    0110 0110 0110 0110 0110 0110 0110 0110 0110 0110 0110 0110 0110 0110 0110 0110

 4    0000 1111 0000 1111 0000 1111 0000 1111 0000 1111 0000 1111 0000 1111 0000 1111
 5    0101 1010 0101 1010 0101 1010 0101 1010 0101 1010 0101 1010 0101 1010 0101 1010
 6    0011 1100 0011 1100 0011 1100 0011 1100 0011 1100 0011 1100 0011 1100 0011 1100
 7    0110 1001 0110 1001 0110 1001 0110 1001 0110 1001 0110 1001 0110 1001 0110 1001

 8    0000 0000 1111 1111 0000 0000 1111 1111 0000 0000 1111 1111 0000 0000 1111 1111
 9    0101 0101 1010 1010 0101 0101 1010 1010 0101 0101 1010 1010 0101 0101 1010 1010
10    0011 0011 1100 1100 0011 0011 1100 1100 0011 0011 1100 1100 0011 0011 1100 1100
11    0110 0110 1001 1001 0110 0110 1001 1001 0110 0110 1001 1001 0110 0110 1001 1001

12    0000 1111 1111 0000 0000 1111 1111 0000 0000 1111 1111 0000 0000 1111 1111 0000
13    0101 1010 1010 0101 0101 1010 1010 0101 0101 1010 1010 0101 0101 1010 1010 0101
14    0011 1100 1100 0011 0011 1100 1100 0011 0011 1100 1100 0011 0011 1100 1100 0011
15    0110 1001 1001 0110 0110 1001 1001 0110 0110 1001 1001 0110 0110 1001 1001 0110

16    0000 0000 0000 0000 1111 1111 1111 1111 0000 0000 0000 0000 1111 1111 1111 1111
17    0101 0101 0101 0101 1010 1010 1010 1010 0101 0101 0101 0101 1010 1010 1010 1010
18    0011 0011 0011 0011 1100 1100 1100 1100 0011 0011 0011 0011 1100 1100 1100 1100
19    0110 0110 0110 0110 1001 1001 1001 1001 0110 0110 0110 0110 1001 1001 1001 1001

20    0000 1111 0000 1111 1111 0000 1111 0000 0000 1111 0000 1111 1111 0000 1111 0000
21    0101 1010 0101 1010 1010 0101 1010 0101 0101 1010 0101 1010 1010 0101 1010 0101
22    0011 1100 0011 1100 1100 0011 1100 0011 0011 1100 0011 1100 1100 0011 1100 0011
23    0110 1001 0110 1001 1001 0110 1001 0110 0110 1001 0110 1001 1001 0110 1001 0110

24    0000 0000 1111 1111 1111 1111 0000 0000 0000 0000 1111 1111 1111 1111 0000 0000
25    0101 0101 1010 1010 1010 1010 0101 0101 0101 0101 1010 1010 1010 1010 0101 0101
26    0011 0011 1100 1100 1100 1100 0011 0011 0011 0011 1100 1100 1100 1100 0011 0011
27    0110 0110 1001 1001 1001 1001 0110 0110 0110 0110 1001 1001 1001 1001 0110 0110

28    0000 1111 1111 0000 1111 0000 0000 1111 0000 1111 1111 0000 1111 0000 0000 1111
29    0101 1010 1010 0101 1010 0101 0101 1010 0101 1010 1010 0101 1010 0101 0101 1010
30    0011 1100 1100 0011 1100 0011 0011 1100 0011 1100 1100 0011 1100 0011 0011 1100
31    0110 1001 1001 0110 1001 0110 0110 1001 0110 1001 1001 0110 1001 0110 0110 1001

32    0000 0000 0000 0000 0000 0000 0000 0000 1111 1111 1111 1111 1111 1111 1111 1111
33    0101 0101 0101 0101 0101 0101 0101 0101 1010 1010 1010 1010 1010 1010 1010 1010
34    0011 0011 0011 0011 0011 0011 0011 0011 1100 1100 1100 1100 1100 1100 1100 1100
35    0110 0110 0110 0110 0110 0110 0110 0110 1001 1001 1001 1001 1001 1001 1001 1001

36    0000 1111 0000 1111 0000 1111 0000 1111 1111 0000 1111 0000 1111 0000 1111 0000
37    0101 1010 0101 1010 0101 1010 0101 1010 1010 0101 1010 0101 1010 0101 1010 0101
38    0011 1100 0011 1100 0011 1100 0011 1100 1100 0011 1100 0011 1100 0011 1100 0011
39    0110 1001 0110 1001 0110 1001 0110 1001 1001 0110 1001 0110 1001 0110 1001 0110

40    0000 0000 1111 1111 0000 0000 1111 1111 1111 1111 0000 0000 1111 1111 0000 0000
41    0101 0101 1010 1010 0101 0101 1010 1010 1010 1010 0101 0101 1010 1010 0101 0101
42    0011 0011 1100 1100 0011 0011 1100 1100 1100 1100 0011 0011 1100 1100 0011 0011
43    0110 0110 1001 1001 0110 0110 1001 1001 1001 1001 0110 0110 1001 1001 0110 0110

44    0000 1111 1111 0000 0000 1111 1111 0000 1111 0000 0000 1111 1111 0000 0000 1111
45    0101 1010 1010 0101 0101 1010 1010 0101 1010 0101 0101 1010 1010 0101 0101 1010
46    0011 1100 1100 0011 0011 1100 1100 0011 1100 0011 0011 1100 1100 0011 0011 1100
47    0110 1001 1001 0110 0110 1001 1001 0110 1001 0110 0110 1001 1001 0110 0110 1001

48    0000 0000 0000 0000 1111 1111 1111 1111 1111 1111 1111 1111 0000 0000 0000 0000
49    0101 0101 0101 0101 1010 1010 1010 1010 1010 1010 1010 1010 0101 0101 0101 0101
50    0011 0011 0011 0011 1100 1100 1100 1100 1100 1100 1100 1100 0011 0011 0011 0011
51    0110 0110 0110 0110 1001 1001 1001 1001 1001 1001 1001 1001 0110 0110 0110 0110

52    0000 1111 0000 1111 1111 0000 1111 0000 1111 0000 1111 0000 0000 1111 0000 1111
53    0101 1010 0101 1010 1010 0101 1010 0101 1010 0101 1010 0101 0101 1010 0101 1010
54    0011 1100 0011 1100 1100 0011 1100 0011 1100 0011 1100 0011 0011 1100 0011 1100
55    0110 1001 0110 1001 1001 0110 1001 0110 1001 0110 1001 0110 0110 1001 0110 1001

56    0000 0000 1111 1111 1111 1111 0000 0000 1111 1111 0000 0000 0000 0000 1111 1111
57    0101 0101 1010 1010 1010 1010 0101 0101 1010 1010 0101 0101 0101 0101 1010 1010
58    0011 0011 1100 1100 1100 1100 0011 0011 1100 1100 0011 0011 0011 0011 1100 1100
59    0110 0110 1001 1001 1001 1001 0110 0110 1001 1001 0110 0110 0110 0110 1001 1001

60    0000 1111 1111 0000 1111 0000 0000 1111 1111 0000 0000 1111 0000 1111 1111 0000
61    0101 1010 1010 0101 1010 0101 0101 1010 1010 0101 0101 1010 0101 1010 1010 0101
62    0011 1100 1100 0011 1100 0011 0011 1100 1100 0011 0011 1100 0011 1100 1100 0011
63    0110 1001 1001 0110 1001 0110 0110 1001 1001 0110 0110 1001 0110 1001 1001 0110
```

the long code. This long code shall be periodic with period $2^{42} - 1$ chips.

Quadrature spreading. The sequences used for the spread in quadrature are shown in Fig. 15.28 at point F. These sequences are periodic with period 2^{15} chips, and the spread polynomials of channel I and Q pilot PN sequences are

$$P_I(x) = x^{15} + x^{13} + x^9 + x^8 + x^7 + x^5 + 1$$

$$P_Q(x) = x^{15} + x^{12} + x^{11} + x^{10} + x^6 + x^5 + x^4 + x^3 + 1$$

which are of period $2^{15} - 1$. The pilot PN sequence repeats every 26.66 ms ($2^{15}/1228800$ s). There are exactly 75 repetitions in every 2 s. Reverse CDMA channel I and Q mapping for an offset QPSK modulation is shown in Fig. 15.29.

Access channel and reverse traffic channel

1. Access channel
   - Time alignment—an access channel frame shall begin only when system time is an integral multiple of 20 ms.
   - Modulation rate—a fixed rate of 4800 bps.
   - The reverse CDMA channel may contain up to 32 access channel numbers, 0 through 31, per supported paging channel (Fig. 15.30a). Each access channel is associated with a single paging channel on the corresponding forward CDMA channel (Fig. 15.30b). The forward CDMA channel structure will be described later.
   - Frame structure:

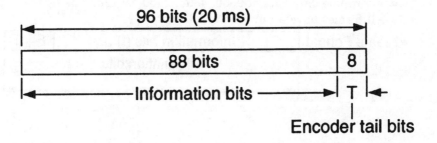

I	Q	Phase
0	0	$\pi/4$
1	0	$3\pi/4$
1	1	$-3\pi/4$
0	1	$-\pi/4$

(a)

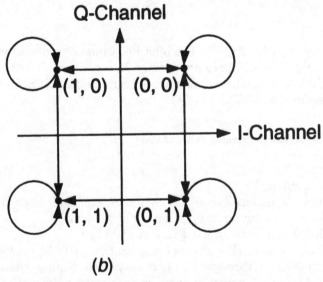

(b)

Figure 15.29 Reverse CDMA channel quadrature spreading. (*a*) Reverse CDMA channel I and Q mapping; (*b*) offset QPSK constellation and phase transition.

2. Reverse traffic channel
   - A variable data rate of 9600, 4800, 2400, or 1200 bps
   - All frames have a duration of 20 ms

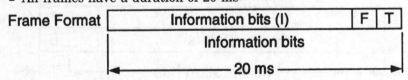

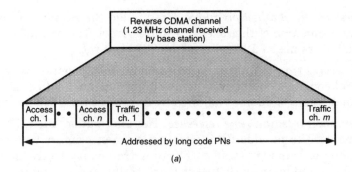

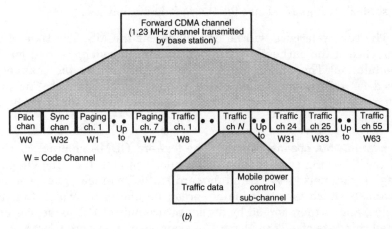

Figure 15.30 CDMA channel structure. (*a*) Example of logical reverse CDMA channels received at a base station; (*b*) example of a forward CDMA channel transmitted by a base station.

Information bits (I)	172 bits (for 9600 bps)
	80 bits (for 4800 bps)
	40 bits (for 2400 bps)
	16 bits (for 1200 bps)
Frame quality indicator (F)	12 bits (for 9600 bps)
(detect errors by CRC)	8 bits (for 4800 bps)
	0 bits (for 2400 bps)
	0 bits (for 1200 bps)
Tail bits (T)	8 bits for all data rates

where the generator polynomials for frame quality indicators are

$$g(x) = x^{12} + x^{11} + x^{10} + x^9 + x^8 + x^4 + x + 1 \qquad \text{(for 9600 bps)}$$

$$g(x) = x^8 + x^7 + x^4 + x^3 + x + 1 \qquad \text{(for 4800 bps)}$$

Reverse traffic channel preamble. Used to aid the BS in performing initial acquisition of the reverse traffic channel. The preamble shall consist of frames of 192 zeros at the 9600-bps rate.

Null reverse traffic channel. Used when no service option is active. It is a keep-alive operation. The null traffic channel data shall consist of a frame of 16 ones followed by 8 zeros at the 1200-bps rate.

Information bits and time reference. The information bits (172 bits) can be used to provide for the transmission of primary traffic and signaling or secondary traffic. Signaling traffic may be transmitted via blank-and-burst with the primary traffic and signaling traffic sharing the frame. Five different information bit structures described in Fig. 15.31 are for the mobile station use.

The time reference will be established at the MS. The time of occurrence of the earliest arriving multipath component is used for demodulation. The time reference from the forward traffic channel is used for the transmit time of the reverse traffic channel. The time reference from the paging channel is used for the transmit time of the access.

Forward CDMA channel signals. The forward CDMA channel consists of the following code channels: the pilot channel, the sync channel, paging channels (1 to 7), and forward traffic channels. They are code channels. Each is orthogonally spread by one of 64 Walsh function codes, and is then spread by a quadrature pair of PN sequences at a fixed chip rate of 1.2288 Mcps. The example of a forward CDMA channel transmitted by a BS is shown in Fig. 15.32. Each traffic channel consists of traffic data and mobile power control subchannels.

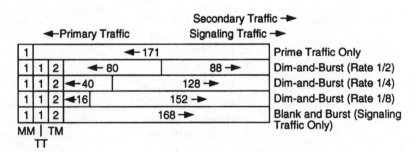

MM - mixed mode bit

TT - traffic type bit

TM - traffic mode bits

Figure 15.31 Information bits for primary traffic and secondary traffic.

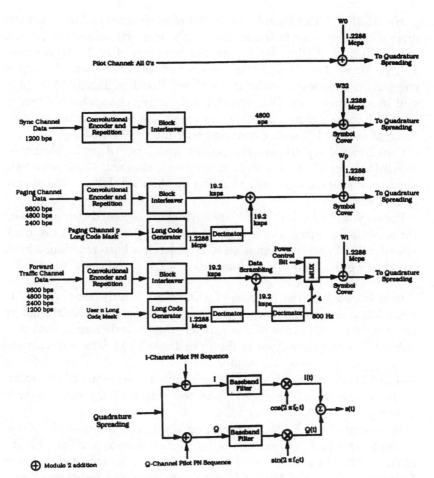

Figure 15.32 Forward CDMA channel structure. (*a*) Modulation; (*b*) quadrature spreading.

Forward CDMA channel structure. The structures of pilot channels, sync channel, paging channel, and forward traffic channel data are shown in Fig. 15.32. There are two parts, modulation and quadrature spreading. In the modulation part, the data rate is at the input.

Data rates at the input

1. The pilot channel sends all 0s at 19.2 kbps rate.
2. The sync channel operates at a fixed rate of 1200 bps.
3. The paging channel supports the fixed data rate at 9600, 4800, or 2400 bps.
4. The forward traffic channel supports variable data rate operation at 9600, 4800, 2400, or 1200 bps.

Modulation.   The modulation of the pilot channel has not used the error-correction prior to transmission. The channel takes each bit and spreads it into a 64-bit Walsh code. The data rate of 19.2 kbps becomes 1.2288 Mcps. The modulation parameters of sync channel, paging channel, and forward traffic channel are listed in Tables 15.7, 15.8, and 15.9, respectively. The sync channel, paging channel, and forward traffic channel are encoded prior to transmission. The rate of convolutional code is 1/2 with constraint length of 9 (9 registers).

Code symbol repetition.   For paging and forward traffic channels, repetition depends on the data rate of each channel. A low data rate needs more repeats in order to make up the modulation symbol rate of 19.2 kbps.

For a sync channel, each encoded symbol is repeated two times and the modulation symbol rate is 4800 sps. The 4800-sps data are modulated with Walsh function code W32 which has been multiplied by 4. In other words, each symbol becomes $4 \times 64 = 256$ cps.

Block interleaving.   The purpose of using block interleaving is to try to avoid burst errors while sending the data through a multipath fading environment. The input of the sync channel interleaver is shown in Table 15.10 and the output is shown in Table 15.11, where the arrows in each column show the sequential order of the data flow. For the forward traffic channel and paging channel, the input of the interleaver is shown in Table 15.12 and the output of the interleaver is shown in Table 15.13.

Data scrambling.   Data scrambling shall be accomplished by performing modulo-2 addition of the interleaver output symbol with the binary value of the long-code PN chip ($2^{42} - 1$); the long-code mask is for privacy. Also, the long-code data rate after passing through two decimators is reduced to 800 Hz, which is used for multiplexer (MUX) timing control. The circuit is shown in Fig. 15.33.

TABLE 15.7   Sync Channel Modulation Parameters

	Data rate, bps	
Parameter	1200	Units
PN chip rate	1.2288	Mcps
Code rate	½	bits/code symbol
Code repetition	2	mod. symbol/code symbol*
Modulation symbol rate	4800	sps
PN chips/modulation symbol	256	PN chips/mod. symbol
PN chips/bit	1024	PN chips/bit

*Each repetition of a code symbol is a modulation symbol.

TABLE 15.8 Paging Channel Modulation Parameters

Parameter	Data rate, bps			Units
	9600	4800	2400	
PN chip rate	1.2288	1.2288	1.2288	Mcps
Code rate	½	½	½	bits/code symbol
Code repetition	1	2	4	mod. symbol/ code symbol*
Modulation symbol rate	19,200	19,200	19,200	sps
PN chips/modulation symbol	64	64	64	PN chips/mod. symbol
PN chips/bit	128	256	512	PN chips/bit

*Each repetition of a code symbol is a modulation symbol.

Power control subchannel. At the rate of one bit every 1.25 ms (i.e., 800 bps), a 0 bit indicator is sent to the MS to increase the mean output power level or a 1 bit is sent to decrease it. There are 16 possible starting positions. Each position corresponds to one of the first 16 modulation symbols. Figure 15.34 indicates the randomization of power control bit positions. The reverse traffic channel sends a bit with 6 Walsh symbols in 1.25 ms. The base station measures signal strength, converts the measured signal strength to a power control bit, and transmit with a 4-bit binary number (levels 0 to 15) by scrambling bits 23, 22, 21, and 20. In Fig. 15.34, the value of bits 23, 22, 21, and 20 is 1011 binary (11 decimal). The power control bit starting position is the eleventh position within 1.25 ms of the seventh slot.

Orthogonal spreading. In the forward channel, each code channel transmits one of 64 Walsh functions at a fixed chip rate of 1.2288 Mcps to provide orthogonal channelization among all code channels on a given forward CDMA channel.

TABLE 15.9 Forward Traffic Channel Modulation Parameters

Parameter	Data rate, bps				Units
	9600	4800	2400	1200	
PN chip rate	1.2288	1.2288	1.2288	1.2288	Mcps
Code rate	½	½	½	½	bits/code symbol
Code repetition	1	2	4	8	mod symbol/code symbol*
Modulation symbol rate	19,200	19,200	19,200	19,200	sps
PN chips/modulation symbol	64	64	64	64	PN chips/mod. symbol
PN chips/bit	128	256	512	1024	PN chips/bit

*Each repetition of a code symbol is a modulation symbol.

TABLE 15.10   Sync Channel Interleaver Input (Array Write Operation)

1	9	17	25	33	41	49	57
1	9	17	25	33	41	49	57
2	10	18	26	34	42	50	58
2	10	18	26	34	42	50	58
3	11	19	27	35	43	51	59
3	11	19	27	35	43	51	59
4	12	20	28	36	44	52	60
4	12	20	28	36	44	52	60
5	13	21	29	37	45	53	61
5	13	21	29	37	45	53	61
6	14	22	30	38	46	54	62
6	14	22	30	38	46	54	62
7	15	23	31	39	47	55	63
7	15	23	31	39	47	55	63
8	16	24	32	40	48	56	64
8	16	24	32	40	48	56	64

TABLE 15.11   Sync Channel Interleaver Output (Array Read Operation)

1	3	2	4	1	3	2	4
33	35	34	36	33	35	34	36
17	19	18	20	17	19	18	20
49	51	50	52	49	51	50	52
9	11	10	12	9	11	10	12
41	43	42	44	41	43	42	44
25	27	26	28	25	27	26	28
57	59	58	60	57	59	58	60
5	7	6	8	5	7	6	8
37	39	38	40	37	39	38	40
21	23	22	24	21	23	22	24
53	55	54	56	53	55	54	56
13	15	14	16	13	15	14	16
45	47	46	48	45	47	46	48
29	31	30	32	29	31	30	32
61	63	62	64	61	63	62	64

TABLE 15.12 Forward Traffic and Paging Channel Interleaver Input (Array Write Operation at 9600 bps)

1	25	49	73	97	121	145	169	193	217	241	265	289	313	337	361
2	26	50	74	98	122	146	170	194	218	242	266	290	314	338	362
3	27	51	75	99	123	147	171	195	219	243	267	291	315	339	363
4	28	52	76	100	124	148	172	196	220	244	268	292	316	340	364
5	29	53	77	101	125	149	173	197	221	245	269	293	317	341	365
6	30	54	78	102	126	150	174	198	222	246	270	294	318	342	366
7	31	55	79	103	127	151	175	199	223	247	271	295	319	343	367
8	32	56	80	104	128	152	176	200	224	248	272	296	320	344	368
9	33	57	81	105	129	153	177	201	225	249	273	297	321	345	369
10	34	58	82	106	130	154	178	202	226	250	274	298	322	346	370
11	35	59	83	107	131	155	179	203	227	251	275	299	323	347	371
12	36	60	84	108	132	156	180	204	228	252	276	300	324	348	372
13	37	61	85	109	133	157	181	205	229	253	277	301	325	349	373
14	38	62	86	110	134	158	182	206	230	254	278	302	326	350	374
15	39	63	87	111	135	159	183	207	231	255	279	303	327	351	375
16	40	64	88	112	136	160	184	208	232	256	280	304	328	352	376
17	41	65	89	113	137	161	185	209	233	257	281	305	329	353	377
18	42	66	90	114	138	162	186	210	234	258	282	306	330	354	378
19	43	67	91	115	139	163	187	211	235	259	283	307	331	355	379
20	44	68	92	116	140	164	188	212	236	260	284	308	332	356	380
21	45	69	93	117	141	165	189	213	237	261	285	309	333	357	381
22	46	70	94	118	142	166	190	214	238	262	286	310	334	358	382
23	47	71	95	119	143	167	191	215	239	263	287	311	335	359	383
24	48	72	96	120	144	168	192	216	240	264	288	312	336	360	384

TABLE 15.13 Forward Traffic and Paging Channel Interleaver Output (Array Read Operation at 9600 bps)

1	9	5	13	3	11	7	15	2	10	6	14	4	12	8	16
65	73	69	77	67	75	71	79	66	74	70	78	68	76	72	80
129	137	133	141	131	139	135	143	130	138	134	142	132	140	136	144
193	201	197	205	195	203	199	207	194	202	198	206	196	204	200	208
257	265	261	269	259	267	263	271	258	266	262	270	260	268	264	272
321	329	325	333	323	331	327	335	322	330	326	334	324	332	328	336
33	41	37	45	35	43	39	47	34	42	38	46	36	44	40	48
97	105	101	109	99	107	103	111	98	106	102	110	100	108	104	112
161	169	165	173	163	171	167	175	162	170	166	174	164	172	168	176
225	233	229	237	227	235	231	239	226	234	230	238	228	236	232	240
289	297	293	301	291	299	295	303	290	298	294	302	292	300	296	304
353	361	357	365	355	363	359	367	354	362	358	366	356	364	360	368
17	25	21	29	19	27	23	31	18	26	22	30	20	28	24	32
81	89	85	93	83	91	87	95	82	90	86	94	84	92	88	96
145	153	149	157	147	155	151	159	146	154	150	158	148	156	152	160
209	217	213	221	211	219	215	223	210	218	214	222	212	220	216	224
273	281	277	285	275	283	279	287	274	282	278	286	276	284	280	288
337	345	341	349	339	347	343	351	338	346	342	350	340	348	344	352
49	57	53	61	51	59	55	63	50	58	54	62	52	60	56	64
113	121	117	125	115	123	119	127	114	122	118	126	116	124	120	128
177	185	181	189	179	187	183	191	178	186	182	190	180	188	184	192
241	249	245	253	243	251	247	255	242	250	246	254	244	252	248	256
305	313	309	317	307	315	311	319	306	314	310	318	308	316	312	320
369	377	373	381	371	379	375	383	370	378	374	382	372	380	376	384

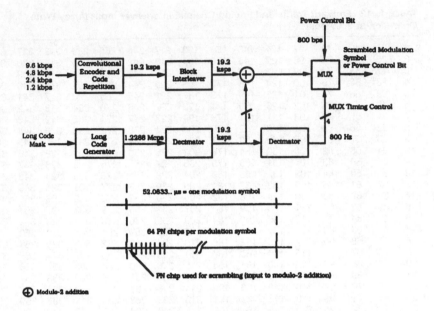

Figure 15.33 Data scrambler function and timing.

PN sequence offset

Pilot channel. A pilot channel is transmitted all times on Walsh function W0 by the base station. Pilot PN sequence offset is used for identifying each base station. Time offset may be revised within a CDMA cellular system.

Sync channel. The sync channel is an encoded, interleaved, spread, and modulated spread signal. The sync channel uses the same pilot PN sequence offset as the pilot channel for a given base station.

Receiver at MS. The MS demodulation process shall perform complimentary operations to the BS modulation process. The MS shall provide a minimum of four processing elements. Three of them are capable of tracking and demodulating multipath components of the forward CDM channel. At least one element shall be a searcher element capable of scanning and estimating the signal strength at each pilot PN sequence offset. The signal strength of the pilot is used to select the desired BS during the idle or initialization stage. Also, the signal strength of the pilot is used for the MS to determine when the handoff shall be requested and which new BS is the candidate. The information on handoff will be sent to the BS via the reverse signaling traffic channel (see Table 15.14). The multiplex option is the same on both the forward traffic channel and the reverse traffic channel.

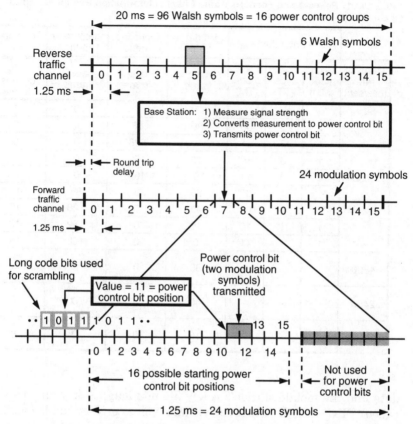

Figure 15.34 Randomization of power control bit positions.

Forward traffic channel frames. There are 14 categories for multiplex option 1. Among those categories, 12 are listed in Table 15.14 and are considered good frames. Categories 9 and 10 are bad frames:

Category 9: 9600-bps frame, primary traffic only, with bit errors

Category 10: Insufficient frame quality

15.3.4 Authentication, encryption, and privacy

Authentication refers to the process by which the base station confirms the identity of the mobile station, i.e., the identical sets of shared secret data. SSD is a 128-bit pattern in the MS. SSD-A consists of 64 bits and SSD-B consists of 64 bits. SSD-A supports the authentication procedure initialized with mobile station specific information, random

TABLE 15.14 Reverse and Forward Traffic Channel Information Bits for Multiplex Option 1

Format bits			Primary traffic	Signaling traffic	Secondary traffic	Categories of
Transmit rate (bits/sec)	Mixed mode (MM)	Traffic type (TT)	Traffic mode (TM)	bits/ frame	bits/ frame	bits/ frame	received traffic channel frame
	'0'	-	-	171	0	0	1
	'1'	'0'	'00'	80	88	0	2
	'1'	'0'	'01'	40	128	0	3
	'1'	'0'	'10'	16	152	0	4
9600	'1'	'0'	'11'	0	168	0	5
*	'1'	'1'	'00'	80	0	88	11
*	'1'	'1'	'01'	40	0	128	12
*	'1'	'1'	'10'	16	0	152	13
*	'1'	'1'	'11'	0	0	168	14
4800	-	-	-	80	0	0	6
2400	-	-	-	40	0	0	7
1200	-	-	-	16	0	0	8

data and the mobile station's A key (64 bits long). A key may be also called PIN. SSD-B supports CDMA voice privacy and message confidentiality.

The purposes for using authentication are (1) MS registration, (2) MS origination, and (3) MS termination. When the information element AUTH in the system parameters overhead message is set to 1 and the MS attempts to register, originate, or terminate, then the auth-signature procedure is executed, and AUTHR is obtained (see Fig. 15.35) and sent with RANDC (eight most significant bits of RAND confirmation) and COUNT to the base station for validation.

Authentication of MS data bursts

1. The BS sends an SSD update message on either the paging channel or the forward traffic channel. In the SSD update message, there is a RANDSSD field which is used for the computation of SSD at the home location register/authentication center (HLR/AUC). (The A key is stored at the MS and the HLR/AUC.) The MS shall then execute the SSD generation procedure by using RANDSSD, ESN, and A key to produce SSD-A-New and SSD-B-New.

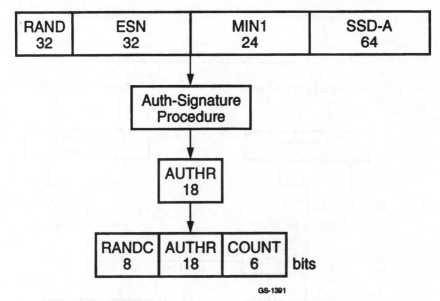

Figure 15.35   Authentication parameters for base station validation.

2. The MS shall select a 32-bit random number RANDBS to BS in a base station challenge order via the access channel or reverse traffic channel.

3. Both BS and MS shall execute an auth-signature procedure by using SSD-A-New and RANDBS, and both obtain an 18-bit AUTHBS.

4. The BS sends the AUTHBS in the base station challenge order confirmation on the paging channel or the forward traffic channel.

5. The MS compares the two AUTHBS, one from its own MS and one from the BS. If the comparison is successful, it sets SSD-A and SSD-B to SSD-A-New and SSD-B-New, respectively. Also, the MS shall send an SSD update confirmation order to the BS indicating the successful comparison. If the comparison is not successful, it discards the two new SSDs and sends an SSD update rejection order to the BS, indicating unsuccessful comparison.

On receipt of the SSD update confirmation order, the BS sets SSD-A and SSD-B to the values received from the HLR/AUC. The SSD update message flow is shown in Fig. 15.36.

Signaling message encryption.   In an effort to protect sensitive subscriber information (such as PIN), the availability of encryption algorithm information is governed under the U.S. International Traffic and Arms Regulation (ITAR) and export administration regulations.

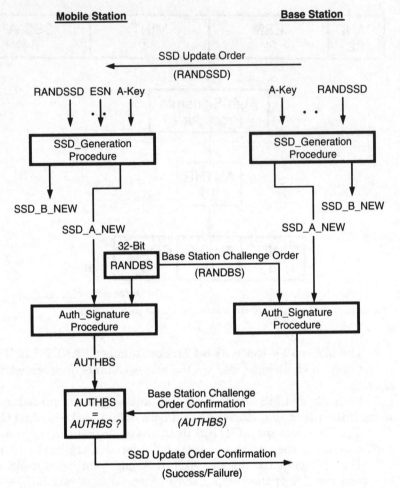

Figure 15.36 SSD update message flow.

Messages shall not be encrypted if authentication is not performed. Signaling message encryption is controlled for each call individually.

Voice privacy. Voice privacy is provided in the CDMA system by means of the private long-code mask used for PN spreading. Voice privacy control is provided on the traffic channels only. All calls are initiated by using the public long-code mask for PN spreading. To initiate a transition to the private or public long-code mask, either the BS or the MS sends a long-code transition request order on the traffic channel.

15.3.5 Malfunction detection

The BS detects the malfunction of an MS by asking the MS to respond to the lock order, lock until power-cycled order, and maintenance required order. This feature identifies the malfunctional MS and prevents the MS from contaminating the CDMA system by sending a signal to disconnect the MS transmit power.

15.3.6 Call processing

MS call processing consists of the following states:

MS initialization state

- The MS selects which system to use.
- It acquires the pilot channel of a CDMA system within 20 ms.
- It obtains system configuration and timing information for a CDMA system.
- It synchronizes its timing to that of a CDMA system.

MS idle state

- The MS shall perform paging channel monitoring procedures. The paging channel is divided into 200-ms slots called *paging channel slots*. Paging and control messages for an MS operating in the non-slotted mode can be received in an array of the paging channel slots. Therefore, the nonslotted mode of operation requires the MS to monitor all slots. An MS operating in the slotted mode generally monitors the paging channel for one or two slots per slot cycle. The MS can control the length of the slot cycle.

- Unless otherwise specified in the requirements for processing a specific message, the MS shall transmit an acknowledgement in response to any message received that is addressed to the MS.

- The MS shall maintain all active registration timers.

The CDMA system supports nine different forms of registration: Autonomous registrations:

1. *Power-up registration.* The mobile station registers when it powers on, switches from using the alternate serving system, or switches from using the analog system.

2. *Power-down registration.* The mobile station registers when it powers off if previously registered in the current serving system.

3. *Timer-based registration.* The mobile station registers when a timer expires.

4. *Distance-based registration.* The mobile station registers when the distance between the current base station and the base station in which it last registered exceeds a threshold.

5. *Zone-based registration.* The mobile station registers when it enters a new zone.

Registrations under different requests:

6. *Parameter-change registration.* The mobile station registers when certain of its stored parameters change.

7. *Ordered registration.* The mobile station registers when the base station requests it.

8. *Implicit registration.* When a mobile station successfully sends an origination message or page response message, the base station can infer the mobile station's location, causing an implicit registration.

9. *Traffic channel registration.* Whenever the base station has registration information for a mobile station that has been assigned to a traffic channel, the base station can notify the mobile station that it is registered.

System access state. The MS sends messages to the BS on the access channel and receives messages from the base station on the paging channel. The entire process of sending one message and receiving acknowledgment for that message is called an *access attempt.* Each transmission in the access attempt is called an *access probe.* The mobile station transmits the same message in each access probe in an access attempt. Each access probe consists of an access channel preamble and an access channel message capsule. There are two types of messages sent on the access channel: a response message and a request message. The access attempt ends after an acknowledgment is received.

MS control on the traffic channel state. The mobile station communicates with the BS using the forward and reverse traffic channels. There are five functions:

1. The MS verifies that it can receive the forward traffic channel and begins transmitting on the reverse traffic channel.

2. The MS waits for an order on an alert with information message.

3. The MS waits for the user to answer the call.

4. The MS's primary service option application exchanges primary traffic packets with the base station.

5. The MS disconnects the call.

15.3.7 Handoff procedures

Types of handoffs. The MS supports four handoff procedures:

1. *Soft handoff.* The MS commences communication with a new base station without interrupting communication with the old base station. Soft handoff means an identical frequency assignment between the old BS and new BS. Soft handoff provides different-site selection diversity to enhance the signal.

2. *CDMA-to-CDMA hard handoff.* The MS transmits between two base stations with different frequency assignments.

3. *CDMA-to-analog handoff.* The MS is directed from a forward traffic channel to an analog voice channel with a different frequency assignment.

4. *Softer handoff.* Handoffs between sectors within a cell.

Pilot sets. The information obtained from the pilot channel is used for the handoff. A pilot is associated with the forward traffic channels in the same forward CDMA channel. A pilot channel is identified by a pilot sequence offset. Each pilot channel is assigned to a particular BS. The MS can obtain four sets of pilot channels:

1. *Active set.* The pilot associated with the forward traffic channels assigned to the MS.

2. *Candidate set.* The pilots that are not in the active set but are received by MS with sufficient strength.

3. *Neighbor set.* The pilots that are not in the active set or the candidate set and are likely candidates for handoff.

4. *Remaining set.* The set in the current system on the current CDMA frequency assignment, excluding the above three sets.

Pilot requirement

1. The base station specified for each of the above pilots sets the search window in which the mobile station is to search for usable multipath components of the pilots in the set.

2. The MS assists the base station in the handoff process by measuring and reporting the strengths of received pilots.

3. A handoff drop timer shall be maintained for each pilot in the active set and candidate set. When the signal strength level is below TDROP (also called T-DROP), T-TDROP is set to zero, i.e., it expires within 100 ms. There are 15 T-TDROP values. The highest value of T-TDROP is 319 s. When the MS receives a signal strength level from the neighboring cell exceeding a given TADD (also called T_ADD) level in decibels, the soft handoff starts. When the MS receives a signal strength level from the home cell below TDROP, the soft handoff ends. The handoff action would take place after the received level from the home cell is below the TDROP. If the time between TADD and TDROP is very short, the T-TDROP time has to be longer. Also, in certain circumstances, it is preferable to reduce the call drops and sacrifice voice quality.

4. The MS shall measure the arrival time for each pilot reported to the base station. The time of the earliest arriving usable multipath component of the pilot is used to measure relative to the MS's time reference in units of PN chips.

5. Soft handoff

   a. All forward traffic channels associated with pilots in the active set of the MS carry modulation symbols identical to those of the power control subchannel. When the active set contains more than one pilot, the MS should provide diversity combining of the associated forward traffic channels. The MS shall provide for differential propagation delays from zero to at least 150 μs.

   b. For reverse traffic channel power control during soft handoff, the handoff direction message identifies sets of forward traffic channels that carry identical closed-loop power control subchannels. A set consists of one or more forward traffic channel transmissions with identical power control information. The MS will obtain at most one power control bit from each set of identical closed-loop power control subchannels. If the power control bits obtained from all sets are equal to 0, the MS shall increase its power; if they are equal to 1, the MS shall decrease its power.

   c. The typical message exchanges between the MS and the BS during handoff are shown in Fig. 15.37. There are seven messages during the soft handoffs. The first message that the MS sends is a pilot strength measurement message when the neighboring pilot strength exceeds TADD. The soft handoff starts. The seventh message is that the MS should move pilot from the active set to the neighboring set and send a handoff completion message. The soft handoff is then completed.

The pilot strength measurement triggered by a candidate pilot is shown in Fig. 15.38. During the soft handoff starts, the two pilots P_1

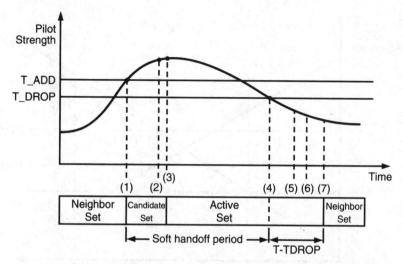

(1) Pilot strength exceeds T_ADD. Mobile station sends a *plot strength measurement message* and transfers pilot to the candidate set.

(2) Base station sends a *handoff direction message*.

(3) Mobile station transfers pilot to the active set and sends a *Handoff completion message*.

(4) Pilot strength drops below T_DROP. Mobile station starts the handoff drop timer.

(5) Handoff drop timer expires. Mobile station sends a *pilot strength measurement message*.

(6) Base station sends a *handoff direction message*.

(7) Mobile station moves pilot from the active set to the neighbor set and sends a *handoff completion message*.

Figure 15.37   Handoff threshold example.

and P_2 are indicated in the active set. There is a P_0 in the candidate set that is stronger than a pilot in the active set only if the difference between their respective strengths is at least T-Comp (level in decibels) as shown in Fig. 15.38.

15.4   Miscellaneous Mobile Systems

In the previous sections, GSM, NA-TDMA, and CDMA were introduced. This setion will briefly introduce other systems such as PHP, PDC, CT-2, DECT, CDPD, PCN, and PCS.

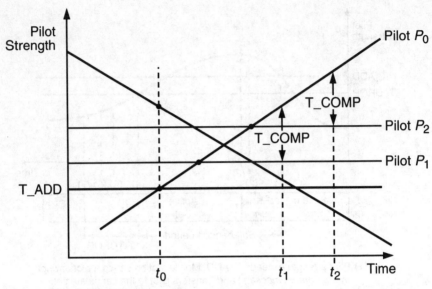

Candidate set: Pilot P_0

Active set: Pilots P_1, P_2

t_0 — *Pilot strength measurement message sent, P_0 > T_ADD*

t_1 — *Pilot strength measurement message sent, P_0 > P_1 + T_COMP*

t_2 — *Pilot strength measurement message sent, P_0 > P_2 + T_COMP*

Figure 15.38 Pilot strength measurements triggered by a candidate pilot.

15.4.1 TDD Systems

Time-division duplexing (TDD) systems are digital systems and use only one carrier to transmit and receive information. There are two kinds of TDD systems (see Fig. 15.39):

- TDD/FDMA—each carrier serves only one user.
- TDD/TDMA—each carrier can have many time slots and each slot can serve one user. Then N transmit time slots can serve N users.

A TDD system is used when only one chunk of spectrum is allocated. In cellular systems, there are two chunks of spectrum, separated by 20 MHz. In each cellular channel, the base transmit frequency and the mobile transmit frequency are 45 MHz apart. Therefore, the separation in frequency between transmitting and receiving is adequate to avoid interference. In TDD there is no separation in frequency between transmitting and receiving, but a separation in time interval.

TDD/FDMA

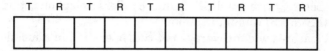

TDD/TDMA

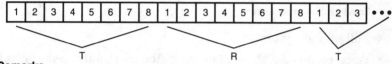

■ Remarks

- All cell sites have to be synchronized in order to eliminate the near-far interference from nieghboring sites
- The guard time in TDD slot would be longer than in regular TDMA slots
- The diversity scheme can be applied at one end to serve both ends
- Do not increase spectrum efficiency from a traffic/capacity point of view

Figure 15.39 Time-division duplexing.

The advantages of TDD are as follows:

1. When only one chunk of spectrum is available, TDD is the best utilization of spectrum.

2. Diversity can be applied at one end (terminal) to serve both ends, since the fading characteristics of one carrier are the same when received at both ends. At the base station, the information on selecting antennas for the space-diversity selective combining receiver can be used to switch to one of two transmitting antennas. Thus the mobile unit (or portable unit) can achieve the same diversity gain with a nondiversity receiver.

One of the concerns is that the TDD system has to be a synchronized system with a master clock. Otherwise, when one BS transmits and another BS receives, equivalent cochannel (co-time slot) interference occurs. Another concern is that the signal structures of TDD and of a frequency duplexing division (FDD) are different. Therefore, the two systems should not coexist in the same area because of their mutual interference.

In this section, several TDD systems will be briefly described.

Personal handy phone (PHP) system[14]. PHP is a wireless communication TDD system which supports personal communication services (PCS). It uses small, low-complexity, lightweight terminals called personal stations (PSs). PHP can be used for public telepoint (a portable phone booth), wireless PBX, home cordless telephone, and walkie-talkie (PS-to-PS) communication.

PHP features wider coverage per cell; operation in a mobile outdoor environment; faster and distributed control of handoff; enhanced authentication, encryption, and privacy; and circuit- and packet-oriented data services. PHP system is also called PHS (personal handy phone system).

Transmission parameters

- Full duplex system
- Voice coder—32 kbps adaptive differential pulse-code modulation (ADPCM).
- Duplexing—TDD. The portable and base units transmit and receive on the same frequency but different time slots. The slot arrangement is shown in Fig. 15.40.
- Multiple access—TDMA-TDD, up to four multiplexed circuits.
- Modulation—$\pi/4$ DQPSK, roll-off rate = 0.5.
- Data rate—192 ksps (or 384 kbps).
- Spectrum allocation—1895 to 1918.1 MHz. This spectrum has been allocated for private and public use.

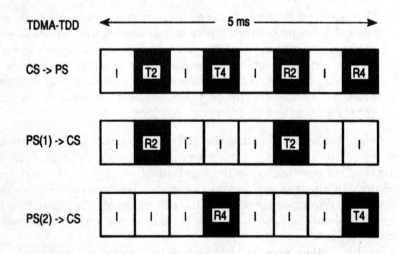

T: transmission R: reception, I: idle, Ti → Ri: Corresponding transmission/reception slot

Figure 15.40 Slot arrangement (corresponding to 32 kbps).

- Carrier frequency spacing—300 kHz.
- Carrier frequency—1895.15 MHz or 1895.15 + N·300 kHz where N is an integer.

Function channel structure The control channel consists of the following logical channels:

1. *Broadcast control channel.* BCCH is a one-way downlink channel for broadcasting control information from CS to PS.
2. *Common control channel.* CCCH sends out the control information for call connection.
   a. *Paging channel.* PCH is a one-way downlink channel.
   b. *Signaling control channel.* SCCH is a bidirectional point-to-point channel.
3. *User packet channel.* UPCH is a bidirectional point-multipoint channel that sends control signal information and user packet data.
4. Associated control channel. ACCH is a bidirectional channel that is associated with the TCH. It carries out control information and user packet data. There is a SACCH and a FACCH.

The traffic channel is a point-to-point bidirectional channel and is used for transmitting user information.

Carrier Structure

Control carrier. A carrier in which only common usage slots can be assigned to study intermittent transmission in a cordless station (CS).

Communications carrier. A carrier in which the user can perform communication through the individual assigned slot. It also can allocate common usage slots for a communications carrier.

Carrier for direct communication between personal stations. A carrier providing direct communication without going through a CS. The connection control and conversation can be carried out on the same slot.

Structure and interfaces of PHP system. The structure of PHP is shown in Fig. 15.41a. The CS is connected to the telecommunications circuit equipment. The interfaces of PHP are shown in Fig. 15.41b. There are three interface points:

Um is the interface point between personal station and cell station or between personal station and personal station. Conforms to the standard.

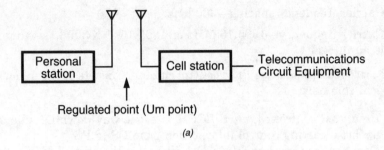

(a)

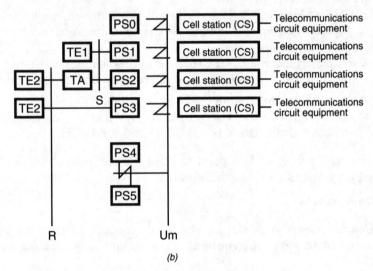

(b)

Figure 15.41 System structure and interface of PHP. (*a*) Structure of personal handy phone system; (*b*) interface points.

R is the interface point between I interface nonconforming terminal and mobile terminal equipment or terminal adapter. Outside scope of the standard.

S is the interface point between I interface conforming terminal or terminal adapter TA and mobile terminal equipment TE.

The five different classes of PS in Fig. 15.41 are defined as follows:

PSO, PS4, PS5	Personal station, including integrated man/machine interface of terminals, etc.
PS1, PS2	Personal station with I interface.
PS3	Personal station without I interface.

Terminal equipment TE1 is with I interface, and TE2 is without I interface. TA is the interface conversion equipment for non–I interface and I interface.

Cordless phone 2 (CT-2).[15] CT-2 was developed by GPT Ltd. in the United Kingdom and was the first TDD system for mobile radio communications. All the other TDD systems such as PHP and DECT adopted CT-2's structure. CT-2 is a portable payphone booth. Calls can be dialed out but not dialed in, and there is no handoff.

System structure. The structure is the same as PHP in Fig. 15.41.

- The CS is called the telepoint or phone zone
- Carrier frequency—864.1 to 868.1 MHz
- Total spectrum—4 MHz
- Channel access—FDMA/TDD
- Number of channels—40
- Channel bandwidth—100-kHz spacing
- Handset output power—1 to 10 mW

Characteristics

- Overall data rate—72 kbps
- Data rate per speech channel—32 kbps
- Speech coding—ADPCM
- Modulation—GMSK

Cordless phone 3 (CT-3). Originally called DCT900, CT-3 was developed by Ericsson as an upgrade from the CT-2 version.

System structure

- Time slots—64
- Two-way call system—call send and call delivery
- Slow-speed handoff
- Caller authentication
- Encryption
- Roaming

Characteristics

- Handset effective radiated power (ERP)—80 mW
- Modulation—filtered minimum-shift keying (MSK)

- Spectrum range—8 MHz
- Frequency—800 to 1000 MHz
- Bandwidth—1 MHz
- RF output power—80 mW peak, 5 mW average
- Number of slots per frame—8
- Overall data rate—640 kbps
- Data rate per speech code—32 kbps
- Speech coding—ADPCM

Digital European cordless telecommunication system (DECT).[16,17]   DECT is a European standard system. It is a CT-2–like system, and the applications are slightly different from the cellular system.

Applications

1. Use the public network to mobile communications within the home and the immediate vicinity.
2. Provide business communications locally. In this case, the PSTN has been replaced by the PBX. The base station has been renamed the radio fixed port (RFP), also called a *cluster controller*. The RFP can be treated as the microcell site.
3. Provide mobile public access. It is a PCS application. The system monitors the location of the active handsets and provides call delivery capability, like a cellular system.
4. Local loop. Using DECT to provide wireless local loops is a cost-effective alternative to running copper wires to residential premises.

System structure

- Duplex method—TDD
- Access method—TDMA
- RF power of handset—10 mW
- Channel bandwidth—1.728 MHz/channel
- Number of carriers—5 (a multiple-carrier system)
- Frequency—1800 to 1900 MHz

Characteristics

- Frame—10 ms
- Time slots—12

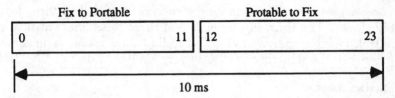

Fix to Portable		Protable to Fix	
0	11	12	23

10 ms

- Bit rate—38.8 kb/slot
- Modulation—GFSK (Gausian FSK)
- Handoff—Yes

15.4.2 Other full-duplexed systems

We have addressed the full-duplexed system such as GSM, NA-TDMA, and CDMA in the previous sections. In this setion PDC, PCN, and PCS will be briefly described.

Personal digital cellular (PDC).[18,19] PDC is a standard system in Japan. The system is a TDMA cellular system operating at 800 MHz and 1.5 GHz which used to be called Japanese digital system (JDC). 1.5 GHz PDC was in service publicly in Osaka in 1994. The PDC network reference model is shown in Fig. 15.42. This system provides nine interfaces among the cellular network. Um is the air interface, which was standardized. Interfaces B, C, D, E, J, K, and H were defined by cellular carriers in Japan. Interface A is an option for operators.

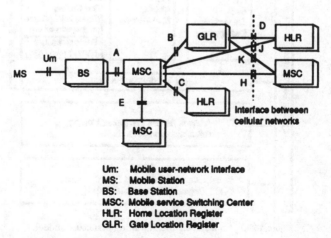

Um: Mobile user-network interface
MS: Mobile Station
BS: Base Station
MSC: Mobile service Switching Center
HLR: Home Location Register
GLR: Gate Location Register

Figure 15.42 PDC network reference model.

Signaling structure. The signaling structure of PDC has three layers (Fig. 15.43). The third layer consists of three functional entities: call control, mobility management, and radio transmission management. In general, the structure of PDC is very similar to that of NA-TDMA.

System structure

- Multiplex access—TDMA
- Number of time slots—3

Characteristics

- 800 MHz

    810—826 MHz
    940—956 MHz

- 1500 MHz

    1429—1441 MHz
    1453—1465 MHz
    1477—1489 MHz
    1561—1513 MHz

- Modulation—π/4, DQPSK
- Speech coder—VSELP

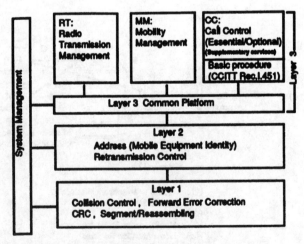

Figure 15.43 Subscriber line signaling structure model.

Personal communication network (PCN).[20]   The PCN system was first initiated by Lord Young in 1988. The characteristics of PCN are as follows:

1. Operational frequency—1.7 to 1.88 GHz (1710–1785 MHz and 1805–1880 MHz)

2. Uses 30 GHz or up for microwave backbone system

3. Covers both small cells and large cells (rural areas)

4. Coverage inside and outside buildings

5. Handover

6. Call delivery

7. Portable handset

8. Uses intelligent network

PCN uses the DCS-1800 system, which is similar to GSM, but upconverts the frequency to 1.7–1.88 GHz. Therefore, the network structure, the signaling structure, and the transmission characteristics are similar between PCN and GSM, but the operational frequencies are different.

Personal communication service.[21]   See Appendix 15.1 for (2) broadband PCS designated in major trading areas (MTA), (1) MTA in the U.S. and (3) basic trading areas (BTA) in the U.S.

Ever since cellular systems were deployed in 1983 in the United States, the rapidly growing wireless communication systems have faced limitations of spectrum utilization. However, the trend is toward personal communications and includes wireless communications for pedestrians and for in-building communications. In 1989 Lord Young of the United Kingdom was promoting a PCN system operating at 1.8 GHz. PCN uses a GSM version of cellular communication. The spectrum range is 1710–1785 MHz and 1805—1880 MHz.

In 1991 the FCC issued a Notice of Proposed Rule Making, Docket No. 90-134, considering allocating spectrum between 1.85 and 2.2 GHz. The FCC has now allocated 120 MHz of spectrum into seven bands as shown in Fig. 15.44 for wideband PCS and narrowband PCS allocated spectrum.

There are now many systems, such as GSM, NADC, CDMA, PCN, and DECT, as described in this chapter, that are being considered as candidates to be adapted as the future PCS system. How to improve the equalizer for the TDMA[28] and reduce the intererence for CDMA are the major concerns. However, users want to carry one unit with

• Wideband PCS — for cellular-like systems

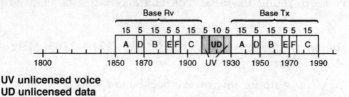

UV unlicensed voice
UD unlicensed data

• Narrowband PCS — for two-way paging systems

Five 50 kHz channels paired with 50 kHz channels

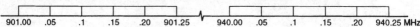

Three 50 kHz channels paired with 12.5 kHz channels

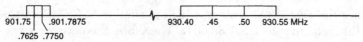

Three 50 kHz unpaired channels

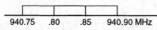

Figure 15.44 FCC's PCS allocation.

which they can place a call to wherever they want and receive calls wherever they may be. Of course, the unit has to be small in size, light in weight, and provide long talking time.

15.4.3 Noncellular systems

There are three systems that can be mentioned in this section: CDPD, MIRS, and satellite mobile systems.

Cellular digital packet data (CDPD) system.[22,27] CDPD is a packet switching system which uses idle voice channels from the cellular system band to carry out traffic. This system can be assigned a dedicated channel or it can hop to idle channels.

In the network reference model of CDPD in Fig. 15.45, acronyms are defined as follows:

SIM   Subscriber identity module

SU   Subscriber unit [also can be indicated as mobile-end station (M-ES)]

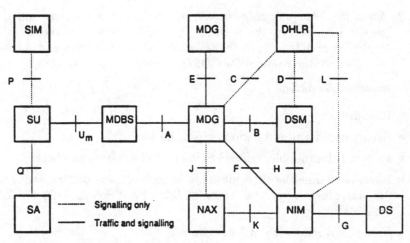

Figure 15.45   CDPD network reference model.

EIR	Equipment identity register
MDBS	Mobile data base station
MDG	Mobile data gateway
NIM	Network interface manager
DHLR	Data home location register
NAX	Network address translator
DSM	Data service manager [also can be indicated as mobile data information station (MD, IS)]
DS	Data service
NG	Network gateway

The network in Fig. 15.45 is self-explanatory. There are two communication interfaces, called *reference points,* identified with a letter (e.g., A or Um). On a given reference point there may be many physical devices to transport the signal such as routes, switches, multiplexers, demultiplexers, and modems.

System operation. The MDBS is collocated at the cellular cell site. DSM is a control center for CDPD. DSM operates independently from the MTSO of cellular systems. In the MDBS, a forward-link logical channel is always on to send the overhead message or send the data to the user. There are two set-ups:

1. *For a dedicated channel setup:* The MS, at the initialization stage, scans the assigned N CDPD channels to lock on a strong CDPD channel. The CDPD channel will be changed while the MS is moving from one cell to another cell.

2. *For a frequency-hopping channel setup:* There is a device called a *sniffer* installed in the MDBS. The sniffer monitors the cellular control channels on both the forward and reverse links and chooses an idle cellular channel for CDPD.

Transmission structure

- Roaming support
- Security and authentication across the airlink
- Forward channel block: Reed-Solomon (63,47) data symbols
- Reverse channel block: 8-bit delay maximum plus dotting sequence (38 bits) plus reverse sync work (22 bits); Reed-Solomon (63,47) data symbols
- Frequency-agile with in-band control
- Channel hopping and dedicated channel
- Cell transfer controlled by M-ES (SU in Fig. 15.45).
- Power control
- AMPS-compatible
- Essentially transparent to AMPS
- Modulation—GMSK (same as GSM)
- Data rate—19.2 kbps
- Link protocol—LAPD
- Point-to-point, broadcast, and multicast delivery

Mobile integrated radio system (MIRS). MIRS is a cellular-like system developed by Motorola to operate at the special mobile radio (SMR) band shown in Fig. 15.46. MIRS system is called by the manufacturers. The system providers call this system ESMR (enhanced SMR). The SMR band is in the Part 90 of FCC CFR in the private sector.

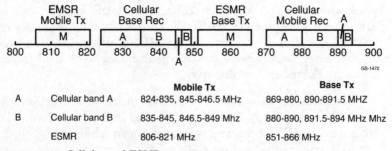

		Mobile Tx	Base Tx
A	Cellular band A	824-835, 845-846.5 MHz	869-880, 890-891.5 MHZ
B	Cellular band B	835-845, 846.5-849 Mhz	880-890, 891.5-894 MHz Mhz
	ESMR	806-821 MHz	851-866 MHz

Figure 15.46 Cellular and ESMR spectrum.

Public information on this system is limited. However, the operational function is the same as the cellular system. Since many existing SMR systems are still in operation, the MIRS frequency channels in the SMR band are restricted depending on the area in which they operate. Therefore, a special frequency assignment arrangement will be imposed in the MIRS system.

A top-level system specification is as follows:

- Full-duplex communication system
- Frequency—806 to 824 MHz (mobile transmitter), 851 to 869 MHz (base transmitter)
- Channel bandwidth—25 KHz
- Multiple access—TDMA
- Number of time slots—6
- Rate of speech coder—4.2 kbps/slot
- Modulation—16 QAM (quadrature AM)
- Speech coder—VSELP
- No equalizer is implemented
- Handoff
- Transmission rate—6.5 kbps/slot
- Forward error correction—3 kbps
- Interleaving

Mobile satellite communications. In the future, mobile satellite communications will enhance terrestrial radio communication, either in rural areas or in global communication to become a broad sense PCS system. Mobile satellite communications is classified by three satellite altitude positions. One uses stationary satellites which are 22,000 mi above the earth. Another uses low-earth-orbit (LEO) satellites which are 400 to 800 mi above the earth. The third uses medium-earth-orbit satellites (MEO) anywhere between the stationary satellite altitude and LEO satellite altitude.

Stationary satellites. To cover the entire earth, only four stationary satellites are required. The required number of stationary satellites is low and the life span is 10 to 15 years. Thus the cost is low. Immarsat[23] is introduced to PCS. However, the time delay due to the long range can be 0.25 s (two ways) and does not include the other delays such as signal processing and network routing. Also, at high latitudes, the stationary satellite has a low elevation angle as seen from the earth.

TABLE 15.15   Comparative Low-Earth-Orbiting Mobile Satellite Service Applications*

System characteristics	Loral/QUALCOMM GLOBALSTAR	Motorola IRIDIUM	TRW ODYSSEY	Constellation ARIES (b)	Ellipsat ELLIPSO (c)
Number of satellites	48	66	12	48	24
Constellation altitude (NM)	750	421	5600	550	1767 x 230
Unique feature	Transponder	Onboard Processing	Transponder	Transponder	Transponder
Circuit capacity (U.S.)	6500	3835	4600	100	1210
Signal modulation	CDMA	TDMA	CDMA	FDMA/CDMA	CDMA
Gateways in U.S.	6	2	2	5	6
Gateway	C-band existing	New Ka-Band	New Ka-band	Unknown	Unknown
Coverage	Global	Global	Global	Global	Northern hemisphere

For instance, in Chicago, the elevation angle is 19°. Under this condition, the buildings and the hills block the direct path between the satellite and the mobile. This causes multipath fading and shadow loss. Also, the transmit power and size of antenna to send the signal up to the satellite from a mobile station is limited.

LEO satellites. Many LEO satellites are required to cover the earth. The Iridium[24] system (Motorola) needs 66 satellites, and Globalstar[25] needs 48 satellites. The average satellite life is 5 years. The cost of utilizing LEO mobile satellite systems is high. The satellite will circle the earth with a period between 1 to 2 h, depending on the satellite altitude. Handoffs between satellite antenna spot beams and between satellites may create difficulties for switching signals in space. However, there is no noticeable delay in talk time, and the mobile transmit power is not an issue. A comparison of LEO mobile satellite service applications among different systems is shown in Table 15.15.

MEO Satellites. MEO satellites may share the advantages and disadvantages of both stationary satellites and LEO satellites. Odyssey Personal Communication Satellite System[26] developed by TRW belongs to MEO satellites.

References

1. *Conference Proceedings,* Digital Cellular Radio Conference, Hagen FRG, October 1988 (21 papers describe the GSM system).
2. Bernard J. T. Mallinder, "An Overview of the GSM System," *Conference Proceedings,* Digital Cellular Radio Conference, Hagen FRG, October 1988.
3. M. Mouly, M. B. Pautet, "The GSM System Mobile Communications," M. Mouly et M. B. Pautet, 49, vue Louis Bruneau, F-91120 Palaisea, France 1992.

4. "European Digital Cellular Telecommunications System (Phase 2): General Description of a GSM Public Land Mobile Network," ETSI, 06921 Sophia Antipolis Cedex, France, October 1993, GSM 01-12.
5. *Conference Proceedings,* Third Nordic Seminar on Digital Land Mobile Radio Communication, September 12–15, 1988, Copenhagen (21 papers describe the GSM system).
6. Cellular System, IS-54 (incorporating EIA/TIA 553), "Dual-Mode Mobile Station-Base Station Compatibility Standard," Electronic Industries Association Engineering Department, PN-2215, December 1989 (NADCA-TDMA system).
7. Cellular System, IS-55, "Recommended Minimum Performance Standards for Mobile Stations," PN-2216, EIA, Engineering Department, December 1989 (NADC-TDMA system).
8. Cellular System, IS-56, "Recommended Minimum Performance Standards for Base Stations," PN-2217, EIA, Engineering Department, December 1989 (NATC-TDMA system).
9. W. C. Y. Lee, "Cellular Operators feel the squeeze," *Telephony,* vol. 214, no. 2, May 30, 1988, pp. 22–23.
10. Cellular System, IS-95, "Dual-Mode Mobile Station-Base Station Wideband Spread Spectrum Compatibility Standard," PN 3118, EIA, Engineering Department, December 1992 (CDMA system).
11. Cellular System, IS-96, "Recommended Minimum Performance Standards for Mobile Stations Supporting Dual-Mode Wideband Spread Spectrum Cellular Base Stations," PN-3119, EIA, Engineering Department, December 1993 (CDMA system).
12. Cellular System, IS-97, "Recommended Minimum Performance Standards for Base Stations Supporting Dual-Mobile Wideband Spread Spectrum Cellular Mobile Stations," PN-3120, EIA, Engineering Department, December 1993 (CDMA system).
13. A. Salmasi and K. S. Gilhousen, "On the System Design Aspects of Code Division Multiple Access Applied to Digital Cellular and Personal Communications Networks," *IEEE VTC 91' Conference Record,* St. Louis, May 19–22, 1991, pp. 57–62.
14. PHS-Personal Handy Phone Standard, Research Development Center for Radio System (RCR), "Personal Handy Phone Standard (PHS)," CRC STD-28, December 20, 1993.
15. Cordless Telephone 2/Common Air Interface (CT2/CAI), "Management of International Telecommunications, MIT 12-850-201, McGraw Hill, Inc. DataPro Information Service Group, Delran, N.J., February 1994.
16. *Digital European Cordless Telecommunications,* Part I, "Overview," DE/RES 3001-1, Common Interface, Radio Equipmemt and Systems, ETS 300 175-1, ETSI, B.P. 152, F-06561 Valbonne Cedex, France, August 1991.
17. Sybo Dijkstra, Frank Owen, "The Case for DECT," *Mobile Communications International,* pp. 60–65, September–October–November 1993.
18. "PDC—Digital Cellular Telecommunication System, RCR STF-27A Version," January 1992, Research & Development Center for Radio System (RCR), Nippon Ericsson K.K.
19. H. Takamura, A. Nakajima, K. Yamamoto, "Network and Signaling Structure Based on Personal Digital Cellular Telecommunication System Concept," VTC '93, *Conference Record,* pp. 922–926.
20. A. R. Potter, "Implementation of PCNs using DCS/800," *IEEE Communications Magazine,* vol. 30, December 1992, pp. 32–37.
21. FCC, "Spectrum Allocation for PCS," FCC 90-314, Sept. 23, 1993.
22. "CDPD—Cellular Digital Packet Data, Cell Plan II Specification," prepared by PCSI, San Diego, CA 92121, January 1992.
23. N. Hart, H. Haugli, P. Poskett, and K. Smith, "Immarsat's Personal Communications System," *Proc. Third International Mobile Satellite Conference,* Pasadena, pp. 303–304, June 16–18, 1993.
24. J. E. Hatlelid and L. Casey, "The Iridium System: Personal Communications Anytime, Anyplace," *Proc. Third International Mobile Satellite Conference,* Pasadena, pp. 285–290, June 16–18, 1993.

25. R. A. Wiedeman, "The Globalstar Mobile Satellite System for Worldwide Personal Communications," *Proc. Third International Mobile Satellite Conference,* Pasadena, pp. 291–296, June 16–18, 1993.
26. C. Spitzer, "Odyssey Personal Communications Satellite System," *Proc. Third International Mobile Satellite Conference,* Pasadena, pp. 291–296, June 16–18, 1993.
27. D. J. Goodman, "Cellular Packet Communications," *IEEE Trans. on Comm.,* vol. 38, August 1990, pp. 1272–1280.
28. J. G. Proakis, "Adaptive Equalization for TDMA Digital Mobile Radio," *IEEE Trans. on VT,* vol. 40, May 1991, pp. 333–341.

APPENDIX 15.1   Major Trading Areas in the United States (*from Rand McNally & Company*)

No.	Major trading area	Abbrev.	Number of basic areas
01	Atlanta	ATL	14
03	Birmingham	BIR	10
05	Boston-Providence	BOS-PRO	14
07	Buffalo-Rochester	BUF-ROC	4
09	Charlotte-Greensboro-Greenville-Raleigh	C-G-G-R	23
11	Chicago	CHI	18
13	Cincinnati-Dayton	CIN-DAY	9
15	Cleveland	CLEV	10
17	Columbus	COL	6
19	Dallas-Fort Worth	DAL-F.W.	22
21	Denver	DEN	12
23	Des Moines-Quad Cities	DES-Q.C.	13
25	Detroit	DET	18
27	El Paso-Albuquerque	ELP-AL	8
29	Honolulu	HON	4
31	Houston	HOU	6
33	Indianapolis	IND	11
35	Jacksonville	JAX	7
37	Kansas City	K.C.	9
39	Knoxville	KNOX	3
41	Little Rock	L.R.	9
43	Los Angeles-San Diego	L.A.-S.D.	7
45	Louisville-Lexington-Evansville	L-L-E	9
47	Memphis-Jackson	MEM-JAK	11
49	Miami-Fort Lauderdale	MIA-F.L.	5
51	Milwaukee	MILW	16
53	Minneapolis-St. Paul	MPLS-S.P.	23
55	Nashville	NASH	3
57	New Orleans-Baton Rouge	N.O.-B.R.	13
59	New York	N.Y.	20
61	Oklahoma City	O.C.	8
63	Omaha	OMA	7
65	Philadelphia	PHIL	11
67	Phoenix	PHOE	7
69	Pittsburgh	PGH	12
71	Portland	POR	9
73	Richmond-Norfolk	RICH-NOR	7
75	St. Louis	ST.L.	12
77	Salt Lake City	S.L.C.	8
79	San Antonio	SANT	6
81	San Francisco-Oakland-San Jose	SF-O-SJ	13
83	Seattle	SEAT	11
85	Spokane-Billings	SPOK-BIL	11
87	Tampa-St. Petersburg-Orlando	T-SP-O	7
89	Tulsa	TUL	4
91	Washington-Baltimore	WASH-BAL	9
93	Wichita	WICH	8
	U. S. Total		487

Broadband PCS Major Trading Area Designations (*from FCC*)

Market no.	Major trading area
M 001	New York
M 002	Los Angeles-San Diego
M 003	Chicago
M 004	San Francisco-Oakland-San Jose
M 005	Detroit
M 006	Charlotte-Greensboro-Greenville-Raleigh
M 007	Dallas-Fort Worth
M 008	Boston-Providence
M 009	Philadelphia
M 010	Washington-Baltimore
M 011	Atlanta
M 012	Minneapolis-St. Paul
M 013	Tampa-St Petersburg-Orlando
M 014	Houston
M 015	Miami-For Lauderdale
M 016	Cleveland
M 017	New Orleans-Baton Rouge
M 018	Cincinnati-Dayton
M 019	St. Louis
M 020	Milwaukee
M 021	Pittsburgh
M 022	Denver
M 023	Richmond-Norfolk
M 024	Seattle (Excluding Alaska)
M 025	Puerto Rico-U.S. Virgin Islands
M 026	Louisville-Lexington-Evansville
M 027	Phoenix
M 028	Memphis-Jackson
M 029	Birmingham
M 030	Portland
M 031	Indianapolis
M 032	Des Moines-Quad Cities
M 033	San Antonio
M 034	Kansas City
M 035	Buffalo-Rochester
M 036	Salt Lake City
M 037	Jacksonville
M 038	Columbus
M 039	El Paso-Albuquerque
M 040	Little Rock
M 041	Oklahoma City
M 042	Spokane-Billings
M 043	Nashville
M 044	Knoxville
M 045	Omaha
M 046	Wichita
M 047	Honolulu
M 048	Tulsa
M 049	Alaska
M 050	Guam-Northern Mariana Islands
M 051	American Samoa

Basic Trading Areas in the United States

No.	Basic trading area
001	Aberdeen, SD
002	Aberdeen, WA
003	Abilene, TX
004	Ada, OK
005	Adrian, MI
006	Albany-Tifton, GA
007	Albany-Schenectady, NY
008	Albuquerque, NM
009	Alexandria, LA
010	Allentown-Bethlehem-Easton, PA
011	Alpena, MI
012	Altoona, PA
013	Amarillo, TX
014	Anchorage, AK
015	Anderson, IN
016	Anderson, SC
017	Anniston, AL
018	Appleton-Oshkosh, WI
019	Ardmore, OK
020	Asheville-Hendersonville, NC
021	Ashtabula, OH
022	Athens, GA
023	Athens, OH
024	Atlanta, GA
025	Atlantic City, NJ
026	Augusta, GA
027	Austin, TX
028	Bakersfield, CA
029	Baltimore, MD
030	Bangor, ME
031	Bartlesville, OK
032	Baton Rouge, LA
033	Battle Creek, MI
034	Beaumont-Port Arthur, TX
035	Beckley, WV
036	Bellingham, WA
037	Bemidji, MN
038	Bend, OR
039	Benton Harbor, MI
040	Big Spring, TX
041	Billings, MT
042	Biloxi-Gulfport-Pascagoula, MS
043	Binghamton, NY
044	Birmingham, AL
045	Bismarck, ND
046	Bloomington, IL
047	Bloomington-Bedford, IN
048	Bluefield, WV
049	Blytheville, AR
050	Boise-Nampa, ID

Basic Trading Areas in the United States

No.	Basic trading area
051	Boston, MA
052	Bowling Green-Glasgow, KY
053	Bozeman, MT
054	Brainerd, MN
055	Bremerton, WA
056	Brownsville-Harlingen, TX
057	Brownwood, TX
058	Brunswick, GA
059	Bryan-College Station, TX
060	Buffalo-Niagara Falls, NY
061	Burlington, IA
062	Burlington, NC
063	Burlington, VT
064	Butte, MT
065	Canton-New Philadelphia, OH
066	Cape Girardeau-Sikeston, MO
067	Carbondale-Marion, IL
068	Carlsbad, NM
069	Casper-Gillette, WY
070	Cedar Rapids, IA
071	Champaign-Urbana, IL
072	Charleston, SC
073	Charleston, WV
074	Charlotte-Gastonia, NC
075	Charlottesville, VA
076	Chattanooga, TN
077	Cheyenne, WY
078	Chicago, IL
079	Chico-Oroville, CA
080	Chillicothe, OH
081	Cincinnati, OH
082	Clarksburg-Elkins, WV
083	Clarksville, TN-Hopkinsville, KY
084	Cleveland-Akron, OH
085	Cleveland, TN
086	Clinton, IA-Sterling, IL
087	Clovis, NM
088	Coffeyville, KS
089	Colorado Springs, CO
090	Columbia, MO
091	Columbia, SC
092	Columbus, GA
093	Columbus, IN
094	Columbus-Starkville, MS
095	Columbus, OH
096	Cookeville, TN
097	Coos Bay-North Bend, OR
098	Corbin, KY
099	Corpus Christi, TX
100	Cumberland, MD

Basic Trading Areas in the United States

No.	Basic trading area
101	Dallas-Fort Worth, TX
102	Dalton, GA
103	Danville, IL
104	Danville, VA
105	Davenport, IA-Moline, IL
106	Dayton-Springfield, OH
107	Daytona Beach, FL
108	Decatur, AL
109	Decatur-Effingham, IL
110	Denver, CO
111	Des Moines, IA
112	Detroit, MI
113	Dickinson, ND
114	Dodge City, KS
115	Dothan-Enterprise, AL
116	Dover, DE
117	Du Bois-Clearfield, PA
118	Dubuque, IA
119	Duluth, MN
120	Dyersburg-Union City, TN
121	Eagle Pass-Del Rio, TX
122	East Liverpool-Salem, OH
123	Eau Claire, WI
124	El Centro-Calexico, CA
125	El Dorado-Magnolia-Camden, AR
126	Elkhart, IN
127	Elmira-Corning-Hornell, NY
128	El Paso, TX
129	Emporia, KS
130	Enid, OK
131	Erie, PA
132	Escanaba, MI
133	Eugene-Springfield, OR
134	Eureka, CA
135	Evansville, IN
136	Fairbanks, AK
137	Fairmont, WV
138	Fargo, ND
139	Farmington, NM-Durango, CO
140	Fayetteville-Springdale-Rogers, AR
141	Fayetteville-Lumberton, NC
142	Fergus Falls, MN
143	Findlay-Tiffin, OH
144	Flagstaff, AZ
145	Flint, MI
146	Florence, AL
147	Florence, SC
148	Fond du Lac, WI
149	Fort Collins-Loveland, CO
150	Fort Dodge, IA

Basic Trading Areas in the United States

No.	Basic trading area
151	Fort Myers, FL
152	Fort Pierce-Vero Beach-Stuart, FL
153	Fort Smith, AR
154	Fort Walton Beach, FL
155	Fort Wayne, IN
156	Fredericksburg, VA
157	Fresno, CA
158	Gadsden, AL
159	Gainesville, FL
160	Gainesville, GA
161	Galesburg, IL
162	Gallup, NM
163	Garden City, KS
164	Glens Falls, NY
165	Goldsboro-Kinston, NC
166	Grand Forks, ND
167	Grand Island-Kearney, NE
168	Grand Junction, CO
169	Grand Rapids, MI
170	Great Bend, KS
171	Great Falls, MT
172	Greeley, CO
173	Green Bay, WI
174	Greensboro-Winston-Salem-High Point, NC
175	Greenville-Greenwood, MS
176	Greenville-Washington, NC
177	Greenville-Spartanburg, SC
178	Greenwood, SC
179	Hagerstown, MD-Chambersburg, PA-Martinsburg, WV
180	Hammond, LA
181	Harrisburg, PA
182	Harrison, AR
183	Harrisonburg, VA
184	Hartford, CT
185	Hastings, NE
186	Hattiesburg, MS
187	Hays, KS
188	Helena, MT
189	Hickory-Lenoir-Morganton, NC
190	Hilo, HI
191	Hobbs, NM
192	Honolulu, HI
193	Hot Springs, AR
194	Houghton, MI
195	Houma-Thibodaux, LA
196	Houston, TX
197	Huntington, WV-Ashland, KY
198	Huntsville, AL
199	Huron, SD
200	Hutchinson, KS

Basic Trading Areas in the United States

No.	Basic trading area
201	Hyannis, MA
202	Idaho Falls, ID
203	Indiana, PA
204	Indianapolis, IN
205	Iowa City, IA
206	Iron Mountain, MI
207	Ironwood, MI
208	Ithaca, NY
209	Jackson, MI
210	Jackson, MS
211	Jackson, TN
212	Jacksonville, FL
213	Jacksonville, IL
214	Jacksonville, NC
215	Jamestown, NY-Warren, PA-Dunkirk, NY
216	Janesville-Beloit, WI
217	Jefferson City, MO
218	Johnstown, PA
219	Jonesboro-Paragould, AR
220	Joplin, MO-Miami, OK
221	Juneau-Ketchikan, AK
222	Kahului-Wailuku-Lahaina, HI
223	Kalamazoo, MI
224	Kallspell, MT
225	Kankakee, IL
226	Kansas City, MO
227	Keene, NH
228	Kennewick-Pasco-Richland, WA
229	Kingsport, TN-Johnson City, TN-Bristol, VA-TN
230	Kirksville, MO
231	Klamath Falls, OR
232	Knoxville, TN
233	Kokomo-Logansport, IN
234	La Crosse, WI-Winona, MN
235	Lafayette, IN
236	Lafayette-New Iberia, LA
237	La Grange, GA
238	Lake Charles, LA
239	Lakeland-Winter Haven, FL
240	Lancaster, PA
241	Lansing, MI
242	Laredo, TX
243	La Salle-Peru-Ottawa-Streator, IL
244	Las Cruces, NM
245	Las Vegas, NV
246	Laurel, MS
247	Lawrence, KS
248	Lawton-Duncan, OK
249	Lebanon-Claremont, NH
250	Lewiston-Moscow, ID

Basic Trading Areas in the United States

No.	Basic trading area
251	Lewiston-Auburn, ME
252	Lexington, KY
253	Liberal, KS
254	Lihue, HI
255	Lima, OH
256	Lincoln, NE
257	Little Rock, AR
258	Logan, UT
259	Logan, WV
260	Longview-Marshall, TX
261	Longview, WA
262	Los Angeles, CA
263	Louisville, KY
264	Lubbock, TX
265	Lufkin-Nacogdoches, TX
266	Lynchburg, VA
267	McAlester, OK
268	McAllen, TX
269	McComb-Brookhaven, MS
270	McCook, NE
271	Macon-Warner Robins, GA
272	Madison, WI
273	Madisonville, KY
274	Manchester-Nashua-Concord, NH
275	Manhattan-Junction City, KS
276	Manitowoc, WI
277	Mankato-Fairmont, MN
278	Mansfield, OH
279	Marinette, WI-Menominee, MI
280	Marion, IN
281	Marion, OH
282	Marquette, MI
283	Marshalltown, IA
284	Martinsville, VA
285	Mason City, IA
286	Mattoon, IL
287	Meadville, PA
288	Medford-Grants Pass, OR
289	Melbourne-Titusville, FL
290	Memphis, TN
291	Merced, CA
292	Meridian, MS
293	Miami-Fort Lauderdale, FL
294	Michigan City-La Porte, IN
295	Middlesboro-Harlan, KY
296	Midland, TX
297	Milwaukee, WI
298	Minneapolis-St. Paul, MN
299	Minot, ID
300	Missoula, MT

Basic Trading Areas in the United States

No.	Basic trading area
301	Mitchell, SD
302	Mobile, AL
303	Modesto, CA
304	Monroe, LA
305	Montgomery, AL
306	Morgantown, WV
307	Mount Pleasant, MI
308	Mount Vernon-Centralia, IL
309	Muncie, IN
310	Muskegon, MI
311	Muskogee, OK
312	Myrtle Beach, SC
313	Naples, FL
314	Nashville, TN
315	Natchez, MS
316	New Bern, NC
317	New Castle, PA
318	New Haven-Waterbury-Meriden, CT
319	New London-Norwich, CT
320	New Orleans, LA
321	New York, NY
322	Nogales, AZ
323	Norfolk, NE
324	Norfolk-Virginia Beach-Newport News-Hampton, VA
325	North Platte, NE
326	Ocala, FL
327	Odessa, TX
328	Oil City-Franklin, PA
329	Oklahoma City, OK
330	Olean, NY-Bradford, PA
331	Olympia-Centralia, WA
332	Omaha, NE
333	Oneonta, NY
334	Opelika-Auburn, AL
335	Orangeburg, SC
336	Orlando, FL
337	Ottumwa, IA
338	Owensboro, KY
339	Paducah-Murray-Mayfield, KY
340	Panama City, FL
341	Paris, TX
342	Parkersburg, WV-Marietta, OH
343	Pensacola, FL
344	Peoria, IL
345	Petoskey, MI
346	Philadelphia, PA-Wilmington, DE-Trenton, NJ
347	Phoenix, AZ
348	Pine Bluff, AR
349	Pittsburgh-Parsons, KS
350	Pittsburgh, PA

Basic Trading Areas in the United States

No.	Basic trading area
351	Pittsfield, MA
352	Plattsburgh, NY
353	Pocatello, ID
354	Ponca City, OK
355	Poplar Bluff, MO
356	Port Angeles, WA
357	Portland-Brunswick, ME
358	Portland, OR
359	Portsmouth, OH
360	Pottsville, PA
361	Poughkeepsie-Kingston, NY
362	Prescott, AZ
363	Presque Isle, ME
364	Providence-Pawtucket, RI-New Bedford-Fall River, MA
365	Provo-Orem, UT
366	Pueblo, CO
367	Quincy, IL-Hannibal, MO
368	Raleigh-Durham, NC
369	Rapid City, SD
370	Reading, PA
371	Redding, CA
372	Reno, NV
373	Richmond, IN
374	Richmond-Petersburg, VA
375	Riverton, WY
376	Roanoke, VA
377	Roanoke Rapids, NC
378	Rochester-Austin-Albert Lea, MN
379	Rochester, NY
380	Rockford, IL
381	Rock Springs, WY
382	Rocky Mount-Wilson, NC
383	Rolla, MO
384	Rome, GA
385	Roseburg, OR
386	Roswell, NM
387	Russellville, AR
388	Rutland-Bennington, VT
389	Sacramento, CA
390	Saginaw-Bay City, MI
391	St. Cloud, MN
392	St. George, UT
393	St. Joseph, MO
394	St. Louis, MO
395	Salem-Albany-Corvallis, OR
396	Salina, KS
397	Salinas-Monterey, CA
398	Salisbury, MD
399	Salt Lake City-Ogden, UT
400	San Angelo, TX

Basic Trading Areas in the United States

No.	Basic trading area
401	San Antonio, TX
402	San Diego, CA
403	Sandusky, OH
404	San Francisco-Oakland-San Jose, CA
405	San Luis Obispo, CA
406	Santa Barbara-Santa Maria, CA
407	Santa Fe, NM
408	Sarasota-Bradenton, FL
409	Sault Ste. Marie, MI
410	Savannah, GA
411	Scottsbluff, NE
412	Scranton-Wilkes-Barre-Hazleton, PA
413	Seattle-Tacoma, WA
414	Sedalia, MO
415	Selma, AL
416	Sharon, PA
417	Sheboygan, WI
418	Sherman-Denison, TX
419	Shreveport, LA
420	Sierra Vista-Douglas, AZ
421	Sioux City, IA
422	Sioux Falls, SD
423	Somerset, KY
424	South Bend-Mishawaka, IN
425	Spokane, WA
426	Springfield, IL
427	Springfield-Holyoke, MA
428	Springfield, MO
429	State College, PA
430	Staunton-Waynesboro, VA
431	Steubenville, OH-Weirton, WV
432	Stevens Point-Marshfield-Wisconsin Rapids, WI
433	Stillwater, OK
434	Stockton, CA
435	Stroudsburg, PA
436	Sumter, SC
437	Sunbury-Shamokin, PA
438	Syracuse, NY
439	Tallahassee, FL
440	Tampa-St. Petersburg-Clearwater, FL
441	Temple-Killeen, TX
442	Terre Haute, IN
443	Texarkana, TX-AR
444	Toledo, OH
445	Topeka, KS
446	Traverse City, MI
447	Tucson, AZ
448	Tulsa, OK
449	Tupelo-Corinth, MS
450	Tuscaloosa, AL

Basic Trading Areas in the United States

No.	Basic trading area
451	Twin Falls, ID
452	Tyler, TX
453	Utica-Rome, NY
454	Valdosta, GA
455	Vicksburg, MS
456	Victoria, TX
457	Vincennes-Washington, IN
458	Visalia-Porterville-Hanford, CA
459	Waco, TX
460	Walla Walla, WA-Pendleton, OR
461	Washington, DC
462	Waterloo-Cedar Falls, IA
463	Watertown, NY
464	Watertown, SD
465	Waterville-Augusta, ME
466	Wausau-Rhinelander, WI
467	Waycross, GA
468	Wenatchee, WA
469	West Palm Beach-Boca Raton, FL
470	West Plains, MO
471	Wheeling, WV
472	Wichita, KS
473	Wichita Falls, TX
474	Williamson, WV-Pikeville, KY
475	Williamsport, PA
476	Williston, MD
477	Willmar-Marshall, MN
478	Wilmington, NC
479	Winchester, VA
480	Worcester-Fitchburgh-Leominster, MA
481	Worthington, MN
482	Yakima, WA
483	York-Hanover, PA
484	Youngstown-Warren, OH
485	Yuba City-Marysville, CA
486	Yuma, AZ
487	Zanesville-Cambridge, OH

16

Intelligent Cell Concept and Applications

16.1 Intelligent Cell Concept

16.1.1 What is the intelligent cell?

In the cellular industry, system capacity is a great issue. As demand for cellular service grows, system operators try to find ways to increase system capacity. Capacity can be increased by reducing the cell sizes. This is called the *conventional microcell approach,* but it does not provide intelligence. When the cell size becomes smaller, the control of interference among the cells becomes harder. Also, the handoff time from the beginning of the initiation to the action completion sometimes may take around 15 s. If a mobile station is moving at a speed of 25 km/h (7 m/s), then the mobile station will travel 105 m in 15 s; at a speed of 50 km/h, the mobile station travels 205 m in 15 s. Since within a microcell of 0.5-km radius the overlapped region for a handoff is very small, then the mobile station is in the overlapped region too short a time for the handoff action to be complete. As a result, the call drops. In a conventional microcell system, interference is hard to control and the handoffs may not have enough time to complete.

The intelligent cell can solve the two problems. The intelligent cell concept can be used not only in microcells but also in regular cells to bring extra capacity to the system.

There are two definitions to describe an intelligent cell. One definition of intelligent cell is that the cell is able to intelligently monitor where the mobile unit or portable unit is and find a way to deliver confined power to that mobile unit. The other definition of intelligent cell is that signals coexist comfortably and indestructibly with the interference in the cell. From the first definition, the intelligent cell is called the *power-delivery intelligent cell,* and from the second definition, it is called the *processing-gain intelligent cell.* The intelligent cell may be a large cell such as a macrocell or a small cell such as a microcell. The intelligent cell increases capacity and improves performance of voice and data transmission. Since personal communication service (PCS) needs vast capacity and high quality, the intelligent cell concept is well-suited to it. Actually, using any means intelligently in a cell to improve the performance of services is what the intelligent cell stands for.

16.1.2 The philosophy of implementing power-delivery intelligent cells

Many different wireless versions of an intelligent cell can be used as long as they can deliver power to the location of the mobile unit. The easiest explanation is the analogy of a person entering a house (Fig. 16.1). In a conventional macrocell or microcell, when a mobile unit enters a cell or a sector, the cell site will cover the power to the entire cell or sector. This is because the cell site does not know where the mobile unit is within the cell or sector. This is just like a house that turns on all the lights when a person enters it.

Delivering power intelligently. In an intelligent macrocell or microcell, when a mobile unit enters a cell or a sector, the cell site covers only a local area, which follows the mobile unit. This is just like a house

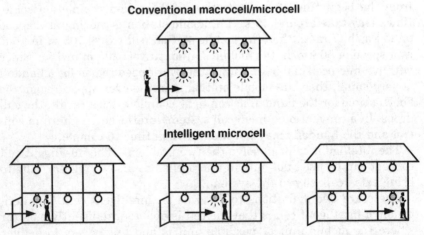

Figure 16.1 Microcell philosophy: energy follows the mobile analogy, light follows the person.

that turns on only the light of the first room a person enters. When the person enters the second room, the light of the first room is turned off and the light of the second room is turned on. Therefore, the light of only one room is on at a time and not the lights in the whole house. When the lights of the entire house A and the lights of the entire house B are on, the two houses should be largely separated in order to avoid the light being seen from one house to the other. If the light of only one room of house A and house B is on, the light that can be seen from one house to the other house is relatively weak. Thus, the distance between the two houses can be much closer.

This same analogy can be applied to a cellular system. In a cellular system, the frequency reuse scheme is implemented for the purpose of increasing spectrum efficiency. If two cochannel cells (cells that use the same frequency) can be placed much closer, then the same frequency channel can be used more frequently in a given geographical area. Thus the finite number of frequency channels can provide many more traffic channels, and both system capacity and spectrum efficiency can be further increased. In order to reduce the separation between two cochannel cells, the power of each cell should be reduced to cover merely one of numerous local areas in a cell if the cell operator is intelligent enough to know in which local area the mobile unit or handset is. Therefore, there are two required conditions:

1. The cell operator has to know where the mobile unit is located. Different resolution methods can be used to locate the mobile unit.
2. The cell operator has to be able to deliver power to that mobile unit. If the power transmitted from the cell site to the mobile unit can be confined in a small area (analogous to the light of a small room turning on when a person enters it), cochannel interference reduces, and the system capacity increases.

The extreme case of interference reduction is to connect the base transmitter and the mobile receiver by a wire. In this case, the wireless communication system becomes a wire line system, and interference is reduced to a minimum.

Radio capacity. In a frequency-reuse system, such as a cellular system, we always use the term *radio capacity* to measure the traffic capacity. The radio capacity m is defined as[1]

$$m = \frac{M}{K} \qquad \text{number of channels/cell} \qquad \text{(for omni cells)} \qquad (16.1\text{-}1)$$

or

$$m = \frac{M}{K \times S} \quad \text{number of channels/sector} \quad \text{(for sector cells)}$$

$$(16.1\text{-}2)$$

where M is the total number of frequency channels, K is the cell reuse factor, and S is the number of sectors. K can be expressed as[2]

$$K = \frac{1}{3}\left(\frac{D}{R}\right)^2 \qquad (16.1\text{-}3)$$

D is the cochannel cell separation and R is the cell radius. Also, according to Figure 16.2a, the relationship between the carrier-to-interference ratio C/I and D/R can be expressed[3] as

$$C/I = \frac{(D/R)^4}{6} \qquad (16.1\text{-}4)$$

Equation (16.1-4) is based on the propagation path loss of 40 dB/dec and omni cells. The radio capacity of omni-cell systems is[3]

$$m = \frac{M}{\sqrt{\dfrac{2}{3}\left(\dfrac{C}{I}\right)}} \quad \text{channels/cell} \qquad (16.1\text{-}5)$$

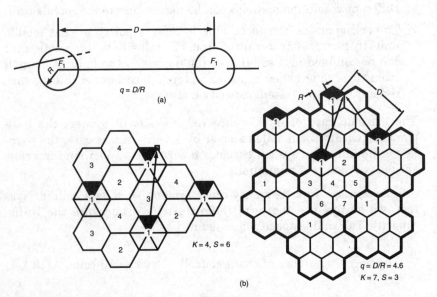

Figure 16.2 The D/R relationship in different sectorial cells.

Any other parameters can be derived from the measures of radio capacity such as erlangs/cell, erlangs/Km², and calls/km².

The normalized radio capacity is

$$\hat{m} = m/B_T \quad \text{channel/cell/spectral band} \quad (16.1\text{-}6)$$

where B_T is the total spectral band. When two systems operate at two different spectral bands such as B_{T_1} and B_{T_2}, then the radio capacities m_1 and m_2 have to be normalized by first using Eq. (16.1-6) to find m_1/B_{T_1} and m_2/B_{T_2} before comparing their radio capacities $\hat{m}_1$ and $\hat{m}_2$.

In a frequency-division multiple-access (FDMA) system, M is the total number of frequency channels, and in a time-division multiple-access (TDMA) system, M is the total number of slot channels. M is a countable number and is fixed, but K is a variable number and depends on cochannel separation D as shown in Fig. 16.2b. However, because of the reuse of frequency channels, more traffic channels are generated. If one frequency channel is used 50 times, then the traffic channel becomes 50 M. Based on a $K = 7$ system, the number of cells will be $50 \times 7 = 350$ cells. The radio capacity $m = M/7$ is measured by the number of frequency channels per cell, which is determined by the cell reuse factor K only, and not by the number of total cells in the system. The radio capacity m increases if K is reduced, provided that the voice quality and data performance are maintained according to the specification.

Implementation of the intelligent cell concept may involve using multiple zones, multiple antenna beams, multiple isolated spaces, or any means of eliminating interference. There are many kinds of intelligent cells as described in the following sections, where we will compare their radio capacities.

16.1.3 Power-delivery intelligent cells

Zone-divided cells. In general there are three kinds of zone-divided cell systems.

Sectorial cells. Sectorial cells are used to reduce interference. Usually sectorial cells are used when the terrain contour in the cells is not flat, causing unevenly distributed interference from other cells. There are two kinds of sectorial cells.

- 7-cell/3-sector reuse system ($K = 7$, $S = 3$)
- 4-cell/6-sector reuse system ($K = 4$, $S = 6$)

In both systems, each sector has a set of unique designated channels. The mobile unit moving from one sector or one cell to another sector or cell requires an intracell handoff. Based on a cluster of either $K = 7$ cells or $K = 4$ cells (see Fig, 16.2b), the radio capacities of these two systems are

$$m_1 = \frac{M}{7 \times 3} = \frac{M}{21}$$

channels/sector (for a cellular system of $K = 7$ and $S = 3$)

$$m_2 = \frac{M}{4 \times 6} = \frac{M}{24}$$

channels/sector (for a cellular system of $K = 4$ and $S = 6$)

Sectorial cells with intracell handoffs do not increase radio capacity, but cochannel interference (Fig. 16.2) can be reduced as shown below:

$$\frac{C}{I} = \frac{R^{-4}}{(D + 0.7R)^{-4} + D^{-4}}$$
$$= 285 \text{ or } 24.5 \text{ dB} \qquad \text{(for } K = 7, S = 3) \qquad (16.1\text{-}7)$$

$$\frac{C}{I} = \frac{R^{-4}}{(D + R)^{-4}} = 395 \text{ or } 26 \text{ dB} \qquad \text{(for } K = 4, S = 6) \qquad (16.1\text{-}8)$$

The above calculation is based on the mobile radio propagation pathloss rule of 40 dB/dec. Because only one or two cochannel interfering sectors are affected, as seen in Fig. 16.2, the C/I will improve by about 7 to 8 dB as compared with $C/I = 18$ dB for a $K = 7$ omni-cell system. When large-size cells are implemented because of the variation of the terrain contour, the sectorial cells should be used to gain this additional margin in decibels to overcome long-term fading. However, more frequent handoffs and larger overlapped regions occur in a $K = 4$ cell than in a $K = 7$ cell.

Intelligent microcells.[4]   When dividing a cell into many zones as in Fig. 16.3, the cell operator knows which zone the mobile unit is in and delivers the radio signal to that zone. When the mobile unit has been assigned a frequency channel for a call, the frequency channel is always associated with that call within the cell. The cell operator simply turns on the new zone site while the mobile unit is entering and turns off the old zone site when it leaves with the assigned frequency channel to the mobile unit unchanged. With this arrangement, we can find the received C/I value at a mobile unit for a scenario having six co-

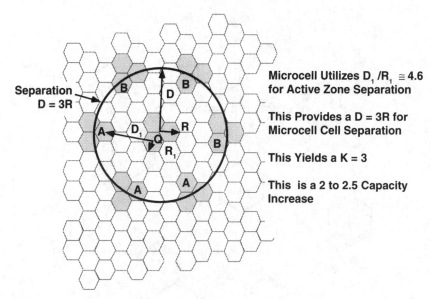

Separation
D = 3R

Microcell Utilizes $D_1/R_1 \cong 4.6$
for Active Zone Separation

This Provides a D = 3R for
Microcell Cell Separation

This Yields a K = 3

This is a 2 to 2.5 Capacity
Increase

Figure 16.3 Intelligent microcell capacity application.

channel interferers at the first tier surrounding the center cell (Fig.
16.4). It is easy to show that the six interferers at the second tier do
not contribute any significant interference to the center cell.[4] Assume,
as shown in Fig. 16.3, that the desired mobile unit is in zone Q of the
center cell and there are three interfering zones marked A and three
interfering zones marked B, one in each interfering cell where the six
interfering mobile units could be. This is a worst-case scenario. In this
case, the voice quality should be maintained at the stated requirement
of $C/I \geq 18$ dB in each zone; that is, the ratio of cochannel zone separation D_1 to zone radius R_1 should be $D_1/R_1 = 4.6$. Based on the six
worst interfering zones one in each cell, with their minimum D_1/R_1
being 4.6, we can find the resultant C/I received by the desired mobile
unit:

$$C/I = \frac{(R_1)^{-4}}{\underbrace{3(4.6R_1)^{-4}}_{\text{zone A}} + \underbrace{3(5.5R_1)^{-4}}_{\text{zone B}}} \approx 100 \text{ or } 20 \text{ dB} \qquad \text{(worst case)}$$

$$(16.1\text{-}9)$$

The average C/I received by the desired mobile unit can be calculated
by taking the probability that each interfering mobile unit would be
located in one of its three zones of an interfering cell. Thus:

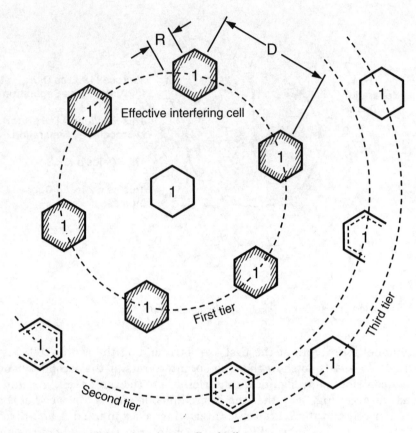

Figure 16.4 Six effective interfering cells of cell 1.

$$C/I = \frac{R^{-4}}{2\left[\dfrac{2}{3}\left(\dfrac{13}{2}R_1\right)^{-4} + \dfrac{1}{3}(4.6R_1)^{-4}\right] + \left[\dfrac{2}{3}(4.6R_1)^{-4} + \dfrac{1}{3}(6R_1)^{-4}\right]}$$

$$+ 2\left[\frac{2}{3}\left(\frac{13}{2}R_1\right)^{-4} + \frac{1}{3}\left(\frac{16}{2}R_1\right)^{-4}\right]$$

$$+ \left[\frac{1}{3}\left(\frac{13}{2}R_1\right)^{-4} + \frac{2}{3}\left(\frac{16}{2}R_1\right)^{-4}\right]$$

$$= 193.4 \text{ or } 22.8 \text{ dB} \quad \text{(average case)} \qquad (16.1\text{-}10)$$

As seen from Eq. (16.1-10), the average C/I almost equals 23 dB, which is 5 dB better than the AMPS voice quality specification. By keeping the minimum $D_1/R_1 = 4.6$ in Fig. 16.3, as a result, the co-channel cell separation D equals $3R$. The value of K can be found from

Eq. (16.1-3) as $K = 3$. The capacity of intelligent microcells is greater than that of AMPS by a factor of $7/3$, or 2.33.

Reuse of sectorial beams with directional antennas. Applying the same intelligent cell concept, we use antenna beams to confine the energy to individual mobile units in the cell. In a $K = 7$ cellular system, each cell has a set of M/K frequency channels. At a cell site, if six sectorial (directional) antennas are used to cover 360° in that cell and if the whole set of frequency channels assigned to the cell is divided into two subsets which are alternating from sector to sector, then there are three cochannel sectors using each subset in a cell, as shown in Fig. 16.5. In this arrangement, we can increase the capacity by 3 times. If N sector beams are reused alternately, the capacity is increased by $N/2$ times the AMPS capacity. This reuse of the sectorial beam scheme can be used in a small-cell system or a large flat-terrain cell system with much less reduction on trunking efficiency. Of course, in reality, the directional antenna front-to-back ratio should be considered to avoid unnecessary interference in the cochannel sectors.

Adaptive antenna array.[5] The antenna pattern can be formed by tracking the mobile unit and nulling the interference. Therefore, if the same frequency channel can be used by N mobile units in a cell, the capacity is Nx. Also, because an adaptive antenna array is used, the antenna beam is able to follow the mobile unit, thus reducing interference. The cell reuse configuration may reduce from $K = 7$ to a smaller K depending on the magnitude of N. If N is large, the N antenna beams operating the same frequency to serve N users within a cell can be treated as they are from an omnidirectional antenna. Therefore, $K = 7$ will remain unchanged. If N is confined in a 120° sector, then the required $C/I = 18$ dB can be used to determine the D/R ratio. Since there are only two interfering cells, as shown in Fig. 16.2, the C/I is expressed approximately as

$$C/I = \frac{R^{-4}}{2D^{-4}} = 63 \text{ or } 18 \text{ dB} \qquad (16.1\text{-}11)$$

Solving Eq. (16.1-11) yields

$$D/R = (126)^{1/4} = 3.35$$

The value of K can be obtained from Eq. (16.1-3) as

$$K = \frac{(3.35)^2}{3} = 3.7$$

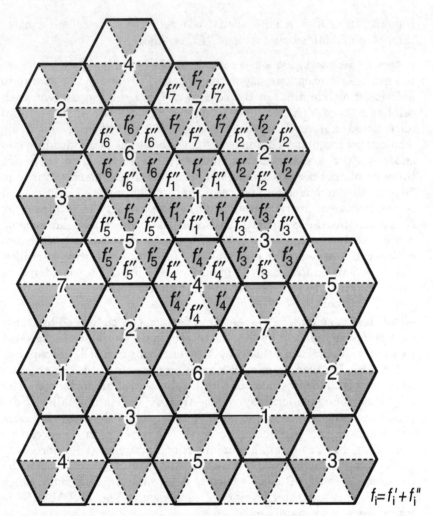

Figure 16.5 Reuse of sectorial beams with directional antennas.

Since the adaptive antenna beam follows the mobile units, cochannel interference is reduced. Also, since sectorization is used, a system with a frequency reuse factor $K = 4$ can be realized.

These adaptive antenna patterns provide a good means of generating multiple cochannel mobile calls on the reverse links. Then with the identical antenna patterns for transmitting and receiving at the cell site, the calls are conducted on the forward links as well. This is because the reciprocity principle works based on Lee's Model,[6] the active region around the mobile unit is defined by a radius of about 100 to 200 wavelengths.[7] The beam angle α received at the cell site is a function of distance R, as shown in Fig. 16.6:

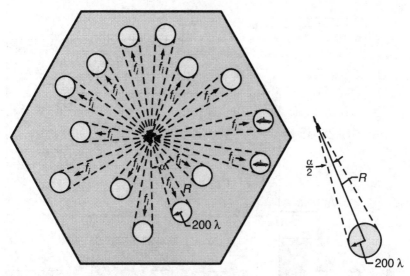

Figure 16.6 Intelligent cell with adaptive antenna-array beams.

$$\alpha = \frac{2 \times 200\lambda}{R} \qquad (16.1\text{-}12)$$

where λ is the wavelength. Usually operating at UHF, the antenna has a beamwidth θ always larger than α. Then the isolation between the two cochannel mobile calls will be measured by θ, not α. Also, the definition of the antenna beamwidth θ is not based on a 3-dB beamwidth but rather on an 18-dB beamwidth. When the two cochannel mobile units move closer within one θ angle, a handoff is initiated. In some proposed systems, the beam nulling is formed between two mobile units. In this case, the nulling angle measured between two mobile units cannot be less than α.

In-building communication. In-building communication needs sufficient traffic channels, but the radio spectrum is limited. We may apply the intelligent cell concept to solve this problem. The number of traffic channels can be increased by treating each floor of a building as a cell. The penetration loss of a radio signal through a reinforced concrete building wall is about 20 dB, and the signal isolation between two adjacent floors is also 20 dB. Therefore let a group of frequency channels M_1 be assigned to the inbuilding use, those same M_1 channels will be reused in every floor of the building, and also in every floor of neighboring buildings (i.e., passive intelligence), as shown in Fig. 16.7.

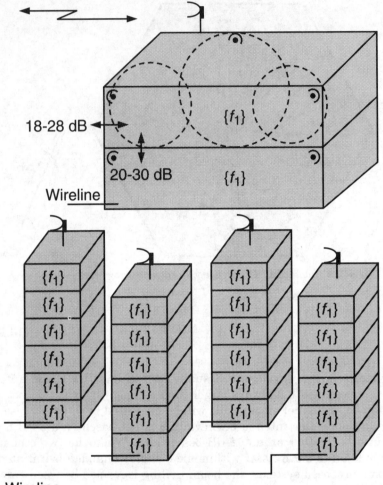

18-28 dB

20-30 dB

Wireline

Figure 16.7 Concept of in-building communication.

Because each floor can be treated as one cell, the radio capacity m_1 of in-building communication can be obtained by setting $K = 1$. Then:

$$m_1 = \frac{M_1}{K} = M_1 \qquad \text{number of channels/floor}$$

Now a 20-floor building has $20 \times M_1$ traffic channels. In case interference occurs between two buildings due to the signal penetration of 4 to 6 dB at the window areas, the intelligent microcell cell with N zones can further reduce the interference. Note that the intelligent cell assumes adequate building shielding. When "active" rather than "passive" intelligence is used, buildings with less isolation can still achieve high efficiency of spectral reuse by self-surveying the amount of signal leakage into the building or between floors.

16.1.4 Processing-gain intelligent cells
($K \rightarrow 1$ system)

Philosophy of implementing processing-gain intelligent cells.   The concept of the processing-gain intelligent cell can be explained by the analogy of many simultaneous conversations in a big hall (Fig. 16.8). The big hall is just like one big radio channel serving all the traffic in an intelligent cell. The conversations of the different parties are the traffic channels. The processing gain is like the size of the hall, which limits the number of persons, hence, the number of conversations, that can be accommodated. Since all the conversations take place in the hall, the speaker level (power control) of each conversation is the key element in this intelligent cell to keep the interference level in each traffic channel down. Also, if the level of each individual conversation can be controlled intelligently, then we can maintain the total interference level and add more conversations. If the power control is not working, the cell is not an intelligent cell. Then the cell will face the

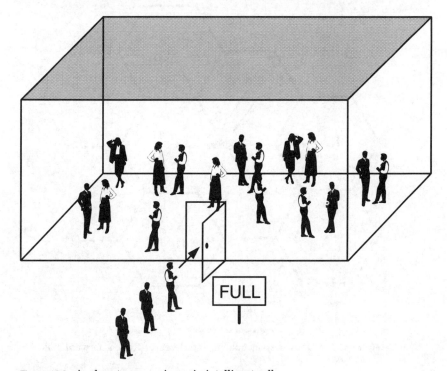

Figure 16.8 Analogy to processing-gain intelligent cell.

cocktail-party syndrome in which no parties can talk except by raising their voices. The processing gain calculation will be shown in the next section.

Direct-sequence CDMA.[8-10] In this code-division multiple-access (CDMA) system, the broadband frequency channel can be reused in every adjacent cell so that K is close to 1, as shown below from our previous definition of Eq. (16.1-3). In a practical sense, D equals $2R$, i.e., all the same available frequencies are used in each cell. Then

$$K = \frac{(D/R)^2}{3} = 1.33 \tag{16.1-13}$$

Radio capacity is also based on Eq. (16.1-1). However, in direct-sequence CDMA (DS-CDMA), K is fixed but M (the total number of available channels) is a variable and depends on the interference situation. For the scenario shown in Fig. 16.9, the interference comes

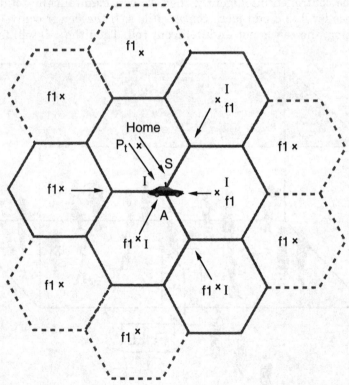

Figure 16.9 CDMA system and its interference (from a forward link scenario).

from the home cell and the adjacent cells and the value of C/I is expressed as

$$\frac{C}{I} = \frac{E_b}{I_0} \times \frac{R_b}{B} = \frac{E_b/I_0}{\text{PG}} \tag{16.1-14}$$

where E_b = energy per bit

I_0 = interference per hertz ($I_0 >> N_0$, where N_0 is thermal noise per hertz)

R_b = information rate

B = bandwidth per channel

PG = processing gain = $10 \log (B/R_b)$ (16.1-15)

In DS-CDMA, since PG is greater than one, C is always smaller than I in Eq. (16.1-14) even if there is only one active user. The processing gain is used to overcome I and determine the number of traffic channels that can be created.

From two sets of given values, E_b/I_0 and R_b/B, we can find the C/I from Eq. (16.1-14). Then, from the scenario of Fig. 16.9, the C/I at the mobile location A can be used to find the number M which is considered a worst case. Assume that the interference level is much higher than the thermal noise level; then

$$\frac{C}{N + I} \rightarrow \frac{C}{I}$$

Assume that in a DS-CDMA system,[8] B = 1.23 MHz and R_b = 9.6 kbps. Then PG = 1.23 MHz/9.6 kbps or 21 dB.

In this system, a voice quality is accepted at a frame error rate FER = 10^{-2}, which typically corresponds to E_b/I_0 = 7 dB. Knowing the values of PG and E_b/I_0, C/I of Eq. (16.1-14) can be obtained:

$$C/I = 7 - 21 = -14 \text{ dB or } 0.03981$$

Find the number of traffic channels m_i in each cell from C/I by:

$$C/I = \frac{\alpha_1(R)^{-4}}{\underbrace{\alpha_1(m_1 - 1) \cdot R^{-4}}_{\text{own cell}} + \underbrace{(\alpha_2 m_2 + \alpha_3 m_3)(R)^{-4}}_{\text{2 adjacent cells}} + \underbrace{\beta \cdot (2R)^{-4}}_{\text{3 interim cells}} + \underbrace{\gamma(2.63R)^{-4}}_{\text{6 distant cells}}}$$

$$= 0.03981 = \frac{1}{25.1}$$

where

$$\beta = \alpha_4 m_4 + \alpha_5 m_5 + \alpha_6 m_6$$

$$\gamma = \sum_{i=7}^{12} \alpha_i m_i \tag{16.1-16}$$

and m_i and α_i are the number of traffic channels and the power level, respectively, in each of the i cells. Solving Eq. (16.1-16), we obtain

$$m_1 = 26.1 - \left[\frac{\alpha_2 m_2 + \alpha_3 m_3}{\alpha_1} \right] - \frac{\beta}{\alpha_1}(2)^{-4} - \frac{\gamma}{\alpha_1}(2.63)^{-4} \tag{16.1-17}$$

Case A. Single cell case ($\alpha_i = 0$ for $i \neq 1$). From Eq. (16.1-17):

$$m_1 = \frac{I}{C} + 1 = 25.1 + 1 = 26.1 \text{ traffic channels/cell}$$

Case B. Identical-cell case. All the cells have the same power and the same number of traffic channels: $m_i = m_j$ and $\alpha_i = \alpha_j$. We may substitute $m_i = m$ and $\alpha_i = \alpha$ into Eq. (16.1-16) or Eq. (16.1-17) and solve for m as follows:

$$26.1 = m[3 + 3 \cdot (2)^4 + 6 \cdot (2.633)^{-4}]$$

$$m = 7.85 \text{ traffic channels/cell}$$

Both traffic channels appearing in Case A and Case B include the overhead channels for sync and set-up, but do not take into consideration the voice activity cycle or sector-reuse factor as used in real commercial systems.[8]

Frequency hopping. A frequency-hopping system can be used as a CDMA system (FH-CDMA). The hopping pattern can be formed as a code sequence. Frequency hopping has been used in the past to overcome enemy jammers in military applications. There are two kinds of hopping, fast frequency hopping and slow frequency hopping.

Slow frequency hopping (SFH) is defined as sending multiple bits on a single hop.[11] Depending on the degree of the enemy's quick reaction, the hopping rate would be adjusted. Fast Frequency Hopping (FFH) is defined as sending a bit on a pseudo-random pattern of frequency channels, then sending the next bit on a different pseudo-random pattern of frequency channels. The multiple frequency channels form a code for one bit which is sent out simultaneously or sequentially. The bandwidth of the channel depends on the transmission bit rate. The scheme of simultaneously sending out the same bit on different frequency channels requires a larger bandwidth for sending each bit. This is another wideband CDMA system. Sending bits over frequency channels sequentially is the conventional FH-CDMA system. The fast frequency hopping CDMA system requires a larger bandwidth than the slow frequency hopping CDMA system. This is because in FFH, one bit requires multiple frequency-hopping channels to be sent out sequentially.

In an FH system, there are two kinds of processing gains. One kind of processing gain is used to measure the power of defeating enemy jammers. It is really focused on minimizing collisions of two carriers occupying a frequency channel at the same time. Then if the desired signal has 1000 channels to hop to avoid the jammer, the processing gain is 30 dB. We can derive this processing gain against jamming from $(C/I)_J$ using Eq. (16.1-14):

$$(C/I)_J = \frac{E_b R_b}{I_0 (BN)} \qquad (16.1\text{-}18)$$

where B is the bandwidth of sending R_b through a single channel (assume that $B = R_b$) and N is the number of available frequency channels to be hopped from. Each channel has the same bandwidth B. The processing gain is

$$\text{PG} = \frac{BN}{R_b} = N \qquad (16.1\text{-}19)$$

From Eq. (16.1-19) we may conclude that FFH and SFH can take the advantage of the processing gain in defeating enemies.

The other kind of processing gain is used to increase radio capacity. It is focused on spreading energy channels such that the interference seen by each bit is near the minimum acceptable performance threshold. The carrier-to-interference ratio $(C/I)_F$ of a frequency-hopping system can be expressed differently from Eq. (16.1-14) as

$$(C/I)_F = \frac{E_b R_b}{I_0 (BF)} = \frac{E_c R_c}{I_0 (BF)} \qquad (16.1\text{-}20)$$

where E_b = energy per bit
R_b = bits per second
E_c = energy per code bit
R_c = number of code bits per second
F = number of frequency channels per bit ($F_b \geq 1$)
B = bandwidth of sending a signal of R_b stream

In a non-FH system or an SFH system, R_c and F are always equal to one because there is no spread spectrum using pseudonoise (PN) coding of the data bits. Then $E_c = E_b$, $F = 1$, and Eq. (16.1-20) becomes Eq. (16.1-14). It has been shown that SFH does not experience processing gain in order to increase radio capacity. The SFH system is more like the Aloha multiple access scheme.[14]

In an FFH system, the processing gain for radio capacity depends on F, the number of frequency channels per bit, as

$$PG = \frac{BF}{R_b} = F \qquad (16.1\text{-}21)$$

Assume that the information rate R_b is 10 kbps and one bit hops among 100 frequencies, $F = 100$, then the total bandwidth required is $BF = 1$ MHz.

Impulses in time domain.[12–13] We may create a spread spectrum system based on impulse position modulation in the time domain. The C_p/I can be obtained as

$$C_p/I = \frac{E_p P_b R_s}{I_0 B} \qquad (16.1\text{-}22)$$

where C_p = carrier power of data stream
 E_p = energy per pulse
 P_b = number of pulses/bit
 R_s = number of bits/s

If $P_b = 1$ then $E_p P_b = E_b$ and Eq. (16.1-22) becomes Eq. (16.1-13). If $P_b > 1$, the impulse position modulation system becomes a spread spectrum system.

The same principle of using DS-CDMA for increasing cellular capacity can be applied to impulse position modulation. When the pulse width of an impulse is less than 1 ns, the advantage of applying impulse position modulation starts to show.

16.1.5 Summary of intelligent cell approaches

The intelligent cell can be described in two ways: (1) it intelligently delivers the signal to the mobile unit; (2) it ensures that the signal indestructibly resides with the interference. Several different methods of increasing traffic capacity by using the intelligent cell concept have been mentioned. These methods are very important for PCSs. Among the methods of forming a power-delivery intelligent cell, the sectorial cells may increase capacity if the cell reuse factor K decreases. Frequency reuse with multiple antenna beams and the microcell with multiple zones apply the intelligent cell concept and can also increase capacity. The method of forming a power delivery intelligent cell can be applied to FDMA and TDMA[15] systems. Among the methods of forming a processing-gain cell, both direct-sequence CDMA[16] and FFH can increase capacity because of their processing gain but SHF cannot.

In a conceptual case, we can narrow the beamwidth θ of an antenna pointing at the mobile unit to a limit; i.e., θ becomes narrower and narrower until $\theta \to 0$ (Fig. 16.10). This is the best case for eliminating interference. When $\theta \to 0$, the wireless line is just like a wire line. Therefore, we conclude that the wire line is the least interference link for the wireless link. If a wire line could be replaced by a wireless line with $\theta \to 0°$, then no radio interference exists. As a result, no limitation in increasing the number of channels has imposed by any radio interference among them.

After learning the capacity issues of the three multiple-access schemes, we may present them in three axes, number of frequency channels (x), number of time slots (y), and number of code sequences (z), as shown in Fig. 16-11a. The shaded regions represent the utilization of the spectrum. The larger the region, the higher the spectrum efficiency. Three subcases, FDMA, TDMA, and CDMA, appear in Fig. 16.11b, c, and d, respectively. The shaded regions in each subcase indicate the degree of spectrum efficiency. There is a different representation of spectrum efficiency in each multiple-access scheme.

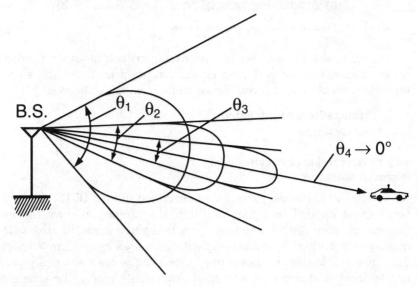

θ-Antenna beamwidth

$\theta_1 > \theta_2 > \theta_3 > \theta_4$ $\qquad\qquad$ $\theta_4 \to 0$

Figure 16.10 The best case of delivering power to the mobile unit in space by antenna.

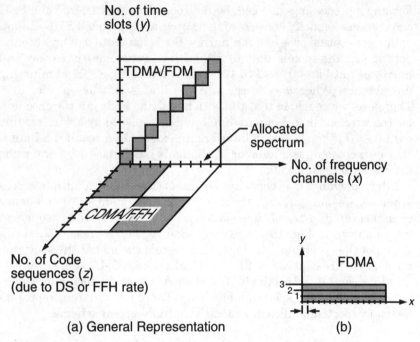

Figure 16.11 Spectrum efficiency representation.

Therefore, the bottom line of the intelligent cell is either to allow no interference to exist during signal reception or to let the signal tolerate a great deal of interference while it is being received.

16.2   Applications of Intelligent[17,18] Microcell Systems

16.2.1   Description of the intelligent microcell operation

An intelligent microcell system was described in Sec. 16.13. This system can be applied to analog and digital systems. To show the improvement over AMPS provided by intelligent microcells, the voice quality is 2 dB better and the capacity increases more than 2 times. The microcell system is shown in Fig. 16.12. The base-site equipment can be located at one zone site or at any remote place. The base site stores a zone selector, a scanning receiver, and a set of radio channels (Fig. 16.12a). The microcell system can be attached to a regular macrocell site (could be a different cellular vendor's equipment), and can be used without any modification on the base-site controller. Also, the microcell system can operate alone if the controller is replaced with an independent microprocessor. Either has to connect to the MTSO.

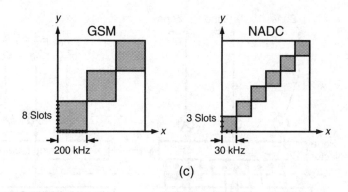

(c)

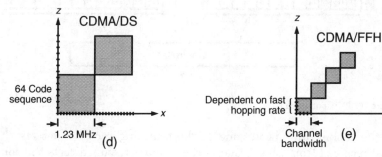

Figure 16.11 (*Continued*)

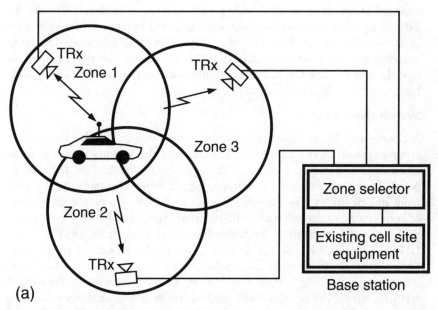

(a)

Figure 16.12 The structure of the microcell system. (*a*) The basic microcell concept; (*b*) modifying the equipment arrangement for the microcell system.

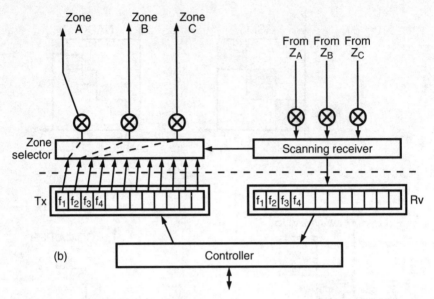

Figure 16.12 (*Continued*)

If the mobile unit is in zone 1, the scanning receiver detects the signal received from zone 1 that is the strongest, then directs the zone selector to switch to zone 1 through a pair of converters (called *translators*). All three zones are receiving the same signal on the same radio channel at any time but only one zone is chosen to transmit the signal. The equipment arrangement for the microcell system at the base site is shown in Fig. 16.12*b*. In this figure, the new components—zone selector, scanning receiver, and converters—above the dotted line are installed to replace the old components such as power amplifiers, combiner, and 800-MHz antennas.

System elements

Converters.   There are three pairs of converters in the three-zone microcell system. Each pair is used to connect the signals from the base site to the designated zone site. The converter (Fig. 16.12*b*) is a broadband device which can upconvert signals before transmitting them, then downconvert signals while receiving them. There are two kinds of converters: microwave and optical-fiber. The microwave type can use different upconverting microwave frequencies, such as 18, 23, or 40 GHz. The higher the frequency, the shorter the reception range. The optical-fiber type is used for upconverting the signal to optical frequency and sending it over fiber cable. Every converter is associated with an amplifier for the weak signals after downconverting back to the cellular frequencies. A converter can carry an average of 20 to 30 cellular channels, depending on the linear range of the converter. In this case, on average, if each zone site will take 20 mobile calls, the total of three zones will have 60 mobile calls. The converters[19] can be either analog or digital regardless of whether the system is analog or digital.

At any zone site, only one converter is needed, as shown in Fig. 16.12. The zone site can physically represent a regular cell site. The converter can be mounted on a utility pole or light pole (Fig. 16.13). Therefore, the size of converters becomes a more important factor for the zone site. The smaller, the lighter, the better.

Scanning receiver. The scanning receiver will scan three zone sites for each frequency. After scanning all the frequencies sequentially, the receiver goes back to the first frequency. The required time of scanning

Figure 16.13 An optical converter (analog conversion) manufactured by Allen Telecomm.

the frequencies and zones is critical. The required time of scanning needs to be short so that more frequencies can be scanned by one scanning receiver.

A scanning receiver for a TDMA system should have the ability to scan not only the frequencies, but also the zones and the time slots. Therefore, a time multiplexing device should be associated with the scanning receiver.

Zone selector. A zone selector is a single switch. It usually switches a signal from one zone to another zone in less than a microsecond. Therefore, no data stream will be affected by this fast switch. In TDMA, the zone selector is able to switch a signal in one time slot of one zone to a designated time slot of another zone.

Areas of application. The intelligent microcell has five applications.

1. *Delivering to extended cells.* Converters can be used to deliver the 800-MHz signal from the base to an extended cell by upconverting the signal to a new frequency for transmission through the air, then downconverting to 800 MHz when the signal reaches a cell where only a converter is installed (Fig. 16.14).

2. *Increasing capacity.* Since the power can be delivered and received intelligently at the mobile unit, the capacity increases.

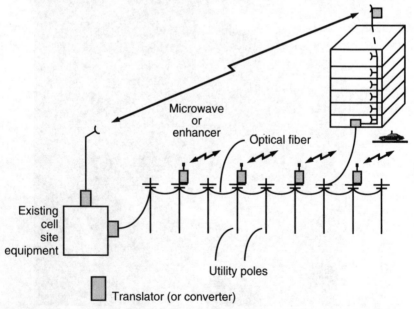

Figure 16.14 Microcell delivery system. The translators are used·to upconvert an 900-MHz GSM signal to an optical signal (or microwave signal) and downconvert an optical signal (or microwave signal) to 900 MHz.

3. *Coverage.* In some areas, the government will not allow regular cell sites to be installed. Under those circumstances, these invisible zone sites can be used to provide the coverage especially in urban areas.[20-22]

4. *Reducing interference.* At some high cell sites, the generated interference to the other low sites becomes a problem. An intelligent microcell can reduce its unnecessary radiated power and the interference it generates.

5. *In-building communication.* The intelligent cell can increase radio capacity many times. A description of this application will appear later.

Advantage of implementing intelligent cells

1. Any number of zones can be included in a microcell. More zones in a microcell can further reduce power and lower the interference.

2. The antennas face inward from the edges of the cell, rather than outward, further reducing the interference.

3. All the zone-site receivers actively receive mobile or portable calls on all frequencies. Since a portable unit's transmit power is usually low, a single cell site receiver has difficulty accepting and maintaining a call, but with three zone sites receiving the same signal from the portable, the reception is much improved. Thus an intelligent microcell is well-suited for PCS terminals.

4. Within a three-zone microcell, no handoff is needed. This arrangement eases the load of switches (MTSO) because of fewer handoffs, and the switches can have more capacity to handle new calls. By reducing handoffs, the intelligent microcell also reduces the number of calls dropped during attempted handoff. In a conventional microcell system, the vehicle often leaves the cell before the handoff action can be completed.

5. The system can be implemented on any existing cellular system. The modification is shown in Fig. 16.12*b*, where a dashed line divides the equipment in a regular cell site from that in a microcell site. Where the power amplifier and channel combiner would be installed in a regular cell site, the scanning receiver, zone selector and the converter are installed instead in a microcell site.

6. The zone site can be moved from one location to another in almost no time. To take an antenna and a converter down from one utility pole and put them up on another utility pole is easy.

7. A fiber-cable network can provide redundancy of connections. If the fiber cable is broken on one end, the signal can be delivered via the network through the other end to the mobile unit.

8. No need to modify the existing cellular subscriber units for the intelligent microcell system.

9. Better voice quality than an analog cellular system.

10. Higher capacity—can be 2.33 times the capacity of the analog system.

Cable cost and converter quality.   In high traffic congestion areas, many zone sites will be installed. The major cost of providing the fiber links is the installation of fiber cable. Usually, the initial cost of the fiber links is high. This situation may lead us to recall that in the beginning of this century many people complained that the initial cost of telephone-wire installation was too high. One hundred years later, we are all benefiting from this telephone network. Since the fiber-cable network will inevitably be the future broadband network, why should we have to be concerned about the initial costs of the installation if it is needed for the future?

The quality of converters is a big challenge. The simplest way of building a converter is to use an analog direct conversion approach. But the required linearity over a broad dynamic range with broadband conversion could make a high-quality analog converter difficult to realize. A digital converter may cost more but may not require linearity, therefore the quality can be better. However, the analog converter would gracefully degrade in performance, but the digital converter would sharply degrade.

16.2.2 Applications to increasing capacity

To verify the increased capacity provided by intelligent microcells, four microcells were deployed in West Los Angeles. Each microcell had three zone sites. The cell reuse pattern was $K = 3$ (Fig. 16.15). Cells 1A and 2B were the cochannel microcells. The measured signal strengths at 1A and 2B are shown in Fig. 16.16a and b respectively. From the measured data, we could calculate the C/I at zone 1A and at zone 2B, as shown in Fig. 16.16c. The result indicates that C/I at each cochannel microcell (1A or 2B) is about 20 dB, as we have expected. Since the cell reuse factor K reduces from $K = 7$ to $K = 3$ while the C/I improves from 18 dB to 20 dB, both the capacity and the voice quality agree with the theoretical prediction.

Deployment along city streets.   Deploying intelligent microcells along city streets is depicted in Fig. 16.17. All the zone sites are located at street crossings. They are lined up diagonally, as shown in the figure. Since each zone site transmits low power, only enough to cover two crossed streets, we may connect the four zone sites to form a microcell

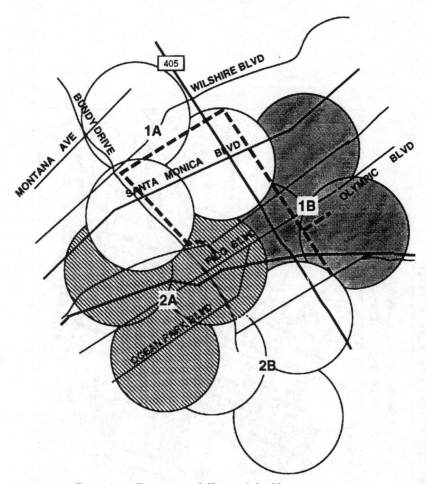

Figure 16.15 Four-microcell pattern and fiber-optic backbone.

using one zone selector. Then, when a vehicle is assigned to one frequency channel f_i, this f_i will be switched from one zone site to another zone site, depending on the zone where the mobile unit is. If the zone selector can serve 20 mobile calls at one time, the same number of calls will be served in this four-zone microcell. For this low-power, intelligent power delivery approach, all the cochannel microcells can be pulled closer. As a result, the capacity is increased.

16.2.3 Applications of coverage provision

There are two applications of the coverage provision: coverage along winding roads and coverage under the ground.

590

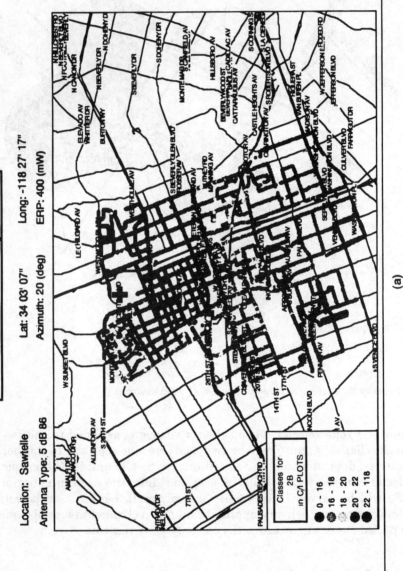

Figure 16.16 (a) Microcell 1A C/I plot (measured). (b) Microcell 2B C/I plot (measured). (c) Microcell C/I plot (calculated).

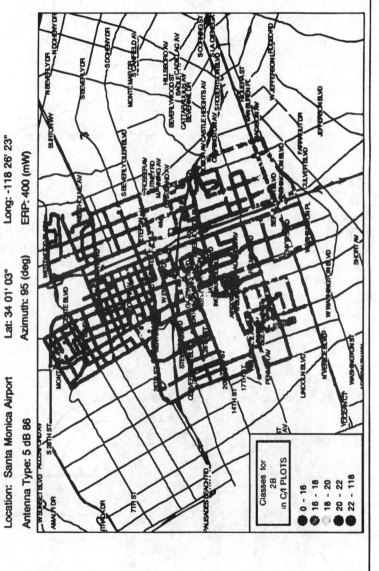

Figure 16.16 (Continued)

591

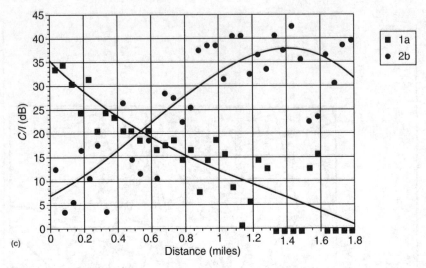

Figure 16.16 (*Continued*)

Coverage along winding roads. Roads like the Pacific Coast Highway, in Malibu, or the road along Santiago Canyon, in Los Angeles, need coverage that stretches up to 15 mi. In Malibu, the zoning requlations prevent the installation of regular cell sites. In Santiago Canyon, the regular cells installed along the winding roads need to be very close for coverage and as a result are very costly. For these reasons, intel-

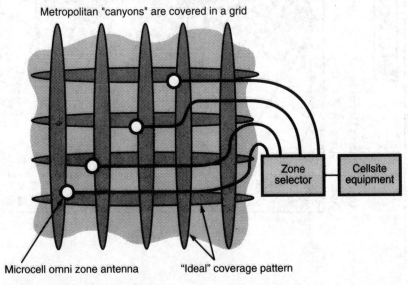

Figure 16.17   Metro "canyon" coverage.

ligent microcells are a good candidate in both cases. In Fig. 16.18, two directional antennas back-to-back are installed on each zone site. Directional antennas and the converters mounted on utility poles form a zone site. Since the traffic is light along these roads, we may use one zone selector in the 15-mi microcell system. Along the road, we number the zone sites along the road 1, 2, 3, 1, 2, 3, repeatedly. In the three-zone selector, all zone sites named zone 1 are connected together, all named zone 2 are connected together, and all named zone 3 are connected together. Then, when a vehicle has been assigned a frequency, f_i and starts to travel from the left end of the road, the left directional antenna of zone 1 receives the mobile signal first, then the right directional antenna on zone 1 receives it later. During this period, all zone 1 sites are on. Because the interference caused by the low-power sites is low, interference is not a concern. The mobile call then passes from zone 1 to zone 2 to zone 3, then back to zone 1 to zone 2 to zone 3, etc. The frequency f_i always stays with this particular mobile call along the 15-mi-long road. No handoff action is needed. This arrangement is suitable for light traffic road conditions.

Coverage under the ground (subway coverage). The intelligent microcell also can be used in subways. In a subway, no interference occurs from the ground cell sites. The three-zone selector can be used to cover a microcell with three subway stations when the stations are very closely separated. Each of the three zones is covered in a different shade pattern in Fig. 16.19. The left-bound train (top) and the right-bound train (bottom) are shared by one zone but covered in different sections. In this case, the two trains stopping at the station will be carried by different zones to ease the traffic.

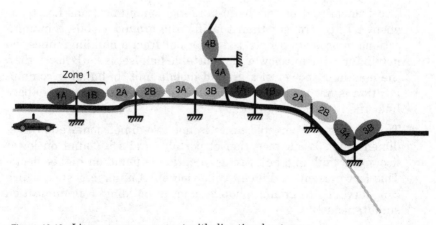

Figure 16.18 Linear coverage concept with directional antennas.

Three station distributed antenna design
Three-zone intelligent cell

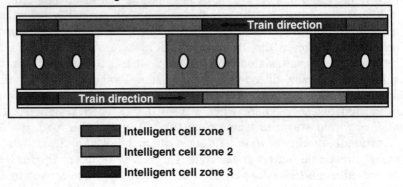

□ Intelligent cell zone 1
□ Intelligent cell zone 2
■ Intelligent cell zone 3

Figure 16.19 Underground communication system.

16.3   In-building Communication

16.3.1   Differences between ground mobile and in-building design

There are many good system design tools which have been developed solely for ground mobile cellular systems. However, no proper design tools have been developed for both ground mobile communications and in-building communications operated within one cellular system. As cellular systems are operated today, in-building communication is provided by transmitting radio signals from cell sites so that they penetrate the building walls to reach portable handsets inside buildings. Several difficulties are raised:

1. The transmitted power for in-building communications has to be about 20 dB stronger than that for the ground mobile communications in order to penetrate into or out from a building. Since the maximum transmit power of a portable handset is 8 dB lower than the maximum power of a ground mobile unit, in-building communication is harder to perform on the reverse link (portable-to-base link) than the forward link.

2. The coverage of portable units is not two-dimensional but three-dimensional. Weak reception of portable units is found on lower floors of a building, but strong reception is found on higher floors. This fact presents a difficult condition for running a system which can serve both ground mobile and in-building communication simultaneously.

3. When a radio channel penetrates from outside into a multifloor building, this particular channel can serve only one user who is located on one floor. The other potential users on different floors cannot use the same channel.

4. In-building communiction needs enormous radio channels which the current cellular system cannot provide.

16.3.2 Natural in-building radio environment

Building penetration. The signal penetrating through the building wall is called the *building penetration*. Building penetration studies have shown that penetration loss depends on geographical areas: 22–28 dB in Tokyo, 18–22 dB in Los Angeles, and 13–17 dB in Chicago. The differences are due to the building's construction. Earthquake resistance is the main factor in constructing buildings in some areas. In earthquake areas, the steel frame of a building is built as a mesh-type structure to resist the vibration of the earthquake. The mesh-type structure causes high penetration loss.

Building height effect. Signal reception is always stronger when the cellular handset is at a higher floor. The floor-height gain is about 2.70 dB/floor, independent of the building construction. In Chicago, the reception on the sixth floor is the same as that on the outside ground level. In Tokyo, with its higher penetration loss because of antiearthquake construction, the reception on the eleventh floor is the same as that on the outside ground level.

Building floor isolation. The signal isolation between floors in a multifloor building is on the average about 20 dB. Within a floor of 150 × 150 feet, the propagation loss due to interior walls, depending on the wall materials, is about 20 dB between the strong and weak areas.

16.3.3 A new in-building communication system

Philosophy of designing a new in-building communication system. After studying the natural in-building radio environment, we found that the building structure is a natural radio shield (Fig. 16.20). Therefore utilizing the shielding advantage is the best policy. This means that a signal should not be forced to penetrate into a building with a high-

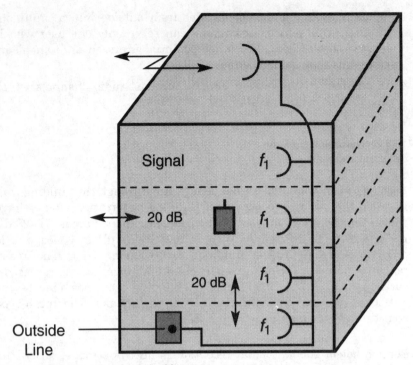

Figure 16.20 Signal within a building.

power radio wave; rather, the signal should be led into the building and distributed onto each floor.

Means of leading the cellular signal into the building. In a cellular system, cellular radios do not need to reside in each building, but rather may be installed in a remote base (Fig. 16.21). This deployment would change the conventional structure of the cell cites and make all the cell sites miniaturized zone sites. There are three methods by which the cellular signal can be led into buildings:

1. Upconvert all cellular signal channels to optical frequency at the base and transmit over optical fibers. When the optical signal reaches the building, downconvert it back to the cellular signals and serve the building users.

2. Upconvert all cellular signal channels to microwave frequency at the base and transmit over a radio link. When the microwave signal reaches the building, downconvert it back to the cellular signals and serve the building users.

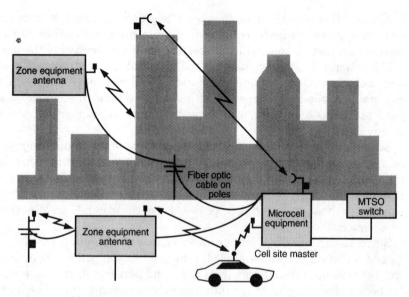

Figure 16.21 Microcell installation concept.

3. Downconvert all cellular signal channels to a 200-MHz UHF signal at the base and transmit it over the cable with low path loss. When the 200-MHz UHF signal reaches the building, upconvert it back to the cellular signals and serve the building users.

16.3.4 In-building system configuration

An in-building system configuration that uses both the natural shielding of the building structure, and a means of leading the cellular signal into the building is illustrated in Fig. 16.7. Calls are sent into the building at upconverted (or downconverted) frequencies to different floors. These lead-in frequencies are converted back to the cellular frequencies as soon as they reach the desired floors.

For a ten-floor building, the same cellular frequency can carry 10 different calls on each of the 10 floors. The capacity is increased 10 times. Also, because of the penetration loss, the same cellular frequencies can be used in neighboring buildings, as shown in Fig. 16.7. Thus the same frequency can be reused many times, and the spectrum utilization is very efficient.

If the building penetration is high, the neighboring buildings will not have cochannel interference. In this case, we can use the intelligent microcell configuration.[2] Each floor will be divided into three or four zones. Only one zone's transmitter is on—the zone where the user is located. Thus the transmitted power is greatly reduced. Within one floor, the user's assigned cellular frequency does not change during the call. The active zone will change, following the location of the user.

If 30 channels are assigned for the in-building users, then the analog cellular system outside the building can still have 365 cellular channel frequencies. These 30 in-building cellular frequencies can generate thousands of calls in an in-building communication system at any time.

Since 30 channel frequencies are designated for the in-building communication only, there is no interference between ground mobile communication and in-building communication. These 30 channel frequencies can be freely reused on each floor in each building without worrying about interference problems with ground mobile communication.

Since the in-building communication in each building can be assigned a station identification (SID) number or the same SID number can be used in all buildings, then the same portable unit can be used for both in-building and the ground mobile communication. The handoff process can be carried out between two communication systems. It would be a benefit to the end user to be able to carry only one portable unit but operate in two systems, the in-building and the ground mobile communication systems.

Handoff between floors can be implemented by assigning a few different channel frequencies from the in-building frequencies to the elevator areas. The handoff will take place by assigning a new elevator area frequency to replace the in-building frequency when the user enters the elevator, and will be handed back to one of the in-building frequencies after the user reaches another floor.

A detailed diagram of an in-building communication system is shown in Fig. 16.22.

16.3.5 A PCS application

Most wireless operators try to divide the market into many segments. Each segment may have many systems in operation such as cellular, in-building, cordless phones, and many others. Each system has to operate its own subscriber units. Thus, there will be many different types of PCS units on the market. However, users (subscribers) want to carry only one portable unit that is small and lightweight and provides long talk time. In addition, they want a PCS that sends out and delivers calls anywhere, anytime.

The new in-building communication system described in Sec. 16.3.4 can meet the end user's needs. It can be used underground, in subways

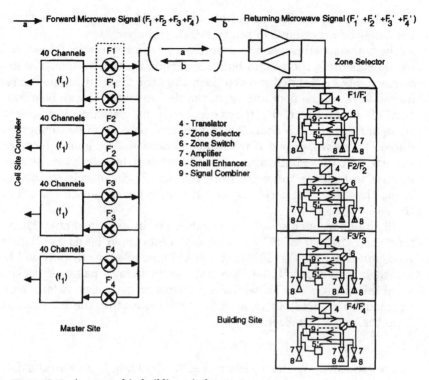

Figure 16.22   A proposed in-building wireless system.

and tunnels. It is a complementary system to the ground mobile macrocell system.

The future PCS unit.   The size and weight of today's cellular portable unit are determined by three factors:

1. The size of human fingers is the limiting factor for the size of keypad (or keyboard).
2. The size of the battery is the limiting factor for long radio path communication and long talk time.
3. The length from the mouth to the ear is the limiting factor for the length of the unit.

Today, voice-activated technology is available to replace the dialing process. A memory device in the unit can have the telephone number recorded by voice to correspond to a person's name. Once the person's name is voiced in to the memory, the corresponding telephone number will be sent out. In this case, not only is no keypad needed, but also the caller does not have to remember the number.

The transmitted power of a portable unit can be drastically reduced by implementing the new in-building communication system. The antenna (probe) is located in each floor, thus the transmitted power required to reach the floor antenna from the portable unit can be a fraction of a milliwatt. Furthermore, with an intelligent microcell configuration on each floor, the transmitted power can be further reduced. Therefore, only a tiny battery is needed. The length between the ear and the mouth also can be shortened by using acoustic vibration to transfer the voice from the mouth via muscle to the ear area. Thus both the earphone and the microphone can be located at the ear.

All the above-mentioned technologies are in existence now. Therefore, soon the size of PCS subscribers' units could be as small as a mechanical pencil (Fig. 16.23). A short simple message display will be attached to the pencil-like PCS unit, as in today's pager. A pocket-sized miniprinter can be carried for lengthier messages. A long message can be printed out in an office by downloading the information from the unit's memory.

Future concerns.   We have shown that the building is a natural shield for radio interference. Also, cellular frequency reuse in each floor and in the neighboring buildings has shown great spectral efficiency. The new in-building communication system is compatible with the existing cellular system and could be a PCS system in the future.

As far as the future PCS in concerned, we may have to explore the two areas shown is Fig. 16.24, intelligent network and radio access.

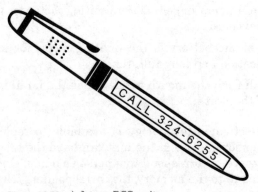

Figure 16.23 A future PCS unit.

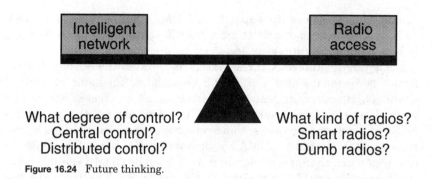

What degree of control?   What kind of radios?
Central control?   Smart radios?
Distributed control?   Dumb radios?

Figure 16.24 Future thinking.

We have to decide whether the intelligent network (IN) should be a central control system or distributed control system. As for radio access, what kind of PCS radio should we have, smart or dumb? If the decision is to use a smart radio, what degree of smartness should the radio have?

16.4 CDMA Cellular Radio Network

16.4.1 System design philosophy

Deploying CDMA systems is like tuning a sophisticated automobile engine. When proper tuning is done, the engine runs very smoothly. But a sophisticated automobile engine needs a sophisticated, computer-aided tuning device, just as a CDMA system needs a computer-aided design tool. As we know, in analog and TDMA systems, capacity increases are a result of the elimination of interference from the desired signal. The signal level of a desired signal is always much stronger than the interference level, say 18 dB or better, for AMPS. However, in a CDMA system, the capacity increase is based on how much interference the desired signal can tolerate. The signal level of a desired signal is always below the interference level. Also, all the users have to share the same radio channel. If one user takes more power than it needs, then the others will suffer and system capacity will be reduced. This scenario is the same as dining in a formal restaurant. The volume of the conversations at every table is low. Therefore, no walls are needed between tables. The guests never feel their conversations are being interrupted by the next table. Therefore, many conversations can occur in the same dining room. This is the concept of CDMA—that all the voice channels are sharing one big radio channel. If people at one table start to raise their voices, the rest of the tables have to either leave or raise their voices too. The former case destroys CDMA. The latter is the so-called cocktail party syndrome which reduces the capacity of CDMA. Neither one is desired. This section addresses how to tune the CDMA cellular radio network in order to tolerate interference.

Designing a uniform CDMA system is comparatively simple. Uniform CDMA means all the cells will be assigned the same number of channels. However, in reality, CDMA systems are not uniform. The voice channels of each cell in a CDMA system are not the same. Because of demographic needs, some cells have more voice channels and some have fewer. Since CDMA has only one radio channel, to generate different voice channels on demand from a single CDMA radio is a big challenge. We would like to describe the challenges by illustrating the design aspects of a CDMA system.

16.4.2 Key elements in designing[23] a CDMA system

The design of a CDMA system is much more complex than the design of a TDMA system. In analog and TDMA systems, the most important key element is C/I. There are two different kinds of C/I. One is the measured C/I which is used to indicate the voice quality in the system. The higher the measured value, the better. The other is the specified C/I [$(C/I)_s$], which is a specified value for a specified cellular system. For example, the $(C/I)_s$ in the AMPS system is 18 dB. Since in analog and TDMA systems, because of spectral and geographical separations, the interference I is much lower than the received signal C, sometimes we can utilize field strength meters to measure C to determine the coverage of each cell. The field strength meter therefore becomes a useful tool in designing the TDMA system. In CDMA all the traffic channels are served solely by a single radio channel in every cell.[1-4] Therefore, in an m-voice channel cell, one of the m traffic channels is the desired channel and the remaining $m - 1$ traffic channels are the interference channels. In this case, the interference is much stronger than the desired channel. Then C/I is hard to obtain by using a signal strength meter. Thus, the key elements in designing a CDMA system are different from those in designing a TDMA system.

Relationship between C/I and FER. In CDMA, the key element is E_b/I_o (energy per bit/power per hertz), which is related to the frame error rate. An acceptable speech quality of a specified vocoder would determine the FER which is related to E_b/I_o at a given vehicle speed. From a system design aspect, we consider the system performance with all the vehicle speeds and environmental conditions and come up with a specified E_b/I_o. Now we can design the CDMA system on the basis of the specified E_b/I_o. The following equation is used:

$$\frac{C}{I} = \left(\frac{E_b}{I_o}\right)\left(\frac{R_b}{B}\right)\eta \qquad (16.4\text{-}1)$$

where R_b is the bits per second, B is the CDMA channel bandwidth, and η is the speech activity cycle in percent. From Eq. (16.4-1), B/R_b is the processing gain (PG), which is known in a given CDMA system. E_b/I_o and η are also known in the system. Then the C/I of each CDMA channel can be obtained. Each coded chanel in CDMA can be treated as a frequency channel in FDMA or TDMA. If the coded channels are sent over a cable transmission medium, the interference among the coded channels can be treated as adjacent channel interference. Due to the nature of channel orthogonality, the interference should be very small. But in the mobile radio environment, because of the multipath wave phenomenon, the orthogonality among the channels cannot be held. Therefore, the processing gain is the only interference protection among the channels.

E_b/I_o always varies in order to meet a specified FER under different conditions. From Eq. (16.4-1) we can find a required $(C/I)_s$ from a specified $(E_b/I_o)_s$ in a worst-case scenario for designing the system. However, the values of $(E_b/I_o)_s$ for the forward link channels and for the reverse-link channels are different because of their different modulation schemes. Therefore, we may have two different requirements for C/I: a $(C/I)_F$ for the forward link channels and a $(C/I)_R$ for the revese link channels.

16.4.3 Uniform cell scenario

Now we try to find the design parameters of each cell for the forward link and the reverse link in a realistic uniform capacity condition.

For the forward link. A worst-case scenario is used to find the relation among the transmitted powers of all cell sites. First we form an equation which relates the C/I received at a mobile location A (Fig. 16.25) to the transmitted powers of all cell sites:

$$C/I = \cfrac{\alpha_1 \cdot R^{-4}}{\underbrace{\alpha(m_1 - 1)R^{-4}}_{\text{self cell}} + \underbrace{(\alpha_2 m_2 + \alpha_3 m_3)R^{-4}}_{\text{2 adjacent cells}}}$$

$$+ \underbrace{\beta(2R)^{-4}}_{\text{3 intermediate cells}} + \underbrace{\gamma(2.633R)^{-4}}_{\text{6 distant cells}} \qquad (16.4\text{-}2)$$

where α_i ($i = 1, 3$) is the transmitted power of each voice channel in the cell and m_i is the number of channels per cell. β and γ are the transmitted powers of the combined adjacent cells at a distance $2R$

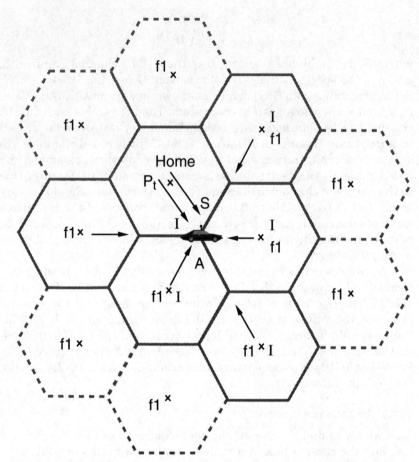

Figure 16.25 CDMA system and its interference (from a forward link scenario).

and 2.633R, respectively. By solving Eq. (16.4-2) we can determine m_i as follows:

$$m_1 = \left(\frac{1}{C/I} + 1\right) - \left[\frac{\alpha_2 m_2 + \alpha_3 m_3}{\alpha_1}\right] - \frac{\beta}{\alpha_1}(2)^{-4} - \frac{\gamma}{\alpha_1}(2.633)^{-4}$$

$$(16.4\text{-}3)$$

Case A. No adjacent cell interference. Let $\alpha_2 = \alpha_3 = \beta = \gamma = 0$ in Eq. (16.4-3). Then

$$m_1 = \frac{1}{(C/I)} + 1 \qquad\qquad (16.4\text{-}4)$$

If the value of C/I obtained from Eq. (16.4-1) is $C/I = -17$ dB, then $m_1 = 51$, the maximum voice channels in a cell.

Case B. No interference other than from the two close-in interfering cells. In Eq. (16.4-3), the third and fourth terms are much smaller in value than the first two terms and therefore can be neglected. Then

$$\alpha_1 = \frac{\alpha_2 m_2 + \alpha_3 m_3}{\dfrac{1}{C/I} + 1 - m_1} \tag{16.4-5}$$

If $C/I = -17$ dB, and the assigned voice channels at three cells are $m_1 = 30$, $m_2 = 25$, and $m_3 = 15$, respectively, then Eq. (16.4-5) becomes:

$$\alpha_1 = \frac{25\alpha_2 + 15\alpha_3}{51 - 30} = 1.19\alpha_2 + 0.714\alpha_3 \tag{16.4-6}$$

Equation (16.4-6) expresses the relationship among α_1, α_2, and α_3.

The total transmitted power P in each cell site is $P_1 = \alpha_1 m_1$, $P_2 = \alpha_2 m_2$, $P_3 = \alpha_3 m_3$. Thus P_1, P_2 and P_3 are the maximum transmitted powers of the three cells. Then Eq. (16.4-5) can be simplified to:

$$\left(\frac{1}{C/I} + 1 \right) \frac{P_1}{m_1} = P_1 + P_2 + P_3 \tag{16.4-7}$$

Following the same derivation steps, we can obtain the following equations:

$$\left(\frac{1}{C/I} + 1 \right) \frac{P_2}{m_2} = P_1 + P_2 + P_3 \tag{16.4-8}$$

$$\left(\frac{1}{C/I} + 1 \right) \frac{P_3}{m_3} = P_1 + P_2 + P_3 \tag{16.4-9}$$

The relationship of the three maximum transmitted powers of the three cells is:

$$\frac{P_1}{m_1} = \frac{P_2}{m_2} = \frac{P_3}{m_3} \tag{16.4-10}$$

Deduced from Eq. (16.4-10), a design criterion which we will use in general for a CDMA system of N cells is

$$\frac{P_i}{m_i} = \frac{P_j}{m_j} = \text{constant} \tag{16.4-11}$$

where i indicates the i cell and j indicates the j cell. Eq. (16.4-11) indicates that the more voice channels generated, the more transmit power is needed. Therefore, either applying power control to the voice channels in a cell such that more voice channels can be provided with a given transmit power, or using fewer channels in a cell such that the transmit power P will decrease, will reduce interference.

For the reverse link. The worst-case scenario (shown in Fig. 16.26) is also used in the reverse link analysis. Assume that all the mobile units traveling in the two adjacent cells will be located at the cell boundary of the home cell. From the reverse link, the powers of the m_1 voice

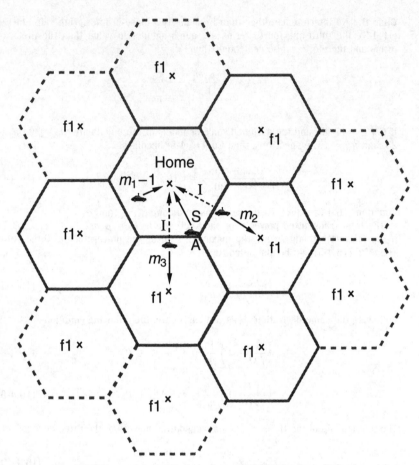

Figure 16.26 CDMA system and its interference (from a reverse link scenario).

signals received at the home site are the same, because of the power control implementation to overcome the near-to-far interference.

Let the received signal from a desired mobile unit at the home cell site be C. Assume that each signal of other m_1 channels received at the home site in Fig. 16.26 is also C. Also, assume that the interference of certain mobile units, say rm_1, from the two adjacent cells comes from the cell boundary. Because of the power control in each adjacent cell, the interference coming from the adjacent cell for each voice channel would roughly be C as received by the home cell site. The received C/I at the desired voice channel can be expressed as:

$$C/I = \frac{C}{(m_1 - 1) \cdot C + r_{12} \cdot m_2 C + r_{13} \cdot m_3 C} \qquad (16.4\text{-}12)$$

$$= \frac{1}{m_1 - 1 + r_{12}m_2 + r_{13}m_3}$$

where r_{12} and r_{13} are a portion of the total number of voice channels in adjacent cells that will interfere with the desired signal at the home cell, which is cell 1.

From Eq. (16.4-12), the worst-case scenario is when:

$$m_1 + r_{12}m_2 + r_{13}m_3 \le \frac{1}{C/I} + 1 \qquad (16.4\text{-}13)$$

Following the same steps, we find:

$$r_{21}m_1 + m_2 + r_{23}m_3 \le \frac{1}{C/I} + 1 \qquad (16.4\text{-}14)$$

$$r_{31}m_1 + r_{32}m_2 + m_3 \le \frac{1}{C/I} + 1 \qquad ((16.4\text{-}15)$$

The value of r depends on the size of the overlapped region in the adjacent cell, and can be reasonably assumed as 1/6 (which is 0.166) if the system is properly designed.

If $C/I = -17$ dB, which is 50, and $r_{12} = r_{13} = 0.166$, then Eq. (16.4-13) becomes:

$$m_1 + 0.166(m_2 + m_3) = 51 \qquad (16.4\text{-}13a)$$

The relationships among the numbers of voice channels in each cell, m_1, m_2, and m_3, are expressed in Eqs. (16.4-13), (16.4-14), and (16.4-15).

Designing a CDMA system. From the reverse-link scenario, we can check to see whether all the conditions expressed in Eqs. (16.4-13), (16.4-14), and (16.4-15) can be met. The main elements in these equations are the demanded voice channels, m_1, m_2, and m_3. For representative values of these terms, we can determine the maximum transmitted power of each cell from the forward link equations, Eqs. (16.4-7) to (16.4-10).

Example 16.1. Given $C/I = 17$ dB and all the r's, $r = r_{ij} = 0.3$:

Case 1: Let the demanded voice channels be $m_1 = 30$, $m_2 = 25$, $m_3 = 15$. Checking the conditions in Eqs. (16.4-13), (16.4-14), and (16.4-15), we find:

$$30 + 0.3\ (25 + 15) = 42 < 51 \qquad \text{(OK)}$$

$$25 + 0.3\ (30 + 15) = 38.5 < 51 \qquad \text{(OK)}$$

$$15 + 0.3\ (30 + 25) = 31.5 < 51 \qquad \text{(OK)}$$

Since the cell sizes of the three cells are the same,

$$\alpha_1 = \alpha_2 = \alpha_3 = CR^{+4}$$

Assume $\alpha_1 = \alpha_2 = \alpha_3 = 100$ mW. Then

$$P_1 = 30 \times 0.1 = 3\ \text{W}$$

$$P_2 = 25 \times 0.1 = 2.5\ \text{W}$$

$$P_3 = 15 \times 0.1 = 1.5\ \text{W}$$

Case 2: Let the demanded voice channels be $m_1 = 40$, $m_2 = 30$, $m_3 = 20$. Checking the conditions in Eqs. (16.4-13), (16.4-14), and (16.4-15), we find:

$$40 + 0.3\ (30 + 20) = 55 > 51 \qquad \text{(does not meet the condition)}$$

$$30 + 0.3\ (40 + 20) = 48 < 51 \qquad \text{(OK)}$$

$$20 + 0.3\ (40 + 30) = 41 < 51 \qquad \text{(OK)}$$

The number of demanded voice channels should be reduced before the system is designed.

16.4.4 Nonuniform cell scenario

Transmit power on the forward link channels. We may first assign the number of voice channels m in each cell according to demographic data. Then we may calculate the total transmit power on the forward link channels in each cell from a worst-case scenario as shown in Fig. 16.27.

All the cell sizes are not the same in a nonuniform CDMA system. We consider only the three cells most affected by the locations of the three vehicles. The vehicles are at the most interference-prone location in each cell.

In this case, the $(C/I)_F$ received at vehicle 1 is

$$(C_1/I_1)_F = \frac{\alpha_1 R_1^{-4}}{(m_1 - 1)\alpha_1 R_1^{-4} + \alpha_2 m_2 R_2^{-4} + \alpha_3 m_3 R_3^{-4} + I_{a_1}} \qquad (16.4\text{-}12)$$

where I_{a1} is the interference coming from other cells outside the three. I_{a1} is usually very small compared to the second and third terms in the denominator and can be neglected.

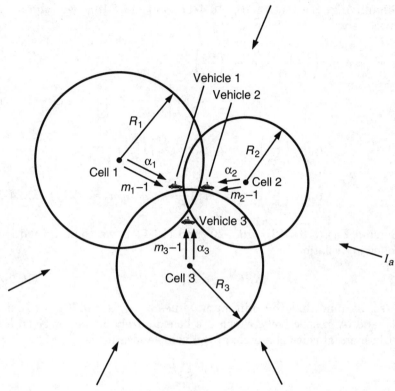

Figure 16.27 The worst-case scenario for forward link channel reception in a non-uniform CDMA system.

The received $(C/I)_F$ at vehicle 2 is

$$(C_2/I_2)_F = \frac{\alpha_2 R_2^{-4}}{(m_2 - 1)\alpha_2 R_2^{-4} + \alpha_1 m_1 R_1^{-4} + \alpha_3 m_3 R_3^{-4} + I_{a2}} \quad (16.4\text{-}17)$$

The received $(C/I)_F$ at vehicle 3 is

$$(C_3/I_3)_F = \frac{\alpha_2 R_2^{-4}}{(m_3 - 1)\alpha_3 R_3^{-4} + \alpha_1 m_1 R_1^{-4} + \alpha_2 m_2 R_2^{-4} + I_{a3}} \quad (16.4\text{-}18)$$

Let

$$(C_1/I_1)_F = (C_2/I_2)_F = (C_3/I_3)_F = (C/I)_F$$

and

$$I_{a1} = I_{a2} = I_{a3} = 0$$

Simplifying Eqs. (16.4-16, 16.4-17 and 16.4-18), we obtain respectively:

$$\alpha_1 m_1 + \alpha_2 m_2 \left(\frac{R_2}{R_1}\right)^{-4} + \alpha_3 m_3 \left(\frac{R_3}{R_1}\right)^{-4}$$

$$= \alpha_1 \left[\frac{1}{(C/I)_F} + 1\right] = \alpha_1 G \quad (16.4\text{-}19)$$

$$\alpha_1 m_1 \left(\frac{R_1}{R_2}\right)^{-4} + \alpha_2 m_2 + \alpha_3 m_3 \left(\frac{R_3}{R_2}\right)^{-4} = \alpha_2 G \quad (16.4\text{-}20)$$

$$\alpha_1 m_1 \left(\frac{R_1}{R_3}\right)^{-4} + \alpha_2 m_2 \left(\frac{R_2}{R_3}\right)^{-4} + \alpha_3 m_3 = \alpha_3 G \quad (16.4\text{-}21)$$

Solving Eqs. (16.4-19), (16.4-20) and (16.4-21) we come up with the following relation:

$$\alpha_1 R_1^{-4} = \alpha_2 R_2^{-4} = \alpha_3 R_3^{-4} \quad (16.4\text{-}22)$$

Also, assume that the minimum values of α_1, α_2, and α_3 will be α_1^0, α_2^0, and α_3^0, respectively, which are based purely on the received level of individual voice channels at the vehicle locations:

$$\alpha_1 \geq \alpha_1^0 = C_0 R_1^{+4} + k$$

$$\alpha_2 \geq \alpha_2^0 = C_0 R_2^{+4} + k \quad (16.4\text{-}23)$$

$$\alpha_3 \geq \alpha_3^0 = C_0 R_3^{+4} + k$$

where C_0 is the required signal level received at the vehicle location and k is a constant related to the antenna heights at the cell sites.

Now the total transmit power of each cell site will be

$$P_1 = m_1 \alpha_1$$

$$P_2 = m_2 \alpha_2 \quad (16.4\text{-}24)$$

$$P_3 = m_3 \alpha_3$$

Transmit power on the reverse link channels. On the reverse link channels, we use the same worst-case scenario (Fig. 16.28). According to the power control algorithm, all the signals will be the same when they reach the cell site. The vehicle 1 signal received at cell site 1 is C_1, the rest of the signals are considered interference.

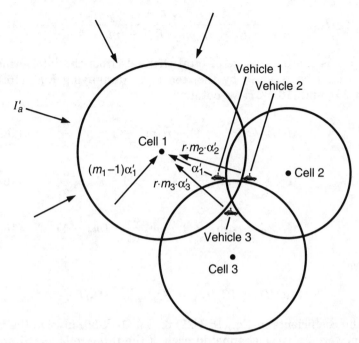

Figure 16.28 The worst-case scenario for reverse-link channel reception in a nonuniform CDMA system.

$$(C_1/I_1)_R \geq \frac{\alpha_1' R_1^{-4}}{(m_1 - 1)\alpha_1' R_1^{-4} + r_{12} m_2 \alpha_2' R_1^{-4} + r_{13} m_3 \alpha_3' R_1^{-4} + I_{a_1}'}$$

$$(16.4\text{-}25)$$

where α_1', α_2', and α_3' are the powers of individual channels transmitted back to their corresponding cell sites, r_{12} and r_{13} are the portion of the total number of voice channels in adjacent cells that will interfere with the desired signal at cell 1, and I_{a_1}' is the interference coming from vehicles in cells other than cell 2 and cell 3. I_{a_1}' is a relatively small value and can be neglected.

Utilizing the same steps, we can derive the following two equations for the two other cases.

Cell site 2 receives the vehicle 2 signal:

$$(C_2/I_2)_R \geq \frac{\alpha_2' R_2^{-4}}{r_{21} m_1 \alpha_1' R_2^{-4} + (m_2 - 1)\alpha_2' R_2^{-4} + r_{23} m_3 \alpha_3' R_2^{-4}} \quad (16.4\text{-}26)$$

Cell site 3 receives the vehicle 3 signal:

$$(C_3/I_3)_R \geq \frac{\alpha_3' R_3^{-4}}{r_{31}m_1\alpha_1'R_3^{-4} + r_{32}m_2\alpha_2'R_3^{-4} + (m_3 - 1)\alpha_3' \cdot R_3^{-4}} \qquad (16.4\text{-}27)$$

where r is the percentage of total channels from the interfering cell that would be received by the, cell site. Simplifying Eqs. (16.4-25), (16.4-26), and (16.4-27) we obtain:

$$(I/C)_R \geq (m_1 - 1) + r_{12}m_2 \frac{\alpha_2'}{\alpha_1'} + r_{13}m_3 \frac{\alpha_3'}{\alpha_1'} \qquad (16.4\text{-}28)$$

$$(I/C)_R \geq r_{21}m_1 \frac{\alpha_1'}{\alpha_2'} + (m_2 - 1) + r_{23}m_3 \frac{\alpha_3'}{\alpha_1'} \qquad (16.4\text{-}29)$$

$$(I/C)_R \geq r_{31}m_1 \frac{\alpha_1'}{\alpha_3'} + r_{32}m_2 \frac{\alpha_2'}{\alpha_3'} + (m_3 - 1) \qquad (16.4\text{-}30)$$

where

$$(C/I)_R = (C_1/I_1)_R = (C_2/I_2)_R = (C_3/I_3)_R$$

All the coefficients in Eq. (16.4-28) to Eq. (16.4-30) involve the transmit power of a voice channel in each of the three cells, α_1', α_2' and α_3'.

For the same reason stated in the derivation of Eq. (16.4-23), the minimum values of α_1', α_2', and α_3' can be defined as follows:

$$\alpha_1' \geq \alpha_1^0 = C_0 R_1^4 + k$$

$$\alpha_2' \geq \alpha_2^0 = C_0 R_2^4 + k \qquad (16.4\text{-}31)$$

$$\alpha_3' \geq \alpha_3^0 = C_0 R_3^4 + k$$

where R_1, R_2, and R_3 are the radii of the three cells and k is a constant related to the antenna heights as the cell sites. We may replace all the α' terms in Eqs. (16.4-28), (16.4-29), and (16.4-30) with the equivalent α^0 terms in Eq. (16.4-31). Equations (16.4-28), (16.4-29) and (16.4-30) become

$$(I/C)_R \geq (m_1 - 1) + r_{12}m_2 \left(\frac{R_2}{R_1}\right)^4 + r_{13}m_3 \left(\frac{R_3}{R_1}\right)^4 \qquad (16.4\text{-}32)$$

$$(I/C)_R \geq r_{21}m_1 \left(\frac{R_1}{R_2}\right)^4 + (m_2 - 1) + r_{23}m_3 \left(\frac{R_3}{R_2}\right)^4 \qquad (16.4\text{-}33)$$

$$(I/C)_R \geq r_{31}m_1 \left(\frac{R_1}{R_3}\right)^4 + r_{32}m_2 \left(\frac{R_2}{R_3}\right)^4 + m_3 - 1 \qquad (16.4\text{-}34)$$

Under the physical condition, the following relationships have to be held. The values m_1, m_2, and m_3 have to be

$$m_1, m_2, m_3 < \frac{1}{(C/I)_R} + 1 \qquad (16.4\text{-}35)$$

which has been derived in Eq. (16.3-4).

Designing a CDMA system.   We first have to check whether all the requirements expressed in Eqs. (16.4-32) to (16.4-32) are met with our given conditions. If they are met, then we can find the transmit powers P_1, P_2, and P_3 from Eq. (16.4-24). Usually, among the three equations, only one dominates. If that one meets the given conditions, the other two will meet them also. The following example addresses this point.

Example 16.2.   Given: $R_1 = 4$ km, $R_2 = 6$ km, $R_3 = 5$ km, and $(C/I)_R = -17$ dB. Also assume that:

$$r_{13} = r_{31} = 0.3$$

$$r_{21} = r_{12} = 0.2$$

$$r_{23} = r_{32} = 0.25$$

Then checking the conditions in Eqs. (16.4-32) through (16.4-34) we obtain:

$$50 \geq m_1 - 1 + 1.0125m_2 + 0.7324m_3 \qquad (16.4\text{-}36)$$

$$50 \geq 0.,0395m_1 + m_2 - 1 + 0.1205m_3 \qquad (16.4\text{-}37)$$

$$50 \geq 0.1229m_1 + 0.5184m_2 + m_3 - 1 \qquad (16.4\text{-}38)$$

Among the three equations, Eq. (16.4-36) should be checked first.
   Let $m_1 = 20$. Then

$$31 \geq 1.0125m_2 + 0.7324m_3 \qquad (16.4\text{-}39)$$

$$50.21 \geq m_2 + 0.1205m_3 \qquad (16.4\text{-}40)$$

$$48.54 \geq 0.5184m_2 + m_3 \qquad (16.4\text{-}41)$$

Among the three equations, the condition of Eq. (16.4-39) is the limiting condition. Also, we may find the conditions for $m_1 = 15$ and $m_1 = 25$ as follows:

$$36 \geq 1.0125m_2 + 0.7324m_3 \qquad (m_1 = 15) \qquad (16.4\text{-}42)$$

$$26 \geq 1.0125m_2 + 0.7324m_3 \qquad (m_1 = 25) \qquad (16.4\text{-}43)$$

The three curves, $m_1 = 15, 20, 25$, are plotted in Fig. 16.29 for various values of m_2 and m_3. From the figure, we may pick three values such as $m_1 = 25$, m_2

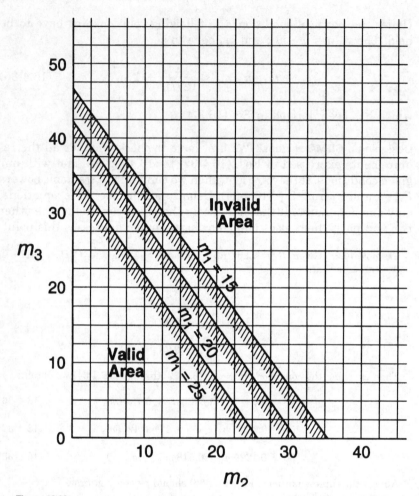

Figure 16.29

= 20, and $m_3 = 7$, and assume that $C = -95$ dBm. Then the transmit power of each voice channel in the respective cell is:

$$\alpha_1 = -95 \text{ dBm} + 40 \log R_1^4 + k \quad \text{(for cell 1)}$$

$$\alpha_2 = -95 \text{ dBm} + 40 \log R_2^4 + k \quad \text{(for cell 2)}$$

$$\alpha_3 = -95 \text{ dBm} + 40 \log R_3^4 + k \quad \text{(for cell 3)}$$

where k is a constant depending on the antenna height.

From Eq. (16.4-31) the total power will be:

$$P_1 = m_1\alpha_1$$

$$P_2 = m_2\alpha_2$$

$$P_3 = m_3\alpha_3$$

References

1. W. C. Y. Lee, *Mobile Cellular Telecommunications Systems,* McGraw Hill, 1989, p. 379.
2. V. H. MacDonald, "The Cellular Concept," *Bell System Technical Journal,* vol. 58, January 1979, pp. 15–42.
3. W. C. Y. Lee, "Spectrum Efficiency in Cellular," *IEEE Transactions of Vehicular Technology,* May 1989, pp. 69–75.
4. W. C. Y. Lee, "Smaller Cell for Greater Performance," *IEEE Communication Magazine,* November 1991, pp. 19–23.
5. M. Cooper and R. Roy, "SDMA Technology—Overview and Development Status," ArrayComm-ID-010, ArrayComm, Inc., Mountain View, Calif.
6. W. C. Y. Lee, "Lee's Model," *IEEE VTS Conference Record 1992,* Denver, May 11, 1992, pp. 343–348.
7. W. C. Y. Lee, *Mobile Communications Engineering,* McGraw Hill, 1982, p. 202.
8. K. Gilhousen, I. Jacobs, R. Padovani, A. Viterbi, L. Weaver, C. Wheatley, "On the Capacity of a Cellular CDMA System," *IEEE Transactions on Vehicular Technology,* Vol. 40, May 1991, pp. 303–312.
9. R. Pickholtz, L. Milstein, D. Schilling, "Spread Spectrum for Mobile Communications," *IEEE Transactions on Vehicular Technology,* Vol. 40, May 1991, pp. 313–322.
10. W. C. Y. Lee, "Overview of Cellular CDMA," *IEEE Transactions on Vehicular Technology,* Vol. 40, May 1991, pp. 291–302.
11. ETSI/TC, "Recommendation GSM 01.02," ETSI/PT12, January 1990.
12. F. Anderson, W. Christensen, L. Fullerton, B. Kortegaard, "Ultra-Wideband Beamforming in Sparse Arrays," *IEEE Proceedings,* vol. 138, no. 4, August 1991, pp. 342–346.
13. R. A. Scholtz, "Multiple Access with Time-Hopping Impulse Modulation," MILCOM 1993, Boston, October 11–14, 1993.
14. R. A. Comroe and D. J. Costello, Jr., "ARQ Schemes for Data Transmission in Mobile Radio Systems," *IEEE Transactions on Vehicular Technology,* vol. VT-33, August 1984, pp. 88–97.
15. K. Raith and J. Uddenfeldt, "Capacity of Digital Cellular TDMA Systems," *IEEE Trans. on Vehicular Technology,* vol. 40, May 1991, pp. 323–332. W. C. Y. Lee, "In-building Telephone Communication System," U.S. patent office number 5,349,631, Sept. 20, 1994.
16. J. C. Liberti, Jr., and T. S. Rappaport, "Analytical Results for Capacity Improvement in CDMA," *IEEE Trans on VT,* vol. 43, August 1994, pp. 680–690.
17. W. C. Y. Lee, "An Innovative Microcell System," *Cellular Business,* December 1991, pp. 42–44.
18. W. C. Y. Lee, "Applying the Intelligent Cell Concept to PCS," *IEEE Trans. on VT,* vol. 43, August 1994, pp. 672–679.
19. Allen Telecomm Co. and 3dBM Co. manufacture analog converters, ADC Kentron manufacture digital converters.
20. D. Parsons, *The Mobile Radio Propagation Channel,* Pentech Press, London, 1992.
21. H. H. Xia, H. L. Bertoni, L. R. Maciel, A. Lindsay-Stewart, and R. Low, "Microcellular Propagation Characteristics for Personal Communications in Urban and Sub-

urban Environments," *IEEE Transactions on Vehicular Technology,* Part II Special Issue on Future PCS Technologies, vol. 43, August 1994.

22. W. C. Y. Lee, Mobile Communications Design Fundamentals, John Wiley & Sons, 1993, section: microcell prediction model.

23. W. C. Y. Lee, "Key Elements in Designing a CDMA System," *IEEE VTC '94 Conference Record,* Stockholm, Sweden, June 8–10, 1994, pp. 1547–1550.

Intelligent Network for Wireless Communications

17.1 Advanced Intelligent Network (AIN)[1-3]

AIN is a network evolving from the intelligent network (IN). It has an independent architecture which allows telecommunication service operators to rapidly create and modify services for both network performance and customers' needs.

In the intelligent network concept, the service providers need more control for new service offerings. The IN is able to separate the specification, creation, and control of telecommunication services from the physical switching network such that the HLR (home location register) and the VLR (visitor location register) are no longer integrated in the MTSO (mobile telephone switching office).

17.1.1 Intelligent network evaluation

In the '60s, crossbar switches were developed and demonstrated to be very reliable switches. However, crossbar switches are mechanical and do not provide the intelligence. Then the No. 1 ESS (electronic switching system) was developed by AT&T and provided stored program control (SPC) capability. In 1965, SPC delivered call waiting and centrex features on No. 1 ESS. In the '70s, ESSs provided intelligent in-network management and maintenance, and offered operations systems (OS), and operation, administration, and maintenance (OA&M). In the '80s, the intelligent network introduced centralized databases and provided the network database services such as 800 toll-free calls and calling card calls. In 1983, the centralized databases were located at the service control point (SCP) to support alternate billing services

(ABS) and 800 calling. The data flow from the physical switches to the SCP is via SS7 (Signaling System No. 7) network. There are several intelligent networks:

IN/1—Functionality is distributed not only at the switches but also at the SS7 network, SCPs and OSs.

IN/2—To expand the switching and SCP capabilities known as *functional components* (FC), and with these expansions to form a new system called the *intelligent peripheral* (IP). It is capable of supporting a wide range of voice and data services.

AIN—The IN evolution began in the '90s at a forum called Multivendor Interaction (MVI) in which 16 vendors participated. AIN was defined by a series of releases. Each release contains additional architecture attributes and capabilities of supporting services.

AIN's network characteristics.   AIN is evolving from IN, and its characteristics are as follows:

- Modular—divided into functional entities
- Uniform—became standard architecture
- Service independent
- Programmable network operable by either user or carrier provider, or both
- Supplier transparent—open system architecture (OSA)
- Capable of rapid introduction of new services
- Accessible for other service providers
- Common channel signaling (CCS)—using out-of-band signaling
- Service logic—invokes AIN service logic programs (SLPs)

AIN uses CCS to deliver the call set-up signaling and the network information. In this case, the traffic channels are never tied up for signaling. For example, if the called party line is detected as busy by the signaling channel, the network would not assign a traffic channel to the calling party. Thus the efficient use of the traffic channels increases. Also the use of CCS can increase the speed of process for call process and information delivery. The CCS network uses digital channels with the SS7 protocol at a rate of 56 kbps.

AIN elements.   An AIN (Fig. 17.1) consists of the following elements:

Service control point. The SCP invokes service logic programs. The common channel signaling network allows the SCP to fully inter-

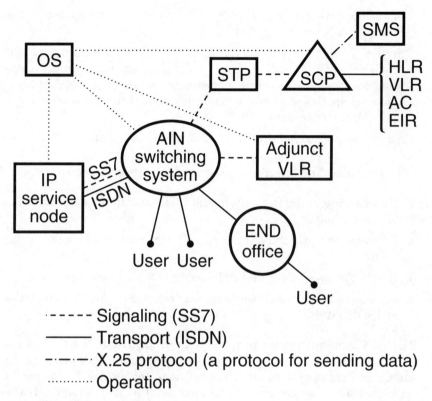

Figure 17.1 AIN system architecture.

connect with AIN switching systems through a signaling transport point (STP). The SCP supports 800 toll-free phone calls, area number calling, or personal location services.

Adjunct system. A direct link to the AIN switching system with a high-speed interface.

Service node. Communicates with AIN switch via the integrated services digital network (ISDN) access link and supports user interaction.

Intelligent peripheral. Controls and manages resources such as voice synthesis announcement, speech recognition, and digit collection.

AIN switch. Routes a call to an IP to ask for a function. When the IP completes the function, it also collects the user's information and sends it to AIN service logic (resides in SCP) via the AIN switch.

Operational system (OS). Provides memory administration, surveillance, network testing, and network traffic management maintenance and operation.

Signaling transport point (STP). The point that interconnects the SCP and AIN switching system.

Service management system (SMS). Provides three functions: (1) provision—creates service order, validation, load record; (2) maintenance—resolves record inconsistency, tests call processing logic, performs special studies; (3) administration—creates service logic, maintains service data.

Service switching point (SSP). Functions as a switch.

AIN interfaces.   The interfaces between AIN network elements are

1. Between the switching system and SCPs or adjunct systems using SS7 signaling.
2. Between the switching system and IPs or service nodes using ISDN.
3. In AIN, between SCP and SMS using the X.25 protocol.
4. Between end users AIN services; may be either conventional analog or ISDN interface.

The AIN general architecture with the indicated AIN interfaces is shown in Fig. 17.1. In cellular system the channel link between the mobile switching system and the user does not use ISDN. Because a 64 kbps ISDN channel needs a bandwidth of 64 kHz for radio transmission. In cellular systems, the data rate of a channel is 16 kbps or less, and needs only a bandwidth of 25 kHz or less. Using less channel bandwidth increases more spectrum efficiency.

17.2   SS7 Network and ISDN for AIN

17.2.1   History of SS7[4]

This is an out-of-band signaling method in which a common data channel is used to convey signaling information related to a large number of trunks (voice and data). Signaling has traditionally supported (1) supervisory functions, e.g., on-hook/off-hook to indicate idle or busy status; (2) addressing function, e.g., called number; and (3) calling information, e.g., dial tone and busy signals. The introduction of electronic processors in switching systems made it possible to provide common channel signaling.

In 1976 common channel interoffice signaling (CCIS) was introduced.CCIS is based on the International Consultative Committee on Telegraphy and Telephony (CCITT) Signaling System No. 6 recommendations and called CCS6. The CCS6 protocol structure was not layered. It was a monolithic structure. The signaling efficiency was high.

In 1980 CCITT first recommended SS7, a signaling system for digital trunks. The layered approach to designing SS7 protocols was being developed for open system interconnection (OSI) data transport. Also, the HDLC (higher-level data link control) bit-oriented protocols had an influence on the development of SS7.

17.2.2 SS7 protocol model

The inefficiencies of layered protocols are far outweighed by their flexibility in realization and management of complex functions. The protocol becomes more aligned with the seven-layer OSI reference model (Fig. 17.2a). The seven layers are physical, data link, network, transport, session, presentation, and application. The SS7 protocol model is shown in Fig. 17.2b for comparison with the OSI model. In SS7, the message transfer part (MTP) provides the OSI layered protocol model as level 1 data service, level 2 link service, and level 3 network service. The full level 3 service is provided by the signaling connection control part (SCCP). SCCP provides an enhanced addressing capability that may be considered as level 3+ or a level close to level 4. Layers 4 to 6 in the OSI model do not exist in the SS7 protocol model. The transaction capabilities application part (TCAP) level and the operations maintenance and administration part (OMAP) level are considered the same as the application part (level 7) in OSI. The application service element (ASE) is at the same level as OMAP. TCAP includes protocols

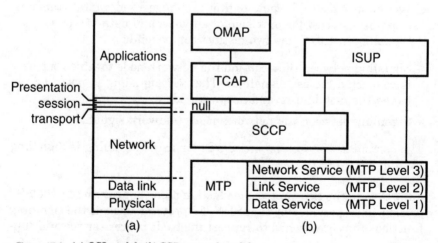

Figure 17.2 (a) OSI model. (b) SS7 protocol model.

and services to perform remote operations. The primary use of TCAP in these networks is for invoking remote procedures in supporting IN services like 800 service. OMAP provides the application protocols and procedures to monitor, coordinate, and control all the network resources which make communication based on SS7 possible. ASE is for the MTP routing verification test (MRVT), which uses the connectionless services of TCAP. MRVT is an important function of OMAP.

17.2.3 SS7 network link deployment for AIN

The SS7 links can provide high-speed service because of the common channel signaling. Based on the connection among all the resource elements, there are six links from A to F.

A link: $\begin{cases} \text{STP} \leftrightarrow \text{SCP} \\ \text{STP} \leftrightarrow \text{SP/SSP} \end{cases}$

B/D and C links   STP ↔ STP

E link   STP ↔ SP/SSP

F link   SP/SSP ↔ SP/SSP

The SS7 network link deployment chart is shown in Fig. 17.3. The interfaces between any two entities are indicated by the letters from A to F.

17.2.4 ISDN[5]

Signaling has evolved with the technology of the telephone. The integrated services digital network (ISDN) is used to integrate all-digital networks in which the same digital exchanges and digital transmission paths are used for provision of all voice and data services.

Signaling in ISDN has two distinct components:

- Signaling between the user and the network node to which the user is connected (access signaling). The SS7 signaling is not used between the mobile user and network node.

- Signaling between the network nodes (network signaling)

The current set of protocol standards for ISDN signaling is Signaling System No. 7 (SS7).

ISDN-UP. In the SS7 protocol model, functions not covered by the SS7 levels will be provided by the ISDN-UP protocol, such as the signaling functions that are needed to support the basic bearer service and sup-

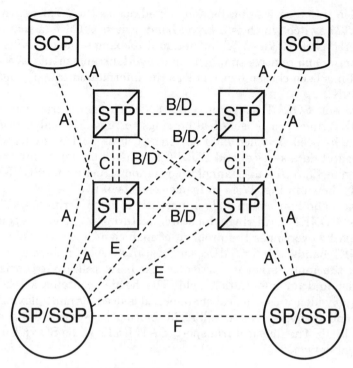

Figure 17.3   SS7 network link deployment.

plementary services for switched voice and data applications in an ISDN environment.

B-ISDN.[6]   The broadband ISDN will support a range of voice, video, data, image, and multimedia services using available resources. These resources include transmission, switching and buffer capacity, and control intelligence. The target is to provide switched services over synchronous optical network/asynchronous transfer mode (SONET/ATM) transport using signaling based on the extended ISDN protocol.

17.2.5   SONET and ATM[7,8]

A fiber-optic transport system named SONET was introduced in 1984. By 1988, as a global transmission standard, synchronous digital hierarchy (SDH) became an umbrella standard that allowed local variations. SDH is a set of interface standards which is common to all SONET equipment. The highest rate for SONET is 155 Mbps. Rates above 155 Mbps can be carried by the ATM cells. All services are divided into a series of cells and routed across the ATM network via an ATM switch which is a broad band switch. ATM is a connection-oriented packet-like switching and multiplexing principle. ATM offers flexibility and relative simplicity in network arrangements as a whole. ATM has been chosen by CCITT as the information transfer mode for B-ISDN.

Present SONET rings use STM (synchronous transfer mode). SONET rings can be self-healing rings to provide reliable transport for high-speed switched data services such as SMDS (switched multimegabit data service) and FDDI (fiber distributed data interface interconnection). A four-channel information structure for SONET uses STM, shown in Fig. 17.4a. A frame consists of five time slots: a framing slot and four channel slots. Recently a SONET ring uses ATM is called SONET/ATM ring which architecture using point-to-point virtue paths overcomes the problem of inefficient and expensive use of SONET bandwidth for SMDS services. In an ATM information structure, the information is carried in fixed-size cells that consist of a header and an information field. The header contains a label that uniquely identifies a logical channel and is used for multiplexing, routing, and switching. A four-cell ATM information structure is shown in Fig. 17.4b. The physical transport of ATM cells is the SONET or SDH optical network.

17.3 AIN for Mobile Communication[9–12]

The AIN for mobile communication has to meet a unique requirement: the call has to reach the mobile station in a required time frame while

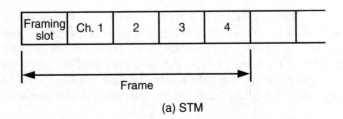

(a) STM

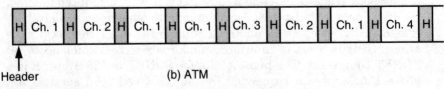

(b) ATM

Figure 17.4 The information structures of (a) STM and (b) ATM.

it is in motion. Therefore, the call processing time or the handoff time has to be within a specific limit, otherwise the mobile station can move out of the coverage area and the call either cannot be connected or will be dropped. Besides, the call has to be delivered to wherever the mobile station is; i.e., the system must have a roaming feature. The system handoff between the two different systems also needs AIN.

Two major directions in AIN for mobile communications are wireless access technology and increased network functionality. Wireless access technology is aiming at low power, light weight, efficiently used spectrum, and low-cost operation and maintenance. The increased network functionality is achieved through the use of SS7 for call control and database transactions. Mobility is the main concern in the connectionless structure of the protocol, which has to be suited to real-time application. The mobile application part (MAP) can be applied to mobile communications. The exchange of data between components of a mobile network to support end user mobility and network call control are taken care of by MAP. The MAP is an application service element. The MAP of CCITT SS7 is shown in Fig. 17.5. TCAP is composed of both the component sublayer and transaction sublayer. The component sublayer provides the exchange of protocol data units, invoking remote operations and reporting their results. The transaction sublayer is responsible for establishing a pseudo-association service for exchange of related protocol data units. The interrogations and transfer of information take place by using the ASE of the MAP and the component sublayer of TCAP. A number of MAP procedures relate to (1) location registration and cancellation, (2) handling of supplementary services, (3) retrieval of subscriber information during call establishment, (4) handoff, and (5) subscriber management including location information request and retrieval. The AIN mobile system architecture is shown in Fig. 17.6, which is the same as Fig. 17.1 except that SSP is replaced by MSC and SMS collocates with the service creation environment (SCE), which defines new features and services.

17.4 Asynchronous Transfer Mode (ATM) Technology[13-22]

ATM technology, because of its flexibility and its support of multimedia traffic, draws much interest and attention. In wire line and wireless communications, we are interested in broadband switches as mentioned in Sec. 17.2.5. The ATM switch can meet our needs. The ATM technology will be described in this section.

The interest in ATM first came from carriers and manufacturers of wide-area networking equipment, and now interest is growing in the application of ATM technology to the local and campus area network-

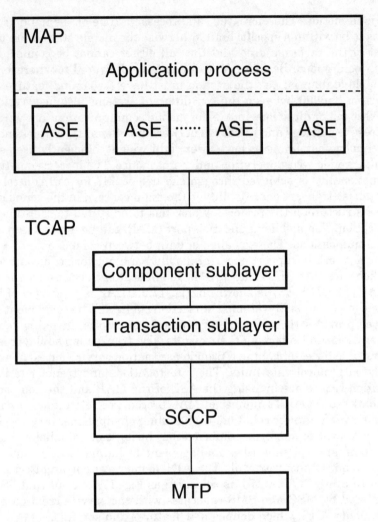

Figure 17.5 Map of CCITT SS7.

ing environment. ATM is designed to support multimedia traffic and is capable of offering seamless integration with wide-area ATM networks, both public and private. Since it offers the benefit of handling the broadband signal channels needed for the increasing volume of data communications traffic, it has been chosen for the switch of B-ISDN. The ATM network concept is shown in Fig. 17.7.

LAN applications. For bringing ATM technology to the customer premises, it must offer LAN-like service for data traffic and be compatible with the existing data communication protocols, application, and

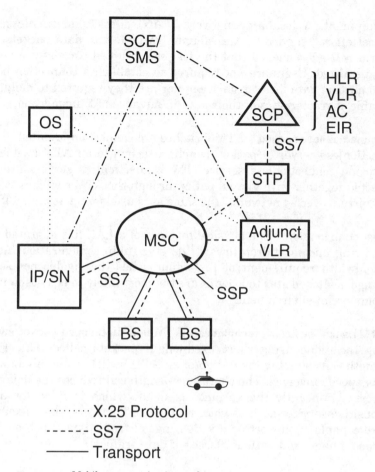

Figure 17.6 Mobile communication architecture.

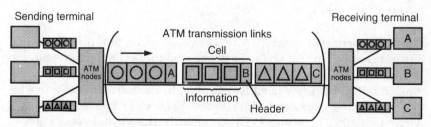

Figure 17.7 ATM network concept.

equipment. A local-area network (LAN) offers connectionless, i.e., "best effort", service for transferring variable size data packets. The term *best effort* means that the lost or corrupted packets are not retransmitted. Users are not required to establish a connection before submitting data for transmission, nor are they required to define the traffic characteristics of their data in advance of transmission.

Connectionless service.   ATM switches are connection-oriented. A connectionless server (a packet switch) attached to an ATM switch can provide connectionless service. The connectionless servers are connected together with virtual paths through the ATM switches to form a "virtual overlay network," the same as is used for narrowband ISDN.

Star configuration.   The physical topology of a LAN has migrated from the ring and multidrop toward the star (hub) configuration. As the bandwidth requirement of LAN approaches the gigabit per second range, switched star topologies are the most likely to be chosen in the commercial environment.

ATM packet-switching techniques.   ATM is a high-speed packet-switching technique using short fixed-length packets called *cells*. Fixed-length cells simplify the design of an ATM switch at the high switching speeds involved. The short fixed-length cell reduces the delay, and most significantly the variance of delay, which is *jitter,* for delay-sensitive services such as voice and video. Therefore, short fixed cells are capable of supporting a wide range of traffic types such as voice, video, image, and various classes of data traffic.

ATM applications

1. ATM multiplexing and switching technologies are used for the B-ISDN.
2. ATM offers LAN, a high-capacity network.
3. ATM's switching technique offers seamless access to private wide-area networking.

Connection-oriented service.   All ATM cells belong to a preestablished virtual connection. All traffic is segmented into cells for transmission across an ATM network. The ATM standard or broadband ISDN defines a cell as having a fixed length of 53 bytes, consisting of a header of 5 bytes and a payload of 48 bytes. Each cell's header contains a virtual channel identifier (VCI) to identify the virtual connection to which the cell belongs. An ATM switch will handle a minimum of several hundred thousand cells per second at every switch port. Each switch port will support a throughput of at least 50 Mbps, while 150 Mbps and 600 Mbps are proposed as standard ports. A switch, if it has more than 100 ports, is considered a large switch. The general structure of an ATM switch is shown in Fig. 17.8. In an ATM switch, cell arrivals are not scheduled. A number of cells from different input ports may simultaneously request the same output port. This event is called *output contention*. A single output port can transmit only one cell at a time. Thus, only one cell can be accepted for transmission and others simultaneously requesting that port must either be buffered or discarded. Therefore, the most significant aspects of the ATM switch design are (1) the topology of the switch fabric, (2) the location of the cell buffers, and (3) the contention resolution mechanism.

Switch fabric.   A switch fabric can be based on time division and space division.

1. *Time division.* All cells flow across a single communication highway shared in common by all input and output ports. The communication highway may be either a shared medium such as a ring or a multidrop bus, or a shared memory as shown in Fig. 17.9. This single shared highway fixes an upper limit on the capacity for a particular implementation.
2. *Space division.* A plurality of paths is provided between the input and output ports. These paths operate concurrently so that many cells may be transmitted across the switch fabric at the same time.

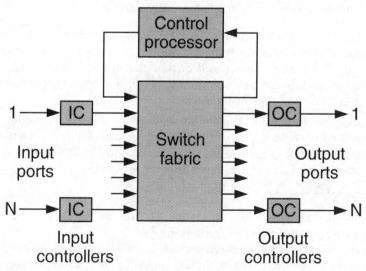

Figure 17.8 General structure of an ATM switch.

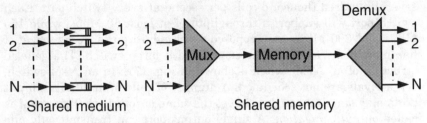

Figure 17.9 Time-division ATM switch fabrics.

Total capacity is measured as follows:

Capacity = path's bandwidth × number of paths.

The upper limit on the total capacity is theoretically unlimited. However, it is restricted by physical implementation constraints, i.e., device capability, connector restrictions, and synchronization considerations for high capacity. There are two approaches:
a. A single-path, self-routing interconnection network is most often proposed for use in ATM switch design.
b. Multiple-path networks are used to improve the performance of a single-path network or to construct large switches from switch modules. Since multiple paths are available between every input-output pair, an algorithm is required to select one of the paths.

Buffering strategies. The buffering strategy is based on whether any cell queries are located within the switch fabric (internally buffered) or outside the switch fabric (externally buffered):

1. *Internal buffering.* A single shared memory switch module may be considered internally buffered when it permits a single buffer to be shared by many input and output parts. This sharing of buffers substantially reduces the number of cell buffers required to support a given switch performance.

2. *External buffering.* Allows the cell queries to be located close to the switch ports that they serve. The absence of cell queries within the switch fabric eases the support of multiple levels of priority across the switch fabric for different classes of traffic.

Contention resolution

1. In an internally buffered switch, contention is handled by placing buffers at the point of contention.
2. In an externally buffered switch, a contention resolution mechanism is required. Three basic actions can be taken once contention is detected.

a. Backpressure. Used in the input-buffered switch design. The cell that cannot be handled at the point of contention will be sent back to the input·buffers.

b. Deflection. This mechanism will route the cells in contention over a path other than the shortest path to the requested destination.

c. Loss. Pure output-buffered designs use a loss mechanism that discards cells that cannot be handled.

17.5   An Intelligent System: Future Public Land Mobile Telecommunication System (FPLMTS)[23,24]

The International Radio Consultative Committee (CCIR) has made recommendations for the third generation of land-mobile systems. The overall objectives of FPLMTS are to provide all services generally available through the fixed network (e.g., voice, fax, and data) to mobile systems. It is intended to provide these services over a wide range of user densities and geographic coverage areas. The frequency allocations were made by the World Administrative Radio Conference of 1992 (WARC '92) in the 1- to 3-GHz frequency range with an amendment of a worldwide co-primary allocation to mobile services over the frequency range 1700 to 2600 MHz. The 230 MHz of spectrum in the following bands is designated for FPLMTS:

1800 to 2025 MHz (140 MHz)

2110 to 2200 MHz (90 MHz)

In addition, the following bands are designated for mobile satellite systems:

1980 to 2010 MHz

2170 to 2200 MHz

Future enhancement

The *land* in FPLMTS refers to the land base station, which can be either a terrestrial or satellite station. Calls within the mobile system are routed to and from the intelligent network, either fixed or mobile via terrestrial or satellite links using at least four kinds of radio interface (R_1, R_2, R_3, R_4). The R_1 interface is used by mobile stations. The R_2 interface is used by indoor and outdoor personal stations (handsets). The R_3 interface is used by mobile stations communicating through a satellite, and the R_4 interface is used by pagers. All mobile calls can be connected either directly to PSTN (public service telephone network) or via a mobile switch. The mobile systems can be either a narrowband or wideband.

The requirement of developing an FPLMT system has been described. Intelligent mobile units, intelligent cells, and the intelligent network will make this intelligent system a reality.

17.6   Wireless Information Superhighway

In 1993, the United States government asked the communications and computer industry to move ahead on building the information superhighway. The information superhighway will provide the ability for many users to frequently send and receive large volumes of information.

There are two types of information superhighways: wireline and wireless. The wireline information superhighway uses optical fiber. The spectral bandwidth of optical fiber is very broad. There is apparently no limitation on assigning a number of broadband channels to the users in the wireline information superhighway system. Therefore, in developing a wireline information superhighway system, the problem is relatively simple. The most difficult task is how the information will get on and off the information superhighway because of the high volume of call traffic. A future advanced intelligent network will be developed for the information superhighway systems.

However, in developing a wireless information superhighway system, the problem is more difficult. Besides the problem of developing the wireline information superhighway, the major difficulty is how to reach and serve customers over radio waves. In the mobile radio environment there is excessive pathloss, multipath fading and dispersive time delay spread as stated in Sec. 1.6. Also, video data transmission over the information superhighway requires a large bandwidth (5 MHz or higher) for each wireless channel. The carrier frequency, which can carry multiple wideband channels, has to be in the range of 20 GHz or above. Therefore, a broadband spectrum is required in order to build this information superhighway. The higher the frequency, the more difficult it is for the radio wave to reach the customer. Thus, mother nature limits the utilization of wide bandwidths for the information superhighway. Mobility in the information superhighway presents another difficulty in wireless.

The means for deploying the broadband channels for the information superhighway systems are by a microwave link and an infrared link. The spectrum of the radio and infrared bands is illustrated in Fig. 17.10.

The advantages of using a microwave system are:

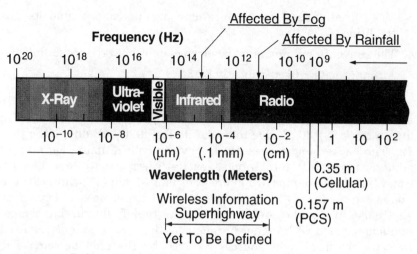

Figure 17.10 Radio and infrared bands.

1. The system does not require line-of-sight conditions for reception.
2. Broadband signal provides less fading.
3. The conventional diversity schemes can further reduce fading.
4. Can apply CDMA scheme for long-range transmission.

The disadvantages of using a microwave system are:

1. The noise floor is relatively high.
2. The propagation loss is high.
3. Attenuation is affected mostly by rainfall.

The advantages of using an infrared link are:

1. Eavesdropping can be prevented.
2. It is highly directional.
3. It provides power conservation.
4. No license is required.
5. It is light weight.

The disadvantages of using an infrared link with today's technology are:

1. For short distances.
2. For mobile communications within rooms using diffused radiation which recreates multiple reflection waves to reach the terminals.

3. Most fixed-to-fixed communications are under the line-of-sight condition.

4. The conventional diversity schemes cannot help reduce fading.

5. Attenuation is affected mostly by fog.

The infrared wavelength is roughly between 0.1 mm to 1 μm. It attenuates heavily by the fog. At a half mile, the loss of over 60 dB is due to thick fog, 30 dB due to moderate fog, and 0 dB due to the thin fog. The background noises due to environment illumination are: (1) sunlight (daylight)—a high noise level at a wavelength less than 1.2 μm; (2) tungsten lamps; (3) fluorescent lamps; and (4) heaters. These noises can produce DC and low frequency photocurrents and generate shot noise in the photodetector. Also, the rapidly fluctuating components associated with higher harmonic of the main frequency emitted by some artificial light sources received by the photodetector. The

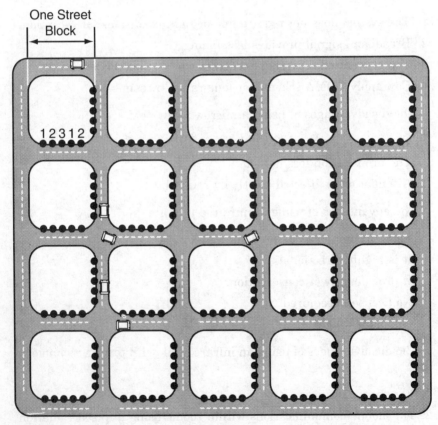

Figure 17.11 The information superhighway structure—along the city streets.

background noise can be reduced by the optical filters. (1) The interference filters (bandpass) which transmission coefficient is less than 40%, optical response changes with angle of incidence of light and cost is high. (2) Absorption filters which noise reduction is 54% for fluorescent tubes, depends on the choice of absorbing material and cost is low.

In the mobile application, we are using diffuse and quasi-diffuse transmission. In diffuse transmission, use sufficient emitted optical power for covering the multipath dispersion, create reflections, and create multi-beams in wireless-transmission configurations. In the quasi-diffuse transmission, the photodetector receives and the laser or LED emits from and to the remote station (RS). Then the common node (CN) and RS are in a line-of-sight condition. All the RS are aiming at CN which is the control center.

Among these difficulties the weather, as mentioned above, introduces undesirable factors to the higher band microwave frequencies or to optical frequencies. Therefore, combining the two systems— the infrared system which is affected by fog, but not affected by rainfall and the microwave system which is affected just the opposite— can achieve multi-propagation-medium diversity. The experiment has been carried out and stated in Sec. 18.4. With that, the airlink for providing the wireless information superhighway system may barely reach 50 meters which is our goal.

The most difficult part of developing the wireless information superhighway is the last 50 meters. Due to the requirements of the wide bandwidth, future technology for the last 50 meters should utilize microwave and infrared, and intelligent cell technologies to obtain the wireless information superhighway. However, more research needs to be conducted in this area. The future wireless information superhighway, thus, has to be hybrid with the wireline information superhighway whose transmission medium uses non-interference wideband optical fiber. For personal mobility and personal communications, the wireless information superhighway will add great value in the entire information superhighway system. We may illustrate an application of using the last 50 meters.

17.6.1 An example for applying the last 50 meters

A wireless information superhighway structure along city streets can be provided as shown in Fig. 17.11. In order to only concentrate on the last 50 meter link, the means of delivering the broadband signal from a distant source will lean on the optical fiber. In the last 50 meters, the microwaves or infrared, and the intelligent microcell can

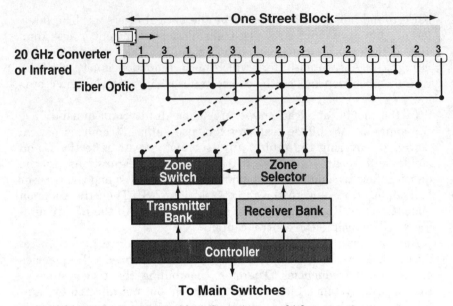

Figure 17.12 Microcell application for information superhighway.

help provide the link along each street block as shown in Fig. 17.12. The signal is delivered via optical fibers to the zone sites along the street. At the zone sites, the signal can be converted to infrared radiation or microwave transmission. The intelligent microcell technology stated in Sec. 16.2 will be applied to convey the signal from the zone site to the mobile unit. The range is within 50 meters. The three-zone selectors detect the signal at each zone. For instance, if the signal at zone 1 is strong, all the zone 1's are turned on. Since the power is very low, no interference would occur at the other neighboring streets. Then the mobile unit traveling along the whole block of the street will keep the same channel, but the zone sites along the street will turn on and turn off as the location of the mobile unit changes. No handoff takes place along the whole street block. This arrangement can ease the load of mobile switching for handoffs and provide more capacity to receive the new calls. Also, this arrangement can reduce the number of zone selectors to only one. This illustration of wireless information superhighway system needs a lot of research effort to make it happen.

References

1. R. B. Robrock II, "The Intelligent Network," *Proc. IEEE,* vol. 75, no. 1, Jan. 1991, pp. 7–20.

2. R. K. Berman, J. H. Brewster, "Perspective on the AIN Architecture," *IEEE Communications Magazine,* vol. 31, February 1993, pp. 27–33.
3. D. A. Pezzutti, "Operations Issues for Advanced Intelligent Networks," *IEEE Communications Magazine,* Feb. 1992, pp. 58–63.
4. A. R. Modarressi, "Signaling System No. 7: A Tutorial," *IEEE Communications Magazine,* July 1990, pp. 19–35.
5. H. Rarig, "ISDN Signal Distribution Network," *IEEE Communications Magazine,* vol. 32, June 1994, pp. 34–38.
6. K. Murano, K. Murakami, E. Iwabuchi, T. Katsuki, H. Ogasawara, "Technologies Towards Broadband ISDN," *IEEE Communications Magazine,* vol. 28, April 1990, pp. 66–70.
7. T. H. Wee, "Cost-Effective Network Evolution," *IEEE Communications Magazine,* Sept. 1993, pp. 64–73.
8. M. Hibino, F. Kaplan, "User Interface Design for SONET Networks," *IEEE Communications Magazine,* vol. 30, August 1992, pp. 24–27.
9. B. Jabbain, "Intelligent Network Concepts in Mobile Communications," *IEEE Communications Magazine,* vol. 31, February 1992, pp. 64–69.
10. A. S. Acampora and M. Naghshinch, "Control and Quality-of-service Provisioning in High-speed Microcellular Network," *IEEE Personal Communications,* vol. no. 2, April, 1994, pp. 36–43.
11. D. J. Goodman, G. P. Pollini, and K. S. Meier-Hellstern, "Network Control for Wireless Communications," *IEEE Communications Magazine,* Dec. 1992, pp. 116–125. W. T. Webb, "Modulation Methods for PCNs," *IEEE Communications Magazine,* Dec. 1992, pp. 90–95.
12. A. Nakajima, M. Eguchi, T. Arita, and H. Takeda, "Intelligent Mobile Communications Network Architecture," *Proceeding of ISS '90,* Stockholm, Sweden, May 1990.
13. I. W. Habib, T. N. Saadawi, "Controlling Flow and avoiding Congestion in Broadband Networks," *IEEE Communications Magazine,* October 1991, vol. 29, pp. 46–53.
14. A. A. Lazar, G. Pacifici, "Control of Resources in Broadband Networks With Quality of Service Guarantees," *IEEE Communications Magazine,* October 1991, vol. 29, pp. 66–73.
15. Y. Inoue, N. Terada, "Granulated Broadband Network," *IEEE Communications Magazine,* April 1994, pp. 56–63.
16. K. Sato, S. Ohta, I. Tokizawa, "Broadband ATM Network Architecture Based on Virtual Paths," *IEEE Transactions on Communication,* August 1990.
17. P. Newman, "ATM Local Area Networks," *IEEE Communications Magazine,* vol. 32, March 1994, pp. 86–98.
18. S. Isaku, M. Ishikura, "ATM Network Architecture for Supporting the Connectionless Service," *Proc. IEEE Infocom,* vol. 2, pp. 796–802, San Francisco, June 1990.
19. K. Kato, T. Shimoe, K. Hajikano, K. Murakami, "Experimental Broadband ATM Switching System," *Proc. Globecom '88,* pp. 1288–1292.
20. J. A. McEachern, "Gigabit Networking on the Public Transmission Network," *IEEE Communications Magazine,* vol. 30, April 1992, pp. 70–78.
21. E. W. Zegma, "Architecture for ATM Switching Systems," *IEEE Communications Magazine,* vol. 31, February 1993, pp. 28–37.
22. P. Condreuse, M. Servel, "Prelude: An Asynchronous Time-Division Switched Network," *Proc. Intl. Communication Conference,* June 1987, pp. 769–773.
23. P. Gardenier, M. Shafi, R. B. Vernall, M. Milner, "Sharing Issues Between FPLMTS and Fixed Services," *IEEE Communications Magazine,* vol. 32, June 1994, pp. 74–78.
24. R. Steele, "The Evolution of Personal Communications," *IEEE Personal Communications,* April 1994, pp. 6–11.

Cellular Related Topics

18.1 Study of a 60-GHz Cellular System

A series of studies[1,2] have explored the implementation of a 60-GHz mobile telephone system using direct line-of-sight transmission along urban streets. Use of the 60-GHz band is encouraged for the following reasons.

1. The 60-GHz band is in the oxygen-absorption range. In this range the attenuation of radio signals in the oxygen-absorption band is two orders of magnitude greater than the attenuation outside the oxygen-absorption band over a 1-km distance. Therefore, the 60-GHz band cannot be used by many applications; however, if it is properly implemented, it can be allocated to a cellular mobile radio system. The FCC may be glad to give this band away.

2. The characteristics of this high-attenuation signal over the propagation path create a natural barrier to cochannel or adjacent-channel interference in the cellular mobile system.

18.1.1 Propagation in the scattered environment

In the UHF range or at X band, multipath signal scattering from vehicles and buildings in the mobile radio environment has frequently caused deep and rapid fading of the received signals. Sometimes, the signal can be received by propagation through scattered or reflected signals, a natural phenomenon. However, when a 60-GHz band is used, line-of-sight propagation is required. A series of experiments at 59.5 GHz were carried out in 1972 in urban areas.[2]

The average output power at the mobile antenna was 40 mW (16 dBm), and the parabola antenna beamwidths were 3° at both the base station and mobile ends. The data were collected on three streets in Red Bank, New Jersey. The streets were chosen because they were representative of the urban mobile environment and were at least 0.5 mi long.

18.1.2   Fixed terminals

The mobile transmitter and the mobile receiver were parked about 0.56 km (0.35 mi) apart on opposite sides of a street. The results for each street are plotted in Fig. 18.1. Fades exceeding 3 dB occurred 5 percent of the time on Monmouth Street, 20 percent of the time on Bridge Street, and 34 percent of the time on Broad Street. The measuring limit was at −52 dBm, which presents a signal fade of 15 dB. Fades exceeding this limit occurred 1 percent of the time on Bridge Street, 2 percent of the time on Monmouth Street, and 3 percent of the time on Broad Street. Fades exceeding 15 dB were caused by large trucks or buses completely blocking the path within 30 m (100 ft) of the transmitter or receiver.

18.1.3   Moving terminal

The mobile receiver was parked at the side of the street, and the mobile transmitter started at least 0.8 km (0.5 mi) from the receiver and

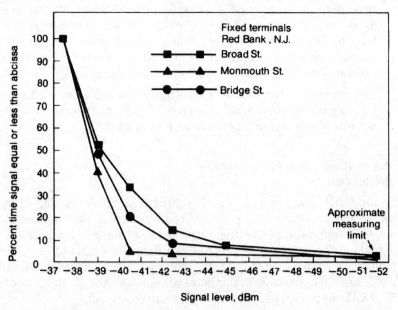

Figure 18.1   Fading statistics from fixed terminals, Red Bank, N.J. (*From Ref. 2.*)

proceeded toward and past the receiver at a constant speed of 24 km/h (15 mi/h). The results are plotted in Fig. 18.2. The fading statistics of all three streets are roughly the same. The fades exceeded 3 dB about 70 percent of the time. The fades exceeding the −52-dBm measuring limit (15 dB fades) occurred 9 percent of the time on Broad and Monmouth Streets but only 5 percent on Bridge Street because the line-of-sight signal was blocked by large trucks or buses. The Doppler frequencies ranged from a few hertz to 170 Hz. When vehicle speed was 15 mi/h the fading rate was 25 Hz. Fades exceeding 10 dB occurred 20 percent of the time. A typical run measured on Monmouth Street is shown in Fig. 18.3.

The experiments at 60 GHz reveal that the amplitude of the scattered signal was small in comparison to the direct signal. The effect of the scattered signal is small. The fading observed was neither deep nor rapid.

18.1.4 System consideration

From the preceding data, we may note the following problems in a 60-GHz mobile radio system: (1) the excessive propagation loss, (2) the pointing mechanism of the antennas, (3) the small separation between repeaters, and (4) the ways of avoiding line-of-sight blockage from large trucks and buses.

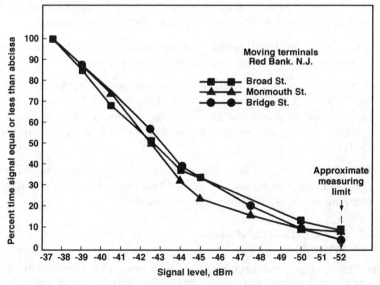

Figure 18.2 Fading statistics from moving terminals, Red Bank, N.J. (*From Ref. 2.*)

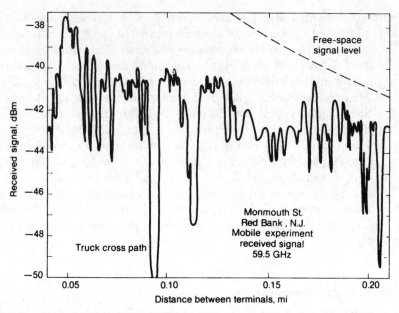

Figure 18.3 Received signal at Monmouth Street, Red Bank, N.J. (*From Ref. 2.*)

18.2 Cellular Fixed Stations

18.2.1 Radios for fixed-station telecommunications

Many alternatives can be used in the installation of radios in fixed-station telecommunications.[3] The cellular fixed station, one of these alternatives, is installed according to certain specifications.

1. To ensure a sufficient number of radio channels, each radio channel is assigned to a customer on a permanent basis.

2. When the number of available (unoccupied) radio channels is limited, two approaches can be used. The trunking approach is used to scan for an available channel and a set-up channel is used to initiate a call, and the call is then transferred to and carried on an assigned voice channel.

3. For cellular fixed ratios, the limited radio channels should be reused. Entire control functions are the same for fixed cellular systems as for other cellular mobile systems.

18.2.2 Two approaches for fixed cellular systems

The fixed cellular concept is not a new one. The implementation of 800-MHz technology has gradually reduced the cost of cellular components, and the fixed cellular system could provide an economic means of providing telephone coverage in a rural area.

There are two approaches to implementation of fixed cellular systems.

1. Use cellular mobile equipment for the fixed cellular stations. It may save the cost of manufacturing a new fixed cellular radio. No new spectrum would be needed.
2. Design a cellular fixed radio.
   a. Operation of this radio could be integrated with local telephone company switching offices (see Fig. 18.4a). However, a new spectrum would be needed.
   b. Operation of this radio could be integrated with cellular mobile switching; in such a case, no new spectrum would be needed.

18.2.3   Special features of fixed cellular systems

There are several advantages.

1. The cost of the radio is low because the logic unit is simpler.
2. Either regular switching or MTSO (mobile telephone switching office) switching can be used.
3. No handoff is needed.
4. Each link can be carefully engineered.
5. A cell with a large area can be served.
   a. Both antenna heights at two ends are high, and most links are line-of-sight links. The propagation loss approaches the free-space loss.
   b. There is no multipath fading. The required C/I for fixed cellular systems is $C/I \geq 10$ dB.
   c. The surroundings are quiet.
   d. Directional antennas can be used at the customer site.
6. A low-cost controller can be provided at either the cell site or the telephone company relay site.
7. The fixed cellular system will expect a longer holding time, and no calls will be dropped.
8. The busy periods (rush hours) in fixed cellular systems usage are different from those in mobile cellular system usage.
9. It is possible to use radio channels for data links between cell site and MTSO.

18.2.4   Design of a cellular fixed system

Because fixed cellular stations cover larger areas, the system can cover a 40-mi-radius cell ($R = 40$ mi) or larger. The frequencies used in the

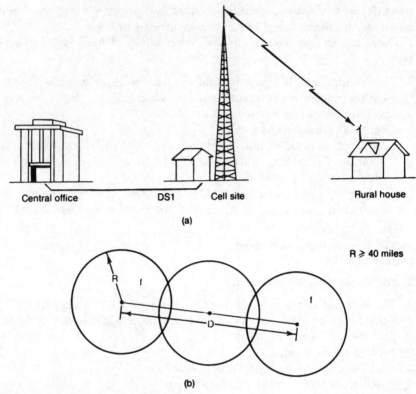

Figure 18.4 Fixed cellular system.

fixed cellular system can be identified as a subset $\{f_i\}$ of the total cellular mobile frequencies, say, 80 channels. The 80 channels need to be divided into only two groups, each with 40 channels. Each cell is assigned one group, and the other group is assigned to an adjacent cell. The separation between cochannel cells is $3R$ (see Fig. 18.1b). The reason for a closed cochannel cell separation is the natural isolation caused by the radio horizon phenomenon. The maximum area of each cell is

$$A = \pi(40)^2 = 5026 \text{ mi}^2$$

With 40 channels per cell, we can serve roughly 800 customers. If the area has to serve more customers, the size of the cells can be decreased.

18.3   Cellular Systems in Rural Service Areas[4]

Cellular mobile systems in the United States, roughly 733 market areas, have been designated as *cellular geographic service areas* (CGSA). Each CGSA can have block A systems and block B systems. The largest 305 cities of the 733 markets are called *metropolitan statistical areas* (MSAs) and the rest (428) are called *rural service areas* (RSAs). RSAs are usually adjacent to MSAs.

How to achieve a flexible and cost-effective cellular system design that will provide adequate RSA coverage but that will not interfere with MSA coverage is a challenging problem. In fact, there are three common obstacles.

1. RSA operators generally do not want to limit their design parameters.
2. MSA operators are afraid of the interference that might result from RSA service.
3. Any cell-site distance-separation specification might be adequate for one system (RSA or MSA) but not for the other.

Two approaches are recommended for controlling cochannel interference and adjacent-channel interference between RSA and MSA systems.

Approach 1—create a buffer zone between the MSA and the RSA. The RSA will be mandated to limit its transmitted power and the antenna height at the RSA cell site so that they are compatible with the MSA specifications. (See Fig. 18.5.)

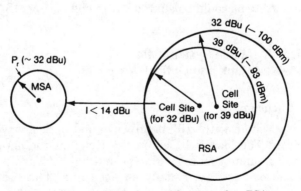

Figure 18.5   Received signal-strength power of an RSA.

Approach 2—let the RSA system designer know both the MSA boundary and the carrier-to-interference ratio at the boundary before planning the system. This way the RSA system designer will know the received signal level at the RSA boundary adjacent to the MSA.

Approach 2 may be more suitable in real situations. Its methodology is as follows.

1. The MSA provides the boundary signal level requirement (level A) to the RSA (see Fig. 18.5).

2. The RSA system designer considers and designs the RSA system in accordance with MSA compatibility requirements.

3. The RSA can implement a signal level of −110 dBm at the boundary of the CGSA if there is no interference with the MSA.

4. The MSA and RSA system designers can consider what specifications will be compatible for both parties.

At the cell site, the RSA operator can install any kind of directional antenna and raise the antenna to maximize the transmitted power, provided the above requirements are satisfied. The directional antenna can also be used to reduce the interference to or from the other site.

There are a few additional considerations.

1. A low-cost antenna mast or an existing mast should be used.

2. Half of the unused voice channels could be used for microwave transmission.

3. The installation of small switching systems should be considered, or these systems could be shared with neighboring MSAs or RSAs.

18.4   Diversity Media System with Millimeter-Wave Link and Optical-Wave Link

18.4.1   Introduction

A diversity system with millimeter-wave and optical-wave links can be achieved.[5] The idea is based on the fact that the optical wave is primarily attenuated by fog[6] and the millimeter wave primarily by rain.[7] Since fog and rain usually do not occur at the same time, a diversity advantage using an optical-wave signal and a millimeter-wave signal may be realized.

For investigating the diversity advantage on these two links (optical wave and millimeter wave) in a metropolitan area, data on the signal attenuation by fog on an optical-wave link and data on the signal attenuation by rain on a millimeter-wave link over the same path simultaneously in that area was collected.

In 1973, an optical-wave (0.9 μm) link between the 88th floor of the Empire State Building and the 58th floor of the Pan American building (a distance of about 0.55 mi) was installed.[8] The data on signal attenuation by fog has been converted to a probability distribution curve (PDC), shown in Fig. 18.6. In addition, there are three other curves, one obtained at Holmdel, New Jersey, and two at the AT&T building in New York City, with different floor heights, also shown in

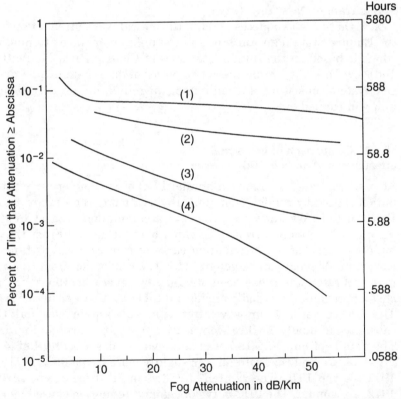

Figure 18.6 Optical signal attenuation by fog at various places: (1) 88th floor of Empire State Building—58th floor of Pan Am Building, New York City, 0.55 mi in distance (one-way transmission); (2) Radio Range Building, Bell Laboratories, Holmdel, N.J., round-trip path is 300 ft; (3) 21st floor of AT&T Building—19th floor of Western Electric Building, New York City, roundtrip path is 240 ft; (4) 4th floor of AT&T Building—3rd floor of Western Electric Building, New York City, round-trip path is 240 ft.

Fig. 18.6 for comparison. We may note that the signal attenuation by fog is drastically changed by the height of the link.

A 3-mm link between the Empire State and Pan Am Buildings, using a 15-mW IMPATT diode as a source, two 30-in parabolic dish antennas (one transmitting and one receiving) for a link gain of 106 dB, and a Schottky-barrier diode for detection, was installed.[9] Limited data were collected to demonstrate that the link was clearly established. The 3-mm signal attenuation caused by rainfall in New York City should be acquired from other sources.

First, we assume that the effect on millimeter-wave signal attenuation of rainfall is small where the link is short. Using this assumption, we can then obtain the 3-mm-wave signal attenuation caused by rainfall over the Empire State–Pan Am link in New York City from some other nearby sources, such as the data collected at Central Park, Manhattan, or New York City.

Now there are two pieces of data: the optical-wave attenuation over the Empire State–Pan Am link and the millimeter-wave attenuation, which is based on the rainfall statistics at Central Park, Manhattan, and New York City. From these two pieces of data, a study of a diversity system consisting of a millimeter-wave link and an optical-wave link will carried out.

18.4.2 Comparison of two signal attenuations from their PDC curves

At first, we have to obtain an annual PDC of millimeter-wave attenuation caused by rainfall from a nearby location of the Empire State–Pan Am link. We know that the rain accumulation data at Central Park can be obtained from U.S. climatological data.[10] From the cumulative rain data, a distribution curve of rain rate versus time for each rainfall event can be generated.[11] The conversion from rain rate to signal attenuation has been obtained by several methods[6,12]; Oguchi's approximate method[12] provides a table that can easily be used. The PDC curves of 3-mm-wave attenuation (dB/km) in New York City calculated annually for three consecutive years are shown in Fig. 18.7. The details of how Fig. 18.7 was obtained are described in Ref. 4. In Fig. 18.7, we find that rainfall was slightly more frequent in 1971 and 1972 than in 1970. The rainfall statistics in all three years, 1970 to 1972, are similar. The PDC curve of signal attenuation caused by rain in 1972 was picked and compared with the PDC curve of signal attenuation caused by fog in our analysis. Comparing Fig. 18.6 with Fig. 18.7, we find that the optical wave is attenuated by fog more frequently than the millimeter wave is attenuated by rain.

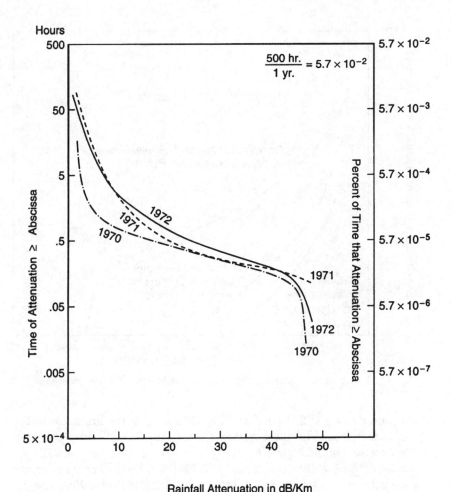

Figure 18.7   100-GHz signal attenuation by rain at Central Park, New York City.

Assuming that fog and rain do not occur simultaneously, then the PDC obtained from a selection-combining diversity signal between the millimeter and the optical wave is

$$P_C = P_r P_f \tag{18.4-1}$$

where P_r and P_f are PDC curves of signal attenuation caused by rainfall and fog respectively. P_r was obtained from Central Park and used for representing the signal attenuation by rain in the Manhattan area. Three selection-combining diversity signals can be obtained from Equation (18.4-1), one at Empire State–Pan Am, one at AT&T 21st

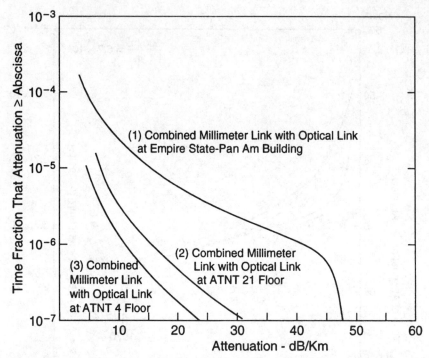

Figure 18.8 A predicted selection-combining signal between two frequencies in New York City.

floor, and one at AT&T 4th floor (Fig. 18.8). All three links are in the Manhattan area. The most advantageous outcome of using this millimeter wave–optical wave diversity scheme can be seen from the selection-combining diversity signal at AT&T 4th floor. We may conclude that a diversity system consisting of a millimeter-wave link and an optical-wave link is suitable for a site at a lower height above the ground.

18.5   Cellular Radio Telecommunications Intersystem Operations[13]

The purpose of having cellular radio telecommunications intersystem operations is to provide to cellular radio telephone subscribers certain services requiring interaction between different cellular systems. Because of the useful and effective services, we need standardized intersystem procedures. The standard is called IS-41B. There are five documents pertaining to the IS-41B as follows:

41.1B Function Overview

41.2B Intersystem Handoff

41.3B Automatic Roaming

41.4B Operative Administration and Maintenance

41.5B Data Communications

Each of these documents identifies its cellular services which require intersystem cooperation.

Standards setting is performed primarily by EIA/TIA with every effort expended to avoid arbitrary restrictions and to encourage innovative new services and capabilities by carriers and/or manufacturers based on market demands and evolving technology.

Restrictions on IS-41B

1. Voice facilities for intersystem handoff are restricted to direct dedicated circuits between pairs of participating systems.

2. When the CSS (cellular subscriber station) is in an alert state, handoffs between systems should not take place.

3. The flow control of data between applications is not provided in the data link and network layer portions of this protocol.

4. Call delivery provided only by means of employing a TLDN (temporary level directory number).

5. Full intersystem functionality does not exist and is being studied for future use.

18.5.1   Intersystem handoff

The intersystem handoff procedure is in the handoff sequence between two different mobile switching centers (MSCs). An intersystem handoff means to switch a CSS telephone call that is in progress on the "serving" MSC to a different MSC.

The dedicated voice circuits between two cooperating systems are used for the purpose of continuing speech transmission after an intersystem handoff is completed. This is currently an IS-41B restriction. A data link connection must be established between the cooperating systems. The physical layer of the data link is referred to as 41.5B. All the voice circuit control signaling such as circuit seizure, release, etc., shall be performed by signaling on the data link. The data link can use either signaling X.25 or SS7.

Intersystem handoff procedures. There are three phases to the intersystem handoff procedure as follows:

1. Location Phase: The serving MSC system makes a handoff measurement request to the candidate MSC systems and determines the target MSC system among them from the handoff measurement reply.

2. Handoff Phase: The servicing MSC initiates a handoff (either on the handoff-forward, handoff-back, or handoff-to-third) to the target MSC, then the target MSC accepts one of these handoff.

3. Release Phase: The initiating MSC releases a handoff request, then the receiving MSC completes the handoff handling.

A call can be engaged in more than two MSCs, then the path minimization is used to keep the number of MSCs involved in a call to a minimum. Then the handoff procedures can be applied to consecutive inter-MSC handoffs for the same CSS call.

18.5.2   Intersystem roaming

The purpose of IS-41.3B is to provide "automatic roaming" as one of several features as follows:

1. Making the identity of the current serving or visited system known to the home system. ·

2. Establishing financial responsibilities for the roaming subscriber. The subscriber is notified of the charge for the equipment rental after each call.

3. Establishing a valid roamer service profile in a visited system can eliminate fraudulent calls.

The dedicated voice circuits between cooperating MSCs are provided for the purpose of delivering an incoming call to the cellular subscriber while in a visited system. A data link connection must be established between the cooperating cellular network elements. The procedures leading to providing routing information of calls to the roaming subscriber in question shall be performed by signaling on the data link.

Automatic roaming procedures. The MSC detecting foreign mobile registration procedure leads to a registration notification that includes autonomous registration, call origination, call termination, or other mechanisms.

If the received message from the CSS cannot be processed, then the VLR or HLR "receiving registration notification invoke" begins. When roaming is not valid, registration cancellation goes into affect. There are several features and requests listed as follows:

Remote feature control

1. MSC detecting remote feature control access
2. VLR receiving remote feature control request invoke
3. HLR receiving remote feature control request invoke

Requests from the CSS's original HLR

1. Location/routing request
2. Call data/routing request
3. Transfer-to-number request
4. Service profile request
5. Qualification request

18.5.3 OA&M (operation, administration and maintenance) and data link

OA&M permits cellular operators to operate, administrate and maintain data communications for roaming, data link, and voice circuits for handoff between two different systems. The functions of OA&M are:

1. The network performance features
2. Abnormal conditions
3. Failure to receive
4. Unreliable roamer data
5. Inter-MSC trunk testing

Data links use the standardized protocols so that a minimum of prior coordination between two mobile systems is necessary to communicate. The OSI protocol is used for mobile application described in Chap. 17. Evaluation of network connectivity, service provided and intersystem capabilities implies that the protocol should support on-line negotiation of protocol features.

References

1. C. L. Ruthruff, "A 60 GHz Cellular System," Microwave Mobile Symposium, Boulder, Colorado, 1974.
2. L. U. Kibler, "A 60 GHz Propagation Measurement," Microwave Mobile Symposium, Boulder, Colorado, 1974.

3. FCC's Notice of Proposed Rulemaking on Establishing Basic Exchange Telecommunications Radio Service," Communications Commission Docket No. 86-495, January 16, 1987.
4. W. C. Y. Lee, "Cellular Technology for Rural Areas," *Communications,* November 1987, p. 78.
5. L. C. Tillotson, "Customer-To-Customer Via Millimeter Wave," private communication, June 26, 1972.
6. C. L. Ruthroff, "Rain Attenuation and Radio Path Design," *Bell System Technical Journal,* vol. 49, pp. 121–136, January 1970.
7. T. S. Chu and D. C. Hogg, "Effects of Precipitation on Propagation at 0.63, 3.5, 10.6 Microns" *Bell System Technical Journal,* May–June 1968, pp. 723–759.
8. H. J. Schulte, "Optical-Wave Link Setup," private communication, 1973.
9. W. C. Y. Lee, "Millimeter-Wave Link Setup," private communication, 1973.
10. U.S. Dept. of Commerce, "Climatological Data, Annual," published once a year.
11. W. C. Y. Lee, "An Approximate Method for Obtaining Rain Rate Statistics for Use in Signal Attenuation Estimating", *IEEE Trans. on Antennas and Propagation,* vol. AP-27, May 1979, pp. 407–413.
12. T. Oguchi, "Attenuation of Electromagnetic Wave Due to Rain With Distorted Raindrops (Part II)," *Journal of Radio Research Lab,* Tokyo, vol. II, pp. 19–44, January 1964.
13. EIA/TIA, "Cellular Radio-Telecommunication Intersystem Operations," Interim Standard, Eng. Dept. EIA, December 1991, distributed by Global Engineering Documents, Irvine, California.

Index

Granglund combiner, 318
Ground reflection angle, 105
Group identification, 75
GSM (group of special mobile), 463
GSM channels, 471
GSM (overview), 485
GSM handover, 483

Handing off the call, 358
Handoff, 26, 65, 74, 283, 386
  creating a, 290
  delaying, 288
  forced, 289
  initiation of, 286
  intersystem, 295, 356
  lost calls due to, 384
  number of, per cell, 286
  power-difference, 292
  probability of, 285
  queuing of, 290, 311, 356
  type of, 283
Handoff blocking, 277
Handoff or handover (HO), 283
  CDMA:
    hard HO, 504, 531
    soft HO, 504, 531
  MAHO, 294
  NA-TDMA, 463
  GSM, 463
Hexagonal-shaped cells, 27
HLR (home location register), 466

Idle stage, 80
Ignition noise, 24
IMPATT diode, 648
IN (intelligent network), 465
Inbuilding communications, 573, 594
  new inbuilding communication system, 595
Incident angle, 11, 104
Induced voltage on a monopole or dipole, 159
Information superhighway, 632
Intelligent cell, 563
  intelligent microcell, 564, 568
  power-delivery, 564
  processing gain, 575
Interference, 189, 221
  caused by portable units, 420
  cochannel, 189
  to mobile units, 420
  noncochannel, 221
  reduction of, 310
Interleaving:
  CDMA, reverse traffic channel, 510
  forward traffic channel, 520

Interleaving (*Cont.*):
  GSM, 476, 477
  NA-TDMA, 497
Intermodulation (IM), 243
IP (intelligent peripheral), 619
IR (infrared), 632
IS-41, 652
IS-54, 503
IS-54 (NA-TDMA, ADC), 486
IS-95, 503
ISDN (integrated service digital network), 619
  B-ISDN, 624
  ISDN-up, 622
IWF (interworking function), 465

Japanese cellular system, 29

Knife-edge diffraction loss, 137, 138

LAN (local area network), 626
Large capacity systems, worldwide, 100, 101
Last 50 meters, 635
Layer Modeling:
  CDMA, 504
  GSM, 468
LCR (level crossing rate), 18, 395
Leaky feeder, 319
Leaky-feeder radio communication, 322
Leaky waveguide, 319
LEO (low earth orbit) satellite, 547
Level-crossing counter, 402
Level crossing rate (LCR), 18, 395
Linear predictive code (LPC), 450–451
Line-of-sight path, 23
Local mean, 13
Locating receiver, 332
Logical channel, 472
Log-normal fading, 13
Long code, 505
Long-distance interference, 254
  overland path, 254
  overwater path, 254
Long-term fading, 13
Low-density small-market design, 339
Lowering antenna height, 202, 236
  in forested area, 203
  on high spot, 202
  in valley, 203
LP (linear predictive code), 450–451, 469

Macrocell prediction formula, 141
MAHO (mobile assistance handoff), 141, 294, 483

ABOUT THE AUTHOR

William C. Y. Lee is Vice President of Technology at PacTel
Cellular, Inc. He was elected a Fellow of the Institute of
Electrical and Electronics Engineers for his distinguished
work in mobile communications, and frequently leads professional seminars on the subject, both in the United States
and abroad. Dr. Lee has written more than 170 articles for
professional journals, three textbooks, and is the author of
Mobile Communications Engineering, also available from
McGraw-Hill.